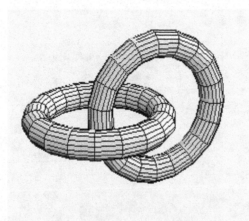

Numerical Methods for
Scientific Computing

by J.H. Heinbockel
Emeritus Professor of Mathematics
Old Dominion University

Order this book online at www.trafford.com
or email orders@trafford.com

Most Trafford titles are also available at major online book retailers.

Print information available on the last page.

ISBN: 978-1-4120-3153-0 (sc)

Trafford rev. 11/09/2018

www.trafford.com
North America & international
toll-free: 1 888 232 4444 (USA & Canada)
fax: 812 355 4082

PREFACE

This is an introductory numerical methods and numerical analysis text. The text presents numerical methods which can be applied to solve many difficult mathematical problems arising in engineering, physics, chemistry and the biological sciences. It contains the basic material needed to pursue an understanding of more advanced topics in numerical analysis.

It is assumed that the reader has completed courses in calculus, differential equations and linear algebra as these courses are prerequisites to the material contained in this text. Many chapters present a review of fundamentals before beginning new material. A summary of the material contained in this text can be discerned from an examination of the table of contents. The final chapters present miscellaneous numerical methods and concepts associated with more advanced topics. Also note that the material presented in the last two chapters covers only elementary concepts and notions from selected topics, and any reader interested in a particular subject area must consult more advanced texts and research papers to pursue and develop further the fundamentals presented.

Many of the numerical methods presented in this text can be obtained as subroutine or function programs written in Fortran, C or C++. Many of these subroutines can be either downloaded as free software or purchased as computational packages. Some of the more general numerical method libraries available are: NAG (Numerical Algorithms Group), IMSL (International Mathematics and Statistical Library), LAPACK (Linear Algebra PACKage), BLAS (Basic Linear Algebra Subprograms) and NETLIB which offers free software of interest to the scientific community.

The material presented can be used for a two semester course on introductory numerical methods and analysis. In many of the exercises presented, there is the introduction of new ideas, and so the beginning student is encouraged to read all of the exercise problems.

The Appendix A contains units of measurements from the Système International d'Unitès along with some selected physical constants. The Appendix B contains solutions to selected exercises. The Appendix C contains chi-square and normal probability tables.

J.H. Heinbockel, 2004

Table of Contents

Numerical Methods for Scientific Computing

Table of Contents

Table of Contents

"Socrates Dialectical Process: The first step is the separation of a subject into its elements. After this, by defining and discovering more about its parts, one better comprehends the entire subject "

Socrates (469-399) BCE[†]

Chapter 1

Basic Concepts

Numerical methods have been around for a long time. However, the usage of numerical methods was limited due to the lengthy hand calculations involved in their implementation. In our current society the application of numerical analysis and numerical methods occurs in just about every field of science and engineering. This is due in part to the rapidly changing digital computer industry. Digital computers have provided a fast computational device for the development and implementation of numerical methods which can handle a variety of difficult mathematical problems. To understand how numerical methods and numerical analysis techniques are developed the reader is required to have knowledge of certain background material from calculus and linear algebra. We begin by reviewing some fundamentals which are used extensively in this text.

Derivative of a Function

The derivative of a continuous function $y = f(x)$ is defined by the limiting process

$$\lim_{h \to 0} \frac{f(x+h) - f(x)}{h} = f'(x) = \frac{dy}{dx},$$

if this limit exits. This limiting process can be represented using the alternative notations

$$\frac{dy}{dx} = \lim_{\Delta x \to 0} \frac{\Delta y}{\Delta x} \quad \text{or} \quad \frac{dy}{dx}\bigg|_{x=x_0} = f'(x_0) = \lim_{x \to x_0} \frac{f(x) - f(x_0)}{x - x_0}.$$

The notation $m = f'(x_0)$ denotes the derivative evaluated at the point x_0. This derivative represents the slope m of the tangent line to the curve $y = f(x)$ which passes through the point $(x_0, f(x_0))$.

[†] BCE ("Before Common Era") replaces B.C. ("Before Christ") usage.

Fundamental Theorem of Calculus

Let $F(x)$ denote any function such that $\frac{dF(x)}{dx} = f(x)$, where $f(x)$ is a continuous function over the domain $a \leq x \leq b$. Divide the domain (a,b) into n equal subintervals of length $\Delta x = \frac{b-a}{n}$. This can be done by defining $a = x_0$ and $b = x_n$ with $x_i = x_0 + i\Delta x$ for $i = 0, 1, 2, \ldots, n$. The resulting numbers

$$a = x_0 < x_1 < x_2 < \cdots < x_{n-1} < x_n = b$$

are then said to partition the interval (a,b) into n equal subintervals. Let c_i denote a number in the ith subinterval, where $x_{i-1} \leq c_i \leq x_i$ for $i = 1, 2, \ldots, n$ and form the sum

$$S_n = \sum_{i=1}^{n} f(c_i)\Delta x = f(c_1)\Delta x + f(c_2)\Delta x + \cdots + f(c_n)\Delta x$$

The fundamental theorem of calculus states that

$$\lim_{n\to\infty} S_n = \lim_{n\to\infty} \sum_{i=1}^{n} f(c_i)\Delta x = \int_a^b f(x)\,dx = F(x)\,]_a^b = F(b) - F(a)$$

represents the area under the curve $y = f(x)$, above the x-axis if $f(x) > 0$, and between the limits $x = a$ and $x = b$.

Note that if $G(x) = \int_a^x f(t)\,dt$, then $\frac{dG(x)}{dx} = f(x)$.

Taylor Series for Functions of a Single Variable

A function $f(x)$ of a single variable x is said to be analytic in the neighborhood of a point $x = x_0$ if it can be represented in a convergent power series of the form

$$f(x) = f(x_0) + f'(x_0)(x - x_0) + \frac{f''(x_0)}{2!}(x - x_0)^2 + \cdots + \frac{f^{(m)}(x_0)}{m!}(x - x_0)^m + \cdots$$
$$f(x) = \sum_{n=0}^{\infty} \frac{f^{(n)}(x_0)}{n!}(x - x_0)^n \tag{1.1}$$

where by definition $0! = 1$ and the zero derivative denotes the function itself so that $f^{(0)}(x_0) = f(x_0)$. For a Taylor series to exist in the neighborhood of a point x_0, one must assume that the function $f(x)$ has continuous derivatives of all orders which can be evaluated at the point x_0.

Some well known Taylor series expansions are

$$e^x = 1 + x + \frac{x^2}{2!} + \frac{x^3}{3!} + \frac{x^4}{4!} + \frac{x^5}{5!} + \cdots$$
$$\sin x = x - \frac{x^3}{3!} + \frac{x^5}{5!} - \frac{x^7}{7!} + \cdots$$
$$\cos x = 1 - \frac{x^2}{2!} + \frac{x^4}{4!} - \frac{x^6}{6!} + \cdots$$

If the Taylor series expansion given by equation (1.1) is truncated after the mth derivative term, then one can write

$$f(x) = f(x_0) + f'(x_0)(x - x_0) + \frac{f''(x_0)}{2!}(x - x_0)^2 + \cdots + \frac{f^{(m)}(x_0)}{m!}(x - x_0)^m + Error \quad (1.2)$$

where the error term is given by

$$Error = \frac{f^{(m+1)}(\xi)}{(m+1)!}(x - x_0)^{m+1}, \qquad x_0 < \xi < x. \tag{1.3}$$

Taylor series expansions of the form

$$f(x_0 + h) = f(x_0) + f'(x_0)h + f''(x_0)\frac{h^2}{2!} + \cdots + f^{(m)}(x_0)\frac{h^m}{m!} + f^{(m+1)}(\xi_1)\frac{h^{m+1}}{(m+1)!}$$

$$f(x_0 - h) = f(x_0) - f'(x_0)h + f''(x_0)\frac{h^2}{2!} + \cdots + (-1)^m f^{(m)}(x_0)\frac{h^m}{m!} + (-1)^{m+1}f^{(m+1)}(\xi_2)\frac{h^{m+1}}{(m+1)!}$$

are used extensively in later chapters.

A continuous function $f(x)$ is said to have a root of multiplicity m if

$$f(x_0) = f'(x_0) = f''(x_0) = \cdots = f^{(m-1)}(x_0) = 0 \quad \text{but} \quad f^{(m)}(x_0) \neq 0 \tag{1.4}$$

That is, a root of multiplicity m is such that the function and its first $(m-1)$ derivatives are zero at $x = x_0$. If $m = 1$, then the root is called a simple root. For example, if, $f(x_0) = 0$, and $f'(x_0) \neq 0$, then x_0 is a simple root. In contrast, the conditions $f(x_0) = 0$, $f'(x_0) = 0$, and $f''(x_0) \neq 0$, imply x_0 is a root of multiplicity 2. Note that a function $f(x)$ which has a root x_0 of multiplicity m has the Taylor series expansion about x_0 of the form

$$f(x_0 + h) = f^{(m)}(x_0)\frac{h^m}{m!}f^{(m+1)}(x_0)\frac{h^{m+1}}{(m+1)!} + \cdots$$

since the function and its first $(m-1)$ derivatives are zero at $x = x_0$.

The Landau Symbol $\mathcal{O}$

The Landau symbol $\mathcal{O}$, sometimes referred to as "big Oh", is used to compare the behavior of one function $f(h)$ with another function $g(h)$ as $h \to 0$. One writes

$$f(h) = \mathcal{O}(g(h)) \quad \text{if} \quad |f(h)| \leq C|g(h)|, \quad C \text{ is a positive constant,}$$

for all h sufficiently small such that $\lim\limits_{h \to 0} \frac{|f(h)|}{|g(h)|} \leq C < \infty$. For example, consider the Taylor series expansion for $\sin x$. One can write

$$\sin x = x - \frac{x^3}{3!} + \mathcal{O}(x^5)$$

since

$$\lim_{x \to 0} \frac{\sin x - x - \frac{x^3}{3!}}{x^5} = \frac{1}{5!} = Constant$$

The Landau symbol $\mathcal{O}$ is used in perturbational methods and numerical methods and is sometimes referred to as an order relation. It will be used throughout this text in the truncation of infinite series to denote the order of the error terms. For example, the Taylor series expansion for $f(x_0 + h)$ when truncated after the second term can be expressed

$$f(x_0 + h) = f(x_0) + f'(x_0)h + \mathcal{O}(h^2)$$

to indicate that the error term is proportional to h^2. One can write

$$Error \le C|h|^2 \quad \text{for any constant } C \text{ satisfying} \quad \frac{f''(\xi)}{2!} \le C.$$

The notation $\mathcal{O}(h^n)$ is used to denote the error being small and behaving like Ch^n, as h gets small, where C is a constant. The statement $Error = \mathcal{O}(h^n)$ is read "the error is of order h^n" and means $\lim_{h \to 0}(Error) = Ch^n$ for some positive constant C.

Taylor Series for Functions of Two Variables

A Taylor series expansion of a function of two variables $f(x, y)$ in the neighborhood of a point (x_0, y_0) can be written in the form

$$
\begin{aligned}
f(x, y) =& f(x_0, y_0) + \frac{\partial f}{\partial x}(x - x_0) + \frac{\partial f}{\partial y}(y - y_0) \\
& + \frac{1}{2!} \left[\frac{\partial^2 f}{\partial x^2}(x - x_0)^2 + 2\frac{\partial^2 f}{\partial x \partial y}(x - x_0)(y - y_0) + \frac{\partial^2 f}{\partial y^2}(y - y_0)^2 \right] + \cdots
\end{aligned}
\tag{1.5}
$$

where all partial derivatives are to be evaluated at the point (x_0, y_0). The above Taylor series expansion can also be written in an operator notation. Define the partial derivative operators

$$D_x f = \frac{\partial f}{\partial x}, \quad D_y f = \frac{\partial f}{\partial y}, \quad D_x^2 f = \frac{\partial^2 f}{\partial x^2}, \quad D_x D_y f = \frac{\partial^2 f}{\partial x \partial y}, \quad D_y^2 f = \frac{\partial^2 f}{\partial y^2}, \quad \text{etc.}$$

and write the Taylor series expansion given by equation (1.5) in the special case where $x = x_0 + h$ and $y = y_0 + k$. One can then write equation (1.5) in the operator form

$$
\begin{aligned}
f(x_0 + h, y_0 + k) =& f(x_0, y_0) + \sum_{n=1}^{\infty} \frac{1}{n!}(hD_x + kD_y)^n f(x, y) \\
f(x_0 + h, y_0 + k) =& f(x_0, y_0) + (hD_x + kD_y)f + \frac{1}{2!}(hD_x + kD_y)^2 f \\
& + \frac{1}{3!}(hD_x + kD_y)^3 f + \frac{1}{4!}(hD_x + kD_y)^4 f + \cdots
\end{aligned}
\tag{1.6}
$$

where all partial derivatives are to be evaluated at the point (x_0, y_0). Note that the operator terms $(hD_x + kD_y)^m f$ can be evaluated by using the binomial expansion.

Example 1-1. (Taylor series.) If $f(x,y)$ and all of its partial derivatives through the nth order are defined and continuous over the rectangular region R defined by $R = \{x, y \mid a \le x \le b, c \le y \le d\}$ and the Taylor series is truncated after the nth derivative terms, then the error term can be calculated from knowledge of Taylor series expansions of a single variable. That is, one can replace x by $x_0 + t(x - x_0)$ and y by $y_0 + t(y - y_0)$ in $f(x,y)$ to obtain a function of the single variable t. One can define

$$\phi(t) = f(x_0 + t(x - x_0), y_0 + t(y - y_0))$$

so that with (x,y) and (x_0, y_0) fixed, the function $\phi(t)$ is a function of a single variable t. Expanding $\phi(t)$ about $t = 0$ gives

$$\phi(t) = \phi(0) + \phi'(0)t + \phi''(0)\frac{t^2}{2!} + \cdots + \phi^{(n)}(0)\frac{t^n}{n!} + Error$$

where

$$Error = \phi^{(n+1)}(t^*)\frac{t^{n+1}}{(n+1)!}, \qquad 0 < t^* < t.$$

Note that at $t = 1$ we have $\phi(1) = f(x,y)$ and when $t = 0$ we have $\phi(0) = f(x_0, y_0)$ so that one can write

$$\phi(1) = f(x,y) = \phi(0) + \phi'(0) + \frac{\phi''(0)}{2!} + \cdots + \frac{\phi^{(n)}(0)}{n!} + Error$$

where for $x - x_0 = h$ and $y - y_0 = k$ we have

$$\phi(0) = f(x_0, y_0)$$

$$\phi'(0) = \left(\frac{\partial f}{\partial x}(x - x_0) + \frac{\partial f}{\partial y}(y - y_0)\right) = (hD_x + kD_y)f$$

$$\phi''(0) = \left(\frac{\partial^2 f}{\partial x^2}(x - x_0)^2 + 2\frac{\partial^2 f}{\partial x \partial y}(x - x_0)(y - y_0) + \frac{\partial^2 f}{\partial y^2}(y - y_0)^2\right) = (hD_x + kD_y)^2 f$$

$$\vdots$$

$$\phi^{(n)}(0) = (hD_x + kD_y)^n f$$

where all partial derivatives are to be evaluated at the point (x_0, y_0). The error term is given by

$$Error\Big|_{t=1} = \frac{\phi^{(n+1)}(t^*)}{(n+1)!} = (hD_x + kD_y)^{n+1} f\Big|_{x=\xi, y=\eta} \qquad \text{for} \quad 0 < t^* < 1,$$

where $\xi = x_0 + t^* h$, $\eta = y_0 + t^* k$ represent some point within the region R.

6

Mean Value Theorem

The mean value theorem states that if $f(x)$ is a continuous function on the closed interval $[a, b]$, then there exists a point ξ, satisfying $a < \xi < b$, such that the slope m_s of the secant line through the points $(a, f(a))$ and $(b, f(b))$ equals the slope of the curve $f(x)$ at $x = \xi$. This can be written and illustrated as follows.

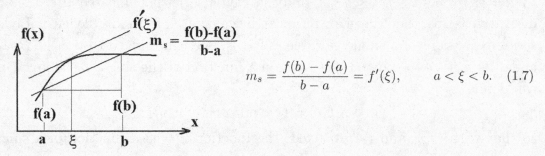

$$m_s = \frac{f(b) - f(a)}{b - a} = f'(\xi), \qquad a < \xi < b. \quad (1.7)$$

Mean Value Theorem for Integrals

The mean value theorem for integrals states the if $f(x)$ is a continuous function and integrable over an interval $[a, b]$, then there exists a value ξ satisfying $a < \xi < b$ such that the average value of the function times the length of the interval from a to b must equal the area under the curve $f(x)$ between a and b. This can be written and illustrated as follows.

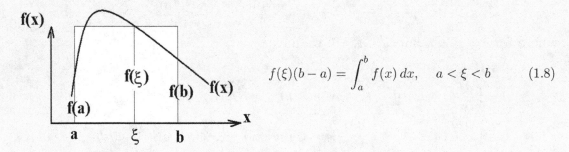

$$f(\xi)(b - a) = \int_a^b f(x)\, dx, \quad a < \xi < b \quad (1.8)$$

The extended mean value theorem for integrals states that if $f(x)$ and $g(x)$ are continuous functions on the closed interval $[a, b]$ and $g(x)$ does not change sign throughout the interval, then there exits a point ξ such that

$$\int_a^b f(x)g(x)\, dx = f(\xi) \int_a^b g(x)\, dx, \qquad a < \xi < b. \quad (1.9)$$

Other forms of this mean value theorem are for the conditions $f(x)$ is positive and monotonic over the interval (a,b) and $g(x)$ is integrable, then one can say there exists at least one value for ξ such that

$$\int_a^b f(x)g(x)\,dx = f(a)\int_a^\xi g(x)\,dx, \qquad a \le \xi \le b.$$

Extreme Value Theorem

The extreme value theorem states that if $f(x)$ is a continuous function over the closed interval $[a,b]$, then there will exist points ξ and η such that $f(\xi)$ is a maximum value of $f(x)$ over the interval and $f(\eta)$ is a minimum value of $f(x)$ over the interval. One can then write

$$minimum = f(\eta) \le f(x) \le f(\xi) = maximum, \qquad \text{for} \quad x \in [a,b].$$

Rolle's Theorem

The Rolle's theorem assumes that $f(x)$ is continuous and differentiable on the closed interval $[a,b]$. One form of Rolle's theorem states that if $f(a) = 0$ and $f(b) = 0$, then there must exist at least one point ξ in the interval such that $f'(\xi) = 0$, $a < \xi < b$.

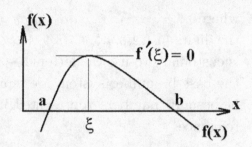

Intermediate Value Theorem

The intermediate value theorem states that if $f(x)$ is continuous on the closed interval $[a,b]$ and there exists a value f_0 such that $f(a) < f_0 < f(b)$, then there exists at least one value ξ such that $f(\xi) = f_0$. In the accompanying figure note that for the f_0 selected there exists more than one value for ξ such that $f(\xi) = f_0$.

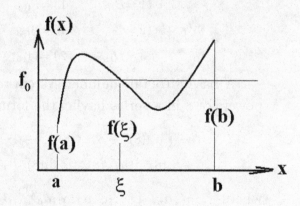

Number Representation

A base 10 (decimal) number system represents a number N in terms of various powers of 10 in a series having the form

$$N = \cdots + \alpha_n(10)^n + \alpha_{n-1}(10)^{n-1} + \cdots + \alpha_3(10)^3 + \alpha_2(10)^2 + \alpha_1(10)^1 + \alpha_0(10)^0$$
$$+ \beta_1(10)^{-1} + \beta_2(10)^{-2} + \beta_3(10)^{-3} + \cdots \tag{1.10}$$

where $\dots, \alpha_n, \alpha_{n-1}, \dots, \alpha_3, \alpha_2, \alpha_1, \alpha_0, \beta_1, \beta_2, \beta_3, \dots$ are coefficients representing one of the digits $\{0, 1, 2, 3, 4, 5, 6, 7, 8, 9\}$. Leading zeros and trailing zeros are not written. For example, the number N=8326.432 in the base ten number system is really a shorthand representation for the number

$$N = 8(10)^3 + 3(10)^2 + 2(10)^1 + 6(10)^0 + 4(10)^{-1} + 3(10)^{-2} + 2(10)^{-3}.$$

A base 16 (hexadecimal) number system represents a number N in terms of various powers of 16 in a series having the form

$$N = \dots + \alpha_n(16)^n + \alpha_{n-1}(16)^{n-1} + \dots + \alpha_3(16)^3 + \alpha_2(16)^2 + \alpha_1(16)^1 + \alpha_0(16)^0$$
$$+ \beta_1(16)^{-1} + \beta_2(16)^{-2} + \beta_3(16)^{-3} + \dots \tag{1.11}$$

where $\dots, \alpha_n, \alpha_{n-1}, \dots, \alpha_3, \alpha_2, \alpha_1, \alpha_0, \beta_1, \beta_2, \beta_3, \dots$ are coefficients representing one of the digits $\{0, 1, 2, 3, 4, 5, 6, 7, 8, 9, A, B, C, D, E, F\}$. (When the base is larger than 10 it is customary to use the letters $A - Z$ to represent the needed digits.) Numbers in the base b number system are represented using a subscript b. Some examples of base 10 numbers represented in the base 16 number system are listed for reference.

$10 = A_{16}$	$14 = E_{16}$	$30 = 1E_{16}$	$70 = 46_{16}$
$11 = B_{16}$	$15 = F_{16}$	$40 = 28_{16}$	$80 = 50_{16}$
$12 = C_{16}$	$16 = 10_{16}$	$50 = 32_{16}$	$90 = 5A_{16}$
$13 = D_{16}$	$20 = 14_{16}$	$60 = 3C_{16}$	$100 = 64_{16}$

A base 8 (octal) number system represents a number N in terms of various powers of 8 in a series having the form

$$N = \dots + \alpha_n(8)^n + \alpha_{n-1}(8)^{n-1} + \dots + \alpha_3(8)^3 + \alpha_2(8)^2 + \alpha_1(8)^1 + \alpha_0(8)^0$$
$$+ \beta_1(8)^{-1} + \beta_2(8)^{-2} + \beta_3(8)^{-3} + \dots \tag{1.12}$$

where $\dots, \alpha_n, \alpha_{n-1}, \dots, \alpha_3, \alpha_2, \alpha_1, \alpha_0, \beta_1, \beta_2, \beta_3, \dots$ are coefficients representing one of the digits $\{0, 1, 2, 3, 4, 5, 6, 7\}$.

A base 2 (binary) number system represents a number N in terms of various powers of 2 in a series having the form

$$N = \dots + \alpha_n(2)^n + \alpha_{n-1}(2)^{n-1} + \dots + \alpha_3(2)^3 + \alpha_2(2)^2 + \alpha_1(2)^1 + \alpha_0(2)^0$$
$$+ \beta_1(2)^{-1} + \beta_2(2)^{-2} + \beta_3(2)^{-3} + \dots \tag{1.13}$$

where $\ldots, \alpha_n, \alpha_{n-1}, \ldots, \alpha_3, \alpha_2, \alpha_1, \alpha_0, \beta_1, \beta_2, \beta_3, \ldots$ are coefficients representing one of the digits $\{0, 1\}$.

Note that a base b number system requires b-digits to be used as coefficients in representing the numbers. For example, the Babylonians of long ago used a base 60 number system which requires 60 symbols to be used as digits. Some conventions from this system that have survived the many centuries is the fact that we have 60 seconds in a minute, 60 minutes in a hour, and 360 degrees in a circle.

Some other names associated with number systems are the following. A base 3 number system is called a ternary system, a base 4 system is called a quaternary system, a base 5 system is called a quinary system, a base 6 number system is called a senary system, a base 7 number system is called a septenary system, a base 9 number system is called a nonary system, a base 11 number system is called a undenary number system, and a base 12 number system is called a duodenary number system.

Number Conversion

To convert a number N from a decimal base to base b it is necessary to calculate the coefficients $\alpha_n, \alpha_{n-1}, \ldots, \alpha_1, \alpha_0, \beta_1, \beta_2, \ldots$ in the base b number system, where $N = \alpha_n b^n + \alpha_{n-1} b^{n-1} + \cdots + \alpha_2 b^2 + \alpha_1 b^1 + \alpha_0 b^0 + \beta_1 b^{-1} + \beta_2 b^{-2} + \cdots$. The integer part of N is denoted $I[N]$ and can be expressed in the factored form

$$I[N] = \alpha_0 + b(\alpha_1 + b(\alpha_2 + b(\alpha_3 + \cdots + b\alpha_n) \cdots))$$

from which one can observe that if the integer part of N is divided by b, then α_0 is the remainder and the quotient is $Q_1 = \alpha_1 + b(\alpha_2 + b(\alpha_3 + \cdots b\alpha_n) \cdots))$. If Q_1 is divided by b, then the remainder is α_1 and the new quotient is $Q_2 = \alpha_2 + b(\alpha_3 + \cdots \alpha_n b) \cdots))$. Continuing this process and saving the remainders $\alpha_0, \alpha_1, \ldots, \alpha_n$ the coefficients for the integer part of N can be determined. The fractional part of N is denoted $F[N]$ and can be expressed in the form

$$F[N] = \beta_1 b^{-1} + \beta_2 b^{-2} + \beta_3 b^{-3} + \cdots$$

from which one can observe that if $F[N]$ is multiplied by b, then there results

$$bF[N] = \beta_1 + \beta_2 b^{-1} + \beta_3 b^{-2} + \cdots$$

so that β_1 is the integer part of $bF[N]$ and the term $\beta_2 b^{-1} + \beta_3 b^{-2} + \cdots$ represents the fractional part of $bF[N]$. Hence if one continues to multiply the resulting fractional parts by b, then one can calculate the coefficients $\beta_1, \beta_2, \beta_3, \ldots$ associated with the fractional part of the base b representation of N.

Example 1-2. (**Number conversion.**)

Convert the number $N = 123.640625$ to a base 2 representation.

Solution: We start with the integer part of N and write $I[N] = 123 = \sum_{i=0}^{n} \alpha_i (2)^i$.
Now divide 123 by 2 and save the remainder R. Continue to divide the resulting quotients and save the remainders as the remainders give us the coefficients $\alpha_0, \alpha_1, \ldots, \alpha_n$ in the base 2 representation. One can construct the following table to find the coefficients

N	I[N/2] =Q	R
123	I[123/2]=61	$1 = \alpha_0$
61	I[61/2]=30	$1 = \alpha_1$
30	I[30/2]=15	$0 = \alpha_2$
15	I[15/2]=7	$1 = \alpha_3$
7	I[7/2]=3	$1 = \alpha_4$
3	I[3/2]=1	$1 = \alpha_5$
1	I[1/2]=0	$1 = \alpha_6$

The integer part of N can now be represented $I[N] = 123 = 1111011_2$ The fractional part of N is written in the form $F[N] = 0.640625 = \sum_{i=1}^{\infty} \beta_i (2)^{-i}$. Now continue to multiply the fractional part by 2 and save the integer part each time. These integer parts represent the coefficients $\beta_1, \beta_2, \ldots$. One can construct the following table for determining the coefficients

N	2N	F[2N]	I[2N]
0.640625	1.28125	0.28125	$1 = \beta_1$
0.28125	0.5625	0.5625	$0 = \beta_2$
0.5625	1.125	0.125	$1 = \beta_3$
0.125	0.25	0.25	$0 = \beta_4$
0.25	0.50	0.50	$0 = \beta_5$
0.50	1.00	0.00	$1 = \beta_6$

The fractional part of N can be represented $F[N] = 0.640625 = 0.101001_2$ and the original number N has the base 2 representation $N = 123.640625 = 1111011.101001_2$.

■

Conversion Between Different Bases

The conversion between octal and binary numbers is accomplished by observing that a binary number represented

$$\cdots \alpha_5 2^5 + \alpha_4 2^4 + \alpha_3 2^3 + \alpha_2 2^2 + \alpha_1 2^1 + \alpha_0 2^0 + \beta_1 2^{-1} + \beta_2 2^{-2} + \beta_3 2^{-3} + \cdots \qquad (1.14)$$

can be represented in the alternate form

$$\cdots + (\alpha_5 2^2 + \alpha_4 2 + \alpha_3)8^1 + (\alpha_2 2^2 + \alpha_2 2 + \alpha_0)8^0 + (2^2\beta_2 + 2\beta_2 + \beta_3)8^{-1} + (2^2\beta_4 + 2\beta_5 + \beta_6)8^{-2} + \cdots \quad (1.15)$$

Observe that if each of the coefficients $\ldots, \alpha_4, \alpha_3, \alpha_2, \alpha_1, \alpha_0, \beta_1, \beta_2, \ldots$ represents one of the digits $\{0,1\}$, then the numbers in each parenthesis represent one of the digits from the set $\{0,1,2,3,4,5,6,7\}$ and so equation (1.15) is an octal representation of the equation (1.14). This indicates that binary numbers can be grouped into sets of three digits beginning at the binary point and moving left and right. Each group of three binary numbers can then be replaced by their octal equivalent. It is convenient to make use of the following table of values to simplify this conversion.

binary	000	001	010	011	100	101	110	111
octal	0	1	2	3	4	5	6	7

Example 1-3.   (Number conversion.)
Convert the binary number 1110101.010110011_2 to an octal representation.
Solution: Start at the binary point and create groups of three in each direction to obtain

$$(001)(110)(101).(010)(110)(011)_2 = 165.263_8$$

Now use the previous binary to octal table of values for conversion of the groups of three from binary to octal. One then obtains the answer given above. ∎

In a similar manner one can show that groupings of four binary digits make conversion from binary to hexadecimal a straightforward matter. For binary to hexadecimal conversion it is convenient to use the following table of values.

binary	0000	0001	0010	0011	0100	0101	0110	0111
hexadecimal	0	1	2	3	4	5	6	7
binary	1000	1001	1010	1011	1100	1101	1110	1111
hexadecimal	8	9	A	B	C	D	E	F

Example 1-4. (**Number conversion.**)

Convert the hexadecimal number $1A.B2_{16}$ to binary.

Solution: One can readily verify that

$$1A.B2_{16} = (0001)(1010).(1011)(0010)_2 = 26.6953125$$

Computer Storage of Numbers

The terminology bits is used as a shorthand notation for binary digits. In computer terminology a word is a group of bits used for storage at a specific location within the computer. The computer word size is determined by the number of bits employed. In the discussions that follow a 32-bit word size will be used for illustrative purposes. The reader will have to make appropriate modifications of the discussions if the word size is larger or smaller.

Digital computers store numbers in a binary format having a fixed number of bits. This discrete size prevents digital computers from representing the continuous aspects of the real number system. This shortcoming will be made apparent shortly. Digital computers usually represent numbers in either a fixed point or floating point format. The fixed point format is reserved for integers which are stored in a binary equivalent format. The fixed point and floating point representation of a number is illustrated in the figure 1-1.

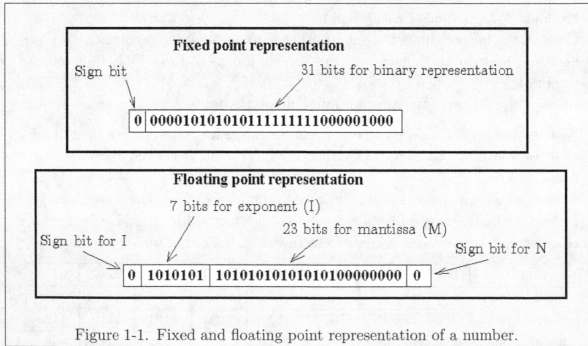

Figure 1-1. Fixed and floating point representation of a number.

The sign bit uses 0 for positive numbers and 1 for negative numbers. An examination of the fixed point representation of an integer reveals that the largest positive integer that can be stored in a 32-bit word size is the equivalent binary number having 31 consecutive 1's.

$$N_{max} = 1111111111111111111111111111111_2 = 2147483647$$

The floating point 32-bit representation of a number N uses a sign bit for N, 23-bits for a mantissa M and 7-bits for an exponent I and a sign bit for I. The number N is calculated from the relation

$$N = \pm M\,(2)^I \tag{1.16}$$

The mantissa M is a binary fraction having the form

$$M = +0.1\beta_2\beta_3\beta_4\cdots\beta_{23} \tag{1.17}$$

where $\beta_1 = 1$ is understood to be the first bit value. This type of floating point number representation is called a normalized floating point binary format. Note that multiplying a binary fraction M by 2^I has the effect of shifting the binary point. If the exponent I is positive, then the binary point is moved I bits to the right. If the exponent I is negative, then the binary point is shifted I bits to the left. A floating point representation like that illustrated in the figure 1-1 can produce the maximum exponent $I = 1111111_2 = 127$. Hence, the exponent I can range over the decimal values

$$-127, -126, \ldots, -1, 0, +1, \ldots, +126, +127.$$

The largest floating point positive number that can be represented in a 32-bit word size is

$$N_{max} = 0.11111111111111111111111_2(2)^{127} = (1 - 2^{-23})(2)^{127}$$

$$N_{max} = 170141163178059628080016879768632819712$$

$$N_{max} = 1.7014116....(10)^{38}.$$

The smallest positive number that can be represented in a 32-bit floating point word size is

$$N_{min} = 0.10000000000000000000000_2(2)^{-127} = \frac{1}{2}(2)^{-127} = 2^{-128}$$

$$N_{min} = \frac{1}{340282366920938463463374607431768211456}$$

$$N_{min} = 2.9387358770557....(10)^{-39}.$$

If you try to store numbers bigger than N_{max} you get a floating point overflow error. Similarly, if you try to store a number less than N_{min} you get a floating point underflow error.

Observe that in representing numbers in floating point format, with a 32-bit word size, there are certain numbers that cannot be represented exactly. For example, the decimal number $0.1 = 1/10$ has no exact finite representation in a binary format. One can verify that the binary representation of $1/10$ is given by the infinite representation

$$\frac{1}{10} = 0.00011001100110011001\overline{1100}_2$$

where $\overline{1100}$ denotes the repetition of the binary numbers 1100 forever. The digital computer floating point representation can only store a finite number of binary digits and so for any attempt to store the number $1/10$ in computer storage there must occur a truncation or chopping off of the infinite digit representation after some fixed number of bits. This illustrates that errors are always introduced in the representation of certain numbers. Consequently, there will always be errors and approximations in calculating with these special numbers if you use a digital computer with a fixed word size. The procedure for replacing a number by its nearest binary digital storage number is called rounding and the resultant error of approximation is called round-off error. If N_t is the true number which is approximated by N_a, then the absolute error of the approximation is defined

$$Absolute\ Error\ =\ |N_t - N_a|.$$

The relative error of the approximation is defined

$$Relative\ Error\ =\ \frac{|N_t - N_a|}{|N_t|}.$$

Note that the relative error is not defined in the special case where $N_t = 0$. The percentage error is defined as 100 times the relative error or

$$Percentage\ Error\ =\ 100 * \frac{|N_t - N_a|}{|N_t|}.$$

The digital machine accuracy ϵ, sometimes called the machine epsilon, is defined as the smallest number ϵ which when added to 1 causes the digital machine to recognize that $1 + \epsilon > 1$. To determine your machine epsilon as a power of 2, you can write a computer program which follows the flow chart illustrated in the figure 1-2.

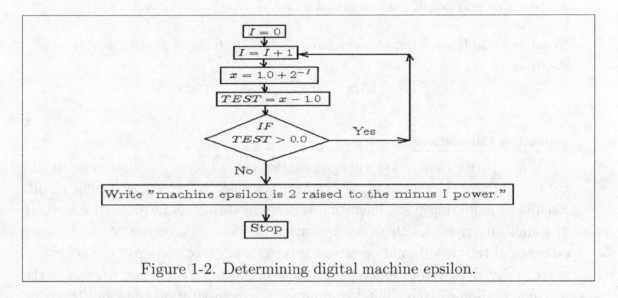

Figure 1-2. Determining digital machine epsilon.

The flowchart illustrated in the figure 1-2 is a suggested method for determining the machine epsilon for your computer. Your program should be run first in single precision and then in double precision.

To illustrate the rounding off of a decimal number consider the symbol e which is used to represent the transcendental number $2.71828182845904523536\ldots$, which has an infinite number of digits. To represent an approximation of e with a finite number of digits we retain as many digits as we desire and then chop off or throw away the remaining digits. This process is called truncation of the number.

The general rule for rounding numbers is as follows: numbers are rounded to cause the least possible error. The general procedure employed to round a number is now illustrated. Let α_i, $i = 1, 2, 3, \ldots$, denote one of the digits 0, 1, 2, 3, 4, 5, 6, 7, 8, 9. Denote by $\alpha_1\alpha_2\alpha_3\ldots\alpha_n\alpha_{n+1}\ldots$ some number, where the decimal point can be anywhere you like if you are careful to fill in any leading or trailing zeros appropriately. To round this number to n significant figures, we throw away all the digits to the right of the nth place to obtain $\alpha_1\alpha_2\ldots\alpha_n$. Let $T_n = \alpha_{n+1}\alpha_{n+2}\ldots$ denote the numbers thrown away. Now if T_n is less than one half of a unit in the nth place, we leave the digit α_n unchanged. If T_n is greater than one half of a unit in the nth place, then add one to the digit α_n. If T_n is exactly one half of a unit in the nth place, then leave the digit α_n unchanged if it is even or add one if it is an odd digit.

Example 1-5. Round the number e to 2,3,4,5 and 6 significant figures.
Solution:

$$2.7, \quad 2.72, \quad 2.718, \quad 2.7183, \quad 2.71828$$

■

Numerical Calculations

When performing numerical computations on a digital computer one should try to minimize the number of operations N needed to achieve a specific result. Each time a floating point operation is performed there is introduced some sort of round off error and these errors accumulate with increasing N. As a rough estimate of the total round off error expected after N operations are performed, one can say that it is probably somewhere between $\sqrt{N}\epsilon$ and $N\epsilon$, where ϵ is the machine epsilon of your digital computer. The computation time and resulting round off error are dependent upon the number and type of calculations being performed.

One example of how to speed up calculations is to always put polynomials in a nested form before evaluating them. For example, the sixth degree polynomial

$$P_6(x) = \alpha_0 x^6 + \alpha_1 x^5 + \alpha_2 x^4 + \alpha_3 x^3 + \alpha_4 x^2 + \alpha_5 x + \alpha_6$$

can be written in the nested form

$$P_6(x) = \alpha_6 + x(\alpha_5 + x(\alpha_4 + x(\alpha_3 + x(\alpha_2 + x(\alpha_1 + \alpha_0 x))))).$$

The nested form does not require the additional operations of calculating x to a power which is required in the unnested form.

Try do avoid calculations where you divide one large number by another large number (∞/∞), or the dividing of a small number by another small number ($0/0$) as these types of calculations lead to errors. One should always scale the variables used in calculations by using some naturally occurring length associated with the problem. Try to use scale factors that keep the variables involved near unity.

If you calculate y as a function of n-independent variables, $y = f(x_1, x_2, \ldots, x_n)$ and each variable x_i has an error Δx_i for $i = 1, \ldots, n$, then one can use a Taylor series expansion to find the error in calculating y. One finds

$$y + \Delta y = f(x_1 + \Delta x_1, x_2 + \Delta x_2, \ldots, x_n + \Delta x_n)$$
$$= f(x_1, x_2, \ldots, x_n) + \frac{\partial f}{\partial x_1}\Delta x_1 + \frac{\partial f}{\partial x_2}\Delta x_2 + \cdots + \frac{\partial f}{\partial x_n}\Delta x_n + h.o.t.$$

where $h.o.t.$ denotes higher order terms which have been neglected. If each Δx_i, $i = 1, \ldots, n$, is small, then one finds the error in calculating y can be approximated by the relation

$$\Delta y = \frac{\partial f}{\partial x_1}\Delta x_1 + \frac{\partial f}{\partial x_2}\Delta x_2 + \cdots + \frac{\partial f}{\partial x_n}\Delta x_n. \tag{1.18}$$

Applied to the operation of multiplication where $y = f(x_1, x_2) = x_1 x_2$ one finds the error associated with multiplication is given by

$$\Delta y = x_2 \Delta x_1 + x_1 \Delta x_2.$$

This implies the relative error is given by

$$\frac{\Delta y}{y} = \frac{\Delta x_1}{x_1} + \frac{\Delta x_2}{x_2}$$

which is a sum of the relative errors of the individual multiplicands.

Infinite Sequence

If to each positive integer n there is assigned a number x_n, then the set of numbers $\{x_1, x_2, x_3, \ldots, x_n, x_{n+1}, \ldots\}$ are said to form an infinite sequence. The sequence is sometimes written in the form $\{x_n\}_{n=1}^{\infty}$. A sequence $\{x_n\}_{n=1}^{\infty}$ is said to converge to a number X, or have a limit X, if for each positive small number ϵ, an integer N can be found such that

$$|x_n - X| < \epsilon \qquad \text{for all} \quad n > N.$$

If the sequence converges, then one can write $\lim\limits_{n \to \infty} x_n = X$. If a limit X does not exist, then the sequence is called a diverging sequence. The above definition for convergence of a sequence implies that for all $n > N$ the sequence values will be trapped within the interval $X - \epsilon < x_n < X + \epsilon$.

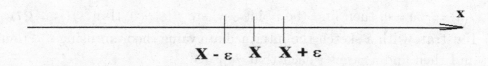

Exercises Chapter 1

▶ **1.**

Find the derivative of the following functions

(a) $\quad f(x) = \sin x^2$ $\qquad$ (b) $\quad f(x) = x^2 \cos x$ $\qquad$ (c) $\quad f(x) = \dfrac{\tan x}{1 + x^2}$

▶ **2.**

Find the derivative of the following functions

(a) $\quad f(x) = \sin^{-1} x^2$ $\qquad$ (b) $\quad f(x) = \cosh x^3$ $\qquad$ (c) $\quad f(x) = x \exp(-x^2)$

▶ **3.**

Evaluate the given integrals

$$I_a = \int_0^{\pi} \sin x \, dx \qquad I_b = \int_0^{\infty} e^{-\alpha x} \, dx \qquad I_c = \int_0^{x} \frac{dt}{1 + t^2}$$

▶ **4.**

Evaluate the given integrals

$$I_a = \int_0^{\pi} x \cos x^2 \, dx \qquad I_b = \int_0^{\infty} x e^{-\alpha x} \, dx \qquad I_c = \int_0^{x} \frac{dt}{\alpha^2 - t^2}$$

▶ **5.** $\quad$ Consider the function $y = f(x) = -\dfrac{2}{3}x^2 + \dfrac{16}{3}x - \dfrac{14}{3}$ over the interval $0 \leq x \leq 8$.

(a) Calculate the slope of the secant line m_s through the points $(2, 10/3)$ and $(5, 16/3)$ and illustrate the mean value theorem by finding the point ξ such that $f'(\xi) = m_s$. Illustrate the situation with a sketch.

(b) Illustrate with a sketch the Rolle's theorem by finding points a and b where $f(a) = f(b) = 0$, then find the point ξ where $f'(\xi) = 0$. Hint: $f(1) = f(7) = 0$.

(c) Illustrate with a sketch the intermediate value theorem using $a = 1$ and $b = 4$ and then find a point ξ such that $f(\xi) = 9/2$.

▶ **6.** (Series.)

The following are tests for convergence of a series $\sum_{n=1}^{\infty} u_n(x)$. A necessary condition for convergence of the series is for $\lim_{n \to \infty} u_n = 0$.

The ratio test: If the limit $\rho = \lim_{n \to \infty} \frac{u_{n+1}}{u_n}$ exists, then the series converges if $\rho < 1$, diverges if $\rho > 1$ and the test fails if $\rho = 1$.

The root test: If the limit $\lim_{n \to \infty} \sqrt[n]{|u_n|} = \rho$ exists, then the series converges if $\rho < 1$, diverges if $\rho > 1$ and the test fails if $\rho = 1$.

Comparison test: If $|u_n| \leq \beta_n$ for all n and if the series $\sum_{n=1}^{\infty} \beta_n$ converges then the series $\sum_{n=1}^{\infty} u_n$ converges absolutely.

Integral test: If $y(x)$ is continuous for all $x \geq x_0$, where x_0 is a constant, and $y(x)$ decreases as x increases such that $\lim_{x \to \infty} y(x) = 0$, and $y(n) = u_n$, then the convergence or divergence of the series $\sum_{n=1}^{\infty} u_n$ is determined by whether the integral $\int_{x_0}^{\infty} y(x) \, dx = \lim_{T \to \infty} \int_{x_0}^{T} y(x) \, dx$ exists or does not exist.

Test the following series for convergence

$$(a) \quad \sum_{n=1}^{\infty} \frac{1}{n^2}, \qquad (b) \quad \sum_{n=1}^{\infty} \frac{1}{n!}, \qquad (c) \quad \sum_{n=1}^{\infty} \frac{(-1)^{n+1}}{(2n-1)!}$$

▶ **7.** (Series.)

Determine the values of x for which the following series converge.

$$(a) \quad \sum_{n=1}^{\infty} \frac{x^n}{2n^2 - n}, \qquad (b) \quad \sum_{n=1}^{\infty} \frac{nx^n}{2^n}, \qquad (c) \quad \sum_{n=1}^{\infty} \frac{1}{nx^{2n}}$$

▶ **8.** (Taylor series.)

Determine the Taylor series expansion of $f(x) = \ln(1-x)$ about the point $x = 0$. Use the ratio test to determine the radius of convergence of the series.

▶ **9.** (Taylor series.)

Determine the Taylor series expansion of $f(x) = \ln(1+x)$ about the point $x = 0$. Use the ratio test to determine the radius of convergence of the series.

▶ **10.** (Taylor series.)

(a) Determine the Taylor series expansion of the function $f(x) = \frac{1}{1-x}$ about the point $x = 0$.

(b) Use the Binomial expansion on $(1-x)^{-1}$ to obtain an equivalent series.

▶ **11.** The numbers π, e and $\sqrt{2}$ truncated to 50 digits are given by

$$3.1415926535897932384626433832795028841971693993751$$

$$2.7182818284590452353602874713526624977572470936999$$

$$1.4142135623730950488016887242096980785696718753769$$

Fill in the following tables representing the percent error in truncating these numbers at various places.

Approximation to π	% Error	Approximation to e	% Error	Approximation to $\sqrt{2}$	% Error
3.000000		2.000000		1.000000	
3.100000		2.700000		1.400000	
3.140000		2.710000		1.410000	
3.141000		2.718000		1.414000	
3.141500		2.718200		1.414200	
3.141590		2.718280		1.414210	
3.141592		2.718281		1.414213	

Comment on accuracy in representing numbers to an excessive number of digits. Whenever possible, use the exact value like π, e or $\sqrt{2}$, which is preferred over decimal approximations.

▶ **12.** (Taylor series.)

(a) Expand $f(x,y) = xy^2$ in a Taylor series about the point $(x_0, y_0) = (1,1)$.

(b) Expand $f(x,y) = xy^2$ in a Taylor series about the point $(x_0, y_0) = (1,2)$.

▶ **13.** (Landau Symbol O.)

Use Taylor series and construct examples that illustrate the following properties of the Landau symbol $\mathcal{O}$.

(a) $\quad \mathcal{O}(h^m) + \mathcal{O}(h^m) = \mathcal{O}(h^m),$

(b) $\quad \begin{array}{l} \mathcal{O}(h^m) + \mathcal{O}(h^n) = \mathcal{O}(h^p) \\ \text{where } p = \min(m,n). \end{array}$

▶ **14.**   The function $y = y(t) = e^{-t/\tau}$ has a parameter τ which is called a time constant. Quantify the statement that after 5 time constants this function is essentially zero. Examine the function at the times $t = \tau, 2\tau, 3\tau, 4\tau, 5\tau, 6\tau$ and determine what "close to zero" means.

▶ **15.**   (**Relative error.**)
By definition a number N_a approximates a number N_t to d significant digits if the relative error satisfies $\dfrac{|N_t - N_a|}{|N_t|} < \dfrac{1}{2}(10)^{-d}$, where d is a positive integer.
(a) Find bounds on N_a if $d = 3$ and N_t is one of the numbers $10, 100, 1000, 10,000$.
(b) Find bounds on N_a if $d = 4$ and N_t is one of the numbers $10, 100, 1000, 10,000$.

▶ **16.**   Let $y = f(x_1, \ldots, x_n) = x_1 x_2 \cdots x_n$ denote a product of n-numbers. If all the numbers have an error ϵ, then use equation (1.18) and find the relative error associated with such a product.

▶ **17.**   Let $y = f(x_1, x_2) = x_1/x_2$ denote the division of two numbers. If both x_1 and x_2 have errors of Δx_1, and Δx_2 respectively, then use equation (1.18) to find the maximum relative error in calculating y. Hint: Δx_1 or Δx_2 can be positive or negative.

▶ **18.**
Represent the numbers $\{100, 10, 1, 0.1, 0.01\}$ in the base 2 number system.

▶ **19.**
Represent the numbers $\{100, 10, 1, 0.1, 0.01\}$ in the base 8 number system.

▶ **20.**
Represent the numbers $\{100, 10, 1, 0.1, 0.01\}$ in the base 16 number system.

▶ **21.**
Represent the numbers $\{100, 10, 1, 0.1, 0.01\}$ in the base 3 number system.

▶ **22.**
(a) Verify that groupings of four binary digits make conversion from binary to hexadecimal a straight forward matter.
(b) Illustrate the results from part (a) by converting the following decimal numbers to binary and then to hexadecimal.

$$\{1024, \quad 512, \quad 256, \quad 128, \quad 64, \quad 32, \quad 16, \quad 8, \quad 4, \quad 2, \quad 1\}$$

▶ **23.** In scientific notation a number of the form $d_1.d_2d_2\ldots d_n \times 10^p$, where d_i, $i = 1,\ldots,n$ are integers, is said to contain n-significant figures. In performing an operation of addition or subtraction using numbers with m-significant figures and n-significant figures, the result can only be accurate to as many decimal places as the least accurate of the numbers involved in the operation. Here the least accurate of the numbers dictates how an answer should be rounded. In performing an operation of multiplication or division using numbers with m-significant figures and n-significant figures, the result is only accurate to the number of significant digits in the least accurate of the numbers involved in the operation.

Let $a = 4.256 \times 10^2$, $b = 1.23 \times 10^{-1}$, $c = 1.2 \times 10^{-2}$ and calculate the following.

(i) $a + b$ (iv) b/c

(ii) $a - b$ (v) a/c

(iii) ab (vi) a/b

▶ **24.** Find the Taylor series expansion of the function $f(x) = \sqrt{x+1}$ about the point $x = 0$.

(a) Truncate the series after three terms and use the resulting polynomial to approximate   (i) $\sqrt{0.5}$ and (ii) $\sqrt{1.5}$

(b) Truncate the series after four terms and use the resulting polynomial to approximate   (i) $\sqrt{0.5}$ and (ii) $\sqrt{1.5}$

(c) Discuss the errors resulting from the approximations in parts (a) and (b).

▶ **25.** Illustrate with a sketch the mean value theorem for integrals. Find a point ξ such that $f(\xi)(b - a) = \displaystyle\int_a^b f(x)\,dx$ for the following cases.

(a) $f(x) = \sin x$ with $a = 0$ and $b = \pi$.

(b) $f(x) = 1/x$ with $a = 1$ and $b = 4$.

▶ **26.** The Greek mathematician Hero (or Heron) of Alexandria (AD 20-62) derived the following formula for the area A of a triangle with sides a, b and c $A = \sqrt{s(s-a)(s-b)(s-c)}$ where $s = \frac{1}{2}(a+b+c)$. Assume that a has an error of ϵ in its measurement, the other values are exact. Use equation (1.18) and find the relative error in calculating A by the above formula if $a = 3$, $b = 4$ and $c = 5$.

► **27.** **(Computer problem–Rounding of numbers)**

Many people have gone to jail for improper rounding of numbers. The rounding of a number is very important in the business world.

Assume you borrow N dollars at $I\%$ interest/year and promise to repay the debt by making periodic monthly payments of R dollars over a period of Y years. This is the amortization method. The monthly interest rate is $i = \frac{I\%}{100*12}$ and the periodic monthly payment is given by $R = \frac{Ni}{1-(1+i)^{-n}}$ where $n = 12 * Y$ is the number of monthly payments. An amortization table might have the following form.

col(1) Payment number	col(2) Outstanding Principal at beginning of each period	col(3) Interest due at end of period	col(4) Payment at end of each period	col(5) Portion of principal reduced by payment
	col(2)-col(5)	col(2)$\times i$	R	col(4)-col(3)
1	20,000.00			
2				
$\vdots$	$\vdots$	$\vdots$	$\vdots$	$\vdots$
36	Last principal due	Interest due	R(last)	
totals		Total Interest Sum of col(3)=S_3	Total Payment Sum of col(4)=S_4	Total Sum of col(5)=S_5

Assume a \$20,000.00 loan over a period of 3 years at a interest rate of 6.735% per year. Fill in the above table such that (1.) All money is rounded to the nearest cent. (2.) The last principal due plus last interest due equals R(last). (3.) The sum of entries in column(5) must equal \$20,000.00. (4) $S_3 + S_4 = S_5$. If you do not round properly your table of values will not add up correctly.

► **28.** Show that a Taylor series expansion can be used to remove the indeterminate forms that result as the numerator and denominator approach zero.

(a) $\displaystyle\lim_{x\to 2} \frac{e^{x-2} - e^{2-x}}{\sin(x-2)}$

(d) $\displaystyle\lim_{x\to 0} \frac{e^x - 1 - x}{\sin^2 x}$

(b) $\displaystyle\lim_{\substack{x\to 0\\ x>0}} \frac{1 - \cos\sqrt{x}}{x}$

(e) $\displaystyle\lim_{x\to 0} \frac{\sin x - x}{x^3}$

(c) $\displaystyle\lim_{x\to 2} \frac{\cos(x-2) - 1}{(x-2)^2}$

(f) $\displaystyle\lim_{x\to 0} \frac{x - \frac{x^3}{3!} - \sin x}{x^5}$

▶ **29.** There are many applications occurring in science and engineering where, in order to obtain a simplified result, one employs the approximations

$$\sin\theta = \tan\theta \approx \theta, \text{ for small angles } \theta.$$

Fill in the following table.

Error of Approximation			
θ (degrees)	θ (radians)	Percent Error in approximation $\sin\theta = \theta$	Percent Error in approximation $\tan\theta = \theta$
1			
2			
3			
4			
5			
6			
7			
8			
9			
10			

▶ **30.** (Error.)

Assume you have a computer which stores the following base 2 approximation of the decimal number 1/10.

$$0.0001100110011001100110_2$$

(a) Let N_a denote the decimal form of the above number. Find N_a

(b) What is the absolute error in the representation of 1/10 ?

(c) What is the relative error in the representation of 1/10 ?

(d) Find the next larger decimal number $N_a + \epsilon_1$ if one bit is added to the binary representation of 1/10?

(e) Find the next smaller decimal number $N_a - \epsilon_2$ if one bit is subtracted from the binary representation of N_a.

(f) Note that $N_a - \epsilon_2, N_a, N_a + \epsilon_1$ are numbers that the digital computer can store. Write a statement concerning the usage of real numbers R_1 and R_2 which satisfy the inequalities $N_a - \epsilon_2 < R_2 < N_a < R_1 < N_a + \epsilon_1$, if these numbers are to be used in a digital computer.

"Socrates dialectical Procedure: For an over all view what is now necessary is the movement of consciousness from knowledge of particular objects to an understanding of general concepts."

Socrates (469-399) BCE

Chapter 2
Equation Solving

This chapter deals with finding solutions of algebraic and transcendental equations of either of the forms

$$f(x) = 0 \qquad \text{or} \qquad f(x) = g(x) \qquad (2.1)$$

where we want to solve for the unknown x. An algebraic equation is an equation constructed using the operations of $+, -, \times, \div$, and possibly root taking (radicals). Rational functions and polynomials are examples of algebraic functions. Transcendental equations in comparison are not algebraic. That is, they contain non-algebraic functions and possibly their inverses functions. Equations which contain either trigonometric functions, inverse trigonometric functions, exponential functions, and logarithmic functions are examples of non-algebraic functions which are called transcendental functions. Transcendental functions also include many functions defined by the use of infinite series or integrals.

Graphical Methods

Confronted with equations having one of the above forms and assuming one has access to a graphical calculator or computer that can perform graphics, then one should begin by plotting graphs of the given functions. If the equation to be solved is of the form $f(x) = 0$, then we plot a graph of $y = f(x)$ over some range of x until we find where the curve crosses the x−axis. Points where $y = 0$ or $f(x) = 0$ are called the roots of $f(x)$ or the zeros of $f(x)$.

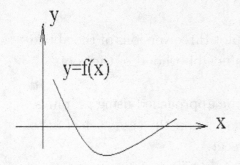

The point (or points) where the given curve $y = f(x)$ crosses the x−axis is where $y = 0$. All such points of intersection then represent solutions to the equation $f(x) = 0$.

Example 2-1. (**Root of algebraic equation.**)

Estimate the solutions of the algebraic equation

$$f(x) = x^3 - \frac{132}{32}x^2 + \frac{28}{32}x + \frac{147}{32} = 0.$$

Solution: We use a computer or calculator and plot a graph of the function $y = f(x)$ and obtain the figure 2-1.

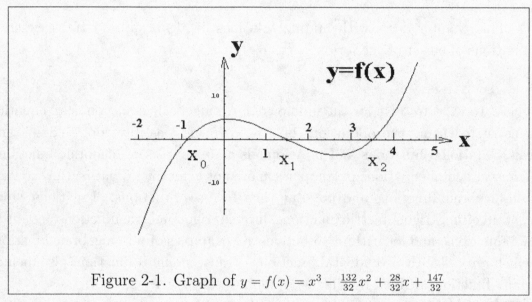

Figure 2-1. Graph of $y = f(x) = x^3 - \frac{132}{32}x^2 + \frac{28}{32}x + \frac{147}{32}$

One can now estimate the solutions of the given equation by determining where the curve crosses the x–axis because these are the points where $y = 0$. Examining the graph in figure 2-1 we can place bounds on our estimates x_0, x_1, x_2 of the solutions. One such estimate is given by

$$-1.0 < x_0 < -0.8$$

$$1.4 < x_1 < 1.6$$

$$3.4 < x_2 < 3.6$$

To achieve a better estimate for the roots one can plot three versions of the above graph which have some appropriate scaling in the neighborhood of the roots.

∎

Finding values for x where $f(x) = g(x)$ can also be approached using graphics. One can plot graphs of the curves $y = f(x)$ and $y = g(x)$ on the same set of axes and then try to estimate where these curves intersect.

Example 2-2. (Root of transcendental equation.)

Estimate the solutions of the transcendental equation

$$x^3 - \frac{132}{32}x^2 + \frac{28}{32}x + \frac{147}{32} = 5\sin x$$

Solution: We again employ a computer or calculator and plot graphs of the functions

$$y = f(x) = x^3 - \frac{132}{32}x^2 + \frac{28}{32}x + \frac{147}{32} \quad \text{and} \quad y = g(x) = 5\sin x$$

to obtain the figure 2-2.

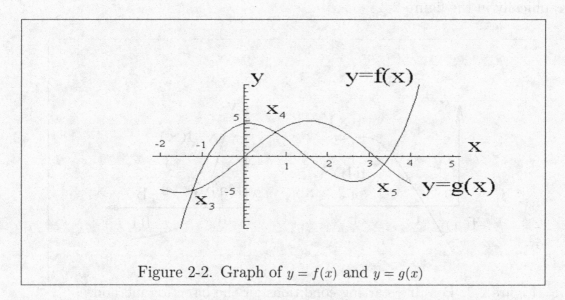

Figure 2-2. Graph of $y = f(x)$ and $y = g(x)$

One can estimate the points where the curve $y = f(x)$ intersects the curve $y = g(x)$. If the curves are plotted to scale on the same set of axes, then one can place bounds on the estimates of the solution. One such set of bounds is given by

$$-1.5 < x_3 < -1.0$$
$$0.5 < x_4 < 1.0$$
$$3.0 < x_5 < 3.5$$

By plotting these graphs over a finer scale one can obtain better estimates for the solutions.

Bisection Method

The bisection method is also known as the method of interval halving. The method assumes that you begin with a continuous function $y = f(x)$ and that you desire to find a root r such that $f(r) = 0$. The method assumes that if you plot a graph of $y = f(x)$, then it is possible to select an interval (a, b) such that at the end points of the interval the values $f(a)$ and $f(b)$ are of opposite sign in which case $f(a)f(b) < 0$. Starting with the above assumptions the intermediate value theorem guarantees that there exists at least one root of the given equation in the interval (a, b). The bisection method is a way of determining the root r to some desired degree of accuracy. The assumed starting situation is illustrated graphically in the figure 2-3.

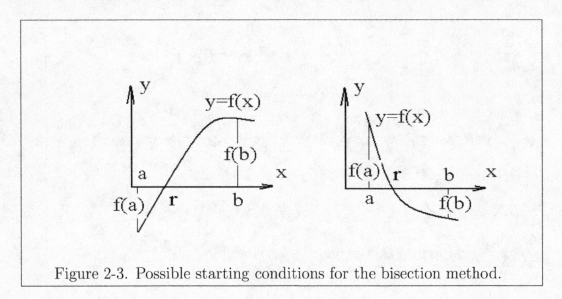

Figure 2-3. Possible starting conditions for the bisection method.

The bisection method generates a sequence of intervals $(a_1, b_1), (a_2, b_2), \ldots, (a_n, b_n)$ which get halved each time. Each interval (a_n, b_n) is determined such that the root r satisfies $a_n < r < b_n$. The bisection method begins by selecting $a_1 = a$ and $b_1 = b$ with $f(a)f(b) < 0$. The midpoint m_1 of the first interval (a_1, b_1) is calculated

$$m_1 = \frac{1}{2}(a_1 + b_1) \tag{2.2}$$

and the curve height $f(m_1)$ is calculated. If $f(m_1) = 0$, then $r = m_1$ is the desired root. If $f(m_1) \neq 0$ then one of the following cases will exist.

Either $f(m_1)f(b_1) < 0$ in which case there is a sign change in the interval (m_1, b_1) or $f(m_1)f(a_1) < 0$ in which case there is a sign change in the interval (a_1, m_1). The new interval (a_2, b_2) is determined from one of these conditions.

(i) If $f(m_1)f(b_1) < 0$, then we select
$a_2 = m_1$ and $b_2 = b_1$.

(ii) If $f(m_1)f(a_1) < 0$, then we select
$a_2 = a_1$ and $b_2 = m_1$.

Whichever case holds, the root r will lie within the new interval (a_2, b_2) which is one-half the size of the previous interval. The figure 2-4 illustrates some possible scenarios that could result in applying the bisection method to find a root of an equation.

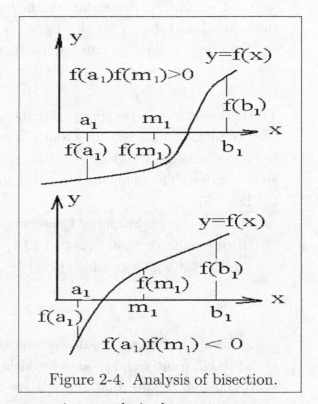

Figure 2-4. Analysis of bisection.

The above process is then repeated as many times as desired to generate new intervals (a_n, b_n) for $n = 3, 4, 5, \ldots$. The bisection method places bounds upon the distance between the nth midpoint m_n and the desired root r. One can define the error of approximation after the nth bisection as

$$\text{Error} = |r - m_n|. \tag{2.3}$$

This produces the following bounds.

$$\text{After first bisection} \quad |r - m_1| < \frac{b-a}{2}, \quad \text{since } r \in (a_1, b_1)$$

$$\text{After second bisection} \quad |r - m_2| < \frac{b-a}{2^2}, \quad \text{since } r \in (a_2, b_2)$$

$$\vdots$$

$$\text{After } n\text{th bisection} \quad |r - m_n| < \frac{b-a}{2^n}, \quad n \geq 1, \quad \text{since } r \in (a_n, b_n).$$

These errors are obtained from the bounds upon the distance between the midpoint m_n and the desired root r. We find the error term associated with the nth step of the iteration procedure for the bisection method will always be less than the initial interval $b - a$ divided by 2^n. If we want the error to be less than some small amount ϵ, then we can require that n be selected such that

$$\text{Error} = |r - m_n| < \frac{b - a}{2^n} < \epsilon. \tag{2.4}$$

The bisection method generates a sequence of midpoint values $\{m_1, m_2, \ldots, m_n, \ldots\}$ used to approximate the true root r. The number of interval halving operations to be performed for a given function $f(x)$ depends upon how accurate you want your solution. The following is a list of some stopping conditions associated with the bisection method.

Stopping Conditions for Bisection Method

(i) If one requires that equation (2.4) be satisfied, then n can be selected as the least integer which satisfies

$$n > \frac{\ln|b - a| - \ln \epsilon}{\ln 2} \tag{2.5}$$

(ii) Given a error ϵ one could continue until $|m_n - m_{n-1}| < \epsilon$. This requires that the two consecutive midpoints be within ϵ of one another.

(iii) One can require that the relative error or percentage error be less than some small amount ϵ. This requires that

$$\frac{|m_n - m_{n-1}|}{|m_n|} < \epsilon \quad \text{or} \quad \frac{|m_n - m_{n-1}|}{|m_n|} \times 100 < \epsilon$$

(iv) The height of the curve $y = f(x)$ is near zero. This requires $|f(m_n)| < \epsilon$ where ϵ is some stopping criteria.

(v) One can arbitrarily select a maximum number of iterations N_{max} and stop the interval halving whenever $n > N_{max}$. One usually selects N_{max} based upon an analysis of equation (2.4).

(vi) The inequality $|r - m_n| \leq |b_n - a_n| < \epsilon$ can be used to define a stopping condition for the error.

Example 2-3.   (Bisection method.)

Find the value of x which satisfies $f(x) = xe^x - 2 = 0$.

Solution: We sketch the given function and select $a_1 = 0$ and $b_1 = 1$ with $f(a_1) = -2$ and $f(b_1) = 0.718$. This type of a problem can be easily entered into a spread sheet program which can do the repetitive calculations quickly. Many free spread sheet programs are available from the internet for those interested. The bisection method produces the following table of values where the error after the n*th* bisection is less than $E = \frac{b-a}{2^n}$.

Bisection method to solve $f(x) = xe^x - 2 = 0$ with $\mid r - m_n \mid < E = \frac{b-a}{2^n}$						
n	a_n	$f(a_n)$	b_n	$m_n = \frac{1}{2}(a_n + b_n)$	$f(m_n)$	E
1	0	-2	1	0.5	-1.1756	0.5
2	0.5	-1.17564	1	0.75	-0.4122	0.25
3	0.75	-0.41225	1	0.875	0.0990	0.125
4	0.75	-0.41225	0.875	0.8125	-0.1690	0.0625
5	0.8125	-0.16900	0.875	0.84375	-0.0382	0.03125
6	0.84375	-0.03822	0.875	0.859375	0.0296	0.015625
7	0.84375	-0.03822	0.859375	0.8515625	-0.0045	0.0078125
8	0.8515625	-0.00453	0.859375	0.85546875	0.0125	0.00390625
9	0.8515625	-0.00453	0.85546875	0.853515625	0.0040	0.001953125
10	0.8515625	-0.00453	0.853515625	0.852539063	-0.0003	0.000976563
11	0.852539063	-0.00029	0.853515625	0.853027344	0.0018	0.000488281
12	0.852539063	-0.00029	0.853027344	0.852783203	0.0008	0.000244141
13	0.852539063	-0.00029	0.852783203	0.852661133	0.0002	0.000122070

Continuing one can achieve the more accurate approximation $r = 0.852605502$.

∎

There can be problems in using the bisection method. In addition to the bisection method being slow, there can be the problem that the initial interval (a, b) is selected too large. If this condition occurs, then there exists the possibility that more than one root exists within the initial interval. Observe that if the starting interval contains more than one root, then the bisection method will find only one of the roots. The good thing about the bisection method is that it always works when the setup conditions are satisfied.

Linear Interpolation

The method of linear interpolation is often referred to as the method of false position or the Latin equivalent "regula falsi". It is a method that is sometimes used in the attempt to speed up the bisection method. The method of linear interpolation is illustrated in the figure 2-5.

Given two points $(a_n, f(a_n))$ and $(b_n, f(b_n))$, where $f(a_n)f(b_n) < 0$, then one can construct a straight line through these points. The point-slope formula can be used to find the equation of the line in figure 2-5. One obtains the equation

$$y - f(b_n) = \left(\frac{f(b_n) - f(a_n)}{b_n - a_n} \right) (x - b_n) \quad (2.6)$$

This line crosses the x–axis at the point $(x_n, 0)$ where

$$x_n = b_n - \left(\frac{f(b_n)}{f(b_n) - f(a_n)} \right) (b_n - a_n).$$

We use the point x_n in place of the midpoint m_n of the bisection method. That is, each iteration begins with end points $(a_n, f(a_n))$ and $(b_n, f(b_n))$ where $f(a_n)$ and $f(b_n)$ are of opposite sign. These points produce a straight line which determines a point x_n by linear interpolation.

A new interval (a_{n+1}, b_{n+1}) is determined by the same procedure used in the bisection method. We calculate $f(x_n)$ and test the sign of $f(x_n)f(a_n)$ in order to determine the new interval (a_{n+1}, b_{n+1}) which contains the desired root r. The method of linear interpolation sometimes has the problem of a slow one-sided approach to the root as illustrated in the figure 2-6.

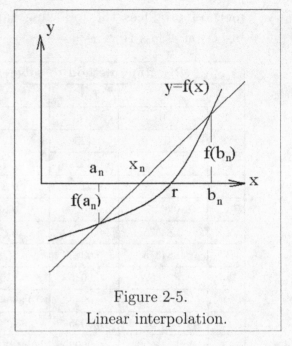

Figure 2-5.
Linear interpolation.

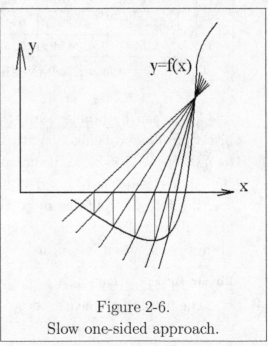

Figure 2-6.
Slow one-sided approach.

Iterative Methods

One can write equations of the form $f(x) = 0$ in the alternative form $x = g(x)$ and then one can define an iterative sequence

$$x_{n+1} = g(x_n) \quad \text{for} \quad n = 0, 1, 2, 3, \ldots \tag{2.7}$$

which can be interpreted as mapping a point x_n to a new point x_{n+1}. One starts with an initial guess x_0 to the solution of $x = g(x)$ and calculates $x_1 = g(x_0)$. The iterative method continues with repeated substitutions into the $g(x)$ function to obtain the values

$$x_2 = g(x_1)$$
$$x_3 = g(x_2)$$
$$\vdots$$
$$x_n = g(x_{n-1})$$
$$x_{n+1} = g(x_n)$$

If the sequence of values $\{x_n\}_{n-1}^{\infty}$ converges to r, then

$$\lim_{n \to \infty} x_{n+1} = r = \lim_{n \to \infty} g(x_n) = g(r)$$

and r is called a fixed point of the mapping. Convergence of the iterative processes is based upon the concept of a contraction mapping. In general, a mapping $x_{n+1} = g(x_n)$ is called a contraction mapping if the following conditions are satisfied.

1. The function $g(x)$ maps all point in a set S_n into a subset S_{n+1} of S_n so that one can write $S_{n+1} \subset S_n$.

2. For $x_n, y_n \in S_n$, with $x_{n+1} = g(x_n)$ and $y_{n+1} = g(y_n)$ both members of the set S_{n+1}, the distance between y_{n+1} and x_{n+1} must be less than the distance between y_n and x_n. This can be expressed

$$\mid y_{n+1} - x_{n+1} \mid \leq K \mid y_n - x_n \mid$$

where K is some constant satisfying $0 \leq K < 1$.

That is, the distance between any two points x_n and y_n belonging to a set S_n is always greater than the distance between the image points x_{n+1} and y_{n+1} belong to the image subset S_{n+1}. The representation of a contraction mapping is illustrated in the figure 2-7 which gives an image showing points from one set being mapped to a smaller set. This is the idea behind a contraction mapping. Each mapping gives a smaller and smaller image set which eventually contracts to a limit point r where $r = g(r)$. This idea can be applied to more general types of mappings.

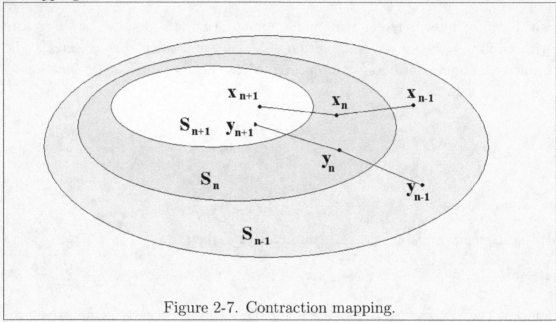

Figure 2-7. Contraction mapping.

In one-dimension, assume that the iterative sequence

$$x_{n+1} = g(x_n) \tag{2.8}$$

converges to a limit r such that

$$r = g(r). \tag{2.9}$$

Subtract the equation (2.9) from the equation (2.8) and write

$$x_{n+1} - r = g(x_n) - g(r) = \left[\frac{g(x_n) - g(r)}{x_n - r} \right] (x_n - r). \tag{2.10}$$

The mean value theorem can now be employed to express the bracketed term in equation (2.10) in terms of a derivative so that

$$\left[\frac{g(x_n) - g(r)}{x_n - r} \right] = g'(\xi_n) \qquad \text{for} \qquad x_n < \xi_n < r. \tag{2.11}$$

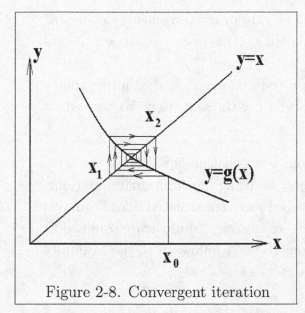

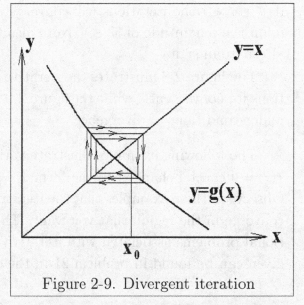

| Figure 2-8. Convergent iteration | Figure 2-9. Divergent iteration |

Substitute the equation (2.11) into the equation (2.10) and then take the absolute value of the terms in the resulting equation. If one defines the distance between x_n and r as the error term $e_n = |x_n - r|$, then the equation (2.10) implies that

$$e_{n+1} = |g'(\xi_n)| \, e_n. \qquad (2.12)$$

Observe that if $|g'(\xi_n)| \le K < 1$, for all values of ξ_n, then the equation (2.12) implies that

$$e_{n+1} \le K \, e_n \qquad (2.13)$$

and so defines a contraction mapping. That is we obtain the following inequalities

$$e_1 \le K \, e_0$$
$$e_2 \le K \, e_1 \le K^2 \, e_0$$
$$e_3 \le K \, e_2 \le K^3 \, e_0$$
$$\vdots$$
$$e_n \le K \, e_{n-1} \le K^n \, e_0.$$

Thus, for $K < 1$ the contraction mapping converges and one can write $\lim_{n \to \infty} x_n = r$. Note that if $K > 1$, the iterations would diverge. We conclude that for $g(x)$ and $g'(x)$ continuous near a root r, then in order for the iterative scheme $x_{n+1} = g(x_n)$ to converge, it is necessary that each x_n be near the root r and $|g'(x)| < 1$ for all x near the root r. This is telling us that if $e_{n+1} \le g'(\xi_n)e_n$ for $n = 1, 2, 3, \ldots$, then if $|g'(\xi_n)| < 1$ the iterations will converge. The rate of convergence is dependent upon the magnitude of $|g'(\xi_n)|$. Note also that if $g'(\xi_n)$ is negative, the errors will alternate in sign.

The figure 2-8 illustrates the iterative procedure for $x_{n+1} = g(x_n)$ under conditions for convergence, while the figure 2-9 illustrates the same iterative procedure under conditions of divergence.

The following examples illustrate various ways that an equation of the form $f(x) = 0$ can be changed to the form $x = g(x)$ by using algebraic manipulations. Observe in these examples that the iterative sequence constructed doesn't always converge in the region that you want. These examples also illustrate some additional problems associated with iterative methods. A follow up to the examples given can be found in problem 21 in the exercises.

Example 2-4.   (Iterative method.)
Write the equation $f(x) = x^3 - 3x^2 + 1 = 0$ in the form $x = g(x)$ and set up an iterative procedure for solving $f(x) = 0$.
Solution: Add x to both sides of the given equation $f(x) = 0$ to obtain

$$x = x^3 - 3x^2 + x + 1.$$

One can then construct the iterative sequence

$$x_{n+1} = g(x_n) = x_n^3 - 3x_n^2 + x_n + 1. \tag{2.14}$$

One problem associated with iterative techniques is a bad initial guess for the root. For example, select $x_0 = 1$ for the starting value in equation (2.14) and observe that $x_1 = 0$ results. The value $x_1 = 0$ substituted into the equation (2.14) gives $x_2 = 1$. This gives the sequence of unending values $\{1, 0, 1, 0, 1, 0, \ldots\}$ which does not converge.

If we plot graphs of $y = f(x) = x^3 - 3x^2 + 1$, $y = g(x) = x^3 - 3x^2 + x + 1$ and $y = g'(x) = 3x^2 - 6x + 1$ we obtain the curves illustrated in the figure 2-10. The $y = f(x)$ curve shows that the roots we desire are approximately $\{-.5, .6, 2.9\}$. The $y = g'(x)$ curve shows that $|g'(x)| < 1$ only for certain values of x. Hence we cannot expect our iterative method to converge.

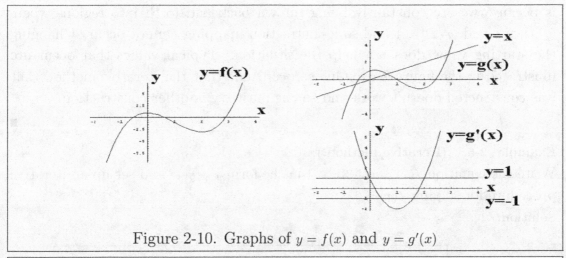

Figure 2-10. Graphs of $y = f(x)$ and $y = g'(x)$

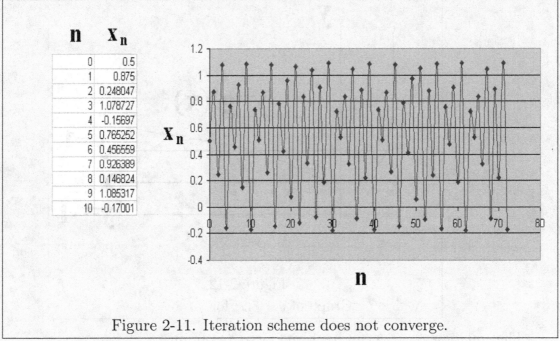

n	x_n
0	0.5
1	0.875
2	0.248047
3	1.078727
4	-0.15697
5	0.765252
6	0.456559
7	0.926389
8	0.146824
9	1.085317
10	-0.17001

Figure 2-11. Iteration scheme does not converge.

Let us try the starting value $x_0 = 0.5$ since this is in a region where $|g'(x)| < 1$. Successive substitutions into the iterative scheme defined by equation (2.14) gives the values

$$x_0 = 0.5, \quad x_1 = 0.875, \quad x_2 = 0.248047, \quad x_3 = 1.07873, \cdots$$

A spread sheet is a good way of calculating the sequence of terms. The figure 2-11 shows that for the starting value selected the series does not converge. This

38

is because we are constantly being thrown back and forth into regions where $|g'(x)| > 1$ and $|g'(x)| < 1$ and so a contraction mapping cannot occur. Changing the starting value does not help the situation. Typical values that occur are illustrated in the figure 2-11. One can conclude that the iterative method that was constructed doesn't work and so one must try another construction.

■

Example 2-5. **(Iterative method.)**
Write the equation $f(x) = x^3 - 3x^2 + 1$ in the form $x = g(x)$ and set up an iterative procedure for solving $f(x) = 0$.
Solution:

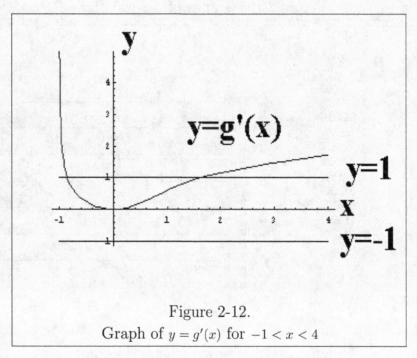

Figure 2-12.
Graph of $y = g'(x)$ for $-1 < x < 4$

The equation $f(x) = 0$ can be solved for x^2 to obtain

$$x^2 = \frac{x^3 + 1}{3}$$

and then one can solve for x to obtain

$$x = g(x) = \left(\frac{x^3 + 1}{3}\right)^{1/2}. \tag{2.15}$$

This gives the iterative procedure

$$x_{n+1} = g(x_n) = \left(\frac{x_n^3 + 1}{3}\right)^{1/2} \tag{2.16}$$

A graph of $g'(x)$ vs x is illustrated in the figure 2-12 and illustrates the region where $|g'(x)| < 1$. We select a starting value of $x_0 = 0.5$ since $g'(0.5) < 1$.

By trial and error we find the iterative scheme only converges to the middle root $r = 0.652704$ for a small selected domain of initial values x_0 where $|g'(x_0)| < 1$. The iterative sequence diverges if the initial starting value is too large. The figure 2-13 illustrates the iterative values obtained.

Selecting $x_0 = 0.5$ produces the iterative values given in the table below.

n	$\mathbf{x_n}$
0	0.5
1	0.612372436
2	0.640218625
3	0.648694262
4	0.651401898
5	0.652279526
6	0.652565308
7	0.652658506
8	0.652688915
9	0.652698838
10	0.652702076
11	0.652703133
12	0.652703478
13	0.652703590
14	0.652703627
15	0.652703639
16	0.652703643
17	0.652703644
18	0.652703644

Figure 2-13. Iterative values

Example 2-6. (Iterative method.)
Write the equation $f(x) = x^3 - 3x^2 + 1$ in the form $x = g(x)$ and set up an iterative procedure for solving $f(x) = 0$.

Solution: The equation $f(x) = 0$ can be multiplied by some continuous function $h(x)$ and then x can be added to both sides of the equation to obtain the equivalent equation $x = x + h(x)f(x)$. This produces the iterative sequence

$$x_{n+1} = g(x_n) = x_n + h(x_n)f(x_n) \tag{2.17}$$

where $g(x) = x + h(x)f(x)$. Suppose the function $h(x)$ is selected such that we force the derivative $g'(x)$ to be less than one in absolute value for all x in the neighborhood of a root r. If this can be accomplished we would expect the iterative procedure to converge for values of x near a root. We differentiate $g(x)$ to obtain

$$g'(x) = 1 + h(x)f'(x) + h'(x)f(x). \tag{2.18}$$

Now if we artificially impose a condition like

$$g'(x) = 1 + h(x)f'(x) + h'(x)f(x) \leq \frac{1}{2}\sin x, \tag{2.19}$$

then $|g'(x)| \leq 1$ for all x. Note that for x near a root r, the term $f(x)$ is approximately zero. Therefore, one can use equation (2.19) and say that for x near a root r, the desired function $h(x)$ has the approximate value

$$h(x) = \frac{\frac{1}{2}\sin x - 1}{f'(x)} = \frac{\frac{1}{2}\sin x - 1}{3x^2 - 6x} = \frac{\sin x - 2}{6x(x - 2)}.$$

We substitute this value for $h(x)$ into the iterative sequence given by equation (2.17) and obtain the following iterations for the initial values selected.

$x_0 = -0.5$	$x_0 = 0.5$	$x_0 = 2.9$
$x_1 = g(x_0) = -0.54132$	$x_1 = g(x_0) = 0.62672$	$x_1 = g(x_0) = 2.88212$
$x_2 = g(x_1) = -0.52983$	$x_2 = g(x_1) = 0.64528$	$x_2 = g(x_1) = 2.87974$
$x_3 = g(x_2) = -0.53267$	$x_3 = g(x_2) = 0.65049$	$x_3 = g(x_2) = 2.87943$
$x_4 = g(x_3) = -0.53194$	$x_4 = g(x_3) = 0.65203$	$x_4 = g(x_3) = 2.87939$
$x_5 = g(x_4) = -0.53213$	$x_5 = g(x_4) = 0.65250$	$x_5 = g(x_4) = 2.87939$
$x_6 = g(x_5) = -0.53208$	$x_6 = g(x_5) = 0.65264$	$x_6 = g(x_5) = 2.87938$
$x_7 = g(x_6) = -0.53209$	$x_7 = g(x_6) = 0.65268$	$x_7 = g(x_6) = 2.87938$
$\vdots$	$\vdots$	$\vdots$
$x_{13} = g(x_{12}) = -0.532088887$	$x_{13} = g(x_{12}) = 0.65270363$	$x_{13} = g(x_{12}) = 2.879385242$

For this example, the initial guesses were all in the neighborhood of a root and so all the iterations converged. ∎

The previous examples illustrate that there are many ways to convert an equation of the form $f(x) = 0$ into the form $x = g(x)$ so that an iterative sequence $x_{n+1} = g(x_n)$ can be constructed. The previous examples illustrate that the constructed iterative sequence is not always a converging sequence. Therefore, one should be careful in modifying equations $f(x) = 0$ to the form $x = g(x)$ to construct an iterative sequence. Also note that the initial starting value can sometimes be very important to insure the iterative procedure converges.

Convergence of Iterative Methods

Let $\{x_n\}_{n=1}^{\infty}$ denote a converging sequence with limit r. Define the error of the nth iterative as $e_n = r - x_n$ for $n = 1, 2, 3, \ldots$. If there exists positive constants α and β which satisfy

$$\lim_{n \to \infty} \frac{|r - x_{n+1}|}{|r - x_n|^\beta} = \lim_{n \to \infty} \frac{|e_{n+1}|}{|e_n|^\beta} = \alpha, \tag{2.20}$$

then β is called the order of the convergence and α is called the asymptotic error constant associated with the converging sequence. For $\beta = 1$, the convergence is said to be linear, in which case the error satisfies $\lim_{n \to \infty} \frac{|e_{n+1}|}{|e_n|} = \alpha$. For $\beta = 2$, the convergence is said to be quadratic, in which case $\lim_{n \to \infty} \frac{|e_{n+1}|}{|e_n|^2} = \alpha$. When the order of the convergence is 2 or higher, one obtains the limit of the sequence very quickly. For example, the inequality

$$|e_n| = |r - x_n| < \epsilon = \frac{1}{2}(10)^{-N} \tag{2.21}$$

means that x_n approximates the root r to N decimal places. Now consider the special case of quadratic convergence with asymptotic error constant of unity, where $|e_{n+1}| \approx |e_n|^2$. One can then say that if $|e_n| < \epsilon$, then $|e_{n+1}| < \epsilon^2 = \frac{1}{4}(10)^{-2N}$. This special case of quadratic convergence illustrates that under certain conditions the number of correct digits in the representation of the solution can be doubled in one iteration. Note that in general the asymptotic error constant is not always in the neighborhood of unity and so if you have quadratic convergence then each iteration only approximately doubles the number of correct digits in the representation of the solution.

Example 2-7. (Comparison of iterative methods.)

Compare the number of iterations associated with two convergent sequences where one sequence has linear convergence and the other sequence has quadratic convergence. Assume the errors e_n, for $n = 0, 1, 2, 3, \ldots$, associated with the linear convergence satisfies $|e_{n+1}| \leq \alpha |e_n|$ and the errors for the sequence with quadratic convergence satisfies $|e_{n+1}| \leq \alpha |e_n|^2$ where α is some positive constant less than one.

Solution: For the assumptions given we form tables for linear and quadratic convergence.

<table>
<tr><td colspan="2">

Linear convergence

$n = 0, \quad |e_1| \leq \alpha |e_0|$

$n = 1, \quad |e_2| \leq \alpha |e_1| \leq \alpha^2 |e_0|$

$n = 2, \quad |e_3| \leq \alpha |e_2| \leq \alpha^3 |e_0|$

$\vdots \qquad \vdots$

$n = m - 1, \quad |e_m| \leq \alpha |e_{m-1}| \leq \alpha^m |e_0|$

How many iterations m are required to achieve an error of ϵ? If we require that

$$|e_m| \leq \alpha^m |e_0| < \epsilon,$$

then m must be selected as the least integer m which satisfies

$$m > \frac{\ln \epsilon - \ln |e_0|}{\ln \alpha}.$$

We assume $|e_0| = 0.5$ and construct tables for values of m associated with various values of ϵ and α.

</td><td colspan="2">

Quadratic convergence

$n = 0, \quad |e_1| \leq \alpha |e_0|^2$

$n = 1, \quad |e_2| \leq \alpha |e_1|^2 \leq \alpha^3 |e_0|^4$

$n = 2, \quad |e_3| \leq \alpha |e_2|^2 \leq \alpha^7 |e_0|^8$

$\vdots \qquad \vdots$

$n = m - 1, \quad |e_m| \leq \alpha |e_{m-1}| \leq \alpha^{2^m - 1} |e_0|^{2^m}$

How many iterations m are required to achieve an error of ϵ? If we require that

$$|e_m| \leq \alpha^{2^m - 1} |e_0|^{2^m} < \epsilon,$$

then m must be selected as the least integer m which satisfies

$$2^m > \frac{\ln \epsilon + \ln \alpha}{\ln(\alpha |e_0|)}.$$

We assume $|e_0| = 0.5$ and construct tables for values of m associated with various values of ϵ and α.

</td></tr>
</table>

Number of Iterations to Achieve Error Bound ϵ Linear Convergence $\mid e_0 \mid = 0.5$			
ϵ	$\alpha = 0.25$	$\alpha = 0.5$	$\alpha = 0.75$
10^{-4}	7	13	30
10^{-5}	8	16	38
10^{-6}	10	19	46
10^{-7}	12	23	54
10^{-8}	13	26	62

Number of Iterations to Achieve Error Bound ϵ Quadratic Convergence $\mid e_0 \mid = 0.5$			
ϵ	$\alpha = 0.25$	$\alpha = 0.5$	$\alpha = 0.75$
10^{-4}	3	3	4
10^{-5}	3	4	4
10^{-6}	3	4	4
10^{-7}	4	4	5
10^{-8}	4	4	5

An examination of the above tables illustrates the dramatic differences between linear and quadratic convergence.

Newton-Raphson Method

The Newton-Raphson method is also an iterative method having the form $x_{n+1} = g(x_n)$ for solving the equation $f(x) = 0$. It is assumed that $f(x)$ is continuous and twice differentiable in the neighborhood of a root. The Newton-Raphson method can be described graphically as follows. In order to find a root r such that $f(r) = 0$, we begin by sketching a graph of the curve $y = f(x)$. A sketch of a representative curve is illustrated in the figure 2-14.

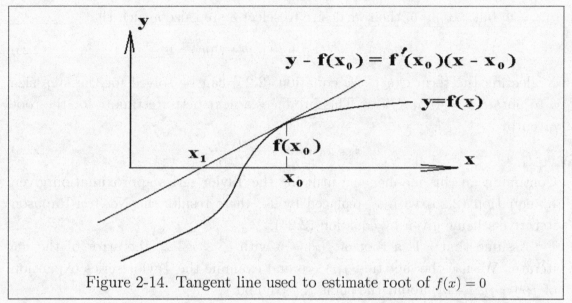

Figure 2-14. Tangent line used to estimate root of $f(x) = 0$

First estimate the root with a selected value x_0, then calculate the height of the curve $f(x_0)$ at the selected point x_0. Also calculate the slope of the tangent line to the curve $y = f(x)$ at the point $(x_0, f(x_0))$ and calculate the equation of the tangent line to the curve at this point. This tangent line is found to be given by

$$y - f(x_0) = f'(x_0)(x - x_0). \tag{2.22}$$

One can now use this tangent line to estimate the root of $f(x) = 0$. The tangent line at $(x_0, f(x_0))$ intersects the x-axis at the point x_1 where $y = 0$. Therefore, by setting $y = 0$ and $x = x_1$ in equation (2.22) one finds that upon solving for x_1 there results

$$x_1 = x_0 - \frac{f(x_0)}{f'(x_0)}. \tag{2.23}$$

Selecting x_1 as the next starting point and going through the same procedure for constructing a tangent line at the point $(x_1, f(x_1))$ suggests the iterative sequence

$$x_{n+1} = x_n - \frac{f(x_n)}{f'(x_n)} \tag{2.24}$$

for $n = 0, 1, 2, \ldots$. This iterative scheme for root finding is known as the Newton-Raphson method.

Another way of approaching the Newton-Raphson method is from the Taylor series expansion of $f(x)$ about a point. If x_0 is the initial guess at the root of $f(x) = 0$, but $f(x_0) \neq 0$, then we desire to select a step size h such that

$$f(x_0 + h) = f(x_0) + f'(x_0)h + \mathcal{O}(h^2) = 0. \tag{2.25}$$

Neglecting the term $\mathcal{O}(h^2)$, the equation (2.25) can be solved for the step size h to obtain $h = -f(x_0)/f'(x_0)$. This produces a next better estimate for the root given by

$$x_1 = x_0 + h = x_0 - \frac{f(x_0)}{f'(x_0)}. \tag{2.26}$$

Continuing in this manner and utilizing the Taylor series approximation given by equation (2.25) with x_0 replaced by x_1, there results the Newton-Raphson iterative scheme given by equation (2.24).

Assume that r is a root of $f(x) = 0$ with $e_n = r - x_n$ the error of the nth iterate. We use the fact that $f(r) = 0$ and examine the Taylor series expansion of $f(r) = f(x_n + e_n)$ about the point x_n. We find

$$0 = f(r) = f(x_n + e_n) = f(x_n) + f'(x_n)e_n + f''(\xi_n)\frac{e_n^2}{2!} \tag{2.27}$$

where the last term is the error term upon truncation of the Taylor series. Note that the error term is evaluated at a point ξ_n which is located somewhere between x_n and r. Assume that the derivative $f'(x_n) \neq 0$, then the equation (2.27) implies that

$$0 = \frac{f(x_n)}{f'(x_n)} + e_n + \frac{f''(\xi_n)}{2f'(x_n)}e_n^2 \tag{2.28}$$

and so the Newton-Raphson iterative method becomes

$$x_{n+1} = x_n - \frac{f(x_n)}{f'(x_n)} = \underbrace{x_n + e_n}_{r} + \frac{f''(\xi_n)}{2f'(x_n)}e_n^2 \tag{2.29}$$

The equation (2.29) implies that

$$\lim_{n \to \infty} \frac{|e_{n+1}|}{|e_n|^2} = \lim_{n \to \infty} \left| \frac{f''(\xi_n)}{2f'(x_n)} \right|. \tag{2.30}$$

If the Newton-Raphson iterative procedure converges and $f'(r) \neq 0$, then the equation (2.30) indicates that the convergence is quadratic with asymptotic error constant $|\frac{f''(r)}{2f'(r)}|$. Hence, when the Newton-Raphson method works it provides the desired root very quickly since it has quadratic convergence.

The Newton-Raphson method is of the form $x_{n+1} = g(x_n)$ with $g(x) = x - \frac{f(x)}{f'(x)}$. For convergence the derivative

$$g'(x) = 1 - \frac{f'(x)f'(x) - f(x)f''(x)}{(f'(x))^2} = \frac{f(x)f''(x)}{(f'(x))^2} \qquad (2.31)$$

must be less then one in absolute value. Hence, for the Newton-Raphson method to converge it is required that

$$|g'(x)| = \left| \frac{f(x)f''(x)}{(f'(x))^2} \right| < 1 \qquad (2.32)$$

for x lying in some interval about the desired root r. Here it is assumed that $f'(x)$ is different from zero for these values of x in the vicinity of the root.

Example 2-8.   (Newton-Raphson iterative method.)
Use the Newton-Raphson iterative method to find the roots of the equation
$$f(x) = x^3 - 3x^2 + 1 = 0.$$
Solution: Here $f'(x) = 3x^2 - 6x$ and the Newton-Raphson iterative method is given by

$$x_{n+1} = x_n - \frac{f(x_n)}{f'(x_n)} = x_n - \frac{x_n^3 - 3x_n^2 + 1}{3x_n^2 - 6x_n}. \qquad (2.33)$$

This produces the following sequences for the selected initial values of $\{-0.5, 0.5, 2.9\}$.

$x_0 = -0.5$	$x_0 = 0.5$	$x_0 = 2.9$
$x_1 = -0.53333333$	$x_1 = 0.666666667$	$x_1 = 2.879693487$
$x_2 = -0.532090643$	$x_2 = 0.652777778$	$x_2 = 2.879385312$
$x_3 = -0.532088886$	$x_3 = 0.652703647$	$x_3 = 2.879385242$

Compare the above iterations with those from the example 2-6 and note how quickly the iterations converged.

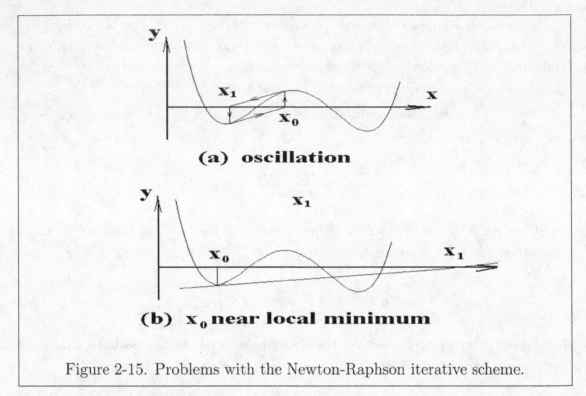

Figure 2-15. Problems with the Newton-Raphson iterative scheme.

There are situations where the Newton-Raphson root finding method fails. One such situation is when the denominator term $f'(x_n) = 0$ and so a division by zero results in applying the iterative sequence. Another problem causing situation is when some iteration gives a point x_n in the neighborhood of a local maximum or minimum of the curve $y = f(x)$. This situation can at times generate a tangent line which produces a next root estimate which is far from the desired root. The new root might even be in a location where $|g'(x)| > 1$ which causes the Newton-Raphson method to diverge. Still another problem arises when the Newton-Raphson method starts to oscillate and so never converges. These later situations are illustrated in the figures 2-15(a)(b).

The Newton-Raphson method usually works fine for functions of a real variable if the initial guess is near a root. However, one must be careful in applying the Newton-Raphson method to functions $f(z)$ of a complex variable $z = x + iy$ where $i^2 = -1$. The study of iterative methods in general, applied to functions of a complex variable z, can become quite complicated. Individuals who pursue this avenue of investigation will be led to a wonderful new area of mathematics called dynamical system theory.

Newton-Raphson Method with Multiple Roots

A function $y = f(x)$ is said to have a root r of multiplicity m if the function and its first $m - 1$ derivatives are all zero at the root r but the mth derivative is different from zero at the root r. If r is root of $f(x) = 0$ with multiplicity m, then one can write

$$f(r) = f'(r) = f''(r) = \cdots = f^{(m-1)}(r) = 0, \quad \text{but} \quad f^{(m)}(r) \neq 0. \qquad (2.34)$$

The graph of $y = f(x)$, for x in the vicinity of a multiple root r, is usually very flat and causes an increase in the interval of error about the root. Therefore, whenever possible it is desirable to work with roots that are simple roots (roots of multiplicity one). Note that if $f(x)$ has a root r of multiplicity m, then its Taylor series expansion about the point r, given by

$$f(x) = f(r) + f'(r)(x - r) + \cdots + \frac{f^{(m-1)}(r)}{(m-1)!}(x - r)^{m-1} + \frac{f^{(m)}(r)}{m!}(x - r)^m + \cdots, \qquad (2.35)$$

reduces to the form

$$f(x) = (x - r)^m \left(\frac{f^{(m)}(r)}{m!} + \frac{f^{(m+1)}(r)}{(m+1)!}(x - r) + \cdots \right) = (x - r)^m h(x) \qquad (2.36)$$

where $h(x)$ is a differentiable function and $h(r) \neq 0$. Let us calculate the ratio $R(x) = f(x)/f'(x)$ in the special case where $f(x)$ is a continuous function having a root r of multiplicity m. In this special case one can write the function and its derivative in the form

$$f(x) = (x - r)^m h(x)$$
$$f'(x) = (x - r)^m h'(x) + m(x - r)^{m-1} h(x)$$

so that the ratio $R(x)$ becomes

$$R(x) = \frac{f(x)}{f'(x)} = \frac{(x - r)^m h(x)}{(x - r)^m h'(x) + m(x - r)^{m-1} h(x)} = \frac{(x - r)h(x)}{(x - r)h'(x) + mh(x)} \qquad (2.37)$$

Alternative to using this ratio in the Newton-Raphson method, one should make note of the fact that at a root r of $f(x) = 0$ we have $R(r) = 0$. One can calculate the derivative $\frac{dR}{dx}$ and show the derivative evaluated at $x = r$ gives $R'(r) = \frac{1}{m} \neq 0$, where m is the multiplicity of the root. This shows that r is just a simple zero of the function $R(x)$. The creation of the function $R(x)$ and using it to find the root r is a way of removing the difficulties associated with having to deal with a multiple root of the original function.

Modifications to Newton-Raphson Method

To find a root r of the equation $f(x) = 0$, we can proceed as we have done previously and take an initial guess x_0 for the root. In general, $f(x_0) \neq 0$ and so we can use a Taylor series expansion to determine a step size h such that $f(x_0 + h) = 0$. We can modify this procedure by including an extra term in the Taylor series and write

$$f(x_0 + h) = f(x_0) + f'(x_0)h + \frac{f''(x_0)}{2!}h^2 + \mathcal{O}(h^3) = 0. \tag{2.38}$$

If the initial guess x_0 is near the root r we expect that the step size h will be small, and so we can neglect the $\mathcal{O}(h^3)$ terms and solve for h from the quadratic equation (2.38). We find

$$h = -\frac{f'(x_0)}{f''(x_0)} \pm \frac{f'(x_0)}{f''(x_0)} \sqrt{1 - 2\frac{f(x_0)f''(x_0)}{(f'(x_0))^2}}.$$

We use the results from equation (2.31) and write the step size in terms of the smaller root

$$h = -\frac{f'(x_0)}{f''(x_0)} \left(1 - \sqrt{1 - 2g'(x_0)}\right) \quad \text{where} \quad g'(x) = \frac{f(x)f''(x)}{(f'(x))^2}. \tag{2.39}$$

We have previously assumed that $|g'(x)| < 1$ for x near the desired root r and so the square root term in equation (2.39) can be approximated if one uses the binomial expansion

$$(1 - 2g'(x_0))^{1/2} = 1 - g'(x_0) - \frac{1}{2}(g'(x_0))^2 - \cdots \tag{2.40}$$

This approximation when substituted into the equation (2.39) gives the step size

$$h = -\frac{f'(x_0)}{f''(x_0)} \left(g'(x_0) + \frac{1}{2}(g'(x_0))^2\right)$$

which simplifies to

$$h = -\frac{f(x_0)}{f'(x_0)} \left(1 + \frac{f(x_0)f''(x_0)}{2(f'(x_0))^2}\right). \tag{2.41}$$

This gives the improved root

$$x_1 = x_0 + h = x_0 - \frac{f(x_0)}{f'(x_0)} \left(1 + \frac{f(x_0)f''(x_0)}{2(f'(x_0))^2}\right). \tag{2.42}$$

One can now use x_1 as a new starting value for a guess at the root and repeat the above procedure. This produces the iterative method

$$x_{n+1} = x_n - \frac{f(x_n)}{f'(x_n)} \left(1 + \frac{f(x_n)f''(x_n)}{2(f'(x_n))^2}\right) \tag{2.43}$$

which is known as Halley's iterative method. Observe that in the special cases where $f''(x)$ and $f(x)$ are small, the product $f(x)f''(x)$ becomes very small, and consequently the equation (2.43) is approximately the Newton-Raphson method.

Another form for Halley's iterative method is obtained by writing the equation (2.39) in the form

$$h = -\frac{f'(x_0)}{f''(x_0)} \cdot \frac{2g'(x_0)}{(1 + \sqrt{1 - 2g'(x_0)})} = \frac{-2f(x_0)}{f'(x_0) + \sqrt{(f'(x_0))^2 - 2f(x_0)f''(x_0)}}. \tag{2.44}$$

This produces the representation for the improved root as

$$x_1 = x_0 + h = x_0 - \frac{2f(x_0)}{f'(x_0) + \sqrt{(f'(x_0))^2 - 2f(x_0)f''(x_0)}}$$

and leads to the iterative method

$$x_{n+1} = x_n - \frac{2f(x_n)}{f'(x_n) + \sqrt{(f'(x_n))^2 - 2f(x_n)f''(x_n)}}. \tag{2.45}$$

Secant Method

A secant line associated with a curve $y = f(x)$ is a line through two points on the curve. Observe that both the bisection method and linear interpolation methods have the restrictions that at each iteration the points $(a_n, f(a_n))$ and $(b_n, f(b_n))$ are such that $f(a_n)f(b_n) < 0$. That is, the ordinates at the end points of any iteration are of opposite sign. The secant method can be thought of as a removal of this restriction and instead considers the construction of a line between any two points on the curve from which the next root approximation is obtained. The figure 2-16 illustrates a secant line to a general curve $y = f(x)$.

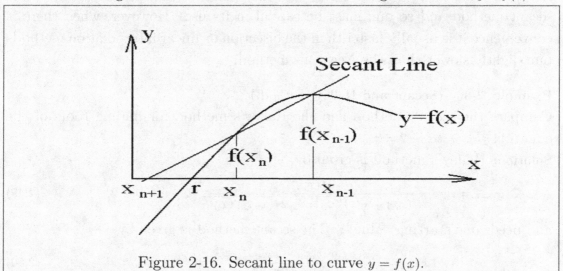

Figure 2-16. Secant line to curve $y = f(x)$.

The secant method for root finding can also be viewed as a modification of the Newton-Raphson method where the procedure avoids the necessity of calculating a derivative. The derivative of the function $y = f(x)$ is obtained as the limit

$$f'(x) = \lim_{h \to 0} \frac{f(x+h) - f(x)}{h} \qquad (2.46)$$

In equation (2.46) make the substitutions $x = x_n$ and $x_n + h = x_{n-1}$, then one can approximate the derivative at the point x_n by the relation

$$f'(x_n) \approx \frac{f(x_{n-1}) - f(x_n)}{x_{n-1} - x_n} = \frac{f(x_n) - f(x_{n-1})}{x_n - x_{n-1}}. \qquad (2.47)$$

Substituting this approximation into the Newton-Raphson iterative method produces the secant iterative method

$$x_{n+1} = x_n - \frac{f(x_n)(x_n - x_{n-1})}{f(x_n) - f(x_{n-1})} \qquad (2.48)$$

This is equivalent to constructing a straight line through the two points $(x_{n-1}, f(x_{n-1}))$ and $(x_n, f(x_n))$ and then determining the point x_{n+1} where this line crosses the x–axis, as illustrated in the figure 2-16. Thus, the secant method can be used to modify both the bisection method and linear interpolation method by selecting two points which are closest to the x–axis. The bisection and linear interpolation methods put bounds on the root because of the change in sign required at each iteration. The secant method requires no such bounds and so is subject to the same type of problems associated with the Newton-Raphson iterative method. There is the possibility of nonconvergence associated with the secant method and so one must be careful in its use. However, when there is convergence it is usually faster than the bisection or linear interpolation methods but slightly slower than the Newton's method.

Example 2-9. (Secant and Halley method.)
Compare the secant method and the Halley's method for finding roots of the equation $f(x) = xe^x - 2 = 0$.
Solution: Halley's method is given by

$$x_{n+1} = x_n - \frac{2f(x_n)}{f'(x_n) + \sqrt{(f'(x_n))^2 - 2f(x_n)f''(x_n)}}, \qquad n = 0, 1, 2, \ldots \qquad (2.49)$$

and needs one starting value x_0. The secant method is given by

$$x_{n+1} = x_n - \frac{f(x_n)(x_n - x_{n-1})}{f(x_n) - f(x_{n-1})}, \qquad n = 1, 2, 3, \ldots \qquad (2.50)$$

and needs two starting values x_0, x_1. We substitute the functional values

$$f(x) = xe^x - 2 \qquad f'(x) = xe^x + e^x, \qquad f''(x) = xe^x + 2e^x$$

and use the starting value $x_0 = 0.5$ for Halley's method and the starting values $x_0 = 0.5$, $x_1 = 0.6$ for the secant method and obtain the table of values given below.

Secant Method			Halley's Method		
x_n	x_{n+1}	$f(x_n)$	x_n	x_{n+1}	$f(x_n)$
0.5		-1.17563936465	0.5	0.86459787245	-1.17563936465
0.6	0.93718587829	-0.90672871977	0.864597872453	0.85260489734	0.05259963616
0.937185878289	0.83533277494	0.3924363448	0.852604897339	0.85260550201	-2.627765E-06
0.835332774936	0.8515050584	-0.07407249156			
0.851505058396	0.85262022205	-0.00477820407			
0.852620222046	0.85260548954	6.3970316E-05			
0.852605489538	0.85260550201	-5.421518E-08			

∎

The secant method requires the construction of a line through two points on a curve. In most cases the previous values calculated will give you the nearest values to the root and so one can select the two previous iterations as the starting value for determining the points to be used for the next iteration of the secant method. However, there can be times when it is not appropriate to select the previous two iterates for use in the secant method. Observe that computer difficulties occur whenever both of the quantities $f(x_n) - f(x_{n-1})$ and $x_n - x_{n-1}$ approach zero. That is, when two small numbers are divided there can be a great loss in significant digits. Also if both $f(x_n)$ and $f(x_{n-1})$ become very small with the same sign, then there is a loss of significant digits when these numbers are subtracted. In such cases large errors in the calculations can result. The secant method is usually quite accurate in regions about the root where a straight line is a good approximation for the function.

Müller's Method

In contrast to the secant method passing a straight line between two initial approximations $(x_0, f(x_0)), (x_1, f(x_1))$, which is then used to approximate where the root is, the Müller method passes a parabola through three initial approximations $(x_0, f(x_0)), (x_1, f(x_1))$ and $(x_2, f(x_2))$ and uses a quadratic equation to estimate the root. The situation is illustrated in the figure 2-17.

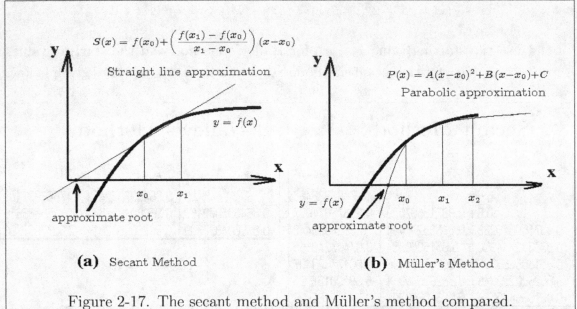

Figure 2-17. The secant method and Müller's method compared.

The parabolic approximation $P(x) = A(x - x_0)^2 + B(x - x_0) + C$ has coefficients A, B, C which are selected so that the parabola passes through the points $(x_0, f(x_0)), (x_1, f(x_1)), (x_2, f(x_2))$. This requires that the following equations hold

$$f(x_0) = C$$

$$f(x_1) = A(x_1 - x_0)^2 + B(x_1 - x_0) + C$$

$$f(x_2) = A(x_2 - x_0)^2 + B(x_2 - x_0) + C.$$

One solves the above equations for the constants A, B, C. The desired root is then approximated by solving $P(x) = A(x - x_0)^2 + B(x - x_0) + C = 0$ to obtain

$$x - x_0 = \frac{-B \pm \sqrt{B^2 - 4AC}}{2A} = \frac{-2C}{B \pm \sqrt{B^2 - 4AC}}.$$

This gives two approximate roots

$$x = x_0 - \frac{2C}{B \pm \sqrt{B^2 - 4AC}}.$$

Selecting the root closest to x_0 requires the sign of the square root be the same as the sign of B. This gives the approximate root

$$x = x_0 - \frac{2C}{B + sign(B)\sqrt{B^2 - 4AC}}.$$

This process can now be continued using the three closest approximations to the root as the initial values used in the construction of the parabola.

Exercises Chapter 2

▶ **1.** Determine the roots of the given equations after first estimating the root using a sketch. State the accuracy of your answers.

$$(a) \qquad x = 2\sin x, \quad x > 0 \qquad\qquad (c) \qquad x = 2\cos x, \quad x > 0$$

$$(b) \qquad 4x = 3x^5, \quad x > 0 \qquad\qquad (d) \qquad \sqrt{x} = 4 - 2x, \quad x > 0$$

▶ **2.** Construct an iterative method of the type specified to find the square root of a real number N. Test your iterative method for $N = 2, 3, 4, 5$ using a starting value of $x_0 = N/2$. Define a quantity to use for the measurement of error and specify the accuracy of your answer.

Hint: Consider $f(x) = x^2 - N = 0$.

    (a) Bisection method   (c) Linear interpolation method

    (b) Newton-Raphson method   (d) Secant method

▶ **3.** Construct an iterative method of the type specified to find the cube root of a real number N. Test your iterative method for $N = 2, 3, 4, 5$ using a starting value of $x_0 = N/2$. Define a quantity to use for the measurement of error and specify the accuracy of your answer.

    (a) Bisection method   (c) Linear interpolation method

    (b) Newton-Raphson method   (d) Secant method

▶ **4.** For the given starting values, determine how many iterations are necessary to find the root of the equation $f(x) = x^3 - 5x^2 + 10 = 0$, by the bisection method, so that your solution has an error less than $\epsilon = 10^{-6}$?

(a) Use the starting values $a_1 = -2$, $f(a_1) = -18$ and $b_1 = -1$, $f(b_1) = 4$.

(b) Use the starting values $a_1 = 1$, $f(a_1) = 6$ and $b_1 = 2$, $f(b_1) = -2$.

(c) Use the starting values $a_1 = 4$, $f(a_1) = -6$ and $b_1 = 5$, $f(b_1) = 10$.

(d) Determine all of the roots to the above equation.

▶ **5.** Determine all the roots of the cubic equation

$$44100x^3 - 44100x^2 + 12809x - 1122 = 0$$

▶ **6.**

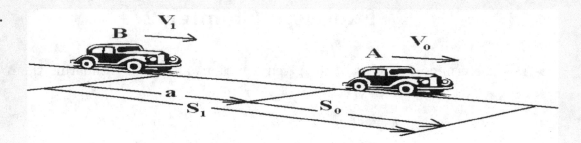

The stopping distance of a car is a function of reaction time and reaction time distance plus braking distance. Stopping under rainy conditions or during night time driving conditions adds to this stopping distance. Data from the department of motor vehicles allows one to construct the following approximate formula for stopping distance S in feet of a vehicle moving with velocity of V miles per hour under rain and night time conditions $S = -2.0 + 2.4V + 0.059V^2 + 0.000028V^3$, for $20 \le V \le 70$.

$V_0 = 40\,\text{mph}$	
a (feet)	V_1 (mph)
10	
20	
30	
40	
50	

You are in car B traveling at V_1 miles per hour and approach the slower moving car A which is traveling at V_0 miles per hour. Assume that when the distance between cars A and B is a-feet, car A suddenly hits its brakes.

(a) Use the distances given in the table and determine the minimum velocity V_1 for an accident to occur if V_0 is 40 miles per hour.

(b) Make up tables similar to the one illustrated for cases where V_0 is greater than 40 mph, say 50 mph and 60 mph.

▶ **7.**   Plot a graph of y vs x over the domain $0 \le x \le 2\pi$, if $\cos x + \cos y - 2\cos(x+y) = 0$, given that initially $y = 2\pi$ when $x = 0$.

▶ **8.**

(a) Graph the curves defined by $x^2 + y^2 - 9 = 0$ and $y - (x-1)^2 - 1 = 0$.

(b) Use a numerical method to determine the points of intersection of the above curves.

▶ **9.**

(a) Graph the curves defined by $x^2 + y^2 - 9 = 0$ and $x - (y-1)^2 = 0$.

(b) Use a numerical method to determine the points of intersection of the above curves.

▶ **10.** **Computer problem**

Fluid is to be stored in a cylindrical tank of radius r and length L. The tank is to be constructed to lie level on its side. A measuring stick is to be inserted into the tank to determine the depth h of the fluid at the center line mark. You are to construct a table of values relating the depth of fluid and the volume of the fluid in the cylindrical tank. You may use the relation for the area $A(h)$ of a segment of a circle of height h which is given by

$$A = A(h) = \frac{1}{2}\pi r^2 - r^2 \sin^{-1}\left(1 - \frac{h}{r}\right) - (r-h)\sqrt{r^2 - (r-h)^2}$$

where r is the radius of the circle. Your table is to have the form illustrated.

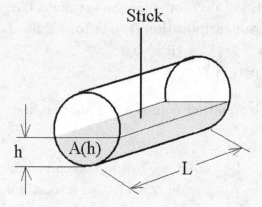

Tank Dimensions $r =$	, $L =$
Volume V (Units)	Depth h (Units)
500	
1000	
1500	
2000	
2500	
3000	
3500	
4000	

(a) Assume $r = 3$ feet and $L = 20$ feet, with the volume given in units of gallons and h is to have units of feet. (Hint: 1 Cubic Foot $=7.48053$ Gallons)
(b) Assume $r = 0.70$ meters and $L = 3$ meters, with volume given in units of liters and h is to have units of meters. (Hint: 1 Cubic Meter $=1000$ Liters)

▶ **11.** Fill in the additional lines to the tables in example 2-7 for the comparison of linear and quadratic convergence.

Number of Iterations to Achieve Error Bound ϵ Linear Convergence $\mid e_0 \mid = 0.5$				Number of Iterations to Achieve Error Bound ϵ Quadratic Convergence $\mid e_0 \mid = 0.5$			
ϵ	$\alpha = 0.25$	$\alpha = 0.5$	$\alpha = 0.75$	ϵ	$\alpha = 0.25$	$\alpha = 0.5$	$\alpha = 0.75$
10^{-9}				10^{-9}			
10^{-10}				10^{-10}			

▶ **12.** Determine all the roots of the cubic equation

$$x^3 - 13.92x^2 + 59.1665x - 76.9411 = 0$$

▶ **13.** In studying the operations of $+, -, *, \div$ associated with electronic computers you will find the process of division is much more complicated than multiplication.

(a) Use the function $f(x) = \dfrac{1}{x} - B$, with B constant, to devise a Newton's method algorithm for approximating the division process of calculating $1/B$. Note the fact that your algorithm only involves multiplication and subtraction. What are the things to avoid with your algorithm?

(b) Assume that $x_i = \frac{1}{N}(1 - \epsilon_i)$ and $x_{i+1} = \frac{1}{N}(1 - \epsilon_{i+1})$. Show that the algorithm in part (a) implies $\epsilon_{i+1} = \epsilon_i^2$. What does this tell you if $|\epsilon_0| < 1$?

(c) To calculate A/B, avoiding the process of division, one can estimate $1/B$ by the algorithm in part (a) and then use multiplication $(1/B)A$ to calculate A/B. Use this method to calculate   (i)   2/3   (ii)   3/7

(d) How accurate are your solutions in part (b)?

▶ **14.** What positive values of x satisfy the equation $\tan x = x$? Find the first five positive roots closest to zero.

▶ **15.** Write the equation $f(x) = e^x - 3x = 0$ in the form $x = g(x)$ and determine an interval (a, b) where the iteration process $x_{n+1} = g(x_n)$ will converge to a solution to the above equation. Find all roots to the above equation.

▶ **16.** Find the cube root of 5 by the bisection method so that the error in your answer is less than $\epsilon = (10)^{-6}$.

▶ **17.** Consider the equation $f(x) = x^3 - 3x^2 + 1 = 0$. Solve for x^3 and then divide both sides of the equation by x^2 and set up an iterative sequence $x_{n+1} = g(x_n) = 3 - \dfrac{1}{x_n^2}$.

(i) Under what conditions will this iterative sequence converge?

(ii) Determine an estimate for $\lim_{n \to \infty}\{x_n\}$ and determine the accuracy of your answer.

▶ **18**

$V = \pi h^2 R - \dfrac{\pi}{3}h^3$

The volume of a spherical cap is given by
$$V = \pi h^2 R - \frac{\pi}{3}h^3,$$
where R is the radius of the sphere and h is the height of the cap. If $R = 6$ meters and $V = 200$ cubic meters, then find h.

▶ **19.** (**Aitken's accelerated convergence**)

The Aitken method of accelerated convergence assumes there is an iterative sequence $x_{n+1} = g(x_n)$ which converges. If r is a fixed point of the iteration, then $r = g(r)$. The error after the (n+1)st iteration is

$$e_{n+1} = r - x_{n+1} = g(r) - g(x_n).$$

The mean value theorem can be applied to the difference on the right-hand side of the error to obtain

$$r - x_{n+1} = g'(\xi_1)(r - x_n). \tag{19a}$$

Replacing n by $n + 1$ gives the result

$$r - x_{n+2} = g'(\xi_2)(r - x_{n+1}). \tag{19b}$$

where ξ_1 and ξ_2 are points near r. The equations (19a) and (19b) imply that if $g'(\xi_1) \approx g'(\xi_2)$, then

$$\frac{r - x_{n+1}}{r - x_{n+2}} = \frac{r - x_n}{r - x_{n+1}}. \tag{19c}$$

(a) Solve the equation (19c) for the root r and show

$$r = x_{n+2} - \frac{(x_{n+2} - x_{n+1})^2}{x_{n+2} - 2x_{n+1} + x_n} \tag{19d}$$

which is known as the Aitken's accelerated convergence formula.

(b) The condition $g'(\xi_1) \approx g'(\xi_2)$ will be true only if x_n is near r. Therefore, equation (19d) is used as an approximation for an improved value for the root r and one can write

$$\text{Improved value} = x_{n+3} = x_{n+2} - \frac{(x_{n+2} - x_{n+1})^2}{x_{n+2} - 2x_{n+1} + x_n} \tag{19e}$$

(c) Consider the sequence of values generated in the figure 2-13. Use the values $x_4 = 0.651401898$, $x_5 = 0.652279526$, $x_6 = 0.652565308$ together with the accelerated convergence formula (19e) to calculate an improved value for the iterates in figure 2-13.

▶ **20.** Set up an iterative scheme for finding the roots of the given equation. After the first three iterations use the Aitken's accelerated convergence formula to approximate the next root. Find all roots in the given interval.

(a) $e^x = x^2$, $-3 < x < 3$,

(b) $xe^{-x} = 0.2$, $0 < x < 4$,

(c) $\tan x = \cos x$, $-1 < x < 1$

58

▶ **21.** Show the equation $f(x) = x^3 - 3x^2 + 1 = 0$ can be written in the iterative form $x_{n+1} = g(x_n)$ where $g(x) = x + \alpha(x^3 - 3x^2 + 1)$ and $g'(x) = 1 + \alpha(3x^2 - 6x)$ where α is a constant.

(a) If we desire a root near $x = -0.5$, what would be a choice for α in order for the iterative technique to converge? Test your answer and determine the root near $x = -0.5$.

(b) If we desire a root near $x = +0.5$, what would be a choice for α in order for the iterative technique to converge? Test your answer and determine the root near $x = 0.5$.

(c) If we desire a root near $x = 2.9$, what would be a choice for α in order for the iterative technique to converge? Test your answer and determine the root near $x = 2.9$.

▶ **22.**

Use Müller's method to find the roots of the given polynomial equations

$$(a) \quad x^3 - 3x^2 + 2 = 0, \qquad\qquad (b) \quad x^3 + 3x^2 - 3 = 0.$$

▶ **23.** Find roots to the following equations

$$(a) \quad \sin x = \cos x, \quad 0 < x < \pi/2 \qquad (b) \quad e^x = 2\cos x, \quad 0 < x < \pi/2$$

▶ **24.** Van der Waals's equation of state for a real gas is given by
$$\left(p + \frac{a}{V^2}\right)(V - b) = RT$$
where T is temperature (K), V is volume (m^3/mol), p is pressure (N/m^2), R is the universal gas constant, $R = 8.31$ $(J/mol\,K)$, and a, b are constants dependent upon the type of gas. For carbon dioxide $a = 0.36$ $(N\,m^4/mol^2)$ and $b = .000043$ (m^3/mol). Find V if $T = 1000$ (K), and $p = 6178.44$ N/m^2.

▶ **25.** Use Halley's method to find $\sqrt{N}$ for

$$N = 2, 3, 5, 6, 7, 8, 10, 11, 12, 13, 14, 15$$

▶ **26.** Find a root of $f(\omega) = \tan\omega + \omega = 0$ for $1.7 < \omega < 3$.

(a) Accurate to within $(10)^{-4}$

(b) Accurate to within $(10)^{-6}$

(c) Accurate to within $(10)^{-8}$

Justify the accuracy of your answers.

"An expert is someone who knows some of the worst mistakes that can be made in his subject, and how to avoid them."

Werner Heisenberg (1901-1976)

Chapter 3

Linear and Nonlinear Systems

In this chapter we investigate direct methods and iterative methods for solving linear systems of equations of the form

$$
\begin{aligned}
E_1: \quad & a_{11}x_1 + a_{12}x_2 + \cdots + a_{1j}x_j + \cdots + a_{1n}x_n = b_1 \\
E_2: \quad & a_{21}x_1 + a_{22}x_2 + \cdots + a_{2j}x_j + \cdots + a_{2n}x_n = b_2 \\
& \quad \vdots \qquad \vdots \qquad \ddots \qquad \vdots \qquad \vdots \\
E_i: \quad & a_{i1}x_1 + a_{i2}x_2 + \cdots + a_{ij}x_j + \cdots + x_{in}x_n = b_i \\
& \quad \vdots \qquad \vdots \qquad \ddots \qquad \vdots \qquad \vdots \\
E_m: \quad & a_{m1}x_1 + a_{m2}x_2 + \cdots + a_{mj}x_j + \cdots + a_{mn}x_n = b_m
\end{aligned}
\tag{3.1}
$$

where the coefficient in the ith row and jth column is denoted by a_{ij} for $i = 1, \ldots, m$ and $j = 1, \ldots n$. These coefficients of the system are assumed to be known and the quantities on the right-hand side b_i for $i = 1, \ldots m$ are also assumed to be known. The equations are labeled $E_1, E_2, \ldots, E_m$ and the problem is to solve for, if possible, the unknowns $x_1, x_2, \ldots, x_n$. This system represents m linear equations from which n unknowns must be determined. The system is referred to as a $m \times n$ (read m by n) linear system of equations since it has m equations with n unknowns, with m not necessarily equal to n.

The above system of equations can be written in alternative forms. One compact form is to write the system of equations as

$$
E_i: \quad \sum_{j=1}^{n} a_{ij}x_j = b_i, \qquad i = 1, 2, \ldots, m
\tag{3.2}
$$

The system of equations (3.1) can also be expressed in the matrix form

$$
A\overline{x} = \overline{b}
\tag{3.3}
$$

where

$$A = [a_{ij}] = \begin{bmatrix} a_{11} & a_{12} & \cdots & a_{1n} \\ a_{21} & a_{22} & \cdots & a_{2n} \\ \vdots & \vdots & \ddots & \vdots \\ a_{m1} & a_{m2} & \cdots & a_{mn} \end{bmatrix}, \quad \overline{\mathbf{x}} = \begin{bmatrix} x_1 \\ x_2 \\ \vdots \\ x_{n-1} \\ x_n \end{bmatrix}, \quad \overline{\mathbf{b}} = \begin{bmatrix} b_1 \\ b_2 \\ \vdots \\ b_m \end{bmatrix}, \tag{3.4}$$

Here it is to be understood that i, j are integers in the range $1 \leq i \leq m$ and $1 \leq j \leq n$ with a_{ij} denoting the matrix element in the ith row and jth column. The matrix A is called a $m \times n$ (read m by n) matrix. The special $n \times 1$ matrix $\overline{\mathbf{x}}$ is called a n-dimensional column vector with x_i, $1 \leq i \leq n$, the element in the ith row. Similarly, the special $m \times 1$ matrix $\overline{\mathbf{b}}$ is called a m-dimensional column vector with b_k denoting the element in the kth row, for $1 \leq k \leq m$. We will examine direct methods of solution and iterative methods for solving such systems.

We also investigate methods for solving nonlinear systems of equations of the form

$$\begin{aligned} E_1: && f_1(x_1, x_2, \ldots, x_n) &= 0 \\ E_2: && f_2(x_1, x_2, \ldots, x_n) &= 0 \\ \vdots && \vdots \\ E_m: && f_m(x_1, x_2, \ldots, x_n) &= 0 \end{aligned} \tag{3.5}$$

where f_i, for $i = 1, \ldots, m$, represent known continuous functions of the variables $x_1, x_2, \ldots, x_n$. The problem is to find, if possible, some numerical procedure for determining the unknown quantities $x_1, x_2, \ldots, x_n$ which satisfy all of the nonlinear equations in the system of equations (3.5). Note that the system of equations (3.1) is a special case of the more general system of equations (3.5). The system of nonlinear equations (3.5) can be written in the vector form

$$\overline{\mathbf{f}}(\overline{\mathbf{x}}) = \overline{0} \quad \text{where} \quad \overline{\mathbf{x}} = \begin{bmatrix} x_1 \\ x_2 \\ \vdots \\ x_n \end{bmatrix} \quad \text{and} \quad \overline{\mathbf{f}}(\overline{\mathbf{x}}) = \begin{bmatrix} f_1(x_1, x_2, \ldots, x_n) \\ f_2(x_1, x_2, \ldots, x_n) \\ \vdots \\ f_m(x_1, x_2, \ldots, x_n) \end{bmatrix} \tag{3.6}$$

with $\overline{0}$ being an m-dimensional column vector of zeros.

We develop several numerical methods which are applicable for solving systems of the above types. We desire to develop numerical methods for solving the above systems of equations in cases where the numbers m and n are large. In some applied problems it is not unusual for the number of unknowns $x_1, x_2, \ldots, x_n$

to be very large, say $n > 10^5$ or $n > 10^6$. For illustrative purposes and examples of the numerical techniques we will use much smaller values for n.

Preliminaries

In dealing with the matrix form associated with the equations (3.2) there are occasions when it is convenient to make use of special matrices and special functions associated with these matrices. Some of these special matrices and functions are as follows.

(i) The $n \times n$ identity matrix $I = [\delta_{ij}]$, where $\delta_{ij} = \begin{cases} 1, & \text{if } i = j \\ 0, & \text{if } i \neq j \end{cases}$. The identity matrix has 1's down the main diagonal and 0's everywhere else. A 3×3 identity matrix has the form $I = \begin{bmatrix} 1 & 0 & 0 \\ 0 & 1 & 0 \\ 0 & 0 & 1 \end{bmatrix}$.

(ii) If A is a $n \times n$ square matrix and there exists a matrix B with the property that $BA = AB = I$, then B is called the inverse of A and is written $B = A^{-1}$. That is, the inverse matrix A^{-1} has the property that $AA^{-1} = A^{-1}A = I$. If A^{-1} exists, then A is said to be nonsingular. If the matrix A does not have an inverse, then the matrix A is said to be singular. In the special case $m = n$ and A is a square matrix having an inverse, then the solution to the matrix system (3.3) can be symbolically obtained by multiplying both sides of the matrix equation (3.3) by the inverse of A to obtain

$$A^{-1}A\overline{\mathbf{x}} = A^{-1}\overline{\mathbf{b}} \quad \text{which simplifies to} \quad \overline{\mathbf{x}} = I\,\overline{\mathbf{x}} = A^{-1}\overline{\mathbf{b}}.$$

This technique for solving a $n \times n$ linear system of equations is only recommended in the case where n is small. This is because the number of multiply and divided operations needed to calculate A^{-1} increases like n^3 and so the inverse matrix calculation becomes very lengthy and burdensome when n is large.

(iii) Associated with a $m \times n$ matrix $A = [a_{ij}]$ is the $n \times m$ transpose matrix denoted by $A^T = [a_{ji}]$ and formed by interchanging the rows and columns of the $m \times n$ matrix A. Note that column vectors such as the $\overline{\mathbf{x}}$ given in equation (3.4) can be expressed as $\overline{\mathbf{x}} = [x_1, x_2, \ldots, x_n]^T$. The transpose is used to determine if a matrix is symmetric. A square matrix is said to be symmetric if $A^T = A$. The matrix product $A^T A$ always produces a square matrix. The transpose operation satisfies the following properties.

(i) $(AB)^T = B^T A^T$ The transpose of a product is the product
of the transpose matrices in reverse order.

(ii) $(A^{-1})^T = (A^T)^{-1}$ When A^{-1} exists, then the transpose of an inverse
is the inverse of the transpose.

(iii) $(A^T)^T = A$ The transpose of the transpose matrix
returns the original matrix.

(iv) A $n \times n$ lower triangular matrix $L = [\ell_{ij}]$ has the form

$$L = \begin{bmatrix} \ell_{11} & & & & \\ \ell_{21} & \ell_{22} & & & \\ \ell_{31} & \ell_{32} & \ell_{33} & & \\ \vdots & \vdots & \vdots & \ddots & \\ \ell_{n1} & \ell_{n2} & \ell_{n3} & \cdots & \ell_{nn} \end{bmatrix} \qquad \ell_{ij} = 0 \text{ for } j > i \qquad (3.7)$$

with all zeros above the main diagonal. In the special case the diagonal elements of a lower triangular matrix are all 1's, then L is called a unit lower triangular matrix.

(v) A $n \times n$ upper triangular matrix $U = [u_{ij}]$ has the form

$$U = \begin{bmatrix} u_{11} & u_{12} & u_{13} & \cdots & u_{1n} \\ & u_{22} & u_{23} & \cdots & u_{2n} \\ & & u_{33} & \cdots & u_{3n} \\ & & & \ddots & \vdots \\ & & & & u_{nn} \end{bmatrix} \qquad u_{ij} = 0 \text{ for } i > j \qquad (3.8)$$

with all zeros below the main diagonal. In the special case the diagonal elements of an upper triangular matrix are all 1's, then U is called a unit upper triangular matrix.

(vi) A $n \times n$ square matrix $A = [a_{ij}]$ with the property that

$$a_{ij} = \begin{cases} 0, & \text{for } i + s \le j, \quad 1 < s < n \\ 0, & \text{for } j + t \le i, \quad 1 < t < n \end{cases} \qquad (3.9)$$

is said to be a banded matrix with band width $w = s + t - 1$. For example, the 5×5 tridiagonal matrix

$$\begin{bmatrix} a_{11} & a_{12} & 0 & 0 & 0 \\ a_{21} & a_{22} & a_{23} & 0 & 0 \\ 0 & a_{32} & a_{33} & a_{34} & 0 \\ 0 & 0 & a_{43} & a_{44} & a_{45} \\ 0 & 0 & 0 & a_{54} & a_{55} \end{bmatrix} \qquad (3.10)$$

is a banded matrix with band width 3. Tridiagonal matrices occur quite frequently. The elements in a tridiagonal matrix are along the diagonal, the subdiagonal and superdiagonal. Tridiagonal system of equations are easily solved. Tridiagonal matrices are sometimes denoted using the notation

$$T = tridiagonal(\vec{a}, \vec{b}, \vec{c})$$

where

$\vec{a} = (a_1, a_2, a_3, a_4, \ldots, a_{n-1}, a_n)$ is vector along the subdiagonal

$\vec{b} = (b_1, b_2, b_3, b_4, \ldots, b_{n-1}, b_n)$ is vector along the diagonal

$\vec{c} = (c_1, c_2, c_3, c_4, \ldots, c_{n-1}, c_n)$ is vector along the superdiagonal

This is a shorthand notation for the tridiagonal matrix

$$T = [t_{ij}] = \begin{bmatrix} b_1 & c_1 & & & & \\ a_2 & b_2 & c_2 & & & \\ & a_3 & b_3 & c_3 & & \\ & & \ddots & \ddots & \ddots & \\ & & & a_{n-1} & b_{n-1} & c_{n-1} \\ & & & & a_n & b_n \end{bmatrix} \quad \text{with} \quad \begin{cases} t_{ij} = 0, & i > j+2 \\ t_{ij} = 0, & j > i+2 \end{cases}$$

Another matrix which is nice to work with is the diagonal matrix with band width one. An example of a 5×5 diagonal matrix is

$$D = \begin{bmatrix} a_{11} & 0 & 0 & 0 & 0 \\ 0 & a_{22} & 0 & 0 & 0 \\ 0 & 0 & a_{33} & 0 & 0 \\ 0 & 0 & 0 & a_{44} & 0 \\ 0 & 0 & 0 & 0 & a_{55} \end{bmatrix} \tag{3.11}$$

(vii) A $n \times n$ matrix A with diagonal elements a_{ii} for $i = 1, 2, \ldots, n$ and satisfying

$$|a_{ii}| \geq \sum_{\substack{j=1 \\ j \neq i}}^{n} |a_{ij}| \quad \text{for} \quad i = 1, 2, \ldots n \tag{3.12}$$

is said to be diagonally dominant.

(viii) A vector norm $\| \cdot \|$ is a mapping from R^n to R which is a measure of distance associated with vectors $\bar{x} = [x_1, x_2, \ldots, x_n]^T$. Vector norms have the following properties:

(a) $\| \bar{x} \| \geq 0$ for all $\bar{x} \in R^n$.

(b) $\| \alpha\bar{x} \| = |\alpha| \| \bar{x} \|$ where α is a scalar $\in R$.

(c) $\| \overline{\mathbf{x}} + \overline{\mathbf{y}} \| \leq \| \overline{\mathbf{x}} \| + \| \overline{\mathbf{y}} \|$ for all $\overline{\mathbf{x}}, \overline{\mathbf{y}} \in R^n$.

(d) $\| \overline{\mathbf{x}} \| = 0$ if and only if $\overline{\mathbf{x}} = 0$.

The Euclidean norm or ℓ_2 norm is used most often and is defined

$$\| \overline{\mathbf{x}} \|_2 = \left[\sum_{i=1}^{n} x_i^2 \right]^{1/2} \tag{3.13}$$

This norm represents the distance of the point $(x_1, x_2, \ldots, x_n)$ from the origin. Other vector norms can be defined so long as they obey the above properties. Two other vector norms used quite frequently are the ℓ_∞ norm and ℓ_p norm defined respectively as

$$\| \overline{\mathbf{x}} \|_\infty = \max_{1 \leq i \leq n} |x_i|$$

$$\| \overline{\mathbf{x}} \|_p = \left[\sum_{i=1}^{n} |x_i|^p \right]^{1/p} \tag{3.14}$$

(ix) A matrix norm $\| A \|$ associated with a $n \times n$ matrix $A = [a_{ij}]$ is any real-valued function $\| \cdot \|$ which satisfies the properties:

(a) $\| A \| \geq 0$

(b) $\| \alpha A \| \leq |\alpha| \| A \|$ where α is a scalar $\in R$

(c) $\| A + B \| \leq \| A \| + \| B \|$, where A, B are $n \times n$ matrices.

(d) $\| AB \| \leq \| A \| \| B \|$

(e) $\| A \| = 0$ if and only if $a_{ij} = 0$ for all i, j values.

The quantity $\| A - B \|$ is used to measure the nearness of two $n \times n$ matrices. Let $\| \cdot \|$ denote a vector norm, then the natural matrix norm of a $n \times n$ matrix A is defined

$$\| A \| = \max_{\| \overline{\mathbf{x}} \| = 1} \| A \overline{\mathbf{x}} \|. \tag{3.15}$$

For example, the ℓ_2 norm of the $n \times n$ matrix A would be represented

$$\| A \|_2 = \max_{\| \overline{\mathbf{x}} \|_2 = 1} \| A \overline{\mathbf{x}} \|_2 \tag{3.16}$$

and the ℓ_∞ norm of A would be represented

$$\| A \|_\infty = \max_{\| \overline{\mathbf{x}} \|_\infty = 1} \| A \overline{\mathbf{x}} \|_\infty \tag{3.17}$$

It can be shown, see the Bronson reference, that if $A = (a_{ij})$ is a $n \times n$ matrix, then

$$\| A \|_\infty = \max_{1 \leq i \leq n} \sum_{j=1}^{n} |a_{ij}|.$$

(x) The characteristic polynomial associated with the real $n \times n$ matrix A is defined in terms of a determinant and given by

$$p(\lambda) = \det[A - \lambda I] \tag{3.18}$$

where λ is a scalar and I is the $n \times n$ identity matrix. The values of λ which satisfy the characteristic equation $p(\lambda) = 0$ are called eigenvalues associated with the matrix A. Nonzero vectors $\overline{x}$ with the property that $A\overline{x} = \lambda\overline{x}$ are called eigenvectors of A corresponding to the eigenvalue λ.

(xi) The spectral radius $\rho(A)$ of the $n \times n$ matrix A is defined

$$\rho(A) = \max |\lambda| \tag{3.19}$$

where λ is an eigenvalue of A. If an eigenvalue is complex with the value $\lambda = \lambda_1 + i\lambda_2$, then $|\lambda| = \sqrt{\lambda_1^2 + \lambda_2^2}$. It can be shown that the spectral radius has the properties

$$(i) \quad \rho(A) \leq \| A \| \quad \text{for any matrix norm } \| \cdot \|.$$

$$(ii) \quad \sqrt{\rho(A^T A)} = \| A \|_2$$

(xii) The condition number $K(A)$ associated with a nonsingular matrix A is defined

$$K(A) = \| A \| \| A^{-1} \| \tag{3.20}$$

where $\| \cdot \|$ is a natural norm. A matrix A is said to be well behaved or well-conditioned if its condition number is close to unity. It is called ill behaved or ill-conditioned if the condition number is far from unity. Great care should be taken when working with ill-conditioned matrices.

Eigenvalues and Eigenvectors

A $n \times n$ square matrix A can be used as an operator to transform a nonzero $n \times 1$ column vector $\vec{x}$. One can imagine an input-output system such as the one illustrated in the figure 3-1.

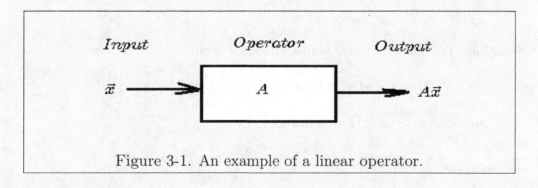

Figure 3-1. An example of a linear operator.

Those nonzero vectors $\vec{x}$ having the special property that the output is proportional to the input must satisfy

$$A\vec{x} = \lambda\vec{x} \tag{3.21}$$

where λ is a scalar proportionality constant. The special nonzero vectors $\vec{x}$ satisfying equation (3.21) are called eigenvectors and the corresponding scalars λ are called eigenvalues. The equation (3.21) is equivalent to the homogeneous system

$$A\vec{x} - \lambda\vec{x} = (A - \lambda I)\vec{x} = \vec{0} \tag{3.22}$$

where I is the $n \times n$ identity matrix. The equation (3.22) has a nonzero solution if and only if

$$\det(A - \lambda I) = |A - \lambda I| = 0. \tag{3.23}$$

The equation (3.23) when expanded is a polynomial equation in λ of degree n having the form

$$C(\lambda) = |A - \lambda I| = (-\lambda)^n + c_{n-1}(-\lambda)^{n-1} + \cdots + c_1(-\lambda) + c_0 = 0 \tag{3.24}$$

which is called the characteristic equation associated with the square matrix A. The solutions of equation (3.24) give the eigenvalues associated with the $n \times n$ square matrix A. For any given eigenvalue λ, the matrix $A - \lambda I$ is a singular matrix such that the homogeneous equations (3.22) produce a nonzero eigenvector $\vec{x}$. Note that if $\vec{x}$ is an eigenvector, then any nonzero constant times $\vec{x}$ is also an eigenvector.

Example 3-1.   (Eigenvalues and Eigenvectors)
Find the eigenvalues and eigenvectors associated with the matrix

$$A = \begin{bmatrix} 1 & 0 \\ 1 & 1 \end{bmatrix}.$$

Solution: We construct the homogeneous system

$$(A - \lambda I)\vec{x} = \vec{0} \quad \text{or} \quad \begin{bmatrix} 1-\lambda & 0 \\ 1 & 1-\lambda \end{bmatrix} \begin{bmatrix} x_1 \\ x_2 \end{bmatrix} = \begin{bmatrix} 0 \\ 0 \end{bmatrix}. \tag{3.25}$$

The characteristic equation is found to be

$$C(\lambda) = \begin{vmatrix} 1-\lambda & 0 \\ 1 & 1-\lambda \end{vmatrix} = (1-\lambda)^2 = 0.$$

The eigenvalues are $\lambda_1 = 1$ and $\lambda_2 = 1$. For $\lambda = 1$, the equation (3.25) reduces to

$$\begin{bmatrix} 0 & 0 \\ 1 & 0 \end{bmatrix} \begin{bmatrix} x_1 \\ x_2 \end{bmatrix} = \begin{bmatrix} 0 \\ 0 \end{bmatrix}.$$

Therefore, $\vec{x} = k \begin{bmatrix} 0 \\ 1 \end{bmatrix}$ is an eigenvector for any nonzero constant k.

∎

Elementary Row Operations

To solve the system of equations (3.1) one is allowed to perform any of the following elementary row operations on the system of equations.

(i) An equation in row i can be multiplied by a nonzero constant α. That is, equation E_i is replaced by the equation αE_i. This is denoted by the notation $(\alpha E_i \rightarrow E_i)$ and is read, "The constant α times equation E_i replaces the equation E_i.

(ii) Equation E_j can be replaced by a multiple of equation E_i added to equation E_j. This can be expressed using the above notation as $(\alpha E_i + E_j \rightarrow E_j)$ where $i \neq j$

(iii) Any two equations can be interchanged. This is denoted by the notation $(E_i \leftrightarrow E_j)$ where $i \neq j$.

Example 3-2.   (Elementary row matrices.)

Row operations performed upon the identity matrix I produces elementary row matrices E. Consider the 3×3 identity matrix $I = \begin{bmatrix} 1 & 0 & 0 \\ 0 & 1 & 0 \\ 0 & 0 & 1 \end{bmatrix}$, then some examples of elementary row matrices are:

(i) Interchanging rows 2 and 3 gives the elementary row matrix $E_1 = \begin{bmatrix} 1 & 0 & 0 \\ 0 & 0 & 1 \\ 0 & 1 & 0 \end{bmatrix}$

(ii) Multiplying row 3 by the scalar 5 gives the elementary row matrix $E_2 = \begin{bmatrix} 1 & 0 & 0 \\ 0 & 1 & 0 \\ 0 & 0 & 5 \end{bmatrix}$

(iii) Multiplying row 1 by 6 and adding the result to row 2 gives the elementary row matrix $E_3 = \begin{bmatrix} 1 & 0 & 0 \\ 6 & 1 & 0 \\ 0 & 0 & 1 \end{bmatrix}$

Note that if $A = \begin{bmatrix} a & b & c \\ d & e & f \\ g & h & i \end{bmatrix}$ is a 3×3 matrix, then

$$E_1 A = \begin{bmatrix} a & b & c \\ g & h & i \\ d & e & f \end{bmatrix}, \quad E_2 A = \begin{bmatrix} a & b & c \\ d & e & f \\ 5g & 5h & 5i \end{bmatrix}, \quad E_3 A = \begin{bmatrix} a & b & c \\ d+6a & e+6b & f+6c \\ g & h & i \end{bmatrix}$$

Observe that the elementary row operations recorded in the matrices E_1, E_2, E_3 have been applied to the matrix A.

■

A shorthand notation for recording the row operations performed upon a linear system of equations is to write down the coefficient matrix A of the linear system $A\overline{x} = \overline{b}$ and then append to the right-hand side of A the column vector $\overline{b}$. The resulting array is called an augmented matrix and is written

$$[A|\overline{b}] = \begin{bmatrix} a_{11} & a_{12} & \cdots & a_{1n} & b_1 \\ a_{21} & a_{22} & \cdots & a_{2n} & b_2 \\ \vdots & \vdots & \ddots & \vdots & \vdots \\ a_{m1} & a_{m2} & \cdots & a_{mn} & b_m \end{bmatrix}.$$

Our objective is to perform row operations upon the resulting array and try to reduce the array A to an upper triangular form. The row operations are then recorded in the augmented column vector. For example, consider the 2×3 augmented array

$$\begin{bmatrix} a_{11} & a_{12} & b_1 \\ a_{21} & a_{22} & b_2 \end{bmatrix},$$

with a_{11} nonzero, where we multiply the first row by $-a_{21}/a_{11}$ and add the result to row 2 to obtain the triangular system

$$\begin{bmatrix} a_{11} & a_{12} & b_1 \\ a_{21} & a_{22} & b_2 \end{bmatrix} \quad \begin{matrix} E_1: \\ -\frac{a_{21}}{a_{11}}E_1 + E_2 \to E_2: \end{matrix} \quad \begin{bmatrix} a_{11} & a_{12} & b_1 \\ 0 & c_{22} & d_2 \end{bmatrix}$$

where $c_{22} = a_{22} - a_{21}a_{12}/a_{11}$ and $d_2 = b_2 - a_{21}b_1/a_{11}$. The nonzero element a_{11} is called a pivot element in the diagonalization process. The resulting upper triangular system can then be solved by back substitution methods.

Gaussian Elimination

The Gaussian elimination method reduces a matrix to upper triangular (or lower triangular) and then uses back substitution to solve for the unknowns. The method is illustrated using an example.

Example 3-3. (Gaussian elimination method.)
Solve the system of equations

$$\begin{aligned} E_1: && 3x_1 - 2x_2 + x_3 &= 8 \\ E_2: && 4x_1 + x_2 - 3x_3 &= 3 \\ E_3: && x_1 + 5x_2 - 4x_3 &= 5 \end{aligned} \tag{3.26}$$

Solution: Let us use basic algebra and solve the above system of equations and observe that all we are doing is performing row operations on the equations. Then for comparison we will perform the same row operations on the augmented matrix array associated with the above system of equations.

Multiply equation E_1 by $-4/3$ and then add the result to equation E_2 to produce a new equation E_2. Using the above notation for row operations on the equations one can write this operation as $(-\frac{4}{3}E_1 + E_2 \rightarrow E_2)$. Next perform the operation $(-\frac{1}{3}E_1 + E_3 \rightarrow E_3)$. That is, a new equation E_3 is created when equation E_1 is multiplied by $-\frac{1}{3}$ and the result added to equation E_3. These row operations change the form of the original set of equations and one obtains the equivalent set of equations

$$
\begin{aligned}
E_1: \quad & 3x_1 - 2x_2 + x_3 = 8 \\
-\frac{4}{3}E_1 + E_2 \rightarrow E_2: \quad & \frac{11}{3}x_2 - \frac{13}{3}x_3 = -\frac{23}{3} \\
-\frac{1}{3}E_1 + E_2 \rightarrow E_3: \quad & \frac{17}{3}x_2 - \frac{13}{3}x_3 = \frac{7}{3}
\end{aligned}
\tag{3.27}
$$

The system of equations (3.27) can be diagonalized by eliminating the variable x_2 from the third row. The row operation $(-\frac{17}{11}E_2 + E_3 \rightarrow E_3)$ is one way to obtain the desired upper triangular system

$$
\begin{aligned}
E_1: \quad & 3x_1 - 2x_2 + x_3 = 8 \\
E_2: \quad & \frac{11}{3}x_2 - \frac{13}{3}x_3 = -\frac{23}{3} \\
-\frac{17}{11}E_2 + E_3 \rightarrow E_3: \quad & \frac{26}{11}x_3 = \frac{156}{11}
\end{aligned}
\tag{3.28}
$$

The original system of equations has been reduced to an equivalent upper triangular form using row operations. Observe that the system of equations (3.26) and (3.28) are equivalent because they have the same solution set. A triangular system such as the equations (3.28) is a form from which the solution can be easily obtained. Start with the last equation from the triangular set and solve for x_3 to obtain $x_3 = 6$. We substitute this value for x_3 into the second equation of the system (3.28) to obtain $x_2 = 5$. One can then substitute the values $x_2 = 5$ and $x_3 = 6$ into the first equation of the system (3.28) and obtain the value $x_1 = 4$. This process is called back substitution. The method of back substitution can be applied to either an upper or lower triangular linear system of equations.

Alternatively, one can write the augmented matrix $[A \mid \overline{b}]$ associated with the given system of linear equations as

$$\begin{bmatrix} 3 & -2 & 1 & 8 \\ 4 & 1 & -3 & 3 \\ 1 & 5 & -4 & 5 \end{bmatrix}$$

and then perform row operations on the array to reduce it to an upper triangular form. The element in column 1 and row 1 must be nonzero to start the process. If this element is zero, then the equations must be rearranged. The nonzero element in the first row, first column of the array is called the pivot element. The numbers in the first row of the array represent the coefficients of the first equation used which is called the pivot equation. In order to obtain zero's below the pivot element we perform the row operations

$$\begin{bmatrix} 3 & -2 & 1 & 8 \\ 4 & 1 & -3 & 3 \\ 1 & 5 & -4 & 5 \end{bmatrix} \quad \begin{array}{c} (E_1 \to E_1) \\ (-\frac{4}{3}E_1 + E_2 \to E_2) \\ (-\frac{1}{3}E_1 + E_3 \to E_3) \end{array} \quad \begin{bmatrix} 3 & -2 & 1 & 8 \\ 0 & 11/3 & -13/3 & -23/3 \\ 0 & 17/3 & -13/3 & 7/3 \end{bmatrix}$$

We now select the $a_{22} = 11/3$ element for the pivot element in the second row and perform row operations to place zero's below this element. This produces the upper triangular form we desire.

$$\begin{bmatrix} 3 & -2 & 1 & 8 \\ 0 & 11/3 & -13/3 & -23/3 \\ 0 & 17/3 & -13/3 & 7/3 \end{bmatrix} \quad \begin{array}{c} (E_1 \to E_1) \\ (E_2 \to E_2) \\ (-\frac{17}{11}E_2 + E_3 \to E_3) \end{array} \quad \begin{bmatrix} 3 & -2 & 1 & 8 \\ 0 & 11/3 & -13/3 & -23/3 \\ 0 & 0 & 26/11 & 156/11 \end{bmatrix} \quad (3.29)$$

The final array on the right-hand side is equivalent to the system (3.28) and the unknowns x_1, x_2, x_3 can now be solved for by back substitution. This process is known as the method of Gaussian elimination.

If additional row operations are performed upon the augmented system (3.29) so that the final result has zeros both above and below the main diagonal, then the method is called the Gauss-Jordan method for solving a system of equations. As an example we modify the system (3.29) with the following row operations

$$\begin{bmatrix} 3 & -2 & 1 & 8 \\ 0 & 11/3 & -13/3 & -23/3 \\ 0 & 0 & 26/11 & 156/11 \end{bmatrix} \quad \begin{array}{c} (\frac{3}{11}E_2 \to E_2) \\ (\frac{11}{26}E_3 \to E_3) \end{array} \quad \begin{bmatrix} 3 & -2 & 1 & 8 \\ 0 & 1 & -13/11 & -23/11 \\ 0 & 0 & 1 & 6 \end{bmatrix} \quad (3.30)$$

followed by the row operations

$$\begin{bmatrix} 3 & -2 & 1 & 8 \\ 0 & 1 & -13/11 & -23/11 \\ 0 & 0 & 1 & 6 \end{bmatrix} \quad \begin{array}{c} (E_1 - E_3 \to E_1) \\ (\frac{13}{11}E_3 + E_2 \to E_2) \end{array} \quad \begin{bmatrix} 3 & -2 & 0 & 2 \\ 0 & 1 & 0 & 5 \\ 0 & 0 & 1 & 6 \end{bmatrix} \quad (2E_2 + E_1 \to E_1) \quad \begin{bmatrix} 3 & 0 & 0 & 12 \\ 0 & 1 & 0 & 5 \\ 0 & 0 & 1 & 6 \end{bmatrix}$$

which produces the solutions $x_1 = 4$, $x_2 = 5$ and $x_3 = 6$.

■

Echelon Form and Rank

The row-reduced echelon form of a matrix is defined as a matrix modified by elementary row operations to a form where the rows and columns of the matrix are such that

(i) Nonzero rows are placed above rows containing all zeros.

(ii) As the elements of the matrix are viewed from left to right, the first nonzero element in a nonzero row is unity.

(iii) If there is a 1 in the ith row and jth column, then each element of column j having a row number greater than i must equal zero.

(iv) If row i is a nonzero row with a 1 in column j, then the number of leading zero's in row $i+1$ must be greater than the number of leading zeros in row i.

The rows of the matrix A can be considered as vectors in a n-dimensional space. The set of all linear combinations of the row vectors from the matrix A forms the row space of A. Similarly, the columns of the matrix A are vectors from a m-dimensional space. The set of all linear combinations of the column vectors from the matrix A forms the column space of A. A theorem from linear algebra states that if A is a $m \times n$ matrix, then the dimension of the row space of A equals the dimension of the column space of A. The rank of the matrix A is the dimension of the row space of A. The row-reduced echelon form of a matrix can be used to determine its rank since the nonzero rows of a row-reduced echelon form produces a basis for the row space of the matrix.

Example 3-4. (Echelon form and row rank.)

Reduce the matrix

$$A = \begin{bmatrix} 3 & 6 & -2 & 7 \\ 2 & 4 & -1 & 6 \\ 3 & 6 & -2 & 8 \\ 4 & 8 & -3 & 8 \end{bmatrix}$$

to echelon form and determine its row rank.

Solution: First one must perform a row operation to place a 1 in the row 1, column 1 position. One way to do this is

$$\begin{bmatrix} 3 & 6 & -2 & 7 \\ 2 & 4 & -1 & 6 \\ 3 & 6 & -2 & 8 \\ 4 & 8 & -3 & 8 \end{bmatrix} \quad (E_1 - E_2 \to E_1) \quad \begin{bmatrix} 1 & 2 & -1 & 1 \\ 2 & 4 & -1 & 6 \\ 3 & 6 & -2 & 8 \\ 4 & 8 & -3 & 8 \end{bmatrix}$$

Now row operations are needed to place zeros below the 1 in column 1.

$$
\begin{bmatrix} 1 & 2 & -1 & 1 \\ 2 & 4 & -1 & 6 \\ 3 & 6 & -2 & 8 \\ 4 & 8 & -3 & 8 \end{bmatrix}
\begin{matrix} \\ (-2E_1 + E_2 \rightarrow E_2) \\ (-3E_1 + E_3 \rightarrow E_3) \\ (-4E_1 + E_4 \rightarrow E_4) \end{matrix}
\begin{bmatrix} 1 & 2 & -1 & 1 \\ 0 & 0 & 1 & 4 \\ 0 & 0 & 1 & 5 \\ 0 & 0 & 1 & 4 \end{bmatrix}
$$

To achieve an echelon form we use row operations applied to the next lower row in order to make the first nonzero element in that row unity. Fortunately, there is already a one as the first nonzero element in row 2 and so we can now use row operations to place zeros below this element.

$$
\begin{bmatrix} 1 & 2 & -1 & 1 \\ 0 & 0 & 1 & 4 \\ 0 & 0 & 1 & 5 \\ 0 & 0 & 1 & 4 \end{bmatrix}
\begin{matrix} \\ \\ (E_2 + E_3 \rightarrow E_3) \\ (E_2 + E_4 \rightarrow E_4) \end{matrix}
\begin{bmatrix} 1 & 2 & -1 & 1 \\ 0 & 0 & 1 & 4 \\ 0 & 0 & 0 & 1 \\ 0 & 0 & 0 & 0 \end{bmatrix} = E
$$

The matrix E is the row-reduced echelon form associated with the given matrix. The matrices A and E are said to be row equivalent. The nonzero rows of the row-reduced echelon form represent a basis for the row space of A. The echelon form shows that the row rank of A equals 3. The row rank and column rank of a $m \times n$ matrix A are identical and so one usually just refers to the rank of the matrix A.

∎

The linear system of equations given by equation (3.4) can be written in terms of the column vectors of the matrix A. We write the matrix equation $A\overline{x} = \overline{b}$ in the form

$$
x_1 \begin{bmatrix} a_{11} \\ a_{21} \\ \vdots \\ a_{m1} \end{bmatrix} + x_2 \begin{bmatrix} a_{12} \\ a_{22} \\ \vdots \\ a_{m2} \end{bmatrix} + \cdots + x_n \begin{bmatrix} a_{1n} \\ a_{2n} \\ \vdots \\ a_{mn} \end{bmatrix} = \begin{bmatrix} b_1 \\ b_2 \\ \vdots \\ b_m \end{bmatrix} \tag{3.31}
$$

If in equation (3.31) all the constants b_i, $(i = 1, \ldots, m)$ are zero, then the system of equations $A\overline{x} = \overline{0}$ is termed a homogeneous system of equations. If at least one of the $b'_i s$ $1 \leq i \leq m$ is different from zero, then the system of equations (3.31) is called a nonhomogeneous system.

The equation (3.31) shows that in order for the equation $A\overline{x} = \overline{b}$ to have a solution, then the vector $\overline{b}$ must be a linear combination of the column vectors of A. In other words $\overline{b}$ must lie in the column space of A. Note that in the special case where $\bar{\mathbf{b}} = \bar{0}$ is the zero vector, the equation (3.31) becomes

$$x_1 \begin{bmatrix} a_{11} \\ a_{21} \\ \vdots \\ a_{m1} \end{bmatrix} + x_2 \begin{bmatrix} a_{12} \\ a_{22} \\ \vdots \\ a_{m2} \end{bmatrix} + \cdots + x_n \begin{bmatrix} a_{1n} \\ a_{2n} \\ \vdots \\ a_{mn} \end{bmatrix} = \begin{bmatrix} 0 \\ 0 \\ \vdots \\ 0 \end{bmatrix} \tag{3.32}$$

which shows that in order for the homogeneous equation $A\bar{\mathbf{x}} = \bar{0}$ to have the trivial solution $\bar{\mathbf{x}} = \bar{0}$, then the column vectors of the matrix A must be linearly independent.

Rank and Solutions

The $m \times n$ system of equations $A\bar{\mathbf{x}} = \bar{\mathbf{b}}$ or the equivalent system given by equation (3.31) can have a unique solution, no solutions or an infinite number of solutions depending upon the rank of the augmented matrix $[A|\bar{\mathbf{b}}]$ and the rank of the matrix A. We make use of the following results from linear algebra.

(i) A unique solution $\bar{\mathbf{x}}$ exists whenever (rank A)=(rank $[A|\bar{\mathbf{b}}]$)= n.

(ii) No solution for $\bar{\mathbf{x}}$ will exist if (rank A)$\neq$ (rank $[A \mid \bar{\mathbf{b}}]$)

(iii) Infinitely many solutions for $\bar{\mathbf{x}}$ will exist whenever the condition (rank A)=(rank $[A \mid \bar{\mathbf{b}}]$)= $r < n$ is satisfied.

Example 3-5. **(Types of solutions.)**
Consider the system of equations

$$\begin{aligned} ax + by &= c \\ dx + ey &= f, \end{aligned} \tag{3.33}$$

where a, b, c, d, e and f are constants. Each equation in the above system represents a straight line. The given system can also be represented in matrix form as

$$\begin{bmatrix} a & b \\ d & e \end{bmatrix} \begin{bmatrix} x \\ y \end{bmatrix} = \begin{bmatrix} c \\ f \end{bmatrix}. \tag{3.34}$$

If we let $x = x_1, \quad y = x_2$ and define

$$\bar{\mathbf{x}} = \text{col}[x_1, x_2], \qquad \bar{\mathbf{b}} = \text{col}[c, f], \qquad A = \begin{bmatrix} a & b \\ d & e \end{bmatrix},$$

then the system of equations (3.33) can be expressed in matrix form as $A\bar{\mathbf{x}} = \bar{\mathbf{b}}$.

In trying to find a solution to equations (3.33), (3.34), we find that various cases can arise:

(i) (The system possesses a unique solution.) Consider the case where the straight lines in equations (3.33) intersect at exactly one point. For example, the system of equations

$$3x - 4y = 2$$

$$x - y = 1$$

has the unique solution $x = 2$ and $y = 1$. Here rank A =rank $[A \mid \overline{b}]$=2.

(ii) (The system possesses no solutions.) Consider the case where the straight lines in equations (3.33) are parallel to one another and do not intersect. For example, the system

$$3x - 4y = 2$$

$$6x - 8y = 24$$

does not possess a common solution point (x, y) because the lines are parallel. Here rank $A = 1$ and rank $[A \mid \overline{b}] = 2$.

(iii) (An infinite number of solutions exist.) Consider the case where the straight lines in equations (3.33) are parallel and they coincide. For example, the system

$$3x - 4y = 2$$

$$6x - 8y = 4.$$

This system of equations possesses the infinite number of solutions given by the parametric representation

$$x = t \qquad y = \frac{1}{4}(3t - 2),$$

where t is a parameter and can have any value. Verification of this solution is done by substituting the values for x and y into the given system of equations. Note that the rank of the augmented matrix and original matrix satisfies the condition rank $A = \text{rank } [A \mid \overline{b}] = 1 < 2$.

■

Example 3-6. (System of equations.)

Solve the system of equations $A\overline{x} = \overline{b}$ given by

$$x_1 \qquad - 4x_3 - 3x_4 - 2x_5 = -5$$

$$x_1 + x_2 - 5x_3 - 2x_4 - x_5 = -3$$

$$x_1 - x_2 - 3x_3 - x_4 - 3x_5 = -7$$

$$x_1 + 2x_2 - 6x_3 - x_4 \qquad = -1$$

Solution The augmented matrix associated with the above system of equations is

$$\left[\begin{array}{ccccc|c} 1 & 0 & -4 & -3 & -2 & -5 \\ 1 & 1 & -5 & -2 & -1 & -3 \\ 1 & -1 & -3 & -4 & -3 & -7 \\ 1 & 2 & -6 & -1 & 0 & -1 \end{array}\right] = [A \mid \mathbf{b}].$$

We now apply row reduction methods to reduce this matrix to an echelon form. For example, by subtracting row 1 from rows 2, 3 and 4 we obtain:

$$\left[\begin{array}{ccccc|c} 1 & 0 & -4 & -3 & -2 & -5 \\ 0 & 1 & -1 & 1 & 1 & 2 \\ 0 & -1 & 1 & -1 & -1 & -2 \\ 0 & 2 & -2 & 2 & 2 & 4 \end{array}\right].$$

Next we add row 2 to row 3, then multiply row 2 by two, and subtract the result from row 4. These calculations produce the echelon form

$$\left[\begin{array}{ccccc|c} 1 & 0 & -4 & -3 & -2 & -5 \\ 0 & 1 & -1 & 1 & 1 & 2 \\ 0 & 0 & 0 & 0 & 0 & 0 \\ 0 & 0 & 0 & 0 & 0 & 0 \end{array}\right]$$

Here $(\text{rank } A) = (\text{rank } [A \mid \mathbf{b}]) = 2$ and the reduced echelon form is equivalent to the system of equations

$$\begin{aligned} x_1 \quad - 4x_3 - 3x_4 - 2x_5 &= -5 \\ x_2 - \quad x_3 + \quad x_4 + \quad x_5 &= 2. \end{aligned}$$

Since the quantities x_3, x_4, and x_5 can be arbitrary, we let

$$x_3 = c_1 \qquad x_4 = c_2 \qquad x_5 = c_3,$$

where c_1, c_2 and c_3 are arbitrary constants, and write the solution as

$$\begin{aligned} x_1 &= 4c_1 + 3c_2 + 2c_3 - 5 \\ x_2 &= \quad c_1 - \quad c_2 - \quad c_3 + 2 \\ x_3 &= \quad c_1 \\ x_4 &= \qquad\qquad c_2 \\ x_5 &= \qquad\qquad\qquad c_3 \end{aligned}$$

The solution can also be expressed in the vector form

$$X = \begin{bmatrix} x_1 \\ x_2 \\ x_3 \\ x_4 \\ x_5 \end{bmatrix} = c_1 \begin{bmatrix} 4 \\ 1 \\ 1 \\ 0 \\ 0 \end{bmatrix} + c_2 \begin{bmatrix} 3 \\ -1 \\ 0 \\ 1 \\ 0 \end{bmatrix} + c_3 \begin{bmatrix} 2 \\ -1 \\ 0 \\ 0 \\ 1 \end{bmatrix} + \begin{bmatrix} -5 \\ 2 \\ 0 \\ 0 \\ 0 \end{bmatrix},$$

where c_1, c_2, and c_3 are arbitrary constants. In general, the number of arbitrary constants N in a complete solution is given by the relation

$$N = \begin{pmatrix} \text{Number of} \\ \text{unknowns} \end{pmatrix} - \begin{pmatrix} \text{Common value of} \\ (\text{rank } A) = (\text{rank} [A \mid \bar{\alpha}]) = r \end{pmatrix}.$$

For the above example we have $N = 5 - 2 = 3$. Also, it should be observed that the column vectors

$$\bar{\mathbf{x}}_1 = \text{col}[4, 1, 1, 0, 0] \qquad \bar{\mathbf{x}}_2 = \text{col}[3, -1, 0, 1, 0] \qquad \bar{\mathbf{x}}_3 = \text{col}[2, -1, 0, 0, 1]$$

are solutions of the homogeneous system $A\bar{\mathbf{x}} = \bar{0}$, whereas the vector

$$\bar{\mathbf{x}}_p = \text{col}[-5, 2, 0, 0, 0]$$

is a particular solution of the nonhomogeneous equation $A\bar{\mathbf{x}} = \bar{\mathbf{b}}$. This example illustrates that there are systems with an infinite number of solutions.

∎

Example 3-7. (**System of equations.**)
Solve the system of equations

$$x_1 + 2x_2 \qquad = 6$$
$$2x_1 + 3x_2 + x_3 = 7$$
$$x_2 - x_3 = 2$$

Solution: We write the augmented matrix associated with this system and find

$$\begin{bmatrix} 1 & 2 & 0 & 6 \\ 2 & 3 & 1 & 7 \\ 0 & 1 & -1 & 2 \end{bmatrix} = [A \mid \bar{\mathbf{b}}].$$

To this matrix we apply row reduction techniques to reduce it to echelon form. We perform the following row operations:

(a) multiply row 1 by -2 and add the results to row 2,

$$\begin{bmatrix} 1 & 2 & 0 & | & 6 \\ 0 & -1 & 1 & | & -5 \\ 0 & 1 & -1 & | & 2 \end{bmatrix}.$$

(b) interchange rows 2 and 3,

$$\begin{bmatrix} 1 & 2 & 0 & | & 6 \\ 0 & 1 & -1 & | & 2 \\ 0 & -1 & 1 & | & -5 \end{bmatrix}.$$

(c) finally, add row 2 to row 3 to get the echelon form

$$\begin{bmatrix} 1 & 2 & 0 & | & 6 \\ 0 & 1 & -1 & | & 2 \\ 0 & 0 & 0 & | & -3 \end{bmatrix}.$$

This reduced form is equivalent to the system of equations

$$x_1 + 2x_2 \qquad = 6$$

$$x_2 - x_3 = 2$$

$$0 = -3 \quad (?)$$

which is clearly inconsistent. For this example we have $(\text{rank } A) = 2$ and $(\text{rank } [A \mid \overline{b}]) = 3$. The ranks of the coefficient matrix and the augmented matrix are different. Therefore, we can conclude that the system of equations has no solution. ∎

Example 3-8. (System of equations.)
Determine the solution of the system of equations

$$3x_1 + 4x_2 = 11$$

$$2x_1 - 3x_2 = -4$$

Solution: Write the augmented matrix associated with the given system of equations as

$$\begin{bmatrix} 3 & 4 & | & 11 \\ 2 & -3 & | & -4 \end{bmatrix} = [A \mid \overline{b}]$$

and apply row operations to reduce this matrix to echelon form. Subtract row 2 from row 1 to obtain

$$\begin{bmatrix} 1 & 7 & | & 15 \\ 2 & -3 & | & -4 \end{bmatrix}.$$

Now multiply row 1 by -2 and add the result to row 2 to obtain

$$\begin{bmatrix} 1 & 7 & | & 15 \\ 0 & -17 & | & -34 \end{bmatrix}.$$

Finally, divide row 2 by -17 to achieve the echelon form

$$\begin{bmatrix} 1 & 7 & | & 15 \\ 0 & 1 & | & 2 \end{bmatrix}.$$

The reduced form is equivalent to the system of equations

$$x_1 + 7x_2 = 15$$

$$x_2 = 2.$$

Here $x_2 = 2$ and $x_1 = 15 - 7x_2 = 1$ is the unique solution. Note that in this example we have (rank A) = (rank $[A \mid \overline{b}]$) = $2 = n$, where $n = 2$ is the number of unknowns in the above system of equations.

■

The Gaussian elimination method works fine for a small system of equations. However, there can be problems in applying the Gauss elimination method to large systems of equations. Some problems that one must prepare for are as follows.

1. Division by zero might occur if a pivot element has a value of zero. If this occurs, then one must rearrange the order of the equations.

2. If a pivot element is small, then divisions might cause overflow errors in the row operation calculations. In such cases one should examine all the equations and place the equation giving the largest magnitude pivot element on the diagonal. This process is called pivoting. Complete pivoting is the process of always placing the largest magnitude pivot element in the correct position. These additional operations are time consuming and are not recommended for very large systems.

3. If the linear system is very large, then a large number of computer operations must be performed. This increases the round off error buildup and introduces errors into the final results.

4. The larger the system the more computer memory is required and more row operations must be performed and consequently the Gauss elimination method is not the fastest method for solving large linear systems.

LU Decomposition

Consider the linear system of equations $A\overline{\mathbf{x}} = \overline{\mathbf{b}}$ where A is a $n \times n$ coefficient matrix. Observe that if it is possible to write the coefficient matrix A as the product of a lower triangular matrix L and an upper triangular matrix U, so that $A = LU$, or

$$\begin{bmatrix} a_{11} & a_{12} & \cdots & a_{1n} \\ a_{21} & a_{22} & \cdots & a_{2n} \\ \vdots & \vdots & \ddots & \vdots \\ a_{n1} & a_{n2} & \cdots & a_{nn} \end{bmatrix} = \underbrace{\begin{bmatrix} \ell_{11} & & & \\ \ell_{21} & \ell_{22} & & \\ \vdots & \vdots & \ddots & \\ \ell_{n1} & \ell_{n2} & \cdots & \ell_{nn} \end{bmatrix}}_{\ell_{ij}=0 \text{ for } j>i} \underbrace{\begin{bmatrix} u_{11} & u_{12} & \cdots & u_{1n} \\ & u_{22} & \cdots & \ell_{2n} \\ & & \ddots & \vdots \\ & & & u_{nn} \end{bmatrix}}_{u_{ij}=0 \text{ for } i>j}$$

then the linear system of equations $A\overline{\mathbf{x}} = \overline{\mathbf{b}}$ can be easily solved using the back substitution methods associated with triangular systems. Here we write $A\overline{\mathbf{x}} = \overline{\mathbf{b}}$ as $LU\overline{\mathbf{x}} = \overline{\mathbf{b}}$, then one can make the substitutions

$$L\overline{\mathbf{y}} = \overline{\mathbf{b}} \qquad \text{and} \qquad U\overline{\mathbf{x}} = \overline{\mathbf{y}} \tag{3.35}$$

and so the original system of equations $A\overline{\mathbf{x}} = \overline{\mathbf{b}}$ is replaced by two triangular systems which are easily solved by back substitution methods. The numerical procedure is to first solve for $\overline{\mathbf{y}}$ and then solve for $\overline{\mathbf{x}}$ by back substitution methods. The factorization of the matrix A into the product LU is called LU-decomposition. The factorization of the matrix A is not unique as is illustrated by the following three methods for decomposition schemes to obtain $A = LU$.

Doolittle's method

This method requires that the matrix L be a unit lower triangular matrix with 1's along the main diagonal.

Crout's method

This method requires that the matrix U be a unit upper triangular matrix with 1's along the main diagonal.

Choleski's method

This method requires that the diagonal elements of L and U be equal with $\ell_{ii} = u_{ii}$ for $i = 1, \ldots, n$.

Example 3-9. (Doolittle's method.)

Solve the system of equations

$$
\begin{array}{rrrrr}
2x_1 & -2x_2 & -x_3 & +x_4 & = & -2 \\
4x_1 & -x_2 & -2x_3 & +3x_4 & = & 16 \\
6x_1 & +6x_2 & -x_3 & +6x_4 & = & 78 \\
10x_1 & +8x_2 & +9x_3 & +5x_4 & = & 146
\end{array}
$$

using Doolittle's method.

Solution: We write the given system of equations in the matrix form $A\overline{\mathbf{x}} = \overline{\mathbf{b}}$ where

$$
A = \begin{bmatrix} 2 & -2 & -1 & 1 \\ 4 & -1 & -2 & 3 \\ 6 & 6 & -1 & 6 \\ 10 & 8 & 9 & 5 \end{bmatrix}, \qquad \overline{\mathbf{x}} = \begin{bmatrix} x_1 \\ x_2 \\ x_3 \\ x_4 \end{bmatrix}, \qquad \overline{\mathbf{b}} = \begin{bmatrix} -2 \\ 16 \\ 78 \\ 146 \end{bmatrix} \tag{3.36}
$$

The Doolittle method requires that we write the matrix A as a product of a unit lower triangular matrix and upper triangular matrix. That is, we write the coefficient matrix in the form $A = LU$ where L has 1's along the main diagonal. This requires that

$$
\begin{bmatrix} 2 & -2 & -1 & 1 \\ 4 & -1 & -2 & 3 \\ 6 & 6 & -1 & 6 \\ 10 & 8 & 9 & 5 \end{bmatrix} = \begin{bmatrix} 1 & 0 & 0 & 0 \\ \ell_{21} & 1 & 0 & 0 \\ \ell_{31} & \ell_{32} & 1 & 0 \\ \ell_{41} & \ell_{42} & \ell_{43} & 1 \end{bmatrix} \begin{bmatrix} u_{11} & u_{12} & u_{13} & u_{14} \\ 0 & u_{22} & u_{23} & u_{24} \\ 0 & 0 & u_{33} & u_{34} \\ 0 & 0 & 0 & u_{44} \end{bmatrix} \tag{3.37}
$$

From the first row vector of L dotted with each of the column vectors from U we find that

$$
u_{11} = 2, \qquad u_{12} = -2, \qquad u_{13} = -1, \qquad u_{14} = 1.
$$

From the second row vector of L dotted with each column vector from U we find the equations

$$
\left.\begin{array}{l} \ell_{21}u_{11} = 4 \\[4pt] \ell_{21}u_{12} + u_{22} = -1 \\[4pt] \ell_{21}u_{13} + u_{23} = -2 \\[4pt] \ell_{21}u_{14} + u_{24} = 3 \end{array}\right\} \quad \text{from which we obtain the values} \qquad \begin{array}{l} \ell_{21} = 2 \\[4pt] u_{22} = 3 \\[4pt] u_{23} = 0 \\[4pt] u_{24} = 1 \end{array}
$$

From the third row vector of L dotted with each column vector from U we obtain the equations

$$
\left.\begin{array}{l} \ell_{31}u_{11} = 6 \\[4pt] \ell_{31}u_{12} + \ell_{32}u_{22} = 6 \\[4pt] \ell_{31}u_{13} + \ell_{32}u_{23} + u_{33} = -1 \\[4pt] \ell_{31}u_{14} + \ell_{32}u_{24} + u_{34} = 6 \end{array}\right\} \quad \text{which produce the values} \qquad \begin{array}{l} \ell_{31} = 3 \\[4pt] \ell_{32} = 4 \\[4pt] u_{33} = 2 \\[4pt] u_{34} = -1 \end{array}
$$

From the fourth row vector of L dotted with each column vector from U one obtains the equations

$$\left.\begin{array}{r}\ell_{41}u_{11} = 10 \\ \ell_{41}u_{12} + \ell_{42}u_{22} = 8 \\ \ell_{41}u_{13} + \ell_{42}u_{23} + \ell_{43}u_{33} = 9 \\ \ell_{41}u_{14} + \ell_{42}u_{24} + \ell_{43}u_{34} + u_{44} = 5\end{array}\right\} \text{ so that } \begin{array}{l}\ell_{41} = 5 \\ \ell_{42} = 6 \\ \ell_{43} = 7 \\ u_{44} = -1\end{array}$$

We can now replace the system $A\bar{\mathbf{x}} = LU\bar{\mathbf{x}} = \bar{\mathbf{b}}$ by the two triangular systems

$$L\bar{\mathbf{y}} = \bar{\mathbf{b}} \qquad\qquad U\bar{\mathbf{x}} = \bar{\mathbf{y}}$$

$$\begin{bmatrix} 1 & 0 & 0 & 0 \\ 2 & 1 & 0 & 0 \\ 3 & 4 & 1 & 0 \\ 5 & 6 & 7 & 1 \end{bmatrix}\begin{bmatrix} y_1 \\ y_2 \\ y_3 \\ y_4 \end{bmatrix} = \begin{bmatrix} -2 \\ 16 \\ 78 \\ 146 \end{bmatrix} \qquad \begin{bmatrix} 2 & -2 & -1 & 1 \\ 0 & 3 & 0 & 1 \\ 0 & 0 & 2 & -1 \\ 0 & 0 & 0 & 1 \end{bmatrix}\begin{bmatrix} x_1 \\ x_2 \\ x_3 \\ x_4 \end{bmatrix} = \begin{bmatrix} y_1 \\ y_2 \\ y_3 \\ y_4 \end{bmatrix} \quad (3.38)$$

These triangular systems are easily solved using back substitution. Solving for $\bar{\mathbf{y}}$ we find $y_1 = -2$, $y_2 = 20$, $y_3 = 4$ and $y_4 = 8$. Solving for $\bar{\mathbf{x}}$ using back substitution gives $x_1 = 2$, $x_2 = 4$, $x_3 = 6$, and $x_4 = 8$.

$\blacksquare$

Example 3-10. (Crout's method.)

Solve the given system of equations by using Crout's method

$$\begin{array}{rrrrrr} 2x_1 & -2x_2 & -x_3 & +x_4 & = & -2 \\ 4x_1 & -x_2 & -2x_3 & +3x_4 & = & 16 \\ 6x_1 & +6x_2 & -x_3 & +6x_4 & = & 78 \\ 10x_1 & +8x_2 & +9x_3 & +5x_4 & = & 146 \end{array}$$

Solution: We again write the system in the matrix form $A\bar{\mathbf{x}} = \bar{\mathbf{b}}$ with elements defined by the equations (3.36). The Crout method of factorization requires that $A = LU$ have the form

$$\begin{bmatrix} 2 & -2 & -1 & 1 \\ 4 & -1 & -2 & 3 \\ 6 & 6 & -1 & 6 \\ 10 & 8 & 9 & 5 \end{bmatrix} = \begin{bmatrix} \ell_{11} & 0 & 0 & 0 \\ \ell_{21} & \ell_{22} & 0 & 0 \\ \ell_{31} & \ell_{32} & \ell_{33} & 0 \\ \ell_{41} & \ell_{42} & \ell_{43} & \ell_{44} \end{bmatrix}\begin{bmatrix} 1 & u_{12} & u_{13} & u_{14} \\ 0 & 1 & u_{23} & u_{24} \\ 0 & 0 & 1 & u_{34} \\ 0 & 0 & 0 & 1 \end{bmatrix} \quad (3.39)$$

where the upper triangular matrix has 1's along the main diagonal. The dot product of the first row vector from L with each column vector from the matrix U produces the equations

$$\left.\begin{array}{l} \ell_{11} = 2 \\ \ell_{11}u_{12} = -2 \\ \ell_{11}u_{13} = -1 \\ \ell_{11}u_{14} = 1 \end{array}\right\} \quad \text{which produces the results} \qquad \begin{array}{l} \ell_{11} = 2 \\ u_{12} = -1 \\ u_{13} = -\dfrac{1}{2} \\ u_{14} = \dfrac{1}{2} \end{array}$$

The dot product of the second row vector from L with each column vector from U produces the equations

$$\left.\begin{array}{l} \ell_{21} = 4 \\ \ell_{21}u_{12} + \ell_{22} = -1 \\ \ell_{21}u_{13} + \ell_{22}u_{23} = -2 \\ \ell_{21}u_{14} + \ell_{22}u_{24} = 3 \end{array}\right\} \quad \text{which gives the results} \qquad \begin{array}{l} \ell_{21} = 4 \\ \ell_{22} = 3 \\ u_{23} = 0 \\ u_{24} = \dfrac{1}{3} \end{array}$$

The dot product of the third row vector from L with each column vector from U gives the equations

$$\left.\begin{array}{l} \ell_{31} = 6 \\ \ell_{31}u_{12} + \ell_{32} = 6 \\ \ell_{31}u_{13} + \ell_{32}u_{23} + \ell_{33} = -1 \\ \ell_{31}u_{14} + \ell_{32}u_{24} + \ell_{33}u_{34} = 6 \end{array}\right\} \quad \text{from which there results} \qquad \begin{array}{l} \ell_{31} = 6 \\ \ell_{32} = 12 \\ \ell_{33} = 2 \\ u_{34} = -\dfrac{1}{2} \end{array}$$

Finally, the dot product of the fourth row vector in L with each of the column vector from U produces the equations

$$\left.\begin{array}{l} \ell_{41} = 10 \\ \ell_{41}u_{12} + \ell_{42} = 8 \\ \ell_{41}u_{13} + \ell_{42}u_{23} + \ell_{43} = 9 \\ \ell_{41}u_{14} + \ell_{42}u_{24} + \ell_{43}u_{34} + \ell_{44} = 5 \end{array}\right\} \quad \text{produces the results} \qquad \begin{array}{l} \ell_{41} = 10 \\ \ell_{42} = 18 \\ \ell_{43} = 14 \\ \ell_{44} = 1 \end{array}$$

We can now replace the matrix system $A\bar{x} = LU\bar{x} = \bar{b}$ by the two triangular matrix systems

$$\overset{\textstyle L\bar{y} = \bar{b}}{\begin{bmatrix} 2 & 0 & 0 & 0 \\ 4 & 3 & 0 & 0 \\ 6 & 12 & 2 & 0 \\ 10 & 18 & 14 & 1 \end{bmatrix} \begin{bmatrix} y_1 \\ y_2 \\ y_3 \\ y_4 \end{bmatrix} = \begin{bmatrix} -2 \\ 16 \\ 78 \\ 146 \end{bmatrix}} \qquad \overset{\textstyle U\bar{x} = \bar{y}}{\begin{bmatrix} 1 & -1 & -1/2 & 1/2 \\ 0 & 1 & 0 & 1/3 \\ 0 & 0 & 1 & -1/2 \\ 0 & 0 & 0 & 1 \end{bmatrix} \begin{bmatrix} x_1 \\ x_2 \\ x_3 \\ x_4 \end{bmatrix} = \begin{bmatrix} y_1 \\ y_2 \\ y_3 \\ y_4 \end{bmatrix}} \qquad (3.40)$$

These triangular systems are easily solved using back substitution. Solving for $\bar{y}$ we find $y_1 = -1$, $y_2 = 20/3$, $y_3 = 2$ and $y_4 = 8$. Solving for $\bar{x}$ using back substitution gives $x_1 = 2$, $x_2 = 4$, $x_3 = 6$, and $x_4 = 8$.

■

Example 3-11. (Choleski's method.)

Solve the given system of equations by using Choleski's method.

$$
\begin{array}{rrrrcr}
2x_1 & -2x_2 & -\ x_3 & +\ x_4 & = & -2 \\
4x_1 & -\ x_2 & -2x_3 & +3x_4 & = & 16 \\
6x_1 & +6x_2 & -\ x_3 & +6x_4 & = & 78 \\
10x_1 & +8x_2 & +9x_3 & +5x_4 & = & 146
\end{array}
$$

Solution: We again write the system in the matrix form $A\bar{x} = \bar{b}$ with elements given by equation (3.36). The Choleski method of factorization requires that $A = LU$ where the diagonal elements of L and U are equal and so we let $u_{ii} = \ell_{ii}$ for $i = 1, 2, 3, 4$. We then obtain the following factorization form

$$
\begin{bmatrix}
2 & -2 & -1 & 1 \\
4 & -1 & -2 & 3 \\
6 & 6 & -1 & 6 \\
10 & 8 & 9 & 5
\end{bmatrix}
=
\begin{bmatrix}
\ell_{11} & 0 & 0 & 0 \\
\ell_{21} & \ell_{22} & 0 & 0 \\
\ell_{31} & \ell_{32} & \ell_{33} & 0 \\
\ell_{41} & \ell_{42} & \ell_{43} & \ell_{44}
\end{bmatrix}
\begin{bmatrix}
\ell_{11} & u_{12} & u_{13} & u_{14} \\
0 & \ell_{22} & u_{23} & u_{24} \\
0 & 0 & \ell_{33} & u_{34} \\
0 & 0 & 0 & \ell_{44}
\end{bmatrix}
$$

The dot product of the first row vector from L with each column vector from U produces the equations

$$
\left.
\begin{aligned}
\ell_{11}^2 &= 2 \\
\ell_{11}u_{12} &= -2 \\
\ell_{11}u_{13} &= -1 \\
\ell_{11}u_{14} &= 1
\end{aligned}
\right\}
\quad \text{which produces the results} \quad
\begin{aligned}
\ell_{11} &= \sqrt{2} \\
u_{12} &= -\sqrt{2} \\
u_{13} &= -\sqrt{2}/2 \\
u_{14} &= \sqrt{2}/2
\end{aligned}
$$

The dot product of the second row vector from L with each column vector from U produces the equations

$$
\left.
\begin{aligned}
\ell_{21}\ell{11} &= 4 \\
\ell_{21}u_{12} + \ell_{22}^2 &= -1 \\
\ell_{21}u_{13} + \ell_{22}u_{23} &= -2 \\
\ell_{21}u_{14} + \ell_{22}u_{24} &= 3
\end{aligned}
\right\}
\quad \text{which gives the results} \quad
\begin{aligned}
\ell_{21} &= 2\sqrt{2} \\
\ell_{22} &= \sqrt{3} \\
u_{23} &= 0 \\
u_{24} &= \sqrt{3}/3
\end{aligned}
$$

The dot product of the third row vector from L with each column vector from U gives the equations

$$\left.\begin{array}{l} \ell_{31}\ell_{11} = 6 \\ \ell_{31}u_{12} + \ell_{32}\ell_{22} = 6 \\ \ell_{31}u_{13} + \ell_{32}u_{23} + \ell_{33}^2 = -1 \\ \ell_{31}u_{14} + \ell_{32}u_{24} + \ell_{33}u_{34} = 6 \end{array}\right\} \text{ from which there results } \begin{array}{l} \ell_{31} = 3\sqrt{2} \\ \ell_{32} = 4\sqrt{3} \\ \ell_{33} = \sqrt{2} \\ u_{34} = -\sqrt{2}/2 \end{array}$$

Finally, the dot product of the fourth row vector in L with each of the column vector from U produces the equations

$$\left.\begin{array}{l} \ell_{41}\ell_{11} = 10 \\ \ell_{41}u_{12} + \ell_{42}\ell_{22} = 8 \\ \ell_{41}u_{13} + \ell_{42}u_{23} + \ell_{43}\ell_{33} = 9 \\ \ell_{41}u_{14} + \ell_{42}u_{24} + \ell_{43}u_{34} + \ell_{44}^2 = 5 \end{array}\right\} \text{ produces the results } \begin{array}{l} \ell_{41} = 5\sqrt{2} \\ \ell_{42} = 6\sqrt{3} \\ \ell_{43} = 7\sqrt{2} \\ \ell_{44} = 1 \end{array}$$

We can now replace the matrix system $A\bar{\mathbf{x}} = LU\bar{\mathbf{x}} = \bar{\mathbf{b}}$ by the two triangular matrix systems

$$L\bar{\mathbf{y}} = \bar{\mathbf{b}}$$

$$\begin{bmatrix} \sqrt{2} & 0 & 0 & 0 \\ 2\sqrt{2} & \sqrt{3} & 0 & 0 \\ 3\sqrt{2} & 4\sqrt{3} & \sqrt{2} & 0 \\ 5\sqrt{2} & 6\sqrt{3} & 7\sqrt{2} & 1 \end{bmatrix} \begin{bmatrix} y_1 \\ y_2 \\ y_3 \\ y_4 \end{bmatrix} = \begin{bmatrix} -2 \\ 16 \\ 78 \\ 146 \end{bmatrix}$$

and
$$U\bar{\mathbf{x}} = \bar{\mathbf{y}}$$

$$\begin{bmatrix} \sqrt{2} & -\sqrt{2} & -\sqrt{2}/2 & \sqrt{2}/2 \\ 0 & \sqrt{3} & 0 & \sqrt{3}/3 \\ 0 & 0 & \sqrt{2} & -\sqrt{2}/2 \\ 0 & 0 & 0 & 1 \end{bmatrix} \begin{bmatrix} x_1 \\ x_2 \\ x_3 \\ x_4 \end{bmatrix} = \begin{bmatrix} y_1 \\ y_2 \\ y_3 \\ y_4 \end{bmatrix}$$

(3.41)

These triangular systems are easily solved using back substitution. Solving for $\bar{\mathbf{y}}$ we find $y_1 = -\sqrt{2}$, $y_2 = 20\sqrt{3}/3$, $y_3 = 2\sqrt{2}$ and $y_4 = 8$. Solving for $\bar{\mathbf{x}}$ using back substitution gives $x_1 = 2$, $x_2 = 4$, $x_3 = 6$, and $x_4 = 8$.

■

There can be problems in using LU-factorization. A problem occurs if a diagonal element in either matrix L or U is zero. In such a case it may be possible to interchange the ordering of the equations to obtain an equivalent system. In the Choleski method there might occur complex numbers in solving the system of equations. That is, one might encounter the square root of a negative number. Some of the advantages of the LU-decomposition approach is that there is less storage of matrix elements and the fact that triangular systems are easily solved by back substitution methods. In the special case the matrix A is symmetric, so that $a_{ij} = a_{ji}$, then the Choleski method produces symmetric L and U matrices. That is, the transpose of L produces U, or the transpose of U produces L. In this special case the number of operations required to perform the LU-decomposition is halved.

Iterative Methods for Solving $A\overline{\mathbf{x}} = \overline{\mathbf{b}}$

The $n \times n$ linear system of equations $A\overline{\mathbf{x}} = \overline{\mathbf{b}}$ can be written in the iterative form

$$\overline{\mathbf{x}}^{(k+1)} = T\overline{\mathbf{x}}^{(k)} + \overline{\mathbf{c}} \quad \text{for} \quad k = 0, 1, 2, \ldots \tag{3.42}$$

where T is a $n \times n$ matrix and $\overline{\mathbf{x}}^{(k)}$, and $\overline{\mathbf{c}}$ are $n \times 1$ column vectors.

Example 3-12. (Jacobi method)
The system of equations

$$\begin{array}{rrrrrl}
4x_1 & -\ x_2 & +\ x_3 & -\ x_4 & =\ 1 \\
x_1 & +5x_2 & -\ x_3 & -2x_4 & =\ 0 \\
x_1 & -2x_2 & +7x_3 & -\ x_4 & =\ 14 \\
x_1 & -3x_2 & -\ x_3 & +8x_4 & =\ 24
\end{array} \tag{3.43}$$

has the solution $x_1 = 1$, $x_2 = 2$, $x_3 = 3$ and $x_4 = 4$. This system can be written in the iterative form of equation (3.42) by solving for the variable x_i in the ith equation. This produces the equations

$$\begin{array}{rlllll}
x_1 & = & \frac{1}{4}x_2 & -\frac{1}{4}x_3 & +\frac{1}{4}x_4 & +\frac{1}{4} \\
x_2 & = & -\frac{1}{5}x_1 & +\frac{1}{5}x_3 & +\frac{2}{5}x_4 & \\
x_3 & = & -\frac{1}{7}x_1 & +\frac{2}{7}x_2 & +\frac{1}{7}x_4 & +2 \\
x_4 & = & -\frac{1}{8}x_1 & +\frac{3}{8}x_2 & +\frac{1}{8}x_3 & +3
\end{array} \tag{3.44}$$

which can be converted to the iterative scheme

$$\begin{bmatrix} x_1 \\ x_2 \\ x_3 \\ x_4 \end{bmatrix}^{(k+1)} = \begin{bmatrix} 0 & 1/4 & -1/4 & 1/4 \\ -1/5 & 0 & 1/5 & 2/5 \\ -1/7 & 2/7 & 0 & 1/7 \\ -1/8 & 3/8 & 1/8 & 0 \end{bmatrix} \begin{bmatrix} x_1 \\ x_2 \\ x_3 \\ x_4 \end{bmatrix}^{(k)} + \begin{bmatrix} 1/4 \\ 0 \\ 2 \\ 3 \end{bmatrix} \tag{3.45}$$

This is called the Jacobi method of forming an iterative system. One usually selects the starting value $\overline{\mathbf{x}}^{(0)} = [0, 0, 0, 0]^T$, but this is not a requirement. Applying this starting value to the system of equations (3.45) one can generate the following table of values.

k	0	1	2	3	4	5	6	7	8	9
$x_1^{(k)}$	0	.25	.5	.843973	.01317	.974231	.986063	.995855	.997786	.999336
$x_2^{(k)}$	0	0	1.55	1.66607	1.9246	1.94681	1.98778	1.99158	1.99803	1.99867
$x_3^{(k)}$	0	2	2.39286	2.83125	2.90086	2.97274	2.98423	2.99562	2.9975	2.9993
$x_4^{(k)}$	0	3	3.21875	3.81786	3.87319	3.97019	3.97987	3.99519	3.99681	3.99922

Continuing the iterations one finds that $\bar{\mathbf{x}}^{(17)} = [1, 2, 3, 4]^T$ and so the iterations converge to the solution.

∎

The system of equations $A\bar{\mathbf{x}} = \bar{\mathbf{b}}$ can be written in the matrix form

$$(D + L + U)\bar{\mathbf{x}} = \bar{\mathbf{b}} \tag{3.46}$$

where D is a diagonal matrix, L is lower triangular and U is upper triangular, then the Jacobi method can be formulated as a matrix equation by writing the system of equations $A\bar{\mathbf{x}} = \bar{\mathbf{b}}$ in the form

$$(D + L + U)\bar{\mathbf{x}} = \bar{\mathbf{b}}$$

$$D\bar{\mathbf{x}} = -(L + U)\bar{\mathbf{x}} + \bar{\mathbf{b}} \tag{3.47}$$

$$\bar{\mathbf{x}} = -D^{-1}(L + U)\bar{\mathbf{x}} + D^{-1}\bar{\mathbf{b}}$$

from which the iterative scheme of equation (3.42) can be written where $T = -D^{-1}(L + U)$ and $\bar{\mathbf{c}} = D^{-1}\bar{\mathbf{b}}$.

Another way to represent the $n \times n$ system of equations $A\bar{\mathbf{x}} = \bar{\mathbf{b}}$ is to write the ith equation in the symbolic form

$$E_i: \qquad a_{i1}x_1 + a_{i2}x_2 + \cdots a_{ii}x_i + \cdots a_{in}x_n = b_i, \qquad i = 1, 2, \ldots n \tag{3.48}$$

If the diagonal elements a_{ii} are different from zero, then one can solve for x_i to obtain

$$x_i = \frac{1}{a_{ii}} \left(\sum_{\substack{j=1 \\ j \neq i}}^{n} -a_{ij}x_j + b_i \right) \quad \text{for} \quad i = 1, 2, \ldots, n. \tag{3.49}$$

This equation can then be used to define the Jacobi iterative sequence

$$x_i^{(k+1)} = \frac{1}{a_{ii}} \left(\sum_{\substack{j=1 \\ j \neq i}}^{n} -a_{ij}x_j^{(k)} + b_i \right) \tag{3.50}$$

for $k = 0, 1, 2, 3, \ldots$

Note the Jacobi iterative method is similar to the iterative method $x_{k+1} = g(x_k)$, introduced in the chapter 2. The Jacobi iterative method employs vectors as the unknowns instead of scalars.

Let $\bar{\mathbf{x}}$ denote the fixed point solution which satisfies the equation

$$\bar{\mathbf{x}} = T\bar{\mathbf{x}} + \bar{\mathbf{c}} \tag{3.51}$$

and let

$$\bar{\mathbf{x}}^{(k+1)} = T\bar{\mathbf{x}}^{(k)} + \bar{\mathbf{c}} \tag{3.52}$$

define the iterates $\{\bar{\mathbf{x}}^{(k)}\}$ for $k = 0, 1, 2, \ldots$. We subtract the equation (3.52) from the equation (3.51) and define the error vector $\bar{\epsilon}_k = \bar{\mathbf{x}} - \bar{\mathbf{x}}^{(k)}$ associated with the kth iteration to show

$$\bar{\mathbf{x}}^{(k+1)} - \bar{\mathbf{x}} = T(\bar{\mathbf{x}}^{(k)} - \bar{\mathbf{x}}) \qquad \text{or} \qquad \bar{\epsilon}_{k+1} = T\bar{\epsilon}_k. \tag{3.53}$$

The $n \times n$ matrix T is called the iteration matrix and in the cases where T has a set of n linearly independent eigenvalues $\lambda_1, \lambda_2, \ldots, \lambda_n$, with corresponding independent eigenvectors $\bar{\mathbf{e}}_1, \bar{\mathbf{e}}_2, \ldots, \bar{\mathbf{e}}_n$, then the eigenvectors constitute a set of basis vectors and so one can express the error $\bar{\epsilon}_0$ associated with the initial guess as

$$\bar{\epsilon}_0 = \sum_{i=1}^{n} c_i \bar{\mathbf{e}}_i \tag{3.54}$$

where $c_1, c_2, \ldots, c_n$ are constants. The equation (3.53) then implies that the errors can be written in the following forms

$$\begin{aligned}
\bar{\epsilon}_1 &= T\bar{\epsilon}_0 = \sum_{i=1}^{n} c_i T\bar{\mathbf{e}}_i = \sum_{i=1}^{n} c_i \lambda_i \bar{\mathbf{e}}_i \\
\bar{\epsilon}_2 &= T\bar{\epsilon}_1 = \sum_{i=1}^{n} c_i \lambda_i T\bar{\mathbf{e}}_i = \sum_{i=1}^{n} c_i \lambda_i^2 \bar{\mathbf{e}}_i \\
&\vdots \\
\bar{\epsilon}_{k+1} &= T\bar{\epsilon}_k = \sum_{i=1}^{n} c_i \lambda_i^k T\bar{\mathbf{e}}_i = \sum_{i=1}^{n} c_i \lambda_i^{k+1} \bar{\mathbf{e}}_i.
\end{aligned} \tag{3.55}$$

Therefore, the error vector will approach zero, as k increases without bound, if and only if each eigenvalue of T satisfies $|\lambda_i| < 1$ for $i = 1, \ldots, n$. It can be shown that the above is equivalent to any of the statements

(i) The matrix T is a convergent matrix.

(ii) The spectral radius of T is less than 1.

(iii) $\lim\limits_{n \to \infty} T^n \overline{\mathbf{x}} = \overline{0}$ for every vector $\overline{\mathbf{x}}$.

(iv) $\lim\limits_{n \to \infty} \| T^n \| = 0$ for all natural norms.

That is, if the iteration matrix T is a convergent matrix, then the successive iterations $\overline{\mathbf{x}}^{(k+1)} = T x^{(k)} + \overline{c}$ can be written as

$$
\begin{aligned}
k = 0 \qquad \overline{\mathbf{x}}^{(1)} &= T\overline{\mathbf{x}}^{(0)} + \overline{c} \\[4pt]
k = 1 \qquad \overline{\mathbf{x}}^{(2)} &= T\overline{\mathbf{x}}^{(1)} + \overline{c} \\
&= T\left(T\overline{\mathbf{x}}^{(0)} + \overline{c}\right) + \overline{c} \\
&= T^2\overline{\mathbf{x}}^{(0)} + (T+I)\overline{c} \\[4pt]
k = 2 \qquad \overline{\mathbf{x}}^{(3)} &= T\overline{\mathbf{x}}^{(2)} + \overline{c} \\
&= T\left(T^2\overline{\mathbf{x}}^{(0)} + (T+I)\overline{c}\right) + \overline{c} \\
&= T^3\overline{\mathbf{x}}^{(0)} + \left(T^2 + T + I\right)\overline{c}
\end{aligned}
\tag{3.56}
$$

$$\vdots$$

$$
k = m \qquad \overline{\mathbf{x}}^{(m+1)} = T^{m+1}\overline{\mathbf{x}}^{(0)} + \left(T^m + T^{m-1} + \cdots + T^2 + T + I\right)\overline{c}
$$

The term $T^{m+1}\overline{\mathbf{x}}^{(0)}$ in equation (3.56) goes to zero if T is a convergent matrix. In the second term of equation (3.56) let S_m denote the matrix sum

$$
\begin{aligned}
S_m &= \qquad\quad T^m + T^{m-1} + \cdots + T^2 + T + I \\
\text{with} \qquad TS_m &= T^{m+1} + T^m + T^{m-1} + \cdots + T^2 + T.
\end{aligned}
\tag{3.57}
$$

When the equations (3.57) are subtracted there results

$$
(I - T)S_m = I - T^{m+1}.
\tag{3.58}
$$

Now if T is a convergent matrix, then

$$
\lim_{m \to \infty}(I - T)S_m = (I - T)\lim_{m \to \infty} S_m = \lim_{m \to \infty}(I - T^{m+1}) = I.
\tag{3.59}
$$

Also observe that if the spectral radius of T satisfies $\rho(T) < 1$, then the spectral radius of $I-T$ must be positive. That is, if λ is an eigenvalue of T, then $1-\lambda$ is an eigenvalue of $I-T$ and from linear algebra if $|\lambda| \le \rho(T) < 1$, the $\rho(I-T) > 0$. This tells us that the matrix $I-T$ is nonsingular and so $(I-T)^{-1}$ exists. The equation (3.59) gives the result that

$$
\lim_{m \to \infty} S_m = (I - T)^{-1}
\tag{3.60}
$$

and consequently one can obtain from the last of the equations (3.56) that

$$\lim_{m \to \infty} \overline{\mathbf{x}}^{(m+1)} = \overline{\mathbf{x}} = (I - T)^{-1}\overline{\mathbf{c}} \tag{3.61}$$

This is, the speed of convergence of the sequence defined by the iterations $\overline{\mathbf{x}}^{(k+1)} = T\overline{\mathbf{x}}^{(k)} + \overline{\mathbf{c}}$, is dependent upon the smallness of the spectral radius $\rho(T)$ associated with the iterative matrix T.

Gauss-Seidel Method

The Gauss-Seidel method is a modification of the Jacobi method. As each component in the iteration vector $\overline{\mathbf{x}}^{(k+1)}$ is calculated, then the most current calculated value is used in the calculation of higher numbered components. That is, after a component is calculated it is used immediately as the most up-dated value in all other equations. In terms of an iterative sequence, the Jacobi iteration given by equation (3.50) is modified to the form

$$x_i^{(k+1)} = \frac{1}{a_{ii}} \left(-\sum_{j=1}^{i-1} a_{ij} x_j^{(k+1)} - \sum_{j=i+1}^{n} a_{ij} x_j^{(k)} + b_i \right), \qquad i = 1, 2, \ldots, n \tag{3.62}$$

where the use of the most updated values are employed.

In matrix form the equation $A\overline{\mathbf{x}} = \overline{\mathbf{b}}$ can be written $(D + L + U)\overline{\mathbf{x}} = \overline{\mathbf{b}}$ from which one can write

$$(D + L)\overline{\mathbf{x}} = \overline{\mathbf{b}} - U\overline{\mathbf{x}}. \tag{3.63}$$

The Gauss-Seidel iterative sequence is defined by

$$(D + L)\overline{\mathbf{x}}^{(k+1)} = \overline{\mathbf{b}} - U\overline{\mathbf{x}}^{(k)}. \tag{3.64}$$

which gives

$$\overline{\mathbf{x}}^{(k+1)} = (D + L)^{-1} \left(-U\overline{\mathbf{x}}^{(k)} + \overline{\mathbf{b}} \right). \tag{3.65}$$

We subtract the equation (3.64) from the equation (3.63) to obtain

$$(D + L)\left(\overline{\mathbf{x}} - \overline{\mathbf{x}}^{(k+1)} \right) = -U\left(\overline{\mathbf{x}} - \overline{\mathbf{x}}^{(k)} \right) \tag{3.66}$$

and then define the error vector $\overline{\mathbf{e}}_k = \overline{\mathbf{x}} - \overline{\mathbf{x}}^{(k)}$ to write equation (3.66) in the form

$$(D + L)\overline{\mathbf{e}}^{(k+1)} = -U\overline{\mathbf{e}}^{(k)} \quad \text{or} \quad \overline{\mathbf{e}}^{(k+1)} = T\overline{\mathbf{e}}^{(k)} \quad \text{with} \quad T = -(D + L)^{-1}U. \tag{3.67}$$

We find that the convergence of the Gauss-Seidel iterative method depends upon the spectral radius of T being less than unity.

Example 3-13. (**Gauss-Seidel method**)

The system of equations

$$
\begin{array}{rrrrr}
4x_1 & - x_2 & + x_3 & - x_4 & = 1 \\
x_1 & +5x_2 & - x_3 & -2x_4 & = 0 \\
x_1 & -2x_2 & +7x_3 & - x_4 & = 14 \\
x_1 & -3x_2 & - x_3 & +8x_4 & = 24
\end{array}
\tag{3.68}
$$

has the solution $x_1 = 1$, $x_2 = 2$, $x_3 = 3$ and $x_4 = 4$. This system can be written in the Gauss-Seidel iterative form

$$
\begin{aligned}
x_1^{(k+1)} &= \qquad\qquad\quad \frac{1}{4}x_2^{(k)} \; - \frac{1}{4}x_3^{(k)} \; + \frac{1}{4}x_4^{(k)} + \frac{1}{4} \\
x_2^{(k+1)} &= -\frac{1}{5}x_1^{(k+1)} \qquad\qquad\; + \frac{1}{5}x_3^{(k)} \; + \frac{2}{5}x_4^{(k)} \\
x_3^{(k+1)} &= -\frac{1}{7}x_1^{(k+1)} + \frac{2}{7}x_2^{(k+1)} \qquad\qquad\; + \frac{1}{7}x_4^{(k)} \; + 2 \\
x_4^{(k+1)} &= -\frac{1}{8}x_1^{(k+1)} + \frac{3}{8}x_2^{(k+1)} + \frac{1}{8}x_3^{(k+1)} \qquad\qquad + 3
\end{aligned}
\tag{3.69}
$$

We employ the starting value of $\overline{\mathbf{x}}^{(0)} = (1,1,1,1)^T$ and obtain the following table of values.

k	0	1	2	3	4	5	6
$x_1^{(k)}$	1	.5	.672	.926	.989	.999	1
$x_2^{(k)}$	1	.5	1.669	1.948	1.994	1.999	2
$x_3^{(k)}$	1	2.214	2.867	2.982	2.998	3	3
$x_4^{(k)}$	1	3.402	3.900	3.988	3.999	4	4

In general the Gauss-Seidel method, when it converges, is faster than the Jacobi method.

■

SOR Methods

The method of successive over-relaxation (SOR) is a modification of the Gauss-Seidel method of iteration. One can treat the updated values given by equation (3.62) as a vector that is to be weighted to achieve a more improved value. That is, we define

$$
y_i^{(k+1)} = \frac{1}{a_{ii}} \left(-\sum_{j=1}^{i-1} a_{ij}x_j^{(k+1)} - \sum_{j=i+1}^{n} a_{ij}x_j^{(k)} + b_i \right), \qquad i = 1, 2, \dots, n
\tag{3.70}
$$

and then define a weight ω to calculate the improved value

$$x_i^{(k+1)} = \omega y_i^{(k+1)} + (1 - \omega) x_i^{(k)}. \tag{3.71}$$

Note that when $\omega = 1$ the Gauss-Seidel method results. The weight factor ω is sometimes referred to as an acceleration factor and the weighting method is sometimes referred to as a relaxation method. Under-relaxation is said to occur for weights in the range $0 < \omega < 1$ used to obtain convergence of nonconvergent Gauss-Seidel systems. Over-relaxation is said to occur for weights $1 < \omega < 2$ used to accelerate convergence of Gauss-Seidel systems. These methods are all referred to as Successive Over-Relaxation or SOR methods. The SOR methods can also be written in a matrix form. We again write A in the form $A = D + L + U$ where D is diagonal, L is lower triangular and U is upper triangular. We write the Gauss-Seidel iterative equation (3.64) in the form

$$D\overline{\mathbf{x}}^{(k+1)} = \overline{\mathbf{b}} - L\overline{\mathbf{x}}^{(k+1)} - U\overline{\mathbf{x}}^{(k)}$$

and define the sequence of values

$$\begin{aligned}
\overline{\mathbf{y}}^{(k+1)} &= D^{-1}\left(\overline{\mathbf{b}} - L\overline{\mathbf{x}}^{(k+1)} - U\overline{\mathbf{x}}^{(k)}\right) \\
\overline{\mathbf{x}}^{(k+1)} &= \omega\overline{\mathbf{y}}^{(k+1)} + (1 - \omega)\overline{\mathbf{x}}^{(k)}
\end{aligned} \tag{3.72}$$

for $k = 0, 1, 2, \ldots$. By eliminating the vector $\overline{\mathbf{y}}^{(k+1)}$ from the equations (3.72) there results

$$\left(I + \omega D^{-1}L\right)\overline{\mathbf{x}}^{(k+1)} = \omega D^{-1}\overline{\mathbf{b}} + \left[(1 - \omega)I - \omega D^{-1}U\right]\overline{\mathbf{x}}^{(k)} \tag{3.73}$$

The equation (3.73) can be written in the form of equation (3.42) by solving for $\overline{\mathbf{x}}^{(k+1)}$. One then obtains

$$\overline{\mathbf{x}}^{(k+1)} = \left(I + \omega D^{-1}L\right)^{-1}\left[(1 - \omega)I - \omega D^{-1}U\right]\overline{\mathbf{x}}^{(k)} + \left(I + \omega D^{-1}L\right)^{-1}\omega D^{-1}\overline{\mathbf{b}}$$

so that $\qquad \overline{\mathbf{x}}^{(k+1)} = T\overline{\mathbf{x}}^{(k)} + \overline{\mathbf{c}}$

where $\qquad T = T(\omega) = \left(I + \omega D^{-1}L\right)^{-1}\left[(1 - \omega)I - \omega D^{-1}U\right]$

and $\qquad \overline{\mathbf{c}} = \left(I + \omega D^{-1}L\right)^{-1}\omega D^{-1}\overline{\mathbf{b}}.$

$$\tag{3.74}$$

If this iterative method converges, then the error vector becomes

$$\overline{\mathbf{e}}^{(k+1)} = T(\omega)\overline{\mathbf{e}}^{(k)}. \tag{3.75}$$

Note that the weight factor ω can now be selected to minimize the spectral radius of T to a value that will produce the fastest convergence.

Example 3-14.   (SOR Method)
Illustrate the SOR method to solve the system of equations

$$2x_1 + x_2 = 10$$

$$x_1 + x_2 = 7$$

which has the exact solutions $x_1 = 3$ and $x_2 = 4$.

Solution: Here $D = \begin{bmatrix} 2 & 0 \\ 0 & 1 \end{bmatrix}$, $L = \begin{bmatrix} 0 & 0 \\ 1 & 0 \end{bmatrix}$ and $U = \begin{bmatrix} 0 & 1 \\ 0 & 0 \end{bmatrix}$ so that one can calculate

$$T = T(\omega) = (I + \omega D^1 L)^{-1} \left[(1-\omega)I - \omega D^{-1}U \right] = \begin{bmatrix} 1-\omega & -\omega/2 \\ \omega^2 - \omega & 1 - \omega + \omega^2/2 \end{bmatrix} \qquad (3.76)$$

and after some algebra one can determine the eigenvalues

$$\lambda_1 = \frac{1}{4}\left(4 - 4\omega + \omega^2 + \omega\sqrt{8 - 8\omega + \omega^2} \right)$$

$$\lambda_2 = \frac{1}{4}\left(4 - 4\omega + \omega^2 - \omega\sqrt{8 - 8\omega + \omega^2} \right) \qquad (3.77)$$

The figure 3-2 illustrates the eigenvalues as a function of ω for $\omega > 1$. The figure illustrates that as ω increases, λ_1 decreases and λ_2 increases up to a value of $\omega = 4 - 2\sqrt{2}$ where the eigenvalues are equal.

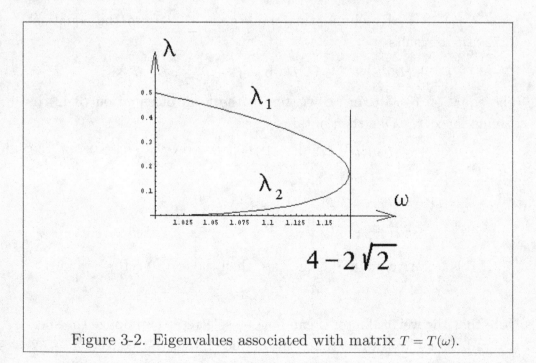

Figure 3-2. Eigenvalues associated with matrix $T = T(\omega)$.

In this example the Gauss-Seidel iteration method

$$x_1^{(k+1)} = -\frac{1}{2}x_2^{(k)} + 5$$
$$x_2^{(k+1)} = -x_1^{(k+1)} + 7$$

(3.78)

is modified to the SOR iteration method

$$y_1^{(k+1)} = -\frac{1}{2}x_2^{(k)} + 5$$
$$x_1^{(k+1)} = \omega y_1^{(k+1)} + (1-\omega)x_1^{(k)}$$
$$y_2^{(k+1)} = -x_1^{(k+1)} + 7$$
$$x_2^{(k+1)} = \omega y_2^{(k+1)} + (1-\omega)x_2^{(k)}$$

(3.79)

which can also be written in the form

$$x_1^{(k+1)} = \omega\left(5 - \frac{1}{2}x_2^{(k)}\right) + (1-\omega)x_1^{(k)}$$
$$x_2^{(k+1)} = \omega\left(7 - x_1^{(k+1)}\right) + (1-\omega)x_2^{(k)}.$$

(3.80)

The equation (3.80) has the equivalent matrix form

$$\left(I + \omega D^{-1}L\right)\bar{\mathbf{x}}^{(k+1)} = \left[(1-\omega)I - \omega D^{-1}U\right]\bar{\mathbf{x}}^{(k)} + \omega D^{-1}\bar{\mathbf{b}}$$

or
$$\begin{bmatrix} 1 & 0 \\ \omega & 1 \end{bmatrix}\begin{bmatrix} x_1^{(k+1)} \\ x_2^{(k+1)} \end{bmatrix} = \begin{bmatrix} 1-\omega & -\omega/2 \\ 0 & 1-\omega \end{bmatrix}\begin{bmatrix} x_1^{(k)} \\ x_2^{(k)} \end{bmatrix} + \begin{bmatrix} 5\omega \\ 7\omega \end{bmatrix}.$$

Using the initial value $\bar{\mathbf{x}}^{(0)} = [1,1]^T$ and $\omega = 4 - 2\sqrt{2}$, the following table can be constructed.

SOR iterations with $\omega = 4 - 2\sqrt{2}$.

k	0	1	2	3	4	5	6	7
$x_1^{(k)}$	1	5.101	3.780	3.206	3.048	3.010	3.002	3.000
$x_2^{(k)}$	1	2.054	3.420	3.858	3.968	3.993	3.999	4.000

In contrast the Gauss-Seidel iterative equations (3.78) takes 20 iterations to converge using 5 digit accuracy.

∎

The Jacobi method and Gauss-Seidel methods, when they converge, will do so slowly. They are not your fastest methods. The SOR method with the appropriate weight factor will generally converge much faster than either the Jacobi or Gauss-Seidel methods. However, for very large linear systems the optimum value for the weight factor is difficult to calculate. Under these conditions one usually tries different values for the weight factor to see the effect on the convergence. How SOR methods can be employed to speed convergence of very large linear systems is still an area of current research.

Observe that the matrix A in the linear system of equations $A\overline{\mathbf{x}} = \overline{\mathbf{b}}$ can be split into the form $A = B + (A - B)$ where B is nonsingular and is called a splitting matrix. The system of equations $A\overline{\mathbf{x}} = \overline{\mathbf{b}}$ can then be rewritten as

$$B\overline{\mathbf{x}} = (B - A)\overline{\mathbf{x}} + \overline{\mathbf{b}}. \tag{3.81}$$

This system can now be used to define the iterative method

$$\overline{\mathbf{x}}^{(k+1)} = B^{-1}(B - A)\overline{\mathbf{x}}^{(k)} + B^{-1}\overline{\mathbf{b}} \tag{3.82}$$

which has the form of equation (3.42) with $T = B^{-1}(B - A)$ the iteration matrix and $\overline{\mathbf{c}} = B^{-1}\overline{\mathbf{b}}$. If $A = D + L + U$, then the special case $B = D$ gives the Jacobi iterative method and the special case $B = D + L$ gives the Gauss-Seidel iterative method. These are just two of many ways to split the coefficient matrix to form iterative methods.

Nonlinear Systems

Consider the problem of determining values of x_1, x_2 such that two nonlinear equations

$$\begin{aligned} f_1(x_1, x_2) &= 0 \\ f_2(x_1, x_2) &= 0 \end{aligned} \tag{3.83}$$

are satisfied simultaneously. One can seek solutions in the neighborhood of an initial point (x_1^0, x_2^0). If $f_1(x_1^0, x_2^0) = 0$ and $f_2(x_1^0, x_2^0) = 0$ we have found the roots. In most cases the initial guess of (x_1^0, x_2^0) produces $f_1(x_1^0, x_2^0) \neq 0$ and $f_2(x_1^0, x_2^0) \neq 0$. We assume that the functions f_1 and f_2 are continuous with derivatives which are also continuous, and that these functions possess Taylor series expansions about the point (x_1^0, x_2^0). These Taylor series expansions have the form

$$\begin{aligned} f_1(x_1, x_2) &= f_1(x_1^0, x_2^0) + \left.\frac{\partial f_1}{\partial x_1}\right]_0 (x_1 - x_1^0) + \left.\frac{\partial f_1}{\partial x_2}\right]_0 (x_2 - x_2^0) + h.o.t. \\ f_2(x_1, x_2) &= f_2(x_1^0, x_2^0) + \left.\frac{\partial f_2}{\partial x_1}\right]_0 (x_1 - x_1^0) + \left.\frac{\partial f_2}{\partial x_2}\right]_0 (x_2 - x_2^0) + h.o.t. \end{aligned} \tag{3.84}$$

where *h.o.t.* stands for higher order terms which are neglected if the differences $\Delta x_1 = x_1 - x_1^0$ and $\Delta x_2 = x_2 - x_2^0$ are small. The terms Δx_1 and Δx_2 denote changes from the initial point (x_1^0, x_2^0) in the x_1 and x_2 directions and the terms

$$\left.\frac{\partial f_1}{\partial x_1}\right]_0, \quad \left.\frac{\partial f_1}{\partial x_2}\right]_0, \quad \left.\frac{\partial f_2}{\partial x_1}\right]_0, \quad \left.\frac{\partial f_2}{\partial x_2}\right]_0 \tag{3.85}$$

denote the partial derivatives evaluated at the initial point (x_1^0, x_2^0). If we set the right-hand side of equation (3.84) equal to zero we obtain the linear system of equations

$$\begin{aligned}
\left.\frac{\partial f_1}{\partial x_1}\right]_0 (x_1 - x_1^0) + \left.\frac{\partial f_1}{\partial x_2}\right]_0 (x_2 - x_2^0) &= -f_1(x_1^0, x_2^0) \\
\left.\frac{\partial f_2}{\partial x_1}\right]_0 (x_1 - x_1^0) + \left.\frac{\partial f_2}{\partial x_2}\right]_0 (x_2 - x_2^0) &= -f_2(x_1^0, x_2^0)
\end{aligned} \tag{3.86}$$

where we can solve the linear system (3.86) to obtain $\Delta x_1, \quad \Delta x_2$, then the values

$$\begin{aligned}
x_1 &= x_1^0 + \Delta x_1 \\
x_2 &= x_2^0 + \Delta x_2
\end{aligned} \tag{3.87}$$

denote the first iteration of values closer to the desired solution. Define the vector quantities

$$\overline{\mathbf{x}}^{(k)} = \begin{bmatrix} x_1^{(k)} \\ x_2^{(k)} \end{bmatrix} \quad \text{and} \quad \overline{\mathbf{f}}(\overline{\mathbf{x}}^{(k)}) = \begin{bmatrix} f_1(\overline{\mathbf{x}}^{(k)}) \\ f_2(\overline{\mathbf{x}}^{(k)}) \end{bmatrix}, \tag{3.88}$$

then the system of equations (3.86) can be written in the matrix iterative form

$$J(\overline{\mathbf{x}}^{(k)}) \left(\overline{\mathbf{x}}^{(k+1)} - \overline{\mathbf{x}}^{(k)} \right) = -\overline{\mathbf{f}}(\overline{\mathbf{x}}^{(k)}) \tag{3.89}$$

for $k = 0, 1, 2, 3, \ldots$, where J is the Jacobian matrix evaluated at $\overline{\mathbf{x}}^{(k)}$

$$J(\overline{\mathbf{x}}^{(k)}) = \left(\frac{\partial f_i}{\partial x_j} \right) = \begin{bmatrix} \frac{\partial f_1(\overline{\mathbf{x}}^{(k)})}{\partial x_1} & \frac{\partial f_1(\overline{\mathbf{x}}^{(k)})}{\partial x_2} \\ \frac{\partial f_2(\overline{\mathbf{x}}^{(k)})}{\partial x_1} & \frac{\partial f_2(\overline{\mathbf{x}}^{(k)})}{\partial x_2} \end{bmatrix}$$

The matrix equation (3.89) can be left multiplied by the inverse Jacobian $J^{-1}(\overline{\mathbf{x}}^{(k)})$ to obtain the equivalent equation

$$\overline{\mathbf{x}}^{(k+1)} = \overline{\mathbf{x}}^{(k)} - J^{-1}(\overline{\mathbf{x}}^{(k)})\overline{\mathbf{f}}(\overline{\mathbf{x}}^{(k)}). \tag{3.90}$$

The iterative scheme given by the equation (3.90) is known as Newton's iterative method for nonlinear systems.

96

In practice the inverse matrix J^{-1} is not calculated because it takes too long. Instead, the matrix form of the equation (3.89) is written as a two-step process where

Step 1 is to solve the linear system $\quad J(\overline{\mathbf{x}}^{(k)})\Delta\overline{\mathbf{x}} = -\overline{\mathbf{f}}(\overline{\mathbf{x}}^{(k)}) \quad$ for $\quad \Delta\overline{\mathbf{x}}$

Step 2 is to calculate the updated values $\quad \overline{\mathbf{x}}^{(k+1)} = \overline{\mathbf{x}}^{(k)} + \Delta\overline{\mathbf{x}}.$

(3.91)

Example 3-15. (Newton's Method)

Solve the nonlinear system of equations

$$f_1(x_1, x_2) = 3x_1 + 4x_2 - 2x_1x_2 = 0$$

$$f_2(x_1, x_2) = x_1 - x_1^2 + \sin x_2 + 4 = 0$$

(3.92)

Solution: The given system of nonlinear equations has the Jacobian matrix

$$J(\overline{\mathbf{x}}) = \begin{bmatrix} \frac{\partial f_1}{\partial x_1} & \frac{\partial f_1}{\partial x_2} \\ \frac{\partial f_2}{\partial x_1} & \frac{\partial f_2}{\partial x_2} \end{bmatrix} = \begin{bmatrix} 3 - 2x_2 & 4 - 2x_1 \\ 1 - 2x_1 & \cos x_2 \end{bmatrix}, \qquad \overline{\mathbf{x}} = \begin{bmatrix} x_1 \\ x_2 \end{bmatrix}.$$

(3.93)

The Newton iterative method is given by equation (3.89) with

$$\overline{\mathbf{f}}(\overline{\mathbf{x}}) = \begin{bmatrix} 3x_1 + 4x_2 - 2x_1x_2 \\ x_1 - x_1^2 + \sin x_2 + 4 \end{bmatrix}, \qquad \overline{\mathbf{x}} = \begin{bmatrix} x_1 \\ x_2 \end{bmatrix}$$

(3.94)

Newtons's method works great if you have some idea of where the roots are. For the given example we can take the first equation and solve for x_2 to obtain

$$x_2 = -3x_1/(4 - 2x_1).$$

(3.95)

This value can be substituted into the second equation to obtain

$$y = x_1 - x_1^2 + \sin\left[\frac{-3x_1}{4 - 2x_1}\right] + 4 = 0 \qquad x_1 \neq 2$$

(3.96)

A graph of y vs x_1 is illustrated in the figure 3-3.

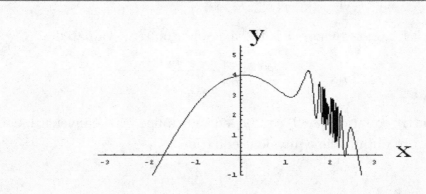

Figure 3-3. Graph of y vs x_1 illustrating four roots.

The figure 3-3 shows that there are four roots to the equation (3.96) . We guess at these roots by examining the graph and select the approximate values of $x_1 \approx -1.7, 2.315, 2.33, 2.6$. The equation (3.95) is then employed to calculate a value of x_2 associated with each x_1 value. This produces the four possible number pairs $(x_1^{(0)}, x_2^{(0)})$ of

$$(-1.7, 0.69), \qquad (2.315, 11.02), \qquad (2.33, 10.6), \qquad (2.6, 6.5)$$

which can be used as starting values for the Newton's method. These starting values produce the following tables.

Table A. Newton's method for nonlinear system.

k	0	1	2	3
$x_1^{(k)}$	-1.7	-1.71085	-1.71082	-1.71082
$x_2^{(k)}$	0.69	0.691565	0.691554	0.691554

Table B. Newton's method for nonlinear system.

k	0	1	2	3	4	5
$x_1^{(k)}$	2.315	2.30497	2.3085	2.309	2.30901	2.30901
$x_2^{(k)}$	11.02	11.3271	11.2233	11.2086	11.2083	11.2083

Table C. Newton's method for nonlinear system.

k	0	1	2	3	4
$x_1^{(k)}$	2.33	2.33261	2.3383	2.3323	2.3323
$x_2^{(k)}$	10.6	10.519	10.5279	10.528	10.528

Table D. Newton's method for nonlinear system.

k	0	1	2	3
$x_1^{(k)}$	2.6	2.60447	2.60448	2.60448
$x_2^{(k)}$	6.5	6.46277	6.46297	6.46297

Observe that nonlinear systems may have more than one solution. If one selects an arbitrary starting value the convergence of Newton's method is not guaranteed. Even in the case where there is convergence, the neighborhood of the fixed point of convergence is not known beforehand. For example, the starting value of $(x_1^{(0)}, x_2^{(0)}) = (1, 1)$ gives the fixed point in Table A after 10 iterations. The starting value of $(x_1^{(0)}, x_2^{(0)}) = (5, 10)$ gives the fixed point in Table D after 4 iterations. A bad initial guess such as $(x_1^{(0)}, x_2^{(0)}) = (1, 3)$ puts the iterations in a loop bouncing around the values $x_1 \approx 2.03$ with $84 < x_2 < 89$ and after 35 iterations there is no convergence. This is because the term $\sin x_2$ oscillates so rapidly in the neighbor hood of $x_1 = 2$. Therefore care should be taken in selecting the starting values associated with systems of nonlinear equations.

Newton's Method for Higher Order Systems

The higher order system of nonlinear equations is treated just like the system of two nonlinear equations. The only difference is that there are more equations and more variables. Consider the system of equations (3.5) in the case where $m = n$. Each equation in this system can be expanded in a Taylor series about some initial point. The vector representation of the Taylor series expansion for the system is written as

$$\bar{\mathbf{f}}(\bar{\mathbf{x}}) = \bar{\mathbf{f}}(\bar{\mathbf{x}}_0 + \Delta \bar{\mathbf{x}}) = \bar{\mathbf{f}}(\bar{\mathbf{x}}_0) + J(\bar{\mathbf{x}}_0)(\bar{\mathbf{x}} - \bar{\mathbf{x}}_0) + h.o.t \tag{3.97}$$

where $h.o.t$ denotes higher order terms which are neglected. Setting equation (3.97) equal to the zero vector and solving for $\bar{\mathbf{x}}$ gives the matrix equation

$$\bar{\mathbf{x}} = \bar{\mathbf{x}}_0 - J^{-1}(\bar{\mathbf{x}}_0)\bar{\mathbf{f}}(\bar{\mathbf{x}}_0) \tag{3.98}$$

where J is the Jacobian matrix

$$J(\bar{\mathbf{x}}) = \begin{bmatrix} \frac{\partial f_1}{\partial x_1} & \frac{\partial f_1}{\partial x_2} & \cdots & \frac{\partial f_1}{\partial x_n} \\ \frac{\partial f_2}{\partial x_1} & \frac{\partial f_2}{\partial x_2} & \cdots & \frac{\partial f_2}{\partial x_n} \\ \vdots & \vdots & \ddots & \vdots \\ \frac{\partial f_n}{\partial x_1} & \frac{\partial f_n}{\partial x_2} & \cdots & \frac{\partial f_n}{\partial x_n} \end{bmatrix} \tag{3.99}$$

and $J^{-1}(\bar{\mathbf{x}})$ is its inverse matrix. The equation (3.98) is modified to produce the iterative method

$$\bar{\mathbf{x}}^{(k+1)} = \bar{\mathbf{x}}^{(k)} - J^{-1}(\bar{\mathbf{x}}^{(k)})\bar{\mathbf{f}}(\bar{\mathbf{x}}^{(k)}) \tag{3.100}$$

for $k = 0, 1, 2, \ldots$ which is called the generalized Newton's method. In the matrix equation (3.100) we use

$$\overline{\mathbf{x}} = [x_1, x_2, \ldots, x_n]^T$$
$$\overline{\mathbf{f}}(\overline{\mathbf{x}}) = [f_1(\overline{\mathbf{x}}), f_2(\overline{\mathbf{x}}), \ldots, f_n(\overline{\mathbf{x}})]^T.$$

Note that for higher ordered systems the inverse Jacobian matrix is not solved for. Instead the system of equations (3.100) is solved by the two step process of first solving the linear system

$$J(\overline{\mathbf{x}}^{(k)})\overline{\mathbf{y}} = -\overline{\mathbf{f}}(\overline{\mathbf{x}}^k) \quad \text{for the vector } \overline{\mathbf{y}}, \tag{3.101}$$

which is a modification of the equation (3.97), then the second step solves for the updated value

$$\overline{\mathbf{x}}^{(k+1)} = \overline{\mathbf{x}}^{(k)} + \overline{\mathbf{y}} \tag{3.102}$$

where $\overline{\mathbf{y}} = \Delta\overline{\mathbf{x}} = \overline{\mathbf{x}}^{(k+1)} - \overline{\mathbf{x}}^{(k)}$. This is the same basic procedure that was used for the two-dimensional equations (3.91) considered previously, but now is extended to higher order nonlinear systems.

The generalized Newton's method is easy to set up on a computer and usually converges quickly if an initial guess is near a root. The disadvantages of Newton's method is that whenever the gradients of the nonlinear functions $f_1, f_2, \ldots, f_n$ are small, then it is possible for the $(k+1)$-st iteration to be far away from the k-th iteration. Also the surfaces generated by the functions $f_1 = 0, f_2 = 0, \ldots, f_n = 0$ may have no point of intersection in common, there may be only one point of intersection in common or there may be many points of intersection in common. For large systems of nonlinear equations it is not known aprori if a solution exists or even how many solutions exists. A graphical illustration of the roots, if they exist, is possible only for one, two and sometimes three-dimensional problems.

Exercises Chapter 3

▶ 1.

Solve the system of equations

$$x_1 + 2x_2 - x_3 = 20$$
$$2x_1 + x_2 - 2x_3 = -20$$
$$3x_1 - 3x_2 + x_3 = 0$$

by using Gaussian elimination.

▶ 2.

Illustrate the Jacobi iterative method to solve the system of equations

$$\begin{bmatrix} 4 & 3 \\ -5 & 7 \end{bmatrix} \begin{bmatrix} x_1 \\ x_2 \end{bmatrix} = \begin{bmatrix} 38 \\ 17 \end{bmatrix}$$

Use $x_1 = 0, x_2 = 0$ for your starting value.

▶ 3.   Illustrate the Gauss Seidel method and solve the system of equations

$$\begin{bmatrix} 2 & 1 \\ 1 & 2 \end{bmatrix} \begin{bmatrix} x_1 \\ x_2 \end{bmatrix} = \begin{bmatrix} 8 \\ 12 \end{bmatrix}$$

Use $x_1 = 0, x_2 = 0$ for your starting value.

▶ 4.   Use Gauss elimination to solve the system of equations

$$x_1 + 2x_2 + 3x_3 = 5$$
$$4x_1 + 10x_2 + 20x_3 = 16$$
$$6x_1 + 16x_2 + 42x_3 = 10$$

▶ 5.   Solve the system of equations

$$x_1 + 2x_2 + 3x_3 = 5$$
$$4x_1 + 10x_2 + 20x_3 = 16$$
$$6x_1 + 16x_2 + 42x_3 = 10$$

using LU decomposition.

▶ **6.** There may be occasions where one has to solve a system of linear equations where the coefficient matrix has elements which vary drastically in magnitude. In such a case it is advisable to scale the coefficients and rearrange the rows of the equations. As an example consider the system of equations

$$\begin{bmatrix} 4 & 5 & 1000 \\ -3 & 1 & 1000 \\ 1 & 2 & 4 \end{bmatrix} \begin{bmatrix} x_1 \\ x_2 \\ x_3 \end{bmatrix} = \begin{bmatrix} 1022 \\ 993 \\ 11 \end{bmatrix} \tag{6a}$$

solved by Gaussian elimination. We shall compare the exact values with what happens when round-off error occurs. In the second column assume that all calculations are done to 4 significant digits. We start by reducing the above system to triangular form.

(a) Verify the following row reduction operations

Exact

$$\begin{bmatrix} 4 & 5 & 1000 & | & 1022 \\ 0 & 19/4 & 1750 & | & 3519/2 \\ 0 & 3/4 & -246 & | & -489/2 \end{bmatrix}$$

$$\begin{bmatrix} 4 & 5 & 1000 & | & 1022 \\ 0 & 19/4 & 1750 & | & 3519/2 \\ 0 & 0 & -9924/19 & | & -9924/19 \end{bmatrix}$$

Exact solutions

$x_1 = 3.000$

$x_2 = 2.000$

$x_3 = 1.000$

Calculations to 4 significant digits

$$\begin{bmatrix} 4 & 5 & 1000 & | & 1022. \\ 0 & 4.75 & 1750 & | & 1759. \\ 0 & 0.75 & -246 & | & -244.5 \end{bmatrix}$$

$$\begin{bmatrix} 4 & 5 & 1000 & | & 1022 \\ 0 & 4.75 & 1750 & | & 1759. \\ 0 & 0 & -522.3 & | & -522.2 \end{bmatrix}$$

Solutions obtained are

$x_1 = 3.0913$

$x_2 = 1.9653$

$x_3 = 0.9998$

(b) Note that if you first scale the elements in the coefficient matrix of equations (6a) by dividing each row by the magnitude of the largest element in the row, one obtains

$$\begin{bmatrix} .004 & .005 & 1.000 \\ -.003 & .001 & 1.000 \\ 0.25 & 0.50 & 1.0 \end{bmatrix} \begin{bmatrix} x_1 \\ x_2 \\ x_3 \end{bmatrix} = \begin{bmatrix} 1.022 \\ 0.993 \\ 2.75 \end{bmatrix} \tag{6b}$$

Now interchange rows and start the row reduction operations with the system

$$\begin{bmatrix} 0.25 & 0.50 & 1.0 \\ .004 & .005 & 1.000 \\ -.003 & .001 & 1.000 \end{bmatrix} \begin{bmatrix} x_1 \\ x_2 \\ x_3 \end{bmatrix} = \begin{bmatrix} 2.75 \\ 1.022 \\ 0.993 \end{bmatrix} \tag{6b}$$

Verify the following row reduction operations

Calculations to 4 significant digits

$$\left[\begin{array}{ccc|c} 0.25 & 0.50 & 1.000 & 2.75 \\ 0 & -.003 & 0.984 & 0.978 \\ 0 & 0.007 & 1.012 & 0.993 \end{array}\right]$$

$$\left[\begin{array}{ccc|c} 0.25 & 0.50 & 1.000 & 2.75 \\ 0 & -.003 & 0.984 & 0.978 \\ 0 & 0 & 3.308 & 3.308 \end{array}\right]$$

Solutions obtained are

$x_1 = 3.000$

$x_2 = 2.000$

$x_3 = 1.000$

▶ **7.** Use Gauss elimination and solve the tridiagonal system of equations

$$
\begin{array}{rcrcrcrcl}
3x_1 & + & x_2 & & & & & = & -2 \\
x_1 & + & 2x_2 & + & x_3 & & & = & 8 \\
& & x_2 & + & 5x_3 & + & x_4 & = & 3 \\
& & & & & + & x_4 & = & 5
\end{array}
$$

▶ **8.** Find the eigenvalues and eigenvectors associates with the matrix

$$A = \begin{bmatrix} 6 & -1 \\ 6 & 1 \end{bmatrix}$$

▶ **9.** Find the eigenvalues and eigenvectors associates with the matrix

$$A = \begin{bmatrix} 7 & -1 \\ 12 & 0 \end{bmatrix}$$

▶ **10.** Find the eigenvalues and eigenvectors associates with the matrix

$$A = \begin{bmatrix} 5 & 2 & -2 \\ -2 & 5 & 2 \\ -2 & 2 & 5 \end{bmatrix}$$

► **11.** Find the eigenvalues and eigenvectors associates with the matrix

$$A = \begin{bmatrix} 4 & 2 & -2 \\ -3 & 4 & 3 \\ -3 & 2 & 5 \end{bmatrix}$$

► **12.** Find all roots of the nonlinear system of equations

$$x_1 + x_2 + \cos \pi x_3 = 3/4$$

$$\cos \pi x_1 + x_2 + x_3 = 11/12$$

$$x_2 + \sin \pi x_3 = 5/12$$

► **13.** Find the points of intersection of the circle $x^2 + y^2 = 1$ with the ellipse $\dfrac{(x-1)^2}{2} + \dfrac{(y-2)^2}{4} = 1$. Hint: Sketch the curves.

► **14.** For the given matrix A find the spectral radius $\rho(A)$ and $\sqrt{\rho(A^T A)}$.

$$A = \begin{bmatrix} -28 & -66 \\ 22 & 49 \end{bmatrix}$$

► **15.** Solve the system of equations $\begin{bmatrix} 4 & 8 \\ 10 & 16 \end{bmatrix} \begin{bmatrix} x_1 \\ x_2 \end{bmatrix} = \begin{bmatrix} 68 \\ 146 \end{bmatrix}$

(a) By Doolittle's method.
(b) By Crout's method.
(c) By Choleski's method.

► **16.** Solve the system of equations $\begin{bmatrix} 16 & 4 & 4 \\ 4 & 10 & 4 \\ 4 & 7 & 8 \end{bmatrix} \begin{bmatrix} x_1 \\ x_2 \\ x_3 \end{bmatrix} = \begin{bmatrix} 100 \\ 94 \\ 271 \end{bmatrix}$

(a) By Doolittle's method.
(b) By Crout's method.
(c) By Choleski's method.

► **17.** Consider the system of equations $\begin{bmatrix} 5 & 1 & -1 \\ 1 & 6 & -1 \\ 0 & 1 & 7 \end{bmatrix} \begin{bmatrix} x_1 \\ x_2 \\ x_3 \end{bmatrix} = \begin{bmatrix} 19/2 \\ 16 \\ 6 \end{bmatrix}$

(a) Write the given system of equations in the form $D\vec{x} = -(L + U)\vec{x} + \vec{b}$.
(b) Multiply by D^{-1} and solve for $\vec{x}$ and then set up an iterative scheme.
(c) What condition must be placed upon $T = -D^{-1}(L + U)$ in order for the iterative scheme to converge?

▶ **18.** Consider the system of equations $\begin{bmatrix} 5 & 1 & -1 \\ 1 & 6 & -1 \\ 0 & 1 & 7 \end{bmatrix} \begin{bmatrix} x_1 \\ x_2 \\ x_3 \end{bmatrix} = \begin{bmatrix} 19/2 \\ 16 \\ 6 \end{bmatrix}$

(a) Write the given system of equations in the form $(D+L)\vec{x} = \vec{b} - U\vec{x}$.

(b) Multiply by $(D+L)^{-1}$ and solve for $\vec{x}$ and then set up an iterative scheme.

(c) What condition must be placed upon $T = -(D+L)^{-1}U$ in order for the iterative scheme to converge?

▶ **19.** Consider the system of equations $\begin{bmatrix} 5 & 1 & -1 \\ 1 & 6 & -1 \\ 0 & 1 & 7 \end{bmatrix} \begin{bmatrix} x_1 \\ x_2 \\ x_3 \end{bmatrix} = \begin{bmatrix} 19/2 \\ 16 \\ 6 \end{bmatrix}$

(a) Write the given system of equations in the form $D\vec{x} = -(L+U)\vec{x} + \vec{b}$.

(b) Multiply by D^{-1} and solve for $\vec{x}$ and then set up an iterative scheme.

(c) Set up a system of iterative equations having the form of equations (3.72).

(c) Experiment with values of ω to achieve convergence of the iterative process.

▶ **20.** Experiment with the matrix equation $\begin{bmatrix} 1 & 2 \\ 3 & 4 \end{bmatrix} \begin{bmatrix} x_1 \\ x_2 \end{bmatrix} = \begin{bmatrix} 5 \\ 11 \end{bmatrix}$ which has the solution $x_1 = 1$ and $x_2 = 2$. Try to find matrices B such that the system can be split into the form

$$B\vec{x} = (B-A)\vec{x} + \vec{b}$$

where the iterative procedure

$$\vec{x}^{(k+1)} = B^{-1}(B-A)\vec{x}^{(k)} + B^{-1}\vec{b}$$

converges quickly.

▶ **21.** Note that the characteristic equation, given by equation (3.24), can be written in the factored form

$$C(\lambda) = (\lambda - \lambda_1)(\lambda - \lambda_2) \cdots (\lambda - \lambda_n) = 0$$

where λ_i for $i = 1, \ldots, n$ are the eigenvalues of A.

(a) Show $C(0) = |A| = c_0 = \lambda_1 \lambda_2 \cdots \lambda_n$ is a product of the eigenvalues.

(b) Show A is singular if any eigenvalue is zero.

▶ **22.** To solve the system of ordinary differential equation

$$\frac{dx_1}{dt} = -2x_1 + x_2$$
$$\frac{dx_2}{dt} = x_1 - 2x_2$$

or $\quad \frac{d\vec{x}}{dt} = A\vec{x}, \quad A = \begin{bmatrix} -2 & 1 \\ 1 & -2 \end{bmatrix}, \quad \vec{x} = \begin{bmatrix} x_1 \\ x_2 \end{bmatrix}$

assume solutions of the form $x_1 = x_1(t) = x_{10}e^{\lambda t}$ and $x_2 = x_2(t) = x_{20}e^{\lambda t}$ where x_{10} and x_{20} are constants.

(a) Show that x_{10}, x_{20} and λ are determined from an appropriate eigenvalue-eigenvector problem.

(b) Solve the above system of equations and find the general solution.

▶ **23.** Illustrate the method of successive over relaxation (SOR) with a weight of $\omega = 3/2$ to solve the system of equations

$$\begin{bmatrix} 3 & 1 \\ 1 & 3 \end{bmatrix} \begin{bmatrix} x_1 \\ x_2 \end{bmatrix} = \begin{bmatrix} 12 \\ 10 \end{bmatrix}$$

Use $x_1 = 0, x_2 = 0$ for your starting values.

▶ **24.**

(a) Factor the matrix $A = \begin{bmatrix} 2 & -1 & 1 \\ 3 & 3 & 9 \\ 3 & 3 & 5 \end{bmatrix}$ into a LU decomposition which has $\ell_{ii} = 1$ for all i.

(b) Use the above LU factorization to solve the system of equations

$$\begin{bmatrix} 2 & -1 & 1 \\ 3 & 3 & 9 \\ 3 & 3 & 5 \end{bmatrix} \begin{bmatrix} x_1 \\ x_2 \\ x_3 \end{bmatrix} = \begin{bmatrix} 27 \\ 87 \\ 55 \end{bmatrix}.$$

▶ **25.** Use Newton's method to find nonzero solutions to the system of equations

$$x^2 + y^2 - x = 0$$
$$x^2 - y^2 - y = 0$$

▶ **26.** Use Gauss elimination to solve the system of equations

$$3x_1 + x_2 + 2x_3 + x_4 = 8$$
$$6x_1 + 5x_3 + x_4 = 27$$
$$3x_1 - 3x_2 + 5x_3 - x_4 = 32$$
$$3x_1 - x_2 + 4x_3 + x_4 = 20$$

▶ **27.** Use LU decomposition to solve the system of equations

$$3x_1 \quad + x_2 + 2x_3 + x_4 = 8$$
$$6x_1 \qquad + 5x_3 + x_4 = 27$$
$$3x_1 - 3x_2 + 5x_3 - x_4 = 32$$
$$3x_1 \quad - x_2 + 4x_3 + x_4 = 20$$

▶ **28.** Use Choleski LU decomposition to solve the system of equations

$$x_1 - x_2 \quad + x_3 \quad + x_4 = 10$$
$$-x_1 + 5x_2 \quad + x_3 \quad + 3x_4 = 40$$
$$x_1 \quad + x_2 + 11x_3 \quad + 6x_4 = 101$$
$$x_1 + 3x_2 \quad + 6x_3 + 22x_4 = 194$$

▶ **29.** Use $\bar{x}^0 = \mathrm{col}(0,0,0)$ as a starting value and solve the given system of equations by Jacobi iteration

$$2x_1 - \quad x_2 + \quad x_3 = 1$$
$$3x_1 + 3x_2 + 9x_3 = 0$$
$$3x_1 + 3x_2 + 5x_3 = 4$$

▶ **30.** Use $\bar{x}^0 = \mathrm{col}(0,0,0)$ as a starting value and solve the given system of equations by the Gauss-Seidel method

$$10x_1 - \quad x_2 \qquad = 9$$
$$-x_1 + 10x_2 - \quad 2x_3 = 7$$
$$-2x_2 + 10x_3 = 8$$

▶ **31.** Find all roots of the nonlinear system of equations

$$x_1^2 + x_2 - 37 = 0$$
$$x_1 - x_2^2 + x_3^2 - 9 = 0$$
$$x_1 + x_2 + x_3 - 9 = 0$$

▶ **32.** (Computer problem)

A tridiagonal system of equations

$$
\begin{pmatrix}
b_1 & c_1 & & & & & & \\
a_2 & b_2 & c_2 & & & & & \\
& a_3 & b_3 & c_3 & & & & \\
& & \ddots & \ddots & \ddots & & & \\
& & & a_i & b_i & c_i & & \\
& & & & \ddots & \ddots & \ddots & \\
& & & & & a_{n-1} & b_{n-1} & c_{n-1} \\
& & & & & & a_n & b_n
\end{pmatrix}
\begin{pmatrix}
x_1 \\ x_2 \\ x_3 \\ \vdots \\ x_i \\ \vdots \\ x_{n-1} \\ x_n
\end{pmatrix}
=
\begin{pmatrix}
d_1 \\ d_2 \\ d_3 \\ \vdots \\ d_i \\ \vdots \\ d_{n-1} \\ d_n
\end{pmatrix}
\tag{32a}
$$

is characterized by a coefficient matrix with nonzero elements along the main diagonal, $b_1, b_2, \ldots, b_n$, the superdiagonal, $c_1, c_2, \ldots, c_{n-1}$ and subdiagonal $a_2, a_3, \ldots, a_n$, with zero elements everywhere else. One can perform row operations on the system of equations (32a) and reduce the system to the form

$$
\begin{pmatrix}
\beta_1 & c_1 & & & & & \\
& \beta_2 & c_2 & & & & \\
& & \beta_3 & c_3 & & & \\
& & & \ddots & \ddots & & \\
& & & & \beta_i & c_i & \\
& & & & & \ddots & \ddots \\
& & & & & & \beta_{n-1} & c_{n-1} \\
& & & & & & & \beta_n
\end{pmatrix}
\begin{pmatrix}
x_1 \\ x_2 \\ x_3 \\ \vdots \\ x_i \\ \vdots \\ x_{n-1} \\ x_n
\end{pmatrix}
=
\begin{pmatrix}
\delta_1 \\ \delta_2 \\ \delta_3 \\ \vdots \\ \delta_i \\ \vdots \\ \delta_{n-1} \\ \delta_n
\end{pmatrix}
\tag{32b}
$$

(a) Show that for nonzero diagonal elements the row operations

$$E_2 - \frac{a_2}{b_1} E_1 \rightarrow E_2$$

$$E_3 - \frac{a_3}{b_2} E_2 \rightarrow E_3$$

$$\vdots \qquad \vdots$$

$$E_i - \frac{a_i}{b_{i-1}} E_{i-1} \rightarrow E_i \quad \text{for } i = 4, 5, \ldots, n$$

reduce the equations (32a) to the set of equations (32b).

(b) Show the system of equations (32b) has a solution set given by

$$x_n = \delta_n / \beta_n$$

$$x_{n-1} = (\delta_{n-1} - c_{n-1} x_n)/\beta_{n-1}$$

$$\vdots$$

$$x_i = (\delta_i - c_i x_{i+1})/\beta_i \quad \text{for } i = n-1, n-2, \ldots, 3, 2, 1$$

(c) Write a computer program to solve a tridiagonal system of equations.

(d) Test your computer program on an $(n \times n)$ tridiagonal system for $n = 4, 5, 6$ where $\vec{a}, \vec{b}, \vec{c}, \vec{d}$ have the patterns

$$\vec{a} = (1, 1, \ldots, 1, 1)$$
$$\vec{b} = (2, 2, \ldots, 2, 2)$$
$$\vec{c} = (1, 1, \ldots, 1, 1)$$
$$\vec{d} = \text{col}(3, 4, 4, \ldots, 4, 4, 3)$$

▶ **33.** **(Computer problem)**

A pentadiagonal system of equations has the form

$$
\begin{pmatrix}
c_1 & d_1 & e_1 \\
b_2 & c_2 & d_2 & e_2 \\
a_3 & b_3 & c_3 & d_3 & e_3 \\
& a_4 & b_4 & c_4 & d_4 & e_4 \\
& & \ddots & \ddots & \ddots & \ddots & \ddots \\
& & & a_i & b_i & c_i & d_i & e_i \\
& & & & \ddots & \ddots & \ddots & \ddots & \ddots \\
& & & & & a_{n-2} & b_{n-2} & c_{n-2} & d_{n-2} & e_{n-2} \\
& & & & & & a_{n-1} & b_{n-1} & c_{n-1} & d_{n-1} \\
& & & & & & & a_n & b_n & c_n
\end{pmatrix}
\begin{pmatrix}
x_1 \\ x_2 \\ x_3 \\ x_4 \\ \vdots \\ x_i \\ \vdots \\ x_{n-2} \\ x_{n-1} \\ x_n
\end{pmatrix}
=
\begin{pmatrix}
f_1 \\ f_2 \\ f_3 \\ f_4 \\ \vdots \\ f_i \\ \vdots \\ f_{n-2} \\ f_{n-1} \\ f_n
\end{pmatrix}
\tag{33a}
$$

with elements $a_{ij} = 0$ for $|i - j| \geq 3$. Pentadiagonal systems of equations can be reduced, using row operations, to the form

$$
\begin{pmatrix}
\beta_1 & \gamma_1 & e_1 \\
& \beta_2 & \gamma_2 & e_2 \\
& & \beta_3 & \gamma_3 & e_3 \\
& & & \beta_4 & \gamma_4 & e_4 \\
& & & & \ddots & \ddots & \ddots \\
& & & & & \beta_i & \gamma_i & e_i \\
& & & & & & \ddots & \ddots & \ddots \\
& & & & & & & \beta_{n-2} & \gamma_{n-2} & e_{n-2} \\
& & & & & & & & \beta_{n-1} & \gamma_{n-1} \\
& & & & & & & & & \beta_n
\end{pmatrix}
\begin{pmatrix}
x_1 \\ x_2 \\ x_3 \\ x_4 \\ \vdots \\ x_i \\ \vdots \\ x_{n-2} \\ x_{n-1} \\ x_n
\end{pmatrix}
=
\begin{pmatrix}
f_1^* \\ f_2^* \\ f_3^* \\ f_4^* \\ \vdots \\ f_i^* \\ \vdots \\ f_{n-2}^* \\ f_{n-1}^* \\ f_n^*
\end{pmatrix}
\tag{33b}
$$

(a) What row operations are needed to reduce the system of equations (33a) to the form of equations (33b)?

(b) How would one write the solutions to the system of equations (33b)?

(c) Write a computer program to solve a pentadiagonal system of equations.

Hint: See problem 32.

"Try to learn something about everything and everything about something"

Thomas Henry Huxley (1825-1895)

Chapter 4

Interpolation and Approximation

The Weierstrass approximation theorem states that a continuous function $f(x)$ over a closed interval $[a, b]$ can be approximated by a polynomial $P_n(x)$, of degree n, such that

$$|f(x) - P_n(x)| \leq \epsilon, \quad x \in [a, b] \tag{4.1}$$

where $\epsilon > 0$ is a small quantity and n is sufficiently large. A polynomial representation is just one way of approximating a function. Approximation theory is concerned with finding various ways to represent a function over an interval and is not restricted to polynomial approximation.

The interpolation problem is the construction of a curve $y(x)$ which passes through a given set of data points (x_i, y_i), for $i = 0, 1, \ldots, n$ where the data points are such that $a = x_0 < x_1 < x_2 < \cdots x_{n-1} < x_n = b$. The constructed curve $y(x)$ can then be used to estimate the values of y at positions x which are between the end points a and b (interpolation) or to estimate the value of y for x exterior to the end points (extrapolation).

Various industrial, business and research organizations routinely collect and analyze data. We shall investigate collected data in the form of two variables which we label x and y. We assume that the data can be labeled in some convenient way and represented in a tabular form. The table 4.1 illustrates one possible way of labeling and representing the data.

x	y
x_0	y_0
x_1	y_1
x_2	y_2
x_3	y_3
$\vdots$	$\vdots$
x_{n-1}	y_{n-1}
x_n	y_n

Table 4.1 Tabular listing of (x,y) data values.

Whenever the data in table 4.1 is such that the independent variable x is evenly spaced, then the difference between any consecutive x-values has a constant value. We denote this constant value by h and write

$$\Delta x = x_{k+1} - x_k = h = \text{constant} \qquad \text{for} \quad k = 0, 1, 2, 3, \ldots.$$

Given a set of $(n+1)$ data pairs (x_i, y_i), $i = 0, \ldots, n$, where the x_i values are equally spaced, we assume these (x,y) data pairs represent selected sample values $(x_i, y(x_i))$ from a continuous function $y(x)$, even though we do not know the function $y(x)$. We will study how to construct a polynomial $P_n(x)$, which satisfies $y_i = P_n(x_i)$ for $i = 0, 1, \ldots, n$. This is called the interpolating polynomial which reproduces the function values at the given points x_i for $i = 0, 1, \ldots, n$. The simplest polynomial interpolation is the straight line through two data points as illustrated in the figure.

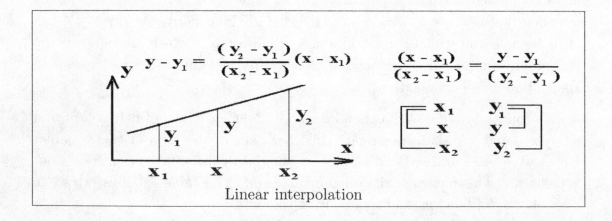

Linear interpolation

The construction of a polynomial function $P_n(x)$ which satisfies $y_i = P_n(x_i)$ for $i = 0, 1, \ldots, n$ has several purposes. First it can be used as an approximation function for reproducing the data values (x_i, y_i) for $i = 0, 1, \ldots, n$. Secondly, polynomial interpolation is said to occur whenever one uses the approximating polynomial $P_n(x)$ to estimate the true y-value for a nontabulated x-value, where $x_0 \leq x \leq x_{n+1}$. Polynomial extrapolation is said to occur whenever one uses the approximating polynomial $P_n(x)$ to estimated the true y-values for x outside the interval $[x_0, x_{n+1}]$. We will use polynomial interpolation in later chapters to develop numerical techniques for differentiation and integration of a function. Polynomial interpolation will also arise in the development of numerical techniques for solving differential equations.

Difference Tables

We shall examine differences in the consecutive y-values associated with the table 4.1 representation of data. Define the first forward differences

$$\Delta y_1 = y_2 - y_1, \quad \Delta y_2 = y_3 - y_2, \quad \cdots \quad \Delta y_i = y_{i+1} - y_i \tag{4.2}$$

and define second forward differences as differences of first forward differences. Second forward differences are written

$$\Delta^2 y_i = \Delta(\Delta y_i) = \Delta(y_{i+1} - y_i) = \Delta y_{i+1} - \Delta y_i$$
$$= (y_{i+2} - y_{i+1}) - (y_{i+1} - y_i) = y_{i+2} - 2y_{i+1} + y_i \tag{4.3}$$

for $i = 1, 2, 3, \ldots$. An $(n+1)$-st ordered forward difference is defined as the difference of n-th ordered forward differences. Alternatively, one can define the stepping operator E defined by

$$E y_i = y_{i+1}, \quad E^2 y_i = y_{i+2}, \quad \ldots, \quad E^n y_i = y_{i+n} \tag{4.4}$$

then the first and higher ordered forward differences can be written in an operator form. For example, since

$$\Delta y_i = y_{i+1} - y_i = E y_i - y_i = (E - 1) y_i \tag{4.5}$$

one can write $\Delta = E - 1$, and so the various forward differences can be expressed

first forward difference $\quad \Delta y_i = (E - 1) y_i = y_{i+1} - y_i$

second forward difference $\quad \Delta^2 y_i = (E - 1)^2 y_i = (E^2 - 2E + 1) y_i = y_{i+2} - 2y_{i+1} + y_i$

third forward difference $\quad \Delta^3 y_i = (E - 1)^3 y_i = y_{i+3} - 3y_{i+2} + 3y_{i+1} - y_i$

$$\vdots$$

n-th forward difference $\quad \Delta^n y_i = (E - 1)^n y_i$

$$\tag{4.6}$$

Using the binomial expansion one can verify that

$$\Delta^n y_i = y_{i+n} - \binom{n}{1} y_{i+n-1} + \binom{n}{2} y_{i+n-2} - \binom{n}{3} y_{i+n-3} + \cdots + (-1)^n y_i \tag{4.7}$$

where

$$\binom{n}{m} = \frac{n!}{m!(n-m)!} \tag{4.8}$$

are the binomial coefficients. One can now append columns of differences to the given data set with equal x-spacing to form a forward difference table. The subscript labeling of the points (x, y) in a difference table is arbitrary in that any point can be labeled (x_0, y_0) and the other points, as well as corresponding differences, are then labeled accordingly. The construction of a representative forward difference table associated with a constant x-value step size is illustrated in the table 4.2.

In the special case $\Delta x = h$ is constant, then the entries in the difference table can be scaled using the transformation

$$s = \frac{x - x_0}{h} \tag{4.9}$$

to obtain the scaled column of integer values listed in the table 4.2.

Table 4.2 Forward Difference Table

s	x	y	Δy	$\Delta^2 y$	$\Delta^3 y$	$\Delta^4 y$	$\Delta^5 y$	$\Delta^6 y$
-2	x_{-2}	y_{-2}						
			Δy_{-2}					
-1	x_{-1}	y_{-1}		$\Delta^2 y_{-2}$				
			Δy_{-1}		$\Delta^3 y_{-2}$			
0	x_0	y_0		$\Delta^2 y_{-1}$		$\Delta^4 y_{-2}$		
			Δy_0		$\Delta^3 y_{-1}$		$\Delta^5 y_{-2}$	
1	x_1	y_1		$\Delta^2 y_0$		$\Delta^4 y_{-1}$		$\Delta^6 y_{-2}$
			Δy_1		$\Delta^3 y_0$		$\Delta^5 y_{-1}$	
2	x_2	y_2		$\Delta^2 y_1$		$\Delta^4 y_0$		$\Delta^6 y_{-1}$
			Δy_2		$\Delta^3 y_1$		$\Delta^5 y_0$	
3	x_3	y_3		$\Delta^2 y_2$		$\Delta^4 y_1$		
			Δy_3		$\Delta^3 y_2$			
4	x_4	y_4		$\Delta^2 y_3$				
			Δy_4					
5	x_5	y_5						

It is assumed that the data used to form the difference table is a discrete sampling from a function $y = y(x)$ which is continuous on some interval $[a, b]$. Therefore, one can apply the Weierstrass approximation theorem to construct an approximation of the function.

Note that if the data set is constructed from some polynomial, then the difference table will have the special property that the nth difference column will be all constants and so all columns of higher order differences will be zero. Whenever this occurs the data can be represented by a nth degree polynomial. Observe that

$$\Delta(x^n) = (x + h)^n - x^n = nhx^{n-1} + \text{lower order terms} \tag{4.10}$$

and for c_0 some nonzero constant one would have

$$\Delta(c_0 x^n) = c_0 nhx^{n-1} + \text{lower order terms.} \tag{4.11}$$

Consider the nth degree polynomial

$$P_n(x) = c_0 x^n + c_1 x^{n-1} + \cdots + c_{n-1} x + c_n \tag{4.12}$$

where $c_0, c_1, \ldots, c_n$ are constants. Taking differences of this polynomial produces

$$\Delta P_n(x) = c_0 nhx^{n-1} + \text{lower order terms.}$$
$$\Delta^2 P_n(x) = c_0 n(n-1)h^2 x^{n-2} + \text{lower order terms}$$
$$\vdots \tag{4.13}$$
$$\Delta^n P_n(x) = c_0 n(n-1)(n-2) \cdots (3)(2)(1)h^n = c_0 n! h^n = \text{constant}$$
$$\Delta^{(n+1)} P_n(x) = 0$$

which demonstrates that for n-th degree polynomials, the nth differences are constant and the $(n+1)$-st differences are zero.

Example 4-1.   (Difference table.)

Form a forward difference table associated with the function $y = y(x) = x^3$ and the x-values 0.2, 0.4, 0.6, 0.8, 1.0, 1.2, 1.4

Solution: We calculate the y-values and then the first, second, third and fourth forward differences to obtain the table 4.3.

		Forward Difference Table			
x	y	Δy	$\Delta^2 y$	$\Delta^3 y$	$\Delta^4 y$
0.2	0.008				
		0.056			
0.4	0.064		0.096		
		0.152		0.048	
0.6	0.216		0.144		0
		0.296		0.048	
0.8	0.512		0.192		0
		0.488		0.048	
1.0	1.000		0.240		0
		0.728		0.048	
1.2	1.728		0.288		
		1.016			
1.4	2.744				

Table 4.3 Table of differences formed from the function $y = x^3$.

■

Whenever data pairs are collected from some sampling of an experiment there is usually errors associated with the data and in such cases the (x, y) data points will not give a forward difference table with a column of all constant values and so the data will not be a polynomial. However, by the Weierstrass approximation theorem, one can replace the true function $y = y(x)$ with some polynomial approximation $P_n(x)$. Let us investigate the construction of various polynomials from a selected set of data pairs.

Interpolating Polynomials

We wish to construct an approximating polynomial $P_n(x)$ which takes on the values $y_0, y_1, \ldots, y_n$ of $y(x)$ at the points $x_0, x_1, \ldots, x_n$ called nodes. If such a polynomial function can be constructed it is called an interpolation polynomial or collocation polynomial. The constructed polynomial function which passes through the given data points can be used to approximate $y(x)$ for any value of x over the interpolation interval (x_0, x_n). If one uses the approximation polynomial to estimate values of $y(x)$ outside the interval (x_0, x_n), then the process is called extrapolation. The process of extrapolation with polynomials is not recommended because polynomials $P_n(x)$ tend to oscillate between the values $y_0, y_1, \ldots, y_n$ when n is large and to diverge outside the interpolation interval. Whenever possible, interpolation is to be preferred over extrapolation when dealing with polynomials of high order.

One can construct an nth degree polynomial $y = P_n(x)$ which passes through $(n+1)$ data points (x_i, y_i) for $i = 0, 1, 2, \ldots, n$ and hence it can be called an interpolating polynomial. The polynomial constructed will be unique. To show uniqueness we employ the fundamental theorem of algebra which states that a polynomial of degree n has exactly n-roots. Now if we assume there are two different polynomials of degree n, say $y = P_n(x)$ and $y = \mathcal{P}_n(x)$ which have the same values at $(n+1)$ data values (x_i, y_i) for $i = 0, 1, 2, \ldots, n$, then the difference function $D(x) = P_n(x) - \mathcal{P}_n(x)$ is at most a polynomial of degree n which has $(n+1)$ zeros. This can only occur if $D(x)$ is identically zero for all x values. Consequently, $P_n(x) = \mathcal{P}_n(x)$ and so the polynomials must be identically the same.

Note that polynomials can be represented in different ways. For example, the second degree polynomial $P_2(x) = x^2$ that passes through the points $(0, 0), (1, 1)$ and $(2, 4)$ is unique, however, its representation is not unique and so the polynomial $P_2(x) = x^2$ can be represented in different ways. Four possible representations are

$$P_2(x) = x(x - 2) + 2x \qquad P_2(x) = \frac{1}{2} + \frac{1}{2}(2x^2 - 1)$$
$$P_2(x) = (x - 1)^2 - 2(x - 1) + 1 \qquad P_2(x) = x + x(x - 1)$$

We now develop methods for construction of nth degree polynomials which collocate with the $(n+1)$ data points $(x_0, y_0), \ldots, (x_n, y_n)$.

Equally Spaced Data

Assume the x-values are equally spaced such that $x_i = x_0 + ih$ for $i = 0, 1, 2, \ldots, n$, then the difference between any consecutive x-values is a constant and one can write $x_m - x_{m-1} = h$, for $m = 1, 2, \ldots, n$. A polynomial representation of the form

$$P_n(x) = c_0 + c_1(x - x_0) + c_2(x - x_0)(x - x_1) + c_3(x - x_0)(x - x_1)(x - x_2) + \cdots + c_n(x - x_0)(x - x_1) \cdots (x - x_{n-1}),$$

where the coefficients $c_0, c_1, \ldots, c_n$ are constants and selected to make the polynomial produce the given data values, is required to satisfy the conditions

$$P_n(x_0) = c_0 = y_0$$
$$P_n(x_1) = c_0 + c_1 h = y_1$$
$$P_n(x_2) = c_0 + c_1 2h + c_2 2h^2 = y_2$$
$$P_n(x_3) = c_0 + c_1 3h + c_2 6h^2 + c_3 6h^3 = y_3$$
$$\vdots$$
$$P_n(x_n) = c_0 + c_1 nh + c_2(n)(n-1)h^2 + \cdots + n! c_n h^n = y_n.$$

Solving for the coefficients $c_0, c_1, c_2, \ldots, c_n$ we find

$$
\begin{aligned}
c_0 &= y_0 \\
c_1 &= \frac{\Delta y_0}{h} = \frac{y_1 - y_0}{h} \\
c_2 &= \frac{\Delta^2 y_0}{2h^2} = \frac{y_2 - 2y_1 + y_0}{2h^2} \\
c_3 &= \frac{\Delta^3 y_0}{3!h^3} = \frac{y_3 - 3y_2 + 3y_1 - y_0}{3!h^3} \\
&\vdots \\
c_n &= \frac{\Delta^n y_0}{n!h^n}
\end{aligned}
\tag{4.14}
$$

This produces the polynomial approximation

$$
\begin{aligned}
P_n(x) =& y_0 + \frac{\Delta y_0}{h}(x - x_0) + \frac{\Delta^2 y_0}{2!h^2}(x - x_0)(x - x_1) + \frac{\Delta^3 y_0}{3!h^3}(x - x_0)(x - x_1)(x - x_2) + \cdots \\
&+ \frac{\Delta^n y_0}{n!h^n}(x - x_0)(x - x_1)(x - x_2)\cdots(x - x_{n-1})
\end{aligned}
\tag{4.15}
$$

which is called Newton's forward interpolation formula. Sometimes referred to as the Newton-Gregory forward interpolating polynomial. In terms of the scaled variable $s = \dfrac{x - x_0}{h}$, which has an integer value corresponding to each x_i data value, the equation (4.15) has the form

$$
\begin{aligned}
P_n(x) =& y_0 + s\Delta y_0 + \frac{s(s - 1)}{2!}\Delta^2 y_0 + \frac{s(s - 1)(s - 2)}{3!}\Delta^3 y_0 + \cdots \\
&+ \frac{s(s - 1)(s - 2)\cdots(s - n + 1)}{n!}\Delta^n y_0
\end{aligned}
\tag{4.16}
$$

which can also be represented in the form

$$
P_n(x) = y_0 + \binom{s}{1}\Delta y_0 + \binom{s}{2}\Delta^2 y_0 + \binom{s}{3}\Delta^3 y_0 + \cdots + \binom{s}{n}\Delta^n y_0
\tag{4.17}
$$

where the binomial coefficients $\binom{s}{i}$ multiplies the ith difference of y_0 from the difference table 4.2. The binomial coefficients $\binom{s}{i}$ represents the number of combination of s elements taken i at a time.

The lozenge diagram[1] illustrated in the figure 4-1 is often constructed as an aid to the representation of various polynomial interpolation formulas. One moves across the lozenge diagram from left to right in a sequence of straight

---

[1] Lozenge refers to the diamond shaped pattern created by diagonal lines.

line paths. The lozenge diagram has certain rules for its use and within the lozenge diagram there are scale factors or coefficients used for construction of a collocation polynomial which is associated with a set of (x, y) data points. The polynomials are constructed as a series of terms and are produced as follows.

1. Move from left to right across the lozenge diagram starting with a value or modified value from the y-column.

2. One moves in a straight line path to the next column of the lozenge diagram. This straight line path can be either diagonally upward, horizontal or diagonally downward.

3. The straight line path points to a difference expression or binomial factor in the next column of the lozenge diagram. The quantity pointed to must be multiplied by a scale factor from the lozenge diagram to produce the next term in the series representing the interpolating polynomial. The scale factors are dependent upon the type of path selected. The following expressions are used as scale factors needed for the representation of the next term in the interpolation polynomial construction

    (a) Use the coefficient below the path
        if you move diagonally upward

    (b) Use the average of the coefficients
        above and below a horizontal path

    (c) Use the coefficient above the path
        if you move diagonally downward

For example, if we start at y_0, then y_0 is a zeroth order polynomial approximation and represents the first term in the polynomial approximation which is given by $P_0(x) = y_0$. If we move horizontally from y_0 in a straight line to the right, we hit the term $\binom{s}{1}$ which must be multiplied by the scale factor $\frac{1}{2}(\Delta y_{-1} + \Delta y_0)$ to produce the first order approximation

$$P_1(x) = y_0 + \frac{1}{2}(\Delta y_{-1} + \Delta y_0)\binom{s}{1}, \qquad s = \frac{x - x_0}{h} \qquad (4.18)$$

Continue moving horizontally to the right we hit the next term $\Delta^2 y_{-1}$ in the lozenge diagram. This term must be multiplied by the scale factor $\frac{1}{2}\left[\binom{s+1}{2} + \binom{s}{2}\right]$

and so we produce the second order polynomial approximation

$$P_2(x) = y_0 + \frac{1}{2}\left(\Delta y_{-1} + \Delta y_0\right)\binom{s}{1} + \frac{1}{2}\Delta^2 y_{-1}\left[\binom{s+1}{2} + \binom{s}{2}\right], \qquad s = \frac{x - x_0}{h} \qquad (4.19)$$

Note that $P_1(x)$ and $P_2(x)$ depend upon the value of the ordinates y_{-1}, y_0 and y_1 centered about the point y_0. By continuing our horizontal path to the right one can construct additional terms to add to the series and so produce higher order approximating polynomials. Observe that by moving further into the lozenge diagram one constructs higher degree interpolating polynomials which include the influence of additional ordinates surrounding the central point y_0.

The lozenge diagram in figure 4-1 can be used to produce the following series in terms of the scaled variable $s = \dfrac{x - x_0}{h}$.

Newton's forward formula (path A-A)

$$P_n(x) = y_0 + \binom{s}{1}\Delta y_0 + \binom{s}{2}\Delta^2 y_0 + \binom{s}{3}\Delta^3 y_0 + \binom{s}{4}\Delta^4 y_0 + \cdots + \binom{s}{n}\Delta^n y_0$$

$$P_n(x) = y_0 + \sum_{i=1}^{n}\binom{s}{i}\Delta^i y_0 \qquad (4.20)$$

Newton's backward formula (path B-B)

$$P_n(x) = y_0 + \binom{s}{1}\Delta y_{-1} + \binom{s+1}{2}\Delta^2 y_{-2} + \cdots + \binom{s+n-1}{n}\Delta^n y_{-n}$$

$$P_n(x) = y_0 + \sum_{i=1}^{n}\binom{s+i-1}{i}\Delta^i y_{-i} \qquad (4.21)$$

Gauss backward formula (path C-C)

$$P_n(x) = y_0 + \binom{s}{1}\Delta y_{-1} + \binom{s+1}{2}\Delta^2 y_{-1} + \binom{s+1}{3}\Delta^3 y_{-2} + \binom{s+2}{4}\Delta^4 y_{-2} + \cdots$$

$$P_{2n}(x) = y_0 + \sum_{i=1}^{n}\left[\binom{s+i-1}{2i-1}\Delta^{2i-1} y_{-i} + \binom{s+i}{2i}\Delta^{2i} y_{-i}\right] \qquad (4.22)$$

Gauss forward formula (path D-D)

$$P_n(x) = y_0 + \binom{s}{1}\Delta y_0 + \binom{s}{2}\Delta^2 y_{-1} + \binom{s+1}{3}\Delta^3 y_{-1} + \binom{s+1}{4}\Delta^4 y_{-2} + \cdots$$

$$P_{2n}(x) = y_0 + \sum_{i=1}^{n}\left[\binom{s+i-1}{2i-1}\Delta^{2i-1} y_{1-i} + \binom{s+i-1}{2i}\Delta^{2i} y_{-i}\right] \qquad (4.23)$$

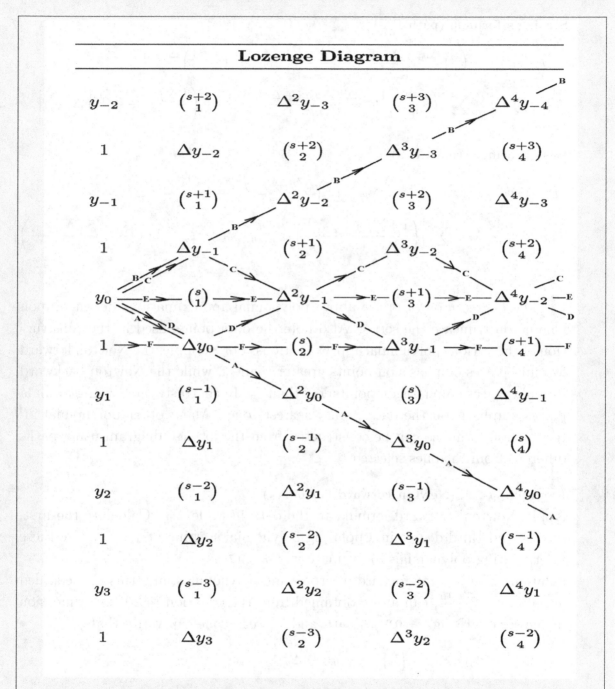

Figure 4-1. Lozenge diagram for polynomial approximation.

Stirling's formula (path E-E)

$$P_n(x) = y_0 + \binom{s}{1}\frac{\Delta y_0 + \Delta y_{-1}}{2} + \frac{\binom{s+1}{2} + \binom{s}{2}}{2}\Delta^2 y_{-1} + \binom{s+1}{3}\frac{\Delta^3 y_{-1} + \Delta^3 y_{-2}}{2} + \cdots$$

$$P_{2n}(x) = y_0 + \sum_{i=1}^{n}\left[\binom{s+i-1}{2i-1}\left(\frac{\Delta^{2i-1}y_{1-i} + \Delta^{2i-1}y_{-i}}{2}\right) + \left(\frac{\binom{s+i}{2i} + \binom{s+i-1}{2i}}{2}\right)\Delta^{2i}y_{-i}\right]$$

$$(4.24)$$

Bessel formula (path F-F)

$$P_n(x) = \frac{y_0 + y_1}{2} + \frac{\binom{s}{1} + \binom{s-1}{1}}{2}\Delta y_0 + \binom{s}{2}\frac{\Delta^2 y_0 + \Delta^2 y_{-1}}{2} + \frac{\binom{s+1}{3} + \binom{s}{3}}{2}\Delta^3 y_1 + \cdots$$

$$P_{2n}(x) = \frac{y_0 + y_1}{2} + \sum_{i=1}^{n}\left[\left(\frac{\binom{s+i-1}{2i-1} + \binom{s+i-2}{2i-1}}{2}\right)\Delta^{2i-1}y_{1-i} + \binom{s+i-1}{2i}\left(\frac{\Delta^{2i}y_{-i} + \Delta^{2i}y_{1-i}}{2}\right)\right]$$

$$(4.25)$$

The degree n or $2n$ in the above polynomial approximations depends upon where you truncate the series. Also note the data points used in the construction of the various polynomial approximations. For example, the Newton forward formula places emphasis on points greater than x_0, while the Newton backward formula places emphasis on points less than x_0. In contrast, the Stirling formula places emphasis on the data points nearest to x_0. Many other polynomial interpolation formulas can be constructed from the lozenge diagram using paths different from the ones selected.

Example 4-2. (Newton Forward Formula .)
Apply Newton's forward formula to the data in table 4.3. Calculate the first, second and third degree interpolating polynomial starting at $x = 0.6$. Use these interpolating polynomials to estimate y at $x = 0.7$.

Solution: The Newton forward interpolating polynomials, in terms of the scaled variable $s = \frac{x - x_0}{h}$, can all be obtained from the equation (4.20) by truncation of the series. Using $h = 0.2$, $x_0 = 0.6$ and $y_0 = 0.216$ one can verify that

$$P_1(x) = y_0 + \binom{s}{1}\Delta y_0 = y_0 + s\,\Delta y_0$$

$$P_2(x) = y_0 + \binom{s}{1}\Delta y_0 + \binom{s}{2}\Delta^2 y_0 = y_0 + s\,\Delta y_0 + s(s-1)\Delta^2 y_0/2$$

$$P_3(x) = y_0 + \binom{s}{1}\Delta y_0 + \binom{s}{2}\Delta^2 y_0 + \binom{s}{3}\Delta^3 y_0$$

$$P_3(x) = y_0 + s\,\Delta y_0 + s(s-1)\Delta^2 y_0/2 + s(s-1)(s-2)\Delta^3 y_0/6$$

where $s = \dfrac{x - 0.6}{0.2} = 5(x - 0.6) = 5x - 3$, $\Delta y_0 = 0.296$, $\Delta^2 y_0 = 0.192$, and $\Delta^3 y_0 = 0.048$. When $x = 0.7$, we have $s = 0.5$ so that the interpolated values are

$$P_1(0.7) = 0.364, \qquad P_2(0.7) = 0.340, \qquad P_3(0.7) = 0.343.$$

Note that the interpolating polynomial $P_3(x)$ reduces to x^3 because this is how we constructed the table 4.3. The figure 4-2 illustrates the data points and interpolating curves.

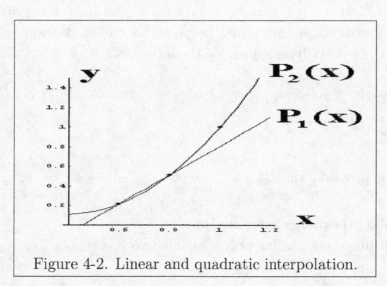

Figure 4-2. Linear and quadratic interpolation.

Unequally Spaced Data

Assume that the data in table 4.1 is not evenly spaced. For these conditions other types of interpolating polynomials are required. The following discussions are applicable to both evenly spaced and unevenly spaced data. The Lagrange polynomials of degree n are defined

$$L_{n,k}(x) = \frac{(x - x_0)(x - x_1) \cdots (x - x_{k-1})(x - x_{k+1}) \cdots (x - x_n)}{(x_k - x_0)(x_k - x_1) \cdots (x_k - x_{k-1})(x_k - x_{k+1}) \cdots (x_k - x_n)}$$

$$L_{n,k}(x) = \prod_{\substack{i=0 \\ i \neq k}}^{n} \frac{(x - x_i)}{(x_k - x_i)} \qquad \text{for} \quad k = 0, 1, 2, \ldots, n \tag{4.26}$$

where the factor $(x - x_k)$ is missing from the product term in the numerator and the term $(x_k - x_k)$ is missing from the denominator. Some textbooks use the notation $\ell_k(x)$ in place of $L_{n,k}(x)$ whenever the degree n is known and no confusion arises. Another notation for representing the Lagrange polynomials is to define the product function $\pi(x) = (x - x_0)(x - x_1) \cdots (x - x_n)$, which involves all products and write the Lagrange polynomial in the form

$$L_{n,k}(x) = \frac{\pi(x)}{(x - x_k)\pi'(x_k)} \tag{4.27}$$

where $\pi'(x_k)$ is the derivative function $\frac{d\pi}{dx}$ evaluated at $x = x_k$.

Observe that the Lagrange polynomials have the property that $L_{n,k}(x_k) = 1$ and $L_{n,k}(x_i) = 0$ for $i = 0, 1, 2, \ldots, k - 1, k + 1, \ldots, n$. The Lagrange polynomials can be used to construct interpolating polynomials called Lagrange interpolating polynomials. These polynomials have the form

$$P_n(x) = y_0 L_{n,0}(x) + y_1 L_{n,1}(x) + \cdots + y_i L_{n,i}(x) + \cdots + y_{n-1} L_{n,n-1}(x) + y_n L_{n,n}(x)$$

$$P_n(x) = \sum_{j=0}^{n} y_j L_{n,j}(x) \tag{4.28}$$

and have the property that at $x = x_i$ we have $P_n(x_i) = y_i$ for $i = 0, 1, \ldots, n$.

Example 4-3.   **(Lagrange interpolation.)**
The straight line interpolating polynomial between two points (x_0, y_0) and (x_1, y_1) is given by

$$P_1(x) = y_0 L_{1,0}(x) + y_1 L_{1,1}(x)$$

where

$$L_{1,0}(x) = \frac{x - x_1}{x_0 - x_1} \quad \text{and} \quad L_{1,1}(x) = \frac{x - x_0}{x_1 - x_0}.$$

Observe that $P_1(x_0) = y_0$ and $P_1(x_1) = y_1$.

∎

Example 4-4.   **(Lagrange interpolation.)**
Consider the unequally spaced data

$$(1.1, 2.1), \quad (2.3, -1.9), \quad (2.7, 4.5), \quad (3.5, 1.3)$$

Construct a Lagrange polynomial which passes through these points.

Solution: We label the given data points

$$(x_0, y_0) = (1.1, 2.1),\ (x_1, y_1) = (2.3, -1.9),\ (x_2, y_2) = (2.7, 4.5),\ (x_3, y_3) = (3.5, 1.3),$$

then the Lagrange polynomial which passes through the above points can be expressed

$$P_3(x) = \sum_{j=0}^{3} y_j L_{3,j}(x) = y_0 L_{3,0}(x) + y_1 L_{3,1}(x) + y_2 L_{3,2}(x) + y_3 L_{3,3}(x)$$

where

$$L_{3,0}(x) = \frac{(x - x_1)(x - x_2)(x - x_3)}{(x_0 - x_1)(x_0 - x_2)(x_0 - x_3)}$$

$$L_{3,1}(x) = \frac{(x - x_0)(x - x_2)(x - x_3)}{(x_1 - x_0)(x_1 - x_2)(x_1 - x_3)}$$

$$L_{3,2}(x) = \frac{(x - x_0)(x - x_1)(x - x_3)}{(x_2 - x_0)(x_2 - x_1)(x_2 - x_3)}$$

$$L_{3,3}(x) = \frac{(x - x_0)(x - x_1)(x - x_2)}{(x_3 - x_0)(x_3 - x_1)(x_3 - x_2)}$$

The Lagrange interpolating polynomial is illustrated in the figure 4-3.

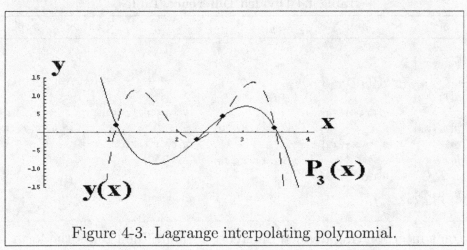

Figure 4-3. Lagrange interpolating polynomial.

Note 1: Many higher ordered polynomials can be made to pass through the given points. What the true function $y = y(x)$ is like is unknown and so when using unequally spaced data points, a spacing that is too wide may miss some important information and consequently the interpolating polynomial can be way off.

Note 2: It is very beneficial if you have some idea as to how your data is suppose to behave. Additional information can suggest other interpolation methods to be used in constructing the proper shape curve to fit the data.

Divided Differences

Associated with nonuniformly spaced data which is ordered in a table, such as in table 4.1, one can define $y[x_i] = y(x_i)$ with first divided difference defined

$$y[x_i, x_{i+1}] = \frac{y(x_{i+1}) - y(x_i)}{x_{i+1} - x_i} \tag{4.29}$$

The second divided difference is defined as a divide difference of the first divided difference and is written

$$y[x_i, x_{i+1}, x_{i+2}] = \frac{y[x_{i+1}, x_{i+2}] - y[x_i, x_{i+1}]}{x_{i+2} - x_i} \tag{4.30}$$

Continuing in this manner the nth divided difference if defined

$$y[x_i, x_{i+1}, x_{i+2}, \ldots, x_{i+n-1}, x_{i+n}] = \frac{y[x_{i+1}, x_{i+2}, \ldots, x_{i+n}] - y[x_i, x_{i+1}, \ldots, x_{i+n-1}]}{x_{i+n} - x_i} \tag{4.31}$$

One can then form the divided difference table 4.4.

Table 4.4 Divided Difference Table

x	$y[\ ]$	$y[,]$	$y[,,]$	$y[,,,]$	$y[,,,,]$	$y[,,,,,]$
x_0	$y[x_0]$					
		$y[x_0, x_1]$				
x_1	$y[x_1]$		$y[x_0, x_1, x_2]$			
		$y[x_1, x_2]$		$y[x_0, x_1, x_2, x_3]$		
x_2	$y[x_2]$		$y[x_1, x_2, x_3]$		$y[x_0, x_1, x_2, x_3, x_4]$	
		$y[x_2, x_3]$		$y[x_1, x_2, x_3, x_4]$		$y[x_0, x_1, x_2, x_3, x_4, x_5]$
x_3	$y[x_3]$		$y[x_2, x_3, x_4]$		$y[x_1, x_2, x_3, x_4, x_5]$	
		$y[x_3, x_4]$		$y[x_2, x_3, x_4, x_5]$		
x_4	$y[x_4]$		$y[x_3, x_4, x_5]$			
		$y[x_4, x_5]$				
x_5	$y[x_5]$					

Newton's interpolating divided difference formula then takes the form

$$P_n(x) = y[x_0] + \sum_{j=1}^{n} y[x_0, x_1, \ldots, x_j](x - x_0)(x - x_1) \cdots (x - x_{j-1}) \tag{4.32}$$

or $\quad P_n(x) = \sum_{j=0}^{n} y[x_0, \ldots, x_j] \prod_{k=0}^{j-1} (x - x_k)$

where by definition

$$\prod_{k=i}^{j} A_k = \begin{cases} A_i A_{i+1} A_{i+2} \cdots A_j & \text{for } i \le j \\ 1 & \text{for } i > j. \end{cases} \tag{4.33}$$

Example 4-5. (Divided difference table.)

Find the Newton interpolating polynomial associated with the data

$$(x_0, y_0) = (1.0, 27/50), \quad (x_1, y_1) = (1.6, 3/4), \quad (x_2, y_2) = (1.9, 22/25), \quad (x_3, y_3) = (2.1, 4/5).$$

Solution: We form the divided difference table

Divided Difference Table

x	$y[]$	$y[,]$	$y[,,]$	$y[,,,]$
1.0	27/50			
		7/20		
1.6	3/4		5/54	
		13/30		-475/297
1.9	22/25		-5/3	
		-2/5		
2.1	4/5			

and obtain the Newton interpolating polynomial

$$P_3(x) = (27/50) + (7/20)(x-1) + (5/54)(x-1)(x-1.6) - (475/297)(x-1)(x-1.6)(x-1.9).$$

This polynomial is illustrated in the figure 4-4.

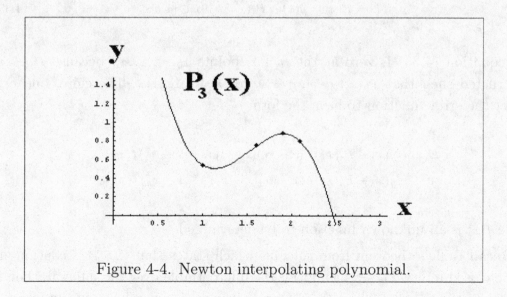

Figure 4-4. Newton interpolating polynomial.

When equal spacing occurs, the relation between forward differences and divided differences is given by

$$y[x_0] = y_0$$

$$y[x_0, x_1] = \frac{y_1 - y_0}{x_1 - x_0} = \frac{\Delta y_0}{h}$$

$$y[x_0, x_1, x_2] = \frac{f[x_1, x_2] - f[x_0, x_1]}{x_2 - x_0} = \frac{\frac{\Delta y_1}{h} - \frac{\Delta y_0}{h}}{2h} = \frac{1}{2!h^2}\Delta^2 y_0$$

and in general

$$y[x_0, x_1, \ldots, x_m] = \frac{1}{m!h^m}\Delta^m y_0 \tag{4.34}$$

Error Term for Polynomial Interpolation

Assume the $(n+1)$ data points (x_i, y_i), for $i = 0, 1, \ldots, n$ represent sampled values from a function $y = y(x)$ for $x \in [a, b]$. We approximate the function $y(x)$ by constructing a nth degree interpolating polynomial $P_n(x)$ and so we can define the error in interpolation as

$$E = E(x) = y(x) - P_n(x) \qquad x \in [a, b] \tag{4.35}$$

The equation (4.35) is zero at the $(n+1)$ points $x_0, x_1, \ldots, x_n$ because $P_n(x)$ is constructed such that $P_n(x_i) = y(x_i) = y_i$ for $i = 0, 1, \ldots, n$. Therefore, one can expect the error function to have the form

$$
\begin{aligned}
E(x) &= y(x) - P_n(x) = (x - x_0)(x - x_1)\cdots(x - x_n)f(x) \\
\text{or} \quad & y(x) - P_n(x) - (x - x_0)(x - x_1)\cdots(x - x_n)f(x) = 0
\end{aligned}
\tag{4.36}
$$

where $f(x)$ is an unknown function to be determined.

Recall Rolle's theorem from calculus which states that if $F(t)$ is continuous on the closed interval $a \leq t \leq b$ and $F(t)$ is differentiable over the interval, then if $t_m < t < t_n \in [a, b]$ and $F(t_m) = F(t_n)$, then there exists a number c in the interval (t_m, t_n) such that $F'(c) = 0$. The situation is illustrated in the figure 4-5

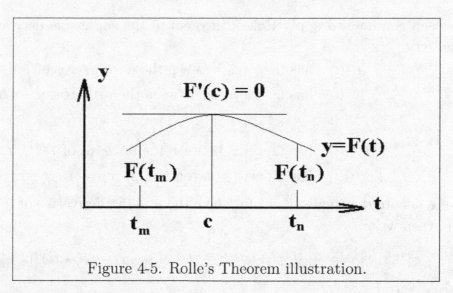

Figure 4-5. Rolle's Theorem illustration.

Consider the function

$$F(t) = y(t) - P_n(t) - (t - x_0)(t - x_1)\cdots(t - x_n)f(x) \qquad a \leq t \leq b \qquad (4.37)$$

which is constructed to have a form similar to equation (4.36). This function is assumed to be continuous and $(n+1)$ times differentiable with respect to the variable t in the closed interval $[a, b]$. In the equation (4.37) the term $P_n(t)$ is a nth degree polynomial and so $\dfrac{d^{n+1}}{dt^{n+1}} P_n(t) = 0$. The equation (4.37) also contains the $(n+1)$st degree polynomial

$$P_{n+1}(t) = (t - x_0)(t - x_1)\cdots(t - x_n) \qquad (4.38)$$

with the property that

$$\frac{d^{n+1}}{dt^{n+1}} P_{n+1}(t) = (n+1)! \qquad (4.39)$$

The function $y(t)$ in equation (4.37) is assumed to be continuous and differentiable $(n+1)$ times. Observe that the function $F(t)$ defined by equation (4.37) is zero for the $(n+2)$ values of t given by

$$x_0, x_1, \ldots, x_n \text{ and } x \qquad (4.40)$$

which are in the interval $[a, b]$. We can then apply the Rolle's theorem to this function and state that

$F'(t)$ has $(n+1)$ zeros between the $(n+2)$ zeros of $F(t)$.

We can continue to apply Rolle's theorem to the successive derivatives of $F'(t)$ and write

$$F''(t) \quad \text{has } (n) \text{ zeros between the } (n+1) \text{ zeros of } F'(t).$$

$$F'''(t) \quad \text{has } (n-1) \text{ zeros between the } (n) \text{ zeros of } F''(t).$$

$$\vdots$$

$$F^{(n)}(t) \quad \text{has } (2) \text{ zeros between the } (3) \text{ zeros of } F^{(n-1)}(t).$$

$$F^{(n+1)}(t) \quad \text{has } (1) \text{ zero between the } (2) \text{ zeros of } F^{(n)}(t).$$

Let ξ denote the single value of t where the $(n+1)$st derivative of F is zero. One can then write

$$F^{(n+1)}(t)\Big|_{t=\xi} = \frac{d^{n+1}}{t^{n+1}}\left(y(t) - P_n(t) - (t - x_0)(t - x_1)\cdots(t - x_n)f(x)\right)\Big|_{t=\xi} = 0$$

which reduces to

$$y^{(n+1)}(\xi(x)) - (n+1)!f(x) = 0 \tag{4.41}$$

The equation (4.41) gives $f(x) = \dfrac{y^{(n+1)}(\xi(x))}{(n+1)!}$ for $a < \xi(x) < b$ and therefore the error term for the polynomial interpolation can be written as

$$E(x) = y(x) - P_n(x) = (x - x_0)(x - x_1)\cdots(x - x_n)\frac{y^{(n+1)}(\xi(x))}{(n+1)!} \qquad a < \xi < b \tag{4.42}$$

or

$$E(x) = y(x) - P_n(x) = \frac{y^{(n+1)}(\xi(x))}{(n+1)!}\pi(x), \qquad \pi(x) = (x - x_0)(x - x_1)\cdots(x - x_n) \tag{4.43}$$

In the special case where the $x_0, x_1, \ldots, x_n$ values are equally spaced, with spacing h, the equation (4.42) can be written in terms of the scaled variable $s = \dfrac{x - x_0}{h}$ as

$$E(x) = y(x) - P_n(x) = \frac{s(s-1)(s-2)\cdots(s-n)}{(n+1)!}h^{n+1}y^{(n+1)}(\xi) \qquad a < \xi, < b$$

or

$$E(x) = \binom{s}{n+1}h^{n+1}y^{(n+1)}(\xi) \qquad \text{where} \quad s = \frac{x - x_0}{h} \tag{4.44}$$

In a similar fashion it can be shown that the error term associated with the Lagrange interpolating polynomial associated with nonuniformly spaced x-values is given by

$$E = E(x) = \frac{y^{(n+1)}(\xi)}{(n+1)!}\pi(x), \qquad \text{where} \quad \pi(x) = (x - x_0)(x - x_1)\cdots(x - x_n). \tag{4.45}$$

Note 1: In using the lozenge diagram to construct a Newton forward interpolating polynomial $P_n(x)$, observe that the next column after the last term used has the form $\binom{s}{n+1}\Delta^{n+1}y_0$. Therefore, if we replace $\Delta^{n+1}y_0$ by $h^{n+1}y^{(n+1)}(\xi)$, we obtain the error term associated with this polynomial approximation.

Note 2: The interpolation polynomial $P_n(x)$ through $(n+1)$ points is unique. Therefore, it doesn't matter what form we select for this polynomial as long as it passes through the same $(n+1)$ points. To get the error term from the lozenge diagram just take the next term after the last in the lozenge diagram pattern (multiplied by the appropriate scale factor) and replace the mth difference term by $h^m y^{(m)}(\xi)$.

Note 3: The above substitution suggests the approximation $y^{(m)}(x) \approx \dfrac{\Delta^m y(x)}{h^m}$ which we shall discover is a valid approximation provided the changes in $\Delta^m y(x)$ remain small.

Interpolation with Piecewise Cubic Splines

Given the $n+1$ data points $(x_0, y_0), (x_1, y_1), \ldots, (x_n, y_n)$ one can plot these points and then connect adjoining points by straight lines. In such a case the resulting sections of straight lines is called a piecewise linear approximation to the function $y(x)$ represented by the data points. This type of approximating function is represented by the straight lines

$$f_k(x) = y_k \left(\frac{x - x_{k+1}}{x_k - x_{k+1}}\right) + y_{k+1}\left(\frac{x - x_k}{x_{k+1} - x_k}\right) \qquad x_k \le x \le x_{k+1} \tag{4.46}$$

for $k = 0, 1, 2, \ldots, n-1$. A representative piecewise linear approximation is illustrated in the figure 4-6.

This type of approximation function has sharp corners where the lines meet and the left-hand derivatives will usually not equal the right-hand derivatives at the points (x_i, y_i) for $i = 1, 2, \ldots, n-1$. These types of functions are not smooth at the points where the lines meet.

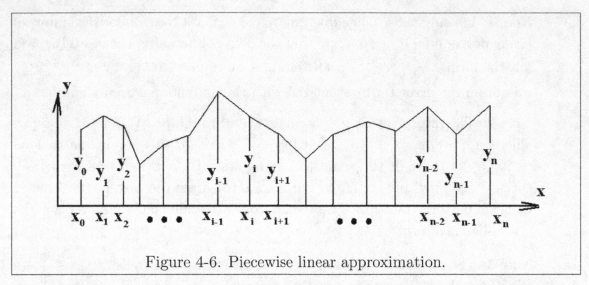

Figure 4-6. Piecewise linear approximation.

The word spline comes from the draftsman drawing device used to help sketch smooth curves connecting a given set of data points. In order to connect the data points by using smooth curves we use cubic polynomial curves which satisfy certain conditions. The resulting cubic polynomials are called cubic splines and are defined as follows.

1. Between each of the points (x_i, y_i) and (x_{i+1}, y_{i+1}) we wish to construct cubic polynomials having the form

$$P_{x_i, x_{i+1}}(x) = a_i(x - x_i)^3 + b_i(x - x_i)^2 + c_i(x - x_i) + d_i \qquad (4.47)$$

for $i = 0, 1, 2, \ldots, n - 1$, where a_i, b_i, c_i, d_i are constants to be determined over each interval (x_i, x_{i+1}). The above cubic polynomial has the first and second derivatives given by

$$
\begin{aligned}
P'_{x_i, x_{i+1}}(x) &= 3a_i(x - x_i)^2 + 2b_i(x - x_i) + c_i \\
P''_{x_i, x_{i+1}}(x) &= 6a_i(x - x_i) + 2b_i
\end{aligned}
\qquad (4.48)
$$

2. The constants a_i, b_i, c_i, d_i in the above cubic polynomial are to be selected such that the following conditions are satisfied.

(i) At $x = x_i$ we want $P_{x_i, x_{i+1}}(x_i)$ to equal y_i. This requires that

$$y_i = P_{x_i, x_{i+1}}(x_i) = d_i \qquad (4.49)$$

(ii) At $x = x_{i+1}$ we want $P_{x_i, x_{i+1}}(x_{i+1})$ to equal y_{i+1}. This requires that

$$y_{i+1} = P_{x_i, x_{i+1}}(x_{i+1}) = a_i h_i^3 + b_i h_i^2 + c_i h_i + d_i \qquad (4.50)$$

where $h_i = x_{i+1} - x_i$ is the length of the ith interval for $i = 0, 1, \ldots, n - 1$.

(iii) We require that there be continuity in the slopes of adjoining curves. That is, we want the slope of the cubic polynomial $P_{x_{i-1},x_i}(x)$ at $x = x_i$ to equal the slope of the polynomial curve $P_{x_i,x_{i+1}}(x)$ at $x = x_i$. Symbolically this is written $P'_{x_{i-1},x_i}(x_i) = P'_{x_i,x_{i+1}}(x_i)$. One can calculate

$$
\begin{aligned}
P'_{x_{i-1},x_i}(x_i) &= 3a_{i-1}(x_i - x_{i-1})^2 + 2b_{i-1}(x_i - x_{i-1}) + c_{i-1} \\
&= 3a_{i-1}h_{i-1}^2 + 2b_{i-1}h_{i-1} + c_{i-1}
\end{aligned} \tag{4.51}
$$

and

$$
P'_{x_i,x_{i+1}}(x_i) = c_i \tag{4.52}
$$

Equating the equations (4.51) and (4.52) gives the slope continuity condition

$$
3a_{i-1}h_{i-1}^2 + 2b_{i-1}h_{i-1} + c_{i-1} = c_i \tag{4.53}
$$

(iv) We let the second derivative values $y_0'', y_1'', y_2'', \ldots, y_{n-2}'', y_{n-1}'', y_n''$ denote unknown quantities and set up a system of equations to solve for the $(n-1)$ values $y_1'', y_2'', \ldots, y_{n-1}''$ which are based upon end point values assigned to either the first derivatives y_0', y_n' or the second derivatives y_0'' and y_n''. This is accomplished as follows. The second derivatives at the end point of the (x_i, x_{i+1}) interval are defined by

$$
\begin{aligned}
p''_{x_i,x_{i+1}}(x_i) &= 2b_i = y_i'' \\
\text{and} \qquad P''_{x_i,x_{i+1}}(x_{i+1}) &= 6a_i(x_{i+1} - x_i) + 2b_i = 6a_ih_i + 2b_i = y_{i+1}''.
\end{aligned} \tag{4.54}
$$

We take the equations (4.54) and solve for the coefficients a_i and b_i in terms of the unknowns y_i'' and y_{i+1}''. This gives the equations

$$
b_i = \frac{y_i''}{2} \qquad \text{and} \qquad a_i = \frac{y_{i+1}'' - y_i''}{6h_i} \tag{4.55}
$$

The equation (4.49) defines the coefficient d_i in terms of the known ordinate value y_i. The coefficient c_i is the remaining coefficient to be expressed in terms of the above unknowns. To do this we substitute the results from equations (4.55) and (4.49) into the equation (4.50) to obtain

$$
y_{i+1} = \left(\frac{y_{i+1}'' - y_i''}{6h_i}\right) h_i^3 + \frac{y_i''}{2}h_i^2 + c_ih_i + y_i \tag{4.56}
$$

and then solve for the coefficient c_i to obtain

$$
c_i = \frac{y_{i+1} - y_i}{h_i} - \frac{y_i''}{2}h_i - \left(\frac{y_{i+1}'' - y_i''}{6}\right) h_i. \tag{4.57}
$$

Now one can use the results from equation (4.57) and substitute the values for c_i and c_{i-1} into the slope continuity equation (4.53). Simplifying the result produces the equation

$$h_{i-1}y_{i-1}'' + 2(h_{i-1} + h_i)y_i'' + h_i y_{i+1}'' = 6\left(\frac{y_{i+1} - y_i}{h_i} - \frac{y_i - y_{i-1}}{h_{i-1}}\right) \tag{4.58}$$

for $i = 1, \ldots n-1$. The equations (4.58) represent $(n-1)$ equations in the $(n-1)$ unknowns $y_1'', \ldots, y_{n-1}''$. provided that one assigns values to the end conditions. Some possible end point conditions are:

(a) The second derivatives are specified at the endpoints. One possible set of assigned values are $y_0'' = 0$ and $y_n'' = 0$. These conditions are referred to as free or natural boundary conditions and the resulting spline is called a natural cubic spline.

(b) The second derivatives are constant near an end point and consequently will take on the values of the second derivative at the nearest point so that $y_0'' = y_1''$ and $y_n'' = y_{n-1}''$.

(c) Assign a linear relation for the second derivative based upon the two nearest end point second derivatives. This gives the values

$$y_0'' = y_1'' - \left(\frac{y_2'' - y_1''}{h_1}\right) h_0$$
$$y_n'' = y_{n-1}'' + \left(\frac{y_{n-1}'' - y_{n-2}''}{h_{n-2}}\right) h_{n-1} \tag{4.59}$$

If we assume a linearity condition for the second derivative, then we have at $x = x_0$ the relation

$$h_1 y_0'' - (h_0 + h_1)y_1'' + h_0 y_2'' = 0 \tag{4.60}$$

and at $x = x_n$ we have

$$h_{n-1}y_{n-2}'' - (h_{n-1} + h_{n-2})y_{n-1}'' + h_{n-2}y_n'' = 0. \tag{4.61}$$

(d) Other types of boundary conditions can be applied. A clamped boundary is said to occur when the first derivatives at the endpoints are specified, for example $P_{x_0,x_1}'(x_0) = y_0'$ and $P_{x_{n-1},x_n}'(x_n) = y_n'$. Another type of boundary condition called a periodic boundary condition occurs when values of the function and derivative are assigned the same values at both the left and right boundaries. These later boundary conditions can be expressed in the form $P_{x_0,x_1}(x_0) = P_{x_{n-1},x_n}(x_n)$ and $P_{x_0,x_1}'(x_0) = P_{x_{n-1},x_n}'(x_n)$.

133

Linearity Condition

The system of equations (4.58), (4.60), and (4.61), associated with a linearity condition for the second derivative at a boundary, can be written in the matrix form $AX = B$ given by

$$
\begin{bmatrix}
h_1 & -\delta_1 & h_0 \\
h_0 & 2\delta_1 & h_1 \\
 & h_1 & 2\delta_2 & h_2 \\
 & & \ddots & \ddots & \ddots \\
 & & & h_{i-1} & 2\delta_i & h_i \\
 & & & & \ddots & \ddots & \ddots \\
 & & & & & h_{n-2} & 2\delta_{n-1} & h_{n-1} \\
 & & & & & h_{n-1} & -\delta_{n-1} & h_{n-2}
\end{bmatrix}
\begin{bmatrix}
y_0'' \\ y_1'' \\ y_2'' \\ \vdots \\ y_{i-1}'' \\ y_i'' \\ y_{i+1}'' \\ \vdots \\ y_{n-2}'' \\ y_{n-1}'' \\ y_n''
\end{bmatrix}
=
\begin{bmatrix}
0 \\ B_1 \\ B_2 \\ \vdots \\ B_{i-1} \\ B_i \\ B_{i+1} \\ \vdots \\ B_{n-2} \\ B_{n-1} \\ 0
\end{bmatrix}
\tag{4.62}
$$

where $\delta_i = h_{i-1}+h_i$ and $B_i = 6\left(\frac{y_{i+1}-y_i}{h_i} - \frac{y_i-y_{i-1}}{h_{i-1}}\right)$ for $i = 1,\ldots,n-1$ and X denotes the column vector of unknowns $\mathrm{col}(y_0'', y_1'', \ldots, y_n'')$. One solves the system of equations (4.62) and constructs the cubic splines

$$
P_{x_i,x_{i+1}}(x) = a_i(x - x_i)^3 + b_i(x - x_i)^2 + c_i(x - x_i) + d_i \qquad x_i \le x \le x_{i+1} \tag{4.63}
$$

for $i = 0,\ldots,n-1$ where

$$
\begin{aligned}
a_i &= \frac{y_{i+1}'' - y_i''}{6h_i} \\
b_i &= \frac{y_i''}{2} \\
c_i &= \frac{y_{i+1} - y_i}{h_i} - \frac{y_i''}{2}h_i - \left(\frac{y_{i+1}'' - y_i''}{6}\right)h_i \\
d_i &= y_i
\end{aligned}
\tag{4.64}
$$

over the respective intervals.

Example 4-6. (Cubic Splines.)

Use cubic splines to construct a smooth curve through the data points

$$
\begin{aligned}
(x_0, y_0) &= (1,1),\ (x_1,y_1) = (2,5),\ (x_2,y_2) = (3,3), \\
(x_3, y_3) &= (4,5),\ (x_4,y_4) = (5,2),\ (x_5,y_5) = (6,3)
\end{aligned}
$$

Solution: Assume a linearity condition with $h = 1 =$ constant, so that the equation (4.62), associated with a linear relation for the second derivative based upon the nearest neighboring second derivatives, becomes

$$\begin{bmatrix} 1 & -2 & 1 & 0 & 0 & 0 \\ 1 & 4 & 1 & 0 & 0 & 0 \\ 0 & 1 & 4 & 1 & 0 & 0 \\ 0 & 0 & 1 & 4 & 1 & 0 \\ 0 & 0 & 0 & 1 & 4 & 1 \\ 0 & 0 & 0 & 1 & -2 & 1 \end{bmatrix} \begin{bmatrix} y_0'' \\ y_1'' \\ y_2'' \\ y_3'' \\ y_4'' \\ y_5'' \end{bmatrix} = \begin{bmatrix} 0 \\ -36 \\ 24 \\ 6 \\ 24 \\ 0 \end{bmatrix}$$

with solution

$$\mathrm{col}\,(y_0'', y_1'', y_2'', y_3'', y_4'', y_5'') = \mathrm{col}\left(\frac{-334}{15}, -6, \frac{154}{15}, \frac{-166}{15}, 4, \frac{286}{15}\right)$$

This gives the cubic splines

$$P_{1,2}(x) = \frac{122}{45}(x-1)^3 - \frac{167}{15}(x-1)^2 + \frac{559}{45}(x-1) + 1$$

$$P_{2,3}(x) = \frac{122}{45}(x-2)^3 - 3(x-2)^2 - \frac{77}{45}(x-2) + 5$$

$$P_{3,4}(x) = -\frac{32}{9}(x-3)^2 + \frac{77}{15}(x-3)^2 + \frac{19}{45}(x-3) + 3$$

$$P_{4,5}(x) = \frac{113}{45}(x-4)^3 - \frac{83}{15}(x-4)^2 + \frac{1}{45}(x-4) + 5$$

$$P_{5,6}(x) = \frac{113}{45}(x-5)^3 + 2(x-5)^2 - \frac{158}{45}(x-5) + 2$$

The figure 4-7 illustrates the cubic spline approximation associated with the above polynomials.

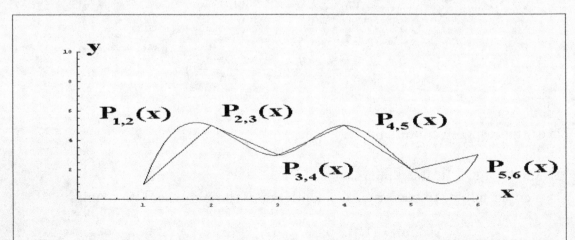

Figure 4-7. Cubic spline approximation with linear relation for y_0'' and y_5''.

For comparison, if we impose the end point conditions $y_0'' = 0$ and $y_5'' = 0$, associated with natural splines, then there results from the equation (4.58) the system of equations

$$\begin{bmatrix} 4 & 1 & 0 & 0 \\ 1 & 4 & 1 & 0 \\ 0 & 1 & 4 & 1 \\ 0 & 0 & 1 & 4 \end{bmatrix} \begin{bmatrix} y_1'' \\ y_2'' \\ y_3'' \\ y_4'' \end{bmatrix} = \begin{bmatrix} -36 \\ 24 \\ 6 \\ 24 \end{bmatrix}$$

with solution

$$y_1'' = \frac{-2520}{209}, \quad y_2'' = \frac{2556}{209}, \quad y_3'' = \frac{-2688}{209}, \quad y_4'' = \frac{1926}{209}.$$

This gives the natural cubic splines

$$P_{1,2}(x) = -\frac{420}{209}(x-1)^3 + \frac{1256}{209}(x-1) + 1$$

$$P_{2,3}(x) = \frac{846}{209}(x-2)^2 - \frac{1260}{209}(x-2)^2 - \frac{4}{209}(x-2) + 5$$

$$P_{3,4}(x) = -\frac{46}{11}(x-3)^3 + \frac{1278}{209}(x-3)^2 + \frac{14}{209}(x-3) + 3$$

$$P_{4,5}(x) = \frac{769}{209}(x-4)^3 - \frac{1344}{209}(x-4)^2 - \frac{52}{209}(x-4) + 5$$

$$P_{5,6}(x) = -\frac{321}{209}(x-5)^3 + \frac{963}{209}(x-5)^2 - \frac{433}{209}(x-5) + 2$$

which are illustrated in the figure 4-8.

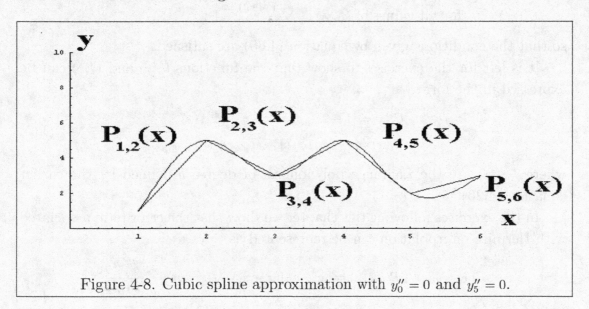

Figure 4-8. Cubic spline approximation with $y_0'' = 0$ and $y_5'' = 0$.

Hermite Interpolation

Whenever one has data $a \le x_0, x_1, \ldots, x_n \le b$ where both the function values y_i and function derivative values y_i' are known at a set of points x_i, $i = 0, 1, 2, \ldots, n$, called nodes, then one can construct a polynomial which matches both the function values and derivative values. This type of polynomial interpolation is called Hermite interpolation or osculating interpolation. The problem is to construct a polynomial $P_{2n+1}(x)$ of degree $2n + 1$ such that

$$P_{2n+1}(x_i) = y_i \quad \text{and} \quad P_{2n+1}'(x_i) = y_i' \tag{4.65}$$

for $i = 0, 1, 2, \ldots, n$.

The Hermite interpolating polynomial of degree $2n + 1$ has the form

$$P_{2n+1}(x) = \sum_{j=0}^{n} [y_i U_i(x) + y_i' V_i(x)] \tag{4.66}$$

with derivative

$$P_{2n+1}'(x) = \sum_{j=0}^{n} [y_i U_i'(x) + y_i' V_i'(x)] \tag{4.67}$$

where $U_i(x)$ and $V_i(x)$ are polynomial functions which satisfy the conditions

$$U_i(x_k) = \begin{cases} 0, & k \ne i \\ 1, & k = i \end{cases} \qquad \begin{aligned} & V_i(x_k) = 0 \quad \text{for all values of } k \\ & V_i'(x_k) = \begin{cases} 0, & k \ne i \\ 1, & k = i \end{cases} \end{aligned} \tag{4.68}$$
$$U_i'(x_k) = 0 \quad \text{for all values of } k$$

so that the conditions given by equation (4.65) are satisfied.

It is left for the exercises to show that the functions $U_i(x)$ and $V_i(x)$ can be expressed in the form

$$\begin{aligned} U_i(x) &= \left[1 - 2L_{n,i}'(x_i)(x - x_i)\right] L_{n,i}^2(x) \\ V_i(x) &= (x - x_i) L_{n,i}^2(x) \end{aligned} \tag{4.69}$$

where $L_{n,i}(x)$ are the Lagrange polynomials of degree n defined by the earlier equation (4.26).

In the exercises following this chapter we show that the error term associated with Hermite interpolation can be represented as

$$E(x) = y(x) - P_{2n+1}(x) = \left[(x - x_0)(x - x_1) \cdots (x - x_n)\right]^2 \frac{y^{(2n+2)}(\xi)}{(2n + 2)!}$$

$$\text{or} \quad E(x) = y(x) - P_{2n+1}(x) = [\pi(x)]^2 \frac{y^{(2n+2)}(\xi)}{(2n + 2)!}, \qquad a < \xi < b. \tag{4.70}$$

Special Functions and Operators

Many special functions and operators are used in numerical methods because their introduction can sometimes simplify a presentation or derivation. In this section we present some selected special functions and operators and illustrate how they can be used to represent polynomial expressions.

Any products of the form

$$A_i A_{i+1} A_{i+2} \cdots A_{i+(n-1)} = \prod_{j=0}^{n-1} A_{i+j}$$

$$\text{or} \qquad A_i A_{i-1} A_{i-2} \cdots A_{i-(n-1)} = \prod_{j=0}^{n-1} A_{i-j} \tag{4.71}$$

are called factorial functions. Here the symbol $\displaystyle\prod_{j=0}^{n-1}$ is used to denote a product of n terms. The following are some examples of factorial functions in terms of the scaled variable $s = \dfrac{x - x_0}{h}$

$$s^{[n]} = \prod_{i=0}^{n-1} (s - i) = s(s-1)(s-2)\cdots(s-(n-1)), \qquad s^{[0]} = 1$$

$$s_{[n]} = \prod_{i=0}^{n-1} (s + i) = s(s+1)(s+2)\cdots(s+(n-1)), \qquad s_{[0]} = 1 \tag{4.72}$$

$$s^{[-n]} = \prod_{i=1}^{n} (s+i)^{-1} = \frac{1}{(s+1)} \frac{1}{(s+2)} \cdots \frac{1}{(s+n)}, \qquad s^{[0]} = 1$$

The binomial coefficient $\dbinom{k}{m}$ is defined

$$\binom{k}{m} = \frac{k!}{m!(k-m)!} = \frac{k(k-1)(k-2)\cdots(k-(m-1))}{m!} = \frac{k^{[m]}}{m!} \tag{4.73}$$

and occurs not only in the binomial expansion

$$(a+b)^n = \sum_{m=0}^{n} \binom{n}{m} a^{n-m} b^m \tag{4.74}$$

but in many other expressions arising in the presentation of numerical methods.

The use of operators can be found in most numerical methods textbooks. These operators help to represent complicated expressions in a simplified compact formalism. The following is a list of selected operators that occur quite frequently.

The forward difference operator Δ is defined

$$\Delta y_k = y_{k+1} - y_k \tag{4.75}$$

with higher order forward differences being defined as differences of lower ordered differences. For example, $\Delta^2 y_k$ and $\Delta^3 y_k$ are calculated as follows

$$\Delta^2 y_k = \Delta(\Delta y_k) = \Delta y_{k+1} - \Delta y_k = y_{k+2} - 2y_{k+1} + y_k$$

$$\text{and} \quad \Delta^3 y_k = \Delta(\Delta^2 y_k) = \Delta^2 y_{k+1} - \Delta^2 y_k = y_{k+3} - 3y_{k+2} + 3y_{k+1} - y_k \tag{4.76}$$

The backward difference operator ∇ is defined

$$\nabla y_k = y_k - y_{k-1} \tag{4.77}$$

with higher ordered differences being defined as differences of the next lowered ordered differences. Similar to the forward difference calculations given by equation (4.76) we calculate the backward differences $\nabla^2 y_k$ and $\nabla^3 y_k$ as follows

$$\nabla^2 y_k = \nabla(\nabla y_k) = \nabla y_k - \nabla y_{k-1} = y_k - 2y_{k-1} + y_{k-2}$$

$$\text{and} \quad \nabla^3 y_k = \nabla(\nabla^2 y_k) = \nabla^2 y_k - \nabla^2 y_{k-1} = y_k - 3y_{k-1} + 3y_{k-2} - y_{k-3}. \tag{4.78}$$

The stepping operator E is defined by

$$Ey_k = y_{k+1} \tag{4.79}$$

with the property $E^m y_k = y_{k+m}$. Some examples of the use of the stepping operator are

$$E^2 y_k = y_{k+2}, \quad E^{-1} y_k = y_{k-1}, \quad E^{-2} y_k = y_{k-2}, \quad E^{1/2} y_k = y_{k+1/2}.$$

One can relate the forward differencing operator Δ and backward differencing operator ∇ and the stepping operator E from the relations

$$\Delta y_k = y_{k+1} - y_k \qquad \nabla y_k = y_k - y_{k-1}$$

$$\Delta y_k = Ey_k - y_k \qquad \nabla y_k = y_k - E^{-1} y_k \tag{4.80}$$

$$\Delta y_k = (E-1)y_k \qquad \nabla y_k = (1 - E^{-1})y_k$$

which implies $\Delta = E - 1$ and $\nabla = 1 - E^{-1}$. We can then use the binomial expansion to write

$$\Delta^n = (E-1)^n = \sum_{j=0}^{n} (-1)^j \binom{n}{j} E^{n-j}$$

so that the nth forward difference can be expressed

$$\Delta^n y_k = \sum_{j=0}^{n} (-1)^j \binom{n}{j} E^{n-j} y_k = \sum_{j=0}^{n} (-1)^j \binom{n}{j} y_{k+n-j}.$$ (4.81)

In a similar fashion it can be verified that

$$\nabla^n y_k = \sum_{j=0}^{n} (-1)^{n+j} \binom{n}{j} E^{j-n} y_k = \sum_{j=0}^{n} (-1)^{n+j} \binom{n}{j} y_{k+j-n}.$$ (4.82)

The central difference operator δ is defined

$$\delta = E^{1/2} - E^{-1/2}$$ (4.83)

with the property

$$\delta y_k = \left(E^{1/2} - E^{-1/2} \right) y_k = y_{k+1/2} - y_{k-1/2}.$$ (4.84)

Higher ordered central differences are defined as differences of next lower ordered differences. For example, $\delta^2 y_k$ and $\delta^3 y_k$ are calculated

$$\begin{aligned}
\delta^2 y_k &= \delta(\delta y_k) = \delta y_{k+1/2} - \delta y_{k-1/2} \\
&= (y_{k+1/2+1/2} - y_{k+1/2-1/2}) - (y_{k+1/2+1/2} - y_{k-1/2-1/2}) \\
&= y_{k+1} - 2y_k + y_{k-1}
\end{aligned}$$

Alternatively one can write

$$\delta^2 = \left(E^{1/2} - E^{-1/2} \right)^2 = E - 2 + E^{-1}$$

so that

$$\delta^2 y_k = (E - 2 + E^{-1}) y_k = y_{k+1} - 2y_k + y_{k-1}.$$

Similarly, one can demonstrate

$$\begin{aligned}
\delta^3 y_k &= \delta(\delta^2 y_k) = (E^{1/2} - E^{-1/2})(y_{k+1} - 2y_k + y_{k-1}) \\
&= y_{k+3/2} - 3y_{k+1/2} + 3y_{k-1/2} - y_{k-3/2}.
\end{aligned}$$ (4.85)

The averaging or mean operator μ is defined

$$\mu = \frac{1}{2} \left(E^{1/2} + E^{-1/2} \right)$$ (4.86)

with the property

$$\mu y_k = \frac{1}{2} \left(y_{k+1/2} + y_{k-1/2} \right). \tag{4.87}$$

One can demonstrate that

$$\mu^2 = \frac{1}{4} \left(E^{1/2} + E^{-1/2} \right)^2 = \frac{1}{4} \left(E + 2 + E^{-1} \right) = 1 + \frac{\delta^2}{4}$$

$$\text{and} \quad \mu\delta = \frac{1}{2} \left(E^{1/2} + E^{-1/2} \right) \left(E^{1/2} - E^{-1/2} \right) = \frac{1}{2}(E - E^{-1}). \tag{4.88}$$

The differential operator D is defined $D = \dfrac{d}{dx}$ with $D^2 = \dfrac{d^2}{dx^2}$, $D^3 = \dfrac{d^3}{dx^3}, \cdots$ denoting operators for higher derivatives. The Taylor series expansion

$$y(x + h) = y(x) + hy'(x) + \frac{h^2}{2!}y''(x) + \cdots + \frac{h^n}{n!}y^{(n)}(x) + \cdots \tag{4.89}$$

can be written in the operator form

$$Ey(x) = \left[1 + hD + \frac{h^2}{2!}D^2 + \cdots + \frac{h^n}{n!}D^n + \cdots \right] y(x) = e^{hD}y(x) \tag{4.90}$$

which implies $E = e^{hD}$ or $D = \dfrac{1}{h}\ln E$.

There are many other relationships that can be derived which interrelate the operators $\Delta, \nabla, \delta, \mu, E$ and D. Some selected relations can be found in the exercises.

The interpolating polynomials of Newton, Gauss, Stirling and Bessel can be written using an operator notation. For example

Newton's forward interpolating polynomial

$$P_n(x) = y_0 + \sum_{m=1}^{n} \frac{s^{[m]}}{m!} \Delta^m y_0 \tag{4.91}$$

is expressed using the forward difference operator Δ.

Newton's backward interpolating polynomial

$$P_n(x) = y_0 + \sum_{m=1}^{n} \frac{s_{[m]}}{m!} \nabla^m y_0 \tag{4.92}$$

is expressed using the backward difference operator ∇.

The **Gauss backward** interpolating polynomial

$$P_{2n}(x) = y_0 + s\delta y_{-1/2} + \sum_{m=1}^{n} \left[\frac{(s+m)^{[2m]}}{(2m)!} \delta^{2m} y_0 + \frac{(s+m)^{[2m+1]}}{(2m+1)!} \delta^{2m+1} y_{-1/2} \right] \tag{4.93}$$

is expressed using the central difference operator δ.

The **Gauss forward** interpolating polynomial

$$P_{2n}(x) = y_0 + s\delta y_{1/2} + \sum_{m=1}^{n} \left[\frac{(s+m-1)^{[2m-1]}}{(2m-1)!} \delta^{2m-1} y_{1/2} + \frac{(s+m-1)^{[2m]}}{(2m)!} \delta^{2m} y_0 \right] \quad (4.94)$$

can also be expressed using the central difference operator δ.

The **Stirling** interpolating polynomial

$$P_{2n}(x) = y_0 + s\mu\delta y_0 + \frac{1}{2}s^2\delta^2 y_0 + \sum_{m=2}^{n} \left[\frac{s f_m(s)}{(2m-1)!} \mu\delta^{2m-1} y_0 + \frac{s^2 f_m(s)}{(2m)!} \delta^{2m} y_0 \right] \quad (4.95)$$

is expressed using the mean operator μ and central difference operator δ where

$$f_m(s) = (s^2 - 1)(s^2 - 4) \cdots (s^2 - (m-1)^2).$$

The **Bessel** interpolating formula

$$P_{2n}(x) = \sum_{m=0}^{n} \left[\frac{(s+m-1)^{[2m]}}{(2m)!} \mu\delta^{2m} y_{1/2} + \frac{(s-1/2)(s+m-1)^{[2m]}}{(2m+1)!} \delta^{2m+1} y_{1/2} \right] \quad (4.96)$$

also is expressed in terms of the operators μ and δ.

In the above formulas $s = \dfrac{x - x_0}{h}$ is a scaled variable. For a derivation of the above formulas see the reference K.S. Kunz.

Rational Functions

A Padé approximation is a method for representing a finite order polynomial

$$f(x) = f_0 + f_1 x + f_2 x^2 + \cdots + f_k x^k \quad (4.97)$$

as a rational function $R_{n,m}(x)$ so that

$$f(x) = R_{n,m}(x) = \frac{P_n(x)}{Q_m(x)} = \frac{p_0 + p_1 x + p_2 x^2 + \cdots p_n x^n}{q_0 + q_1 x + q_2 x^2 + \cdots + q_m x^m} \quad (4.98)$$

where m and n are arbitrary positive integers, k is selected such that $k = n + m$ and q_0 must be different from zero. For simplicity we let $q_0 = 1$. From experience it has been found that the rational functions with polynomials of nearly the same degree or of equal degree generally give better approximations. Therefore, one usually selects either $n = m$ or $n = m + 1$ for the best results.

An arbitrary function $f(x)$ which has a Maclaurin series

$$f(x) = f_0 + f_1 x + f_2 x^2 + \cdots + f_j x^j + \cdots = \sum_{\ell=0}^{\infty} f_\ell x^\ell \qquad (4.99)$$

can be approximated by a rational function $R_{n,m}(x)$ by truncating the Maclaurin series at the kth term, where $k = n + m$. The truncated Maclaurin series produces a polynomial $P_k(x)$ of degree k. The constants $p_0, p_1, \ldots, p_n, q_1, q_2, \ldots, q_m$, in the rational function approximation, represent $n + 1 + m = k + 1$ unknowns that must be determined. To determine these constants require that

$$P_k(x) - R_{n,m}(x) = \frac{\left(\sum_{\ell=0}^{n+m} f_\ell x^\ell \right) \left(\sum_{\ell=0}^{m} q_\ell x^\ell \right) - \sum_{\ell=0}^{n} p_\ell x^\ell}{\sum_{\ell=0}^{m} q_\ell x^\ell} = \mathcal{O}(x^{k+1}) \qquad (4.100)$$

That is, expand the equation

$$(f_0 + f_1 x + \cdots + f_k x^k)(1 + q_1 x + \cdots + q_m x^m) - (p_0 + p_1 x + \cdots + p_n x^n) \qquad (4.101)$$

and set the coefficients of the terms $\{1, x, x^2, \ldots, x^k\}$ all equal to zero. This gives $k + 1$ equations in $k + 1$ unknowns for determining the p and q coefficients. Expanding the equation (4.101) and equating the appropriate coefficients equal to zero produces the following $(n + m + 1)$ equations in the $n + m + 1$ unknowns which determine the coefficients $p_0, p_1, \ldots, p_n, q_1, \ldots, q_m$.

$$
\begin{aligned}
p_0 &= f_0 \\
p_1 &= f_0 q_1 + f_1 \\
p_2 &= f_0 q_2 + f_1 q_1 + f_2 \\
&\vdots \\
p_n &= f_{n-m} q_m + f_{n-m+1} q_{m-1} + \cdots + f_{n-1} q_1 + f_n
\end{aligned}
\qquad (4.102)
$$

and

$$
\begin{aligned}
f_{n-m+1} q_m + f_{n-m+2} q_{m-1} + \cdots + f_{n+1} &= 0 \\
f_{n-m+2} q_m + f_{n-m+3} q_{m-1} + \cdots + f_{n+2} &= 0 \\
&\vdots \\
f_n q_m + f_{n+1} q_{m-1} + \cdots + f_{n+m} &= 0
\end{aligned}
\qquad (4.103)
$$

All the coefficients f_j, $j = 0, 1, \ldots, k$ are assumed known and so in the set of equations (4.103) we first solve for the unknowns $q_1, \ldots, q_m$ and then these values are substituted into the equations (4.102) to produce the values $p_0, p_1, \ldots, p_n$.

Example 4-7. (Rational function)

Approximate the function $f(x) = e^x$ by the rational approximation

$$f(x) = e^x = R_{3,2}(x) = \frac{p_0 + p_1 x + p_2 x^2 + p_3 x^3}{1 + q_1 x + q_2 x^2}$$

Solution We use the Maclaurin series expansion for e^x and truncate it after the x^5 term and set the result equal to $R_{3,2}(x)$. This gives

$$e^x \approx 1 + x + \frac{x^2}{2!} + \frac{x^3}{3!} + \frac{x^4}{4!} + \frac{x^5}{5!} = P_5(x) = R_{3,2}(x).$$

We then require that $P_5(x) - R_{3,2}(x) = \mathcal{O}(x^6)$. This gives

$$\frac{(1 + x + \frac{x^2}{2!} + \frac{x^3}{3!} + \frac{x^4}{4!} + \frac{x^5}{5!})(1 + q_1 x + q_2 x^2) - (p_0 + p_1 x + p_2 x^2 + p_3 x^3)}{1 + q_1 x + q_2 x^2} = \mathcal{O}(x^6).$$

Expanding the numerator terms and equating the coefficients of $\{1, x, x^2, x^3, x^4, x^5\}$ equal to zero gives the six equations

$$p_0 = 1$$
$$p_1 = q_1 + 1$$
$$p_2 = q_2 + q_1 + 1/2$$
$$p_3 = q_2 + \frac{q_1}{2} + 1/6$$
$$\frac{q_2}{2} + \frac{q_1}{6} + 1/24 = 0$$
$$\frac{q_2}{6} + \frac{q_1}{24} + 1/120 = 0$$

Solving the last two equations gives $q_1 = -2/5$ and $q_2 = 1/20$. These values are substituted into the top equations producing the values $p_0 = 1$, $p_1 = 3/5$, $p_2 = 3/20$, $p_3 = 1/60$. This gives the rational approximation

$$e^x \approx R_{3,2}(x) = \frac{1 + \frac{3}{5}x + \frac{3}{20}x^2 + \frac{1}{60}x^3}{1 - \frac{2}{5}x + \frac{1}{20}x^2} = \frac{x^3 + 9x^2 + 36x + 60}{3x^2 - 24x + 60}. \tag{4.104}$$

Sketches of the function e^x and the Padé approximation $R_{3,2}(x)$ are given in the figure 4-9 over the range $0 \le x \le 3$. Observe that the approximation is good in the neighborhood of $x = 0$ where the curves lie on top of one another.

144

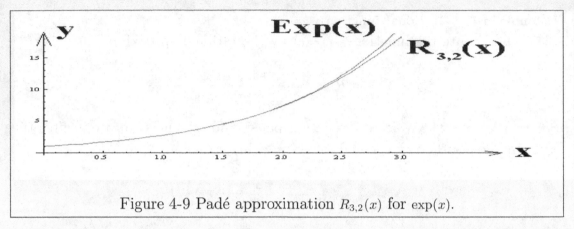

Figure 4-9 Padé approximation $R_{3,2}(x)$ for $\exp(x)$.

Continued Fractions

A continued fraction has the form

$$f = a_0 + \cfrac{b_1}{a_1 + \cfrac{b_2}{a_2 + \cfrac{b_3}{a_3 + \cfrac{b_4}{a_4 + \cfrac{b_5}{a_5 + \cdots}}}}} \tag{4.105}$$

where the number of coefficients $a_0, a_1, \ldots, b_1, b_2, \ldots$ can be finite or infinite. In the continued fraction representation given by equation (4.105) the coefficients can be constants or functions of x. One can use the shorthand notation

$$f = a_0 + \frac{b_1 \mid}{\mid a_1} + \frac{b_2 \mid}{\mid a_2} + \cdots + \frac{b_n \mid}{\mid a_n} + \cdots$$

for representing a continued fractions if the notation saves a lot of writing space.

Continued fractions of the form

$$f = a_0 + \cfrac{1}{a_1 + \cfrac{1}{a_2 + \cfrac{1}{a_3 + \cfrac{1}{a_4 + \cfrac{1}{a_5 + \cdots}}}}} \tag{4.106}$$

are called regular continued fractions and are sometimes represented using the list notation

$$f = [a_0; a_1, a_2, a_3, \ldots]. \tag{4.107}$$

Continued fractions have been around for a long time and can be found in many scientific applications. In 1655 J. Wallis discovered an iterative scheme for evaluating f. His iterative scheme can be written as follows. Set

$$A_{-1} = 1, \quad A_0 = a_0, \quad B_{-1} = 0, \quad B_0 = 1$$

and for $j = 1, 2, 3, \ldots$ define

$$A_j = a_j A_{j-1} + b_j A_{j-2}, \quad B_j = a_j B_{j-1} + b_j B_{j-2} \tag{4.108}$$

or the matrix equivalent

$$\begin{bmatrix} A_j \\ B_j \end{bmatrix} = \begin{bmatrix} A_{j-1} & A_{j-2} \\ B_{j-1} & B_{j-2} \end{bmatrix} \begin{bmatrix} a_j \\ b_j \end{bmatrix}, \tag{4.109}$$

then the ratio $f_n = A_n/B_n$ represents f when the continued fraction is truncated at the b_n/a_n term. If the limit $\lim_{n \to \infty} f_n$ exists, then f is called a convergent continued fraction. Note that when $j = 1$, one obtains

$$A_1 = a_1 A_0 + b_1 A_{-1} = a_1 a_0 + b_1$$

$$B_1 = a_1 B_0 + b_1 B_{-1} = a_1$$

so that

$$f_1 = \frac{A_1}{B_1} = \frac{a_1 a_0 + b_1}{a_1} = a_0 + \frac{b_1}{a_1}.$$

For $j = 2$ we find

$$A_2 = a_2 A_1 + B_2 A_0 = a_2(a_1 a_0 + b_1) + b_2 a_0$$

$$B_2 = a_2 B_1 + b_2 B_0 = a_2 a_1 + b_2$$

so that

$$f_2 = \frac{A_2}{B_2} = \frac{a_0 a_1 a_2 + a_0 b_2 + a_2 b_1}{a_1 a_2 + b_2} = a_0 + \frac{a_2 b_1}{a_1 a_2 + b_2} = a_0 + \cfrac{b_1}{a_1 + \cfrac{b_2}{a_2}}$$

and the general result $f_n = \frac{A_n}{B_n}$ can be proved by induction.

Functions $f(x)$ which have the Maclaurin series expansion

$$f(x) = f_0 + f_1 x + f_2 x^2 + f_3 x^3 + \cdots + f_n x^n + \cdots,$$

where all the coefficients $f_0, f_1, \ldots$ are nonzero, have continued fraction expansions of the form

$$f(x) = a_0 + \cfrac{x}{a_1 + \cfrac{x}{a_2 + \cfrac{x}{a_3 + \cfrac{x}{a_4 + \cfrac{x}{a_5 + \cdots}}}}}$$

where the coefficients a_i, for $i = 0, 1, 2, \ldots$ can be defined in terms of the functions

$$f_{i+1}(x) = \frac{x}{f_i(x) - a_i},$$

where $f_0(x) = f(x)$ and $a_0 = f_0(0)$. One finds that $a_i = f_i(0)$ for $i = 1, 2, 3, \ldots$. The truncated Maclaurin series of order m will then match the truncated continued fraction expansion through the coefficients a_m. One can verify the continued fraction expansion

$$e^x = 1 + \cfrac{x}{1 + \cfrac{x}{-2 + \cfrac{x}{-3 + \cfrac{x}{2 + \cfrac{x}{5 + \cfrac{x}{-2 + \cfrac{x}{-7 + \cfrac{x}{2 + \cfrac{x}{9 + \cdots}}}}}}}}}$$

where the $a_0; a_1, a_2, \ldots$ coefficients have the pattern $[1; 1, -2, -3, 2, 5, -2, -7, 2, 9, \ldots]$ where $a_{2n} = 2(-1)^n$ and $a_{2n+1} = (2n+1)(-1)^n$ for $n = 1, 2, 3, \ldots$.

Example 4-8. (Continued fraction)

Express the square root of a number as a continued fraction.

Solution: Let

$$\sqrt{N} = a_0 + \frac{1}{x} \tag{4.110}$$

so that $x = \dfrac{1}{\sqrt{N} - a_0}$, then one can write

$$x = \frac{1}{\sqrt{N} - a_0} = \frac{\sqrt{N} + a_0}{N - a_0^2} = \frac{a_0 + \frac{1}{x} + a_0}{N - a_0^2} = \frac{1}{N - a_0^2}\left(2a_0 + \frac{1}{x}\right) \tag{4.111}$$

Given a value for N we can select a_0 such that a_0^2 is the closest square to N. For example, if $N = 2$, we select $a_0 = 1$ so that equations (4.110) and (4.111) can be expressed

$$\sqrt{2} = 1 + \frac{1}{x} \qquad \text{and} \qquad x = 2 + \frac{1}{x}. \tag{4.112}$$

From these two equations we can construct the continued fraction

$$\sqrt{2} = 1 + \frac{1}{x} = 1 + \cfrac{1}{2 + \frac{1}{x}} = 1 + \cfrac{1}{2 + \cfrac{1}{2 + \cfrac{1}{2 + \frac{1}{x}}}}.$$

The list notation for square root of 2 then has a periodic pattern for the coefficients and one can write

$$\sqrt{2} = [1; 2, 2, 2, 2, \ldots].$$

Consider the case $N = 41$. If we select $a_0 = 6$, the equations (4.110) and (4.111) can be expressed

$$\sqrt{41} = 6 + \frac{1}{x} \quad \text{and} \quad x = \frac{12}{5} + \frac{1}{5x} \quad \text{or} \quad 5x = 12 + \frac{1}{x}. \tag{4.113}$$

Note that one can write the second equation of (4.113) in the form

$$x = \frac{12 + \frac{1}{x}}{5} = 2 + \frac{2}{5} + \frac{1}{5x} = 2 + \frac{2x+1}{5x} \quad \text{or} \quad x - 2 = \frac{2x+1}{5x}. \tag{4.114}$$

Observe that

$$x = 2 + \frac{1}{\dfrac{5x}{2x+1}} = 2 + \frac{1}{2 + \dfrac{x-2}{2x+1}} = 2 + \frac{1}{2 + \dfrac{1}{5x}}. \tag{4.115}$$

The equation (4.113) allows one to express the result (4.115) in the form

$$x = 2 + \cfrac{1}{2 + \cfrac{1}{12 + \cfrac{1}{x}}} \tag{4.116}$$

This gives the periodic list representation for $\sqrt{41}$ in the form

$$\sqrt{41} = [6; 2, 2, 12, 2, 2, 12, 2, 2, 12, \ldots].$$

∎

As a final note, rational functions can be calculated much more efficiently if represented as a continued fraction. For example, the rational approximation for e^x given by equation (4.104) can be expressed in continued fraction form using long division. One finds

$$
\begin{aligned}
R_{3,2}(x) &= \frac{x^3 + 9x^2 + 36x + 60}{3x^2 - 24x + 60} = \frac{x}{3} + \frac{17}{3} + \frac{152x - 280}{3x^2 - 24x + 60} \\
&= \frac{x}{3} + \frac{17}{3} + \frac{8(19x - 35)}{3x^2 - 24x + 60} = \frac{x}{3} + \frac{17}{3} + \frac{8}{\dfrac{3x^2 - 24x + 60}{19x - 35}} \\
&= \left(\frac{x}{3} + \frac{17}{3}\right) + \frac{8}{\left(\dfrac{3x}{19} - \dfrac{351}{361}\right) + \dfrac{9375/361}{19x - 35}}
\end{aligned}
\tag{4.117}
$$

In general, one will find that the continued fraction representation is the more efficient representation for speed of computation in calculating rational functions.

Parametric Representations

An examination of the data

x	1	3	3	2	4
y	1	2	4	3	3

reveals that associated with the value $x = 3$ there are multiple values for y. Multiple valued functions are usually not used in science and engineering because they can produced results which are wrong.[†]

If there is a reason for the multiplicity, then one way around the difficulty of dealing with multiple-valued functions is to represent the x- and y-values by using parametric representations which are single-valued. This is accomplished by introducing a parameter τ which varies over some domain ($\tau_0 \le \tau \le \tau_n$) and then constructing polynomials to represent the data values in the form $x = x(\tau)$ and $y = y(\tau)$. For example, we append to the given data set equal spaced values of τ ranging from 0 to 4 to obtain the table

τ	0	1	2	3	4
x	1	3	3	2	4
y	1	2	4	3	3

We can now use Lagrange polynomials to construct $x = x(\tau)$ and $y = y(\tau)$ to obtain

$$x = x(\tau) = 1 + \frac{31}{12}\tau - \frac{1}{8}\tau^2 - \frac{7}{12}\tau^3 + \frac{1}{8}\tau^3$$

$$y = y(\tau) = 1 - \frac{17}{6}\tau + \frac{37}{6}\tau^2 - \frac{8}{3}\tau^3 + \frac{1}{3}\tau^3$$

[†] Consider the proof that the weight of a fly equals the weight of an elephant. Let E denote the weight of the elephant and F the weight of the fly, with $C = (E + F)/2$ the average weight.

(1)	$2C = E + F$	$E + F$ is twice the average weight.
(2)	$2C(E - F) = (E + F)(E - F)$	Multiply both sides of equation (1) by $(E - F)$
(3)	$2CE - 2CF = E^2 - F^2$	Use distributive laws and expand equation (2)
(4)	$F^2 - 2CF = E^2 - 2CE$	Add $F^2 - 2CE$ to both sides of equation (3)
(5)	$F^2 - 2CF + C^2 = E^2 - 2CE + C^2$	Add C^2 to both sides of equation (4)
(6)	$(F - C)^2 = (E - C)^2$	Factor equation (5)
(7)	$F - C = E - C$	Take square root of both sides of equation (6)
(8)	$F = E$	Add C to both sides of equation (7)

Which gives the result that the weight of a fly equals the weight of an elephant. You know the result is wrong–but do you know which step in the above argument is wrong?

We can now plot an (x, y) graph of the data using the parametric representations to obtain the figure 4-10.

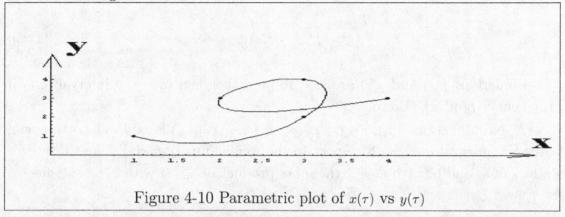

Figure 4-10 Parametric plot of $x(\tau)$ vs $y(\tau)$

One does not have to use equal spacing for the parameter values that are appended to the data. The distances assigned to the τ spacing can affect the shape of the curve. The τ spacing is an additional parameter one can use for data fitting if the shape of the original curve is known.

Having only a limited number of data points one cannot say what the exact shape of the curve through the data points should look like. Alternatively, one can use splines through selected data points (x_i, τ_i) and (y_i, τ_i) to generate piecewise parametric curves to represent the data.

Orthogonal Functions

The inner product of two real functions $f(x)$ and $g(x)$, of a single real variable x, is a weighted integral over a domain $a \le x \le b$, where $f(x)$ and $g(x)$ are defined. The inner product of f, g is defined

$$(f, g) = \int_a^b r(x) f(x) g(x) \, dx \tag{4.118}$$

where $r(x)$ is a weight function which is never negative. The inner product of a function with itself is called a norm squared and is denoted by the notation

$$(f, f) = \| f \|^2 = \int_a^b r(x) f^2(x) \, dx. \tag{4.119}$$

It is assumed that the functions f and g are nonzero, bounded and such that the resulting inner product integral exists. It is also assumed that the norm squared is nonzero unless $f(x)$ is identically zero for all x over the interval $[a, b]$. A function

$f(x)$ is called square integrable on the interval (a, b) with respect to the weight function $r(x)$ whenever

$$\| f \|^2 = \int_a^b r(x) f^2(x)\, dx < +\infty. \tag{4.120}$$

Two functions $f(x)$ and $g(x)$ are said to be orthogonal over the interval (a, b) if their inner product is zero.

A set of functions $\{\phi_1(x), \phi_2(x), \ldots, \phi_n(x), \ldots, \phi_m(x), \ldots\}$ is said to be orthogonal over the interval (a, b) with respect to the weight function $r(x)$ if for all integer values of m and n, with $m \neq n$, the inner product of $\phi_m(x)$ with $\phi_n(x)$ satisfies

$$(\phi_m, \phi_n) = \int_a^b r(x) \phi_m(x) \phi_n(x)\, dx = 0, \qquad m \neq n. \tag{4.121}$$

If the set of functions $\{\phi_n(x)\}$ is an orthogonal set, then one can write

$$(\phi_m, \phi_n) = \| \phi_n \|^2 \delta_{mn} = \begin{cases} 0, & m \neq n \\ \| \phi_n \|^2, & m = n \end{cases}, \quad \text{where} \quad \delta_{mn} = \begin{cases} 0, & m \neq n \\ 1, & m = n \end{cases}$$

where m, n are integers. The symbol δ_{mn} is called the Kronecker delta. The Kronecker delta has the value of unity when the integers m and n are equal and a value of zero when m and n are unequal.

Example 4-9.   (Orthogonal functions)
The set of functions $\{\phi_n(x) = \sin \frac{n\pi x}{L}\}$ for $n = 1, 2, 3, \ldots$ is a set of orthogonal functions over the interval $(0, L)$ with respect to the weight function $r(x) = 1$. One can verify that

$$(\phi_n, \phi_m) = (\sin \frac{n\pi x}{L}, \sin \frac{m\pi x}{L}) = \int_0^L \sin \frac{n\pi x}{L} \sin \frac{m\pi x}{L}\, dx = 0, \quad m \neq n$$

$$(\phi_n, \phi_n) = \| \phi_n \|^2 = \int_0^L \sin^2 \frac{n\pi x}{L} = \frac{L}{2}, \quad \text{for} \quad n = 1, 2, 3 \ldots \qquad \blacksquare$$

Example 4-10.   (Orthogonal functions)
The set of functions $\{\phi_0(x) = 1, \phi_n(x) = \cos \frac{n\pi x}{L}\}$ for $n = 1, 2, 3, \ldots$ is an orthogonal set of functions over the interval $(0, L)$ with respect to the weight function $r(x) = 1$. One can verify that

$$(\phi_0, \phi_n) = \int_0^L \cos\frac{n\pi x}{L}\, dx = 0, \quad \text{for } n = 1, 2, 3, \ldots$$

$$(\phi_n, \phi_m) = \int_0^L \cos\frac{n\pi x}{L} \cos\frac{m\pi x}{L}\, dx = 0, \quad m \neq n$$

$$(\phi_0, \phi_0) = \|\phi_0\|^2 = \int_0^L dx = L$$

$$(\phi_n, \phi_n) = \|\phi_n\|^2 = \int_0^L \cos^2\frac{n\pi x}{L}\, dx = \frac{L}{2}.$$

These results can be summarized by writing

$$(\phi_n, \phi_m) = \left(\cos\frac{n\pi x}{L}, \cos\frac{m\pi x}{L}\right) = \begin{cases} 0, & m \neq n \\ \frac{L}{2}, & m = n \neq 0 \\ L, & m = n = 0 \end{cases}$$

Note 1: The indexing for a set of orthogonal functions $\phi_n(x)$ sometimes begins with $n = 0$ rather than $n = 1$.

Note 2: Whenever one or more of the functions in the set of orthogonal functions has a norm squared different from the other values, it is best to separate that function (or functions) from the others. In the above example, note the norm squared values $\|\phi_0\|^2 = L$ and $\|\phi_n\|^2 = \frac{L}{2}$ for $n = 1, 2, 3, \ldots$.

■

Example 4-11. (Orthogonal functions)

The set of functions $\{1, \cos\frac{n\pi x}{L}, \sin\frac{n\pi x}{L}\}$ is an orthogonal set over the interval $(-L, L)$ with respect to the weight function $r(x) = 1$. One can verify the inner products

$$(1, \sin\frac{n\pi x}{L}) = 0, \qquad (1, \cos\frac{n\pi x}{L}) = 0 \qquad \text{for} \quad n = 1, 2, 3, \ldots$$

$$(\sin\frac{n\pi x}{L}, \sin\frac{m\pi x}{L}) = 0, \qquad (\cos\frac{n\pi x}{L}, \cos\frac{m\pi x}{L}) = 0, \quad \text{for} \quad n \neq m$$

$$(\cos\frac{n\pi x}{L}, \sin\frac{n\pi x}{L}) = 0, \quad \text{for all } n, m \text{ values.}$$

One finds the norm squared values

$$\|1\|^2 = 2L, \quad \|\cos\frac{n\pi x}{L}\|^2 = L, \quad \|\sin\frac{n\pi x}{L}\|^2 = L.$$

■

If a function $f(x)$, is well defined on the interval (a,b), it can be represented as a series of orthogonal functions $\{\phi_n(x)\}$ having the form

$$f(x) = \sum_{n=n_0}^{\infty} c_n \phi_n(x) \qquad a < x < b \tag{4.122}$$

where c_n are constants. This type of representation is called a generalized Fourier series and the coefficients c_n are called the Fourier coefficients. Here n_0 is zero or one, depending upon the indexing of the orthogonal set. For illustrative purposes assume the indexing begins with $n_0 = 1$, then the series given by equation (4.122) can be expanded and written as

$$f(x) = c_1 \phi_1(x) + c_2 \phi_2(x) + \cdots + c_k \phi_k(x) + \cdots \tag{4.123}$$

Take the inner product of both sides of equation (4.123) with the function $\phi_k(x)$ to obtain

$$(f, \phi_k) = c_1(\phi_1, \phi_k) + c_2(\phi_2, \phi_k) + \cdots + c_k(\phi_k, \phi_k) + \cdots \tag{4.124}$$

and note that because the set $\{\phi_n(x)\}$ is an orthogonal set, then all the inner products are zero except for the inner product $(\phi_k, \phi_k) = \| \phi_k \|^2$. Thus, the Fourier coefficients can be found by calculating an inner product divided by a norm squared which can be written

$$c_k = \frac{(f, \phi_k)}{\| \phi_k \|^2} = \frac{\int_a^b r(x) f(x) \phi_k(x)\, dx}{\int_a^b r(x) \phi_k^2(x)\, dx} \quad \text{for} \quad k = 1, 2, 3, \ldots \tag{4.125}$$

Example 4-12. (Fourier series)

Periodic functions $f(x)$ defined on the interval $(-L, L)$ have the Fourier trigonometric series representations

$$f(x) = a_0 + \sum_{n=1}^{\infty} \left(a_n \cos \frac{n\pi x}{L} + b_n \sin \frac{n\pi x}{L} \right)$$

where the Fourier coefficients are given by

$$a_0 = \frac{(f, 1)}{\| 1 \|^2} = \frac{1}{2L} \int_{-L}^{L} f(x)\, dx$$

$$a_n = \frac{(f, \cos \frac{n\pi x}{L})}{\| \cos \frac{n\pi x}{L} \|^2} = \frac{1}{L} \int_{-L}^{L} f(x) \cos \frac{n\pi x}{L}\, dx, \quad n = 1, 2, 3, \ldots$$

$$b_n = \frac{(f, \sin \frac{n\pi x}{L})}{\| \sin \frac{n\pi x}{L} \|^2} = \frac{1}{L} \int_{-L}^{L} f(x) \sin \frac{n\pi x}{L}\, dx, \quad n = 1, 2, 3, \ldots$$

■

The table 4.5 lists some additional orthogonal sets which are used not only in numerical methods and analysis but in many other areas of mathematics.

Name	Orthogonal functions $\{\phi_n(x)\}$	Interval (a,b)	weight function $r(x)$	Norm squared $\|\phi_n\|^2$
Fourier sine	$\{\sin\frac{n\pi x}{L}\}$	$(0,L)$	1	$L/2$
Fourier cosine	$\{1,\cos\frac{n\pi x}{L}\}$	$(0,L)$	1	$L \quad n=0$ $L/2 \quad n\neq0$
Fourier trigonometric	$\{1,\sin\frac{n\pi x}{L},\cos\frac{n\pi x}{L}\}$	$(-L,L)$	1	$\|1\|^2=2L$ $\|\cos\frac{n\pi x}{L}\|^2=L$ $\|\sin\frac{n\pi x}{L}\|^2=L$
Legendre	$\{P_n(x)\}$	$(-1,1)$	1	$\frac{2}{2n+1}$
Chebyshev first kind	$\{T_n(x)\}$	$(-1,1)$	$\frac{1}{\sqrt{1-x^2}}$	$\pi \quad n=0$ $\pi/2 \quad n\neq0$
Chebyshev second kind	$\{U_n(x)\}$	$(-1,1)$	$\sqrt{1-x^2}$	$\pi/2$
Laguerre	$\{L_n(x)\}$	$(0,\infty)$	e^{-x}	1
Hermite	$\{H_n(x)\}$	$(-\infty,\infty)$	e^{-x^2}	$\sqrt{\pi}\,2^n\,n!$

Table 4.5 Orthogonal sets

A more complete list of orthogonal functions and their properties can be found in the Abramowitz and Stegum reference.

Example 4-13. (Generating functions)

A function $G(x,t)$ which can be expanded in a power series, in the variable t, having the form

$$G(x,t) = \sum_{n=0}^{\infty} c_n\phi_n(x)t^n \tag{4.126}$$

where c_n are constants, is called a generating function for the set of functions $\{\phi_n(x)\}$. The trigonometric functions have the following generating functions.

$$\frac{1-t\cos\theta}{1-2t\cos\theta+t^2} = \sum_{n=0}^{\infty}(\cos n\theta)\,t^n, \qquad \frac{t\sin\theta}{1-2t\cos\theta+t^2} = \sum_{n=1}^{\infty}(\sin n\theta)\,t^n, \qquad \theta=\frac{\pi x}{L}$$

Other generating functions are:

$$\frac{1}{\sqrt{1-2xt+t^2}} = \sum_{n=0}^{\infty} P_n(x)\, t^n = 1 + xt + \frac{1}{2}(3x^2 - 1)\, t^2 + \frac{1}{2}(5x^3 - 3x)\, t^3 + \cdots$$

$$\frac{1-xt}{1-2xt+t^2} = \sum_{n=0}^{\infty} T_n(x)\, t^n = 1 + xt + (2x^2 - 1)t^2 + (4x^3 - 3x)t^3 + \cdots$$

$$\frac{1}{1-2xt+t^2} = \sum_{n=0}^{\infty} U_n(x)\, t^n = 1 + 2xt + (4x^2 - 1)t^2 + (8x^3 - 4x)t^3 + \cdots$$

$$\frac{1}{1-t}\exp\left(\frac{-xt}{1-t}\right) = \sum_{n=0}^{\infty} L_n(x)\, t^n = 1 + (1-x)t + \frac{1}{2}(x^2 - 4x + 2)t^2 + \frac{1}{6}(-x^3 + 9x^2 - 18x + 6)t^3 + \cdots$$

$$\exp\left(2xt - t^2\right) = \sum_{n=0}^{\infty} H_n(x)\, \frac{t^n}{n!} = 1 + 2xt + (4x^2 - 2)\frac{t^2}{2!} + (8x^3 - 12x)\frac{t^3}{3!} + \cdots$$

The above generating functions can be used to define the following functions.

The **Legendre functions** $P_n(x)$

$$P_0(x) = 1, \quad P_1(x) = x, \quad P_2(x) = \frac{1}{2}(3x^2 - 1), \quad P_3(x) = \frac{1}{2}(5x^3 - 3x), \quad \ldots$$

The **Chebyshev functions of the first kind** $T_n(x)$

$$T_0(x) = 1, \quad T_1(x) = x, \quad T_2(x) = 2x^2 - 1, \quad T_3(x) = 4x^3 - 3x, \quad \ldots$$

The **Chebyshev functions of the second kind** $U_n(x)$

$$U_0(x) = 1, \quad U_1(x) = 2x, \quad U_2(x) = 4x^2 - 1, \quad U_3(x) = 8x^3 - 4x, \quad \ldots$$

The **Laguerre polynomials** $L_n(x)$

$$L_0(x) = 1, \quad L_1(x) = 1 - x, \quad L_2(x) = \frac{1}{2}(x^2 - 4x + 2), \quad L_3(x) = \frac{1}{6}(-x^3 + 9x^2 - 18x + 6), \quad \ldots$$

The **Hermite polynomials** $H_n(x)$

$$H_0(x) = 1, \quad H_1(x) = 2x, \quad H_2(x) = 4x^2 - 2, \quad H_3(x) = 8x^3 - 12x, \quad \ldots$$

An equation which relates two or more members of a set $\{\phi_n(x)\}$ is called a recurrence relation. The above orthogonal sets have the recurrence relations

Legendre	$(n+1)P_{n+1}(x) = (2n+1)xP_n(x) - nP_{n-1}(x)$
Chebyshev first kind	$T_{n+1}(x) = 2xT_n(x) - T_{n-1}(x)$
Chebyshev second kind	$U_{n+1}(x) = 2xU_n(x) - U_{n-1}(x)$
Laguerre	$(n+1)L_{n+1}(x) = (2n+1-x)L_n(x) - nL_{n-1}(x)$
Hermite	$H_{n+1}(x) = 2xH_n(x) - 2nH_{n-1}(x)$

for $n = 1, 2, 3, \ldots$.

Example 4-14.

Expand the function $f(x) = x^3 + 3x + 1$ as a series involving the Chebyshev polynomials $T_n(x)$.

Solution: If

$$f(x) = c_0 T_0(x) + c_1 T_1(x) + c_2 T_2(x) + c_3 T_3(x) + \cdots$$

where $c_0, c_1, \ldots$ are constants, then one can make use of the orthogonality of the Chebyshev polynomials and write

$$c_n = \frac{(f, T_n)}{\| T_n \|^2} = \frac{\int_{-1}^{1} \frac{1}{\sqrt{1-x^2}} f(x) T_n(x)\, dx}{\int_{-1}^{1} \frac{1}{\sqrt{1-x^2}} T_n^2(x)\, dx}$$

for $n = 0, 1, 2, 3, \ldots$. One finds that

$$c_0 = 1, \qquad c_1 = 15/4, \quad c_2 = 0, \qquad c_3 = 1/4$$

and $c_n = 0$ for integer values of $n > 3$. This gives the representation

$$f(x) = x^3 + 3x + 1 = T_0(x) + \frac{15}{4} T_1(x) + \frac{1}{4} T_3(x).$$

■

Example 4-15.

The generating function $G(x,t) = (1 - 2xt + t^2)^{-1/2} = \sum_{n=0}^{\infty} P_n(x) t^n$ can be differentiated with respect to t to obtain $\frac{\partial G}{\partial t} = (1 - 2xt + t^2)^{-3/2}(x - t) = \sum_{n=0}^{\infty} n P_n(x) t^{n-1}$ which can be written in the alternative form

$$(x - t) \sum_{n=0}^{\infty} P_n(x) t^n = (1 - 2xt + t^2) \sum_{n=0}^{\infty} n P_n(x) t^{n-1}.$$

Expand the above equation and then equate like powers of t^n to obtain the recurrence formula

$$(n + 1) P_{n+1}(x) = (2n + 1) x P_n(x) - n P_{n-1}(x).$$

■

Exercises Chapter 4

▶ **1.** Find a polynomial of degree three which passes through the (x, y) data points

$$(-1, 2), \ (1, -4), \ (3, 6), \ (5, 10).$$

▶ **2.** Given the (x, y) data points

$$(1, -3), \ (2, 4), \ (3, 23), \ (4, 60), \ (5, 121), \ (6, 212)$$

(a) Form a difference table for the given data.

(b) Form a divided difference table for the given data.

(c) Find a polynomial of degree two which passes through the points where $x = 2, 3, 4$.

(d) Estimate the y-value at $x = 2.5$, and state clearly how you obtained your estimate.

▶ **3.** Determine the natural cubic spline that interpolate the data

$$y(0) = 3, \ y(1) = 2, \ y(2) = 3$$

▶ **4.** Determine the clamped cubic spline that interpolates the data

$$y(0) = 2, \ y(1) = 4, \ y(2) = 3 \quad \text{using} \quad y'(0) = 0 \text{ and } y'(2) = 0.$$

▶ **5.** Approximate the function $y(x) = e^{-x}$ by a natural cubic spline over the interval $0 \le x \le 1$ using data points at $x = 0, .25, .5, .75$ and 1.0.

▶ **6.** Show in two different ways that

$$\begin{aligned} f(x) &= 3(x-1)(x-2)(x+1) \\ g(x) &= 3x^3 - 6x^2 - 3x + 6 \\ h(x) &= 3(x-2)^3 + 12(x-2)^2 + 9(x-2) \end{aligned}$$

represent the same polynomial.

▶ **7.** Given the difference table

x	$y = y(x)$	Δy	$\Delta^2 y$	$\Delta^3 y$
1	7			
		——		
2	27		——	
		52		
3	79		50	——
		102		
4	181		——	——
		——		
5	351		——	——
		——		
6	607			

(a) Fill in the missing entries. (9 numbers are missing.)

(b) Find the polynomial of degree two which passes through the points where $x = 2, 3, 4$ by using the Newton-Gregory forward formula.

(c) Find the polynomial of degree one which passes through the points where $x = 2, 3$.

(d) Estimate the value of y at $x = 2.5$ using your previous results.

▶ **8.** Fill in the missing terms in the difference table below.

x	$y = y(x)$	Δy	$\Delta^2 y$	$\Delta^3 y$	$\Delta^4 y$	$\Delta^5 y$
0	-1					
		4				
1	3		4			
		8				
2	11		22	——		
		30			——	
3	41		——	——		——
		——			——	
4	——		——			——
		——		-54		
5	——		-68		——	
		-36				
6	83					

(a) What is the minimum degree polynomial that will pass through all seven of the above data points.

(b) Find the polynomial of degree two which passes the points where $x = 2, 3, 4$.

(c) Find the polynomial of degree two which passes through the points where $x = 4, 5, 6$

(d) Estimate the value of y at $x = 2.5$ by two different methods.

▶ **9.** Let us examine the error associated with polynomial interpolation of a function $f(x)$ over an interval $[a, b]$. We have shown that the error associated with the interpolation of a nth degree polynomial over $n + 1$ nodes $\in [a, b]$ is given by

$$Error = |f(x) - P_n(x)| \leq \frac{|f^{(n+1)}(\xi)|}{(n+1)!}(x - x_0)(x - x_1)(x - x_2) \cdots (x - x_n). \qquad (9a)$$

Is there anyway to control this error? We can't control the $(n + 1)$st derivative or value of ξ or n, the only quantities we can control is the selection of the node points $x_0, x_1, \ldots, x_n$. How do we select the node points to minimize the maximum value of $\prod(x) = (x - x_0)(x - x_1) \cdots (x - x_n)$? The resulting polynomial is called the minimax polynomial for the interval $[a, b]$. The Russian mathematician Chebyshev considered the special case where $[a, b]$ is the interval $[-1, 1]$, he could then answer the above question by employing Chebyshev polynomials. He found that the maximum value of $\prod(x)$ is minimized when the nodes $x_0, x_1, \ldots, x_n$ are selected as the zero's of the $(n + 1)$st Chebyshev polynomial $T_{n+1}(x)$ which are called Chebyshev nodes. These zero's are given by $\xi_{i+1} = \cos \frac{2i + 1}{2(n + 1)} \pi$ for $i = 0, 1, 2, \ldots, n$. To convert the Chebyshev nodes from the interval $[-1, 1]$ to the interval $[a, b]$ the transformation $x = \left(\frac{a+b}{2}\right) + \left(\frac{b-a}{2}\right) \xi$ is employed. The following problem is an example illustrating the difference between polynomial interpolation using equally spaced nodes with that using the Chebyshev nodes.

(a) Approximate the function $y = f(x) = 5e^{-\frac{1}{2}(x-2)^2}$ over the interval $-1 \leq x \leq 5$ by a fourth degree polynomial using the equally spaced nodes

$$\{-1, 3/2, 2, 7/2, 5\} = \{x_0, x_1, x_2, x_3, x_4\}$$

by the Lagrange interpolating polynomial

$$P_4(x) = f(x_0)L_{4,0}(x) + f(x_1)L_{4,1}(x) + f(x_2)L_{4,2}(x) + f(x_3)L_{4,3}(x) + f(x_4)L_{4,4}(x) \qquad (9b)$$

Calculate the Lagrange interpolating polynomials needed for this approximation.

(b) Show the Chebyshev nodes are given by

$$x_0 = 2 + 3 \cos \pi/10,$$
$$x_1 = 2 + 3 \cos 3\pi/10,$$
$$x_2 = 2 + 3 \cos 5\pi/10,$$
$$x_3 = 2 + 3 \cos 7\pi/10,$$
$$x_4 = 2 + 3 \cos 9\pi/10 \qquad (9c)$$

(c) The resulting polynomial approximation using the Chebyshev nodes is denoted $PC_4(x)$ and is obtained from equation (9b) by replacing the equally spaced nodal points by the Chebyshev nodal points. Verify the following figure 4-11. Using the figure 4-11, show where the Chebyshev points are.

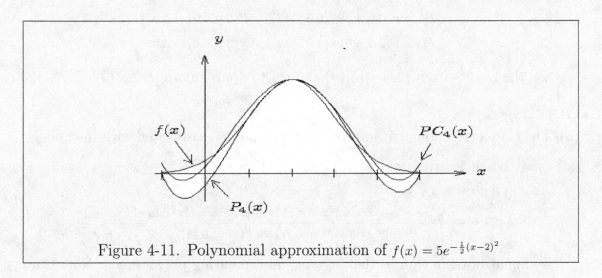

Figure 4-11. Polynomial approximation of $f(x) = 5e^{-\frac{1}{2}(x-2)^2}$

▶ **10.   Two-dimensional interpolation**
Given the set of values

$$F_{11} = F(x_1, y_1), \quad F_{21} = F(x_2, y_1),$$
$$F_{12} = F(x_1, y_2), \quad F_{22} = F(x_2, y_2)$$

(a)   Show

$$F(x_1+\alpha h, y_1+\beta k) = (1-\alpha-\beta)F_{11}+\alpha F_{21}+\beta F_{12}+\mathcal{O}(h^2)$$

(b)   Show that

$$F(x_1 + \alpha h, y_1 + \beta k) = (1-\alpha)(1-\beta)F_{11} + \alpha(1-\beta)F_{21} + \beta(1-\alpha)F_{12} + \alpha\beta F_{22} + \mathcal{O}(h^2)$$

(c)   **(Kriging interpolation)**   Sketch the following situation. Define the distances of the point $(x_3, y_3) = (x_1+\alpha h, y_1+\beta k)$ from the points $(x_1, y_1), (x_1, y_2), (x_2, y_2), (x_2, y_1)$ to be d_1, d_2, d_3, d_4 respectively. For example, let $d_1 = \sqrt{(x_3-x_1)^2 + (y_3-y_1)^2}$ with similar expressions for the other distances. Define the weights

$$W_1 = d_2 d_3 d_4/W, \quad W_2 = d_1 d_3 d_4/W, \quad W_3 = d_1 d_2 d_4/W, \quad W_4 = d_1 d_2 d_3/W$$

160

where $W = d_2d_3d_4 + d_1d_3d_4 + d_1d_2d_4 + d_1d_2d_3$. The interpolated function value for $F(x_3, y_3) = F_{33}$ is then given by the weighted sum.

$$F_{33} = W_1F_{11} + W_2F_{12} + W_3F_{22} + W_4F_{21}.$$

(d) Given the values

$$F_{11} = F(2.0, 4.0) = 6.0, \quad F_{21} = F(3.0, 4.0) = 5.0,$$

$$F_{12} = F(2.0, 5.0) = 4.0, \quad F_{22} = F(3.0, 5.0) = 3.0$$

use the results from parts (a),(b) and (c) to approximate $F(2.3, 4.7)$.

▶ **11.**

(a) The Gauss trigonometric interpolation polynomial associated with the nodes $x_0, x_1, \ldots, x_{2n}$ is given by $y(x) = \sum_{i=0}^{2n} y_i \zeta_i(x)$ where

$$\zeta_i(x) = \frac{\sin \frac{1}{2}(x - x_0) \cdots \sin \frac{1}{2}(x - x_{i-1}) \sin \frac{1}{2}(x - x_{i+1}) \cdots \sin \frac{1}{2}(x - x_{2n})}{\sin \frac{1}{2}(x_i - x_0) \cdots \sin \frac{1}{2}(x_i - x_{i-1}) \sin \frac{1}{2}(x_i - x_{i+1}) \cdots \sin \frac{1}{2}(x_i - x_{2n})}$$

is similar in definition to the Lagrange interpolating polynomials. Note the Gauss interpolating polynomial has the property that $y(x_i) = y_i$.

Compare the Gauss trigonometric interpolating polynomial and Lagrange interpolating polynomial associated with the data points $(1, 1)$, $(2, 4)$, $(3, 6)$, $(4, 3)$, $(5, 5)$.

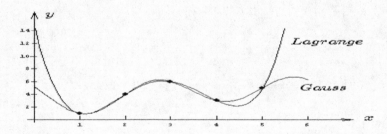

(b) The Hermite periodic interpolation formula is similar to the Gauss formula but uses the functions

$$\zeta_i(x) = \frac{\sin (x - x_0) \cdots \sin (x - x_{i-1}) \sin (x - x_{i+1}) \cdots \sin (x - x_{2n})}{\sin (x_i - x_0) \cdots \sin (x_i - x_{i-1}) \sin (x_i - x_{i+1}) \cdots \sin (x_i - x_{2n})}$$

for $i = 0, 1, \ldots, n$. Find the Hermite periodic interpolation formula associated with the given data.

▶ **12.** Find a rational function approximation $R_{3,2}(x)$ for the function $f(x) = \sin x$ and plot the result over the interval $-2\pi \le x \le 2\pi$.

▶ **13.** For the difference table given, construct the following polynomials.

			Forward Difference Table						
s	x	y	Δy	$\Delta^2 y$	$\Delta^3 y$	$\Delta^4 y$	$\Delta^5 y$	$\Delta^6 y$	$\Delta^7 y$
	1.25	2.0							
			0.5						
	1.50	2.5		0.0					
			0.5		−0.9				
	1.75	3.0		−0.9		0.6			
			−0.4		−0.3		3.0		
	2.0	2.6		−1.2		3.6		−12.0	
			−1.6		3.3		−9.0		27.1
	2.25	1.0		2.1		−5.4		15.1	
			0.5		−2.1		6.1		
	2.50	1.5		0.0		0.7			
			0.5		−1.4				
	2.75	2.0		−1.4					
			−0.9						
	3.00	1.1							

(a) Newton's forward formula using $(1.25, 2.0) = (x_0, y_0)$ as the starting value.

(b) Newton's backward formula using $(3.0, 1.1) = (x_0, y_0)$ as the starting value.

(c) Gauss backward formula using $(2.25, 1.0) = (x_0, y_0)$ as the starting value.

(d) Gauss forward formula using $(2.0, 2.6) = (x_0, y_0)$ as the starting value.

(e) Stirling formula using $(2.0, 2.6) = (x_0, y_0)$ as the starting value.

(f) Bessel formula using $(2.0, 2.6) = (x_0, y_0)$ as the starting value.

(g) Are any of the above polynomials equal? Which ones? Justify your answer.

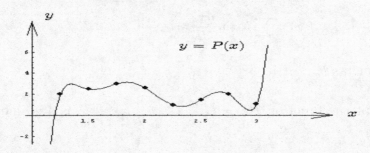

▶ **14.** Find a rational function approximation $R_{4,3}(x)$ for the function $f(x) = \cos x$ and plot the result over the interval $-2\pi \le x \le 2\pi$.

▶ **15.**

Find and plot parametric curves representing the data values and show you get the script letter j.

τ	0	1	2	3	4	5
x	1	2	1.5	1	1	2.5
y	3	4	2.5	1	2	3.5

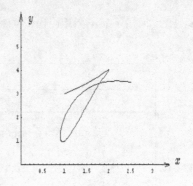

▶ **16.** Represent the function $f(x) = x^3 - 4x^2 + x$ in the following ways.

(a) As a series of Legendre functions $f(x) = c_0 P_0(x) + c_1 P_1(x) + c_2 P_2(x) + \cdots$
Find the constants $c_0, c_1, c_2, \ldots$

(b) As a series of Chebyshev functions of the first kind
$f(x) = c_0 T_0(x) + c_1 T_1(x) + c_2 T_2(x) + \cdots$
Find the constants $c_0, c_1, c_2, \ldots$

(c) As a series of Chebyshev functions of the second kind
$f(x) = c_0 U_0(x) + c_1 U_1(x) + c_2 U_2(x) + \cdots$
Find the constants $c_0, c_1, c_2, \ldots$

(d) As a series of Laguerre functions $f(x) = c_0 L_0(x) + c_1 L_1(x) + c_2 L_2(x) + \cdots$
Find the constants $c_0, c_1, c_2, \ldots$

(e) As a series of Hermite functions $f(x) = c_0 H_0(x) + c_1 H_1(x) + c_2 H_2(x) + \cdots$
Find the constants $c_0, c_1, c_2, \ldots$

▶ **17.** Represent the function $f(x) = x$ as a sine series over the interval $(-L, L)$

$$f(x) = \sum_{n=1}^{\infty} b_n \sin \frac{n\pi x}{L} \qquad -L < x < L$$

Find the coefficients b_n.

▶ **18.** Represent the function $f(x) = \begin{cases} x, & 0 < x < L \\ -x, & -L < x < 0 \end{cases}$ as a cosine series over the interval $(-L, L)$.

$$f(x) = a_0 + \sum_{n=1}^{\infty} a_n \cos \frac{n\pi x}{L} \qquad -L < x < L$$

Find the coefficients a_0, a_n.

▶ **19.** Consider the functions

$$U_i(x) = \left[1 - 2L'_{n,i}(x_i)(x - x_i)\right] L^2_{n,i}(x)$$

$$V_i(x) = (x - x_i)L^2_{n,i}(x)$$

where $L_{n,i}(x)$ are the Lagrange polynomials of degree n.

(a) Verify that $L^2_{n,i}(x)$ is of degree $2n$ and $U_i(x)$ and $V_i(x)$ are of degree $2n+1$.

(b) Verify the derivative formulas

$$U'_i(x) = \left[1 - 2L'_{n,i}(x_i)(x - x_i)\right] 2L'_{n,i}(x)L_{n,i}(x) - 2L'_{n,i}(x_i) \left[L^2_{n,i}(x)\right]^2$$

$$V'_i(x) = (x - x_i)2L_{n,i}(x)L'_{n,i}(x) + [L_{n,i}(x)]^2$$

(c) Verify that

$$U_i(x_k) = \begin{cases} 0, & k \neq i \\ 1, & k = i \end{cases} \qquad\qquad V_i(x_k) = 0 \quad \text{for all values of } k$$

$$U'_i(x_k) = 0 \quad \text{for all values of } k \qquad V'_i(x_k) = \begin{cases} 0, & k \neq i \\ 1, & k = i \end{cases}$$

(c) Verify that the Hermite polynomial

$$P_{2n+1}(x) = \sum_{j=0}^{n} [y_i U_i(x) + y'_i V_i(x)]$$

satisfies the conditions $P_{2n+1}(x_i) = y_i$ and $P'_{2n+1}(x_i) = y'_i$.

▶ **20.** Find the Hermite polynomial $y = h(x)$ which connects the curves $y = g(x) = 10 + x^2$ and $y = f(x) = 5 - \frac{1}{2}x^2$ and passes through the points $(1, 9/2)$ and $(2, 14)$. Note that the Hermite polynomial has the same slopes where the curves touch.

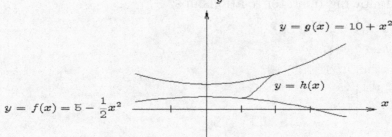

▶ **21.** Hermite interpolation error term.

(a) Show that the Hermite interpolation polynomial error term over the interval $a \leq x_0 < x < x_n \leq b$ is of the form

$$E(x) = y(x) - P_{2n+1}(x) = C[\pi(x)]^2$$

where C is a constant and $\pi(x) = (x - x_0)(x - x_1) \cdots (x - x_n)$.

Hint: Under what conditions must $E(x)$ be identically zero?

(b) Construct the function $F(t) = y(t) - P_{2n+1}(t) - C[\pi(t)]^2$ which is zero for the $n + 2$ values $t = x_0, x_1, \ldots, x_n$ and x. Show also that $F'(t)$ must equal zero for the $n + 1$ values $x_0, \ldots, x_n$ and must have an additional $n + 1$ zeros different from $x_0, \ldots, x_n$ for a total of $2n + 2$ zeros over the interval $[a, b]$.

(c) Show that by a continued application of Rolle's theorem, the $(2n + 2)$ derivative of F, given by $F^{(2n+2)}(t)$, must have at least one zero on the interval $[a, b]$. Call this zero ξ.

(d) Show that the error term can be expressed

$$E(x) = y(x) - P_{2n+1}(x) = \frac{y^{(2n+2)}(\xi)}{(2n+2)!}[\pi(x)]^2$$

▶ **22.** **Operator notation**

Verify the following operator relationships:

(a) $\Delta = hD + \dfrac{h^2 D^2}{2!} + \dfrac{h^3 D^3}{3!} + \cdots = e^{hD} - 1$

(b) $\Delta = \dfrac{\delta^2}{2} + \delta\sqrt{1 + \dfrac{\delta^2}{4}}$

(c) $\nabla = -\dfrac{\delta^2}{2} + \delta\sqrt{1 + \dfrac{\delta^2}{4}}$

▶ **23.** **Operator notation.**

Verify the following operator relationships:

(a) $\delta y_{1/2} = \Delta y_0 = \nabla y_1$

(b) $\Delta^n y_k = \delta^n y_{k+n/2}$

(c) $\mu\delta = \sinh(hD)$

(d) $\nabla = 1 - e^{-hD}$

▶ **24.** **Kriging interpolation.** Kriging interpolation was developed by South African geologist D.G. Krige. It is a weighted prediction method for variables which have a continuous distribution. Kriging interpolation is now widely used in many scientific disciplines. See for example problem 10(c).

(a) One type of Kriging interpolation calculates the distances d_i of an interpolation quantity Q at point P from the known values of Q_i at the points P_i, for $i = 1, \ldots, n$, and then calculates the weights $w_i = \prod_{\substack{j=1 \\ j \neq i}}^{n} d_j$ where the distance d_i is omitted from the products defining w_i for $i = 1, \ldots, n$. The interpolated quantity is then defined by

$$Q = \frac{w_1 Q_1 + w_2 Q_2 + \cdots + w_n Q_n}{w_1 + w_2 + \cdots + w_n}.$$

(a) (One dimensional interpolation) For the data points

$$P_1 = (1,3), \quad P_2 = (2,4), \quad P_3 = (3,2)$$

calculate the Kriging interpolation function

$$y = y(x) = \frac{w_1(x)y_1 + w_2(x)y_2 + w_3(x)y_3}{w_1(x) + w_2(x) + w_3(x)}$$

where the distances are defined by $d_i = |x - x_i|$ for $i = 1, 2, 3$ and the weights are given by

$$w_1 = d_2 d_3, \quad w_2 = d_1 d_3, \quad w_3 = d_1 d_2.$$

Show that Kriging interpolation gives exact values in that it reproduces the data. Using graphics compare the Kriging interpolation with any other kind of interpolation. Comment upon the differences you observe. Which interpolant is better? Define "better".

(b) (Two dimensional interpolation) For the temperatures

$$T_1 = 100, \quad T_2 = 50, \quad T_3 = 75$$

associated with the points $P_1 = (1,3), P_2 = (2,4), P_3 = (3,2)$, define the distances of the point $P = (x,y)$ from the points P_i as $d_i = \sqrt{(x - x_i)^2 + (y - y_i)^2}$, for $i = 1, 2, 3$, and then construct the Kriging interpolant

$$T = T(x,y) = \frac{w_1 T_1 + w_2 T_2 + w_3 T_3}{w_1 + w_2 + w_3}, \qquad w_i = w_i(x,y) \quad i = 1, 2, 3$$

Show that the Kriging interpolation reproduces the data exactly.

(c) (Three dimensional interpolation) For the vectors

$$\vec{V}_1 = \hat{i} + 3\hat{j}, \quad \vec{V}_2 = 2\hat{i} + 4\hat{j}, \quad \vec{V}_3 = 3\hat{i} + 2\hat{j},$$

associated with the points $P_1 = (1, 3, 2)$, $P_2 = (2, 4, 1)$, $P_3 = (3, 2, 3)$, define the distances $d_i = \sqrt{(x - x_i)^2 + (y - y_i)^2 + (z - z_i)^2}$ for $i = 1, 2, 3$ and then construct the Kriging interpolation function

$$\vec{V} = \vec{V}(x, y, z) = \frac{w_1 \vec{V}_1 + w_2 \vec{V}_2 + w_3 \vec{V}_3}{w_1 + w_2 + w_3}, \qquad w_i = w_i(x, y, z), \quad i = 1, 2, 3$$

Show that the Kriging interpolation function reproduces the data exactly.

(d) The Kriging weights are an inverse distance weight in that points closest to the interpolation point has the highest weight. These weights can be defined by

$$w_i = \frac{\frac{1}{d_i^\alpha}}{\sum_{j=1}^{n} \frac{1}{d_j^\alpha}}, \qquad \alpha > 0, \qquad i = 1, 2, \dots, n \tag{24a}$$

Show that the weights in parts (a)(b) and (c) above correspond to $\alpha = 1$ in equation (24a) and $n = 3$.

(e) Try different weights in parts (a)(b) and (c) based upon different values of α in equation (24a). Are the data values reproduced exactly for these new weights? Do different weights affect the shape of the approximating curve or surface?

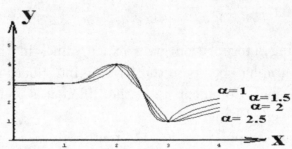

(f) Use the weights from part (d) in part (a) with $\alpha = 1, 1.5, 2, 2.5$ and plot the Kriging interpolation function

$$y = y(x) = \frac{w_1(x) y_1 + w_2(x) y_2 + w_3(x) y_3}{w_1(x) + w_2(x) + w_3(x)}$$

where the distances are defined by $d_i = |x - x_i|$ for $i = 1, 2, 3$.

"You observe that in the life of the intellect there is also a law of inertia. Everything continues to move along its old rectilinear path, and every change, every transition to new and modern ways, meets strong resistance."

Fleix Klein (1849-1925)

Chapter 5
Curve Fitting

Collected data points (x_i, y_i) for $i = 0, 1, 2, \ldots, n$ can be plotted on some type of coordinate paper and then some type of equation can be constructed that will represent the data points in an analytical fashion to produce a graphical (x, y)-representation of the data. For example, in the previous chapter we showed how one can construct polynomial functions which pass through the data points. In this chapter we will not restrict ourselves to polynomials which pass through the data points. We consider instead how to construct equations which represent the data in some "best" way. The numerical procedure for constructing a curve representing the data points is called curve fitting and the equation or equations for representing the curve is called an empirical curve fit. In curve fitting the more experience you have in recognizing certain "basic curve shapes" the easier it becomes to construct an empirical curve representing the data. The figure 5-1 illustrates certain basic curve shapes that the reader should learn to recognize.

Special Graph Paper

The straight line is the easiest curve to recognize and consequently many special types of graph paper are constructed in order to transform special types of curves into a straight line. For example, take the natural logarithm of the exponential curves $y = \alpha e^{mx}$ to obtain

$$\ln y = \ln \alpha + mx. \tag{5.1}$$

Now let $Y = \ln y$ and $b = \ln \alpha$ to obtain

$$Y = mx + b. \tag{5.2}$$

This demonstrates that data of the form of an exponential curve will plot as a straight line on semi-log paper where one axes is logarithmic. Various examples of semi-log paper are illustrated in the figures 5-2, 5-3 and 5-4.

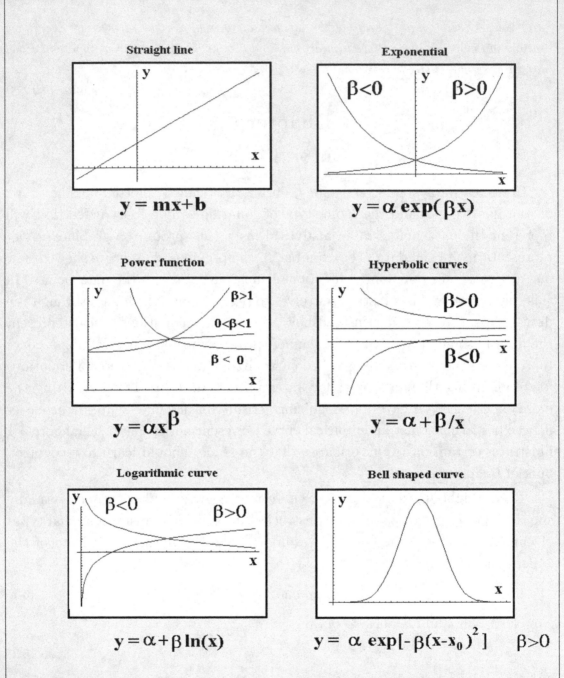

Figure 5-1. Some selected basic curve shapes $\alpha > 0$.

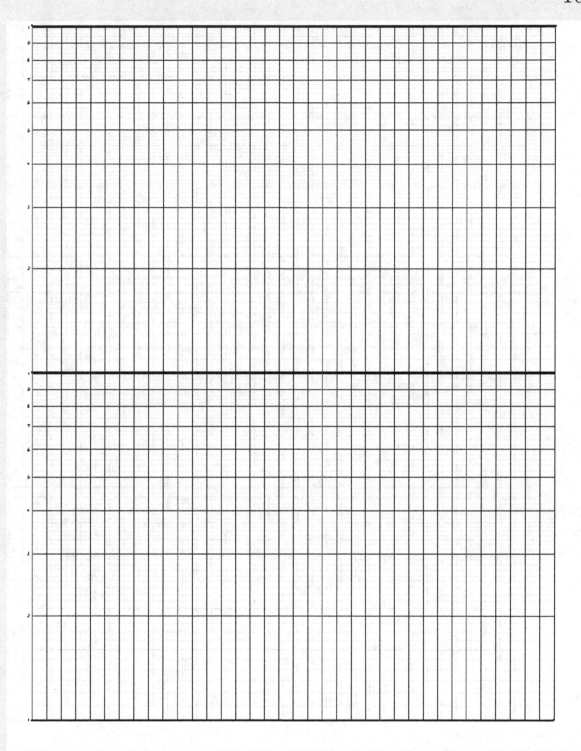

Figure 5-2. 2-Cycle, 34 division Semi-log paper.

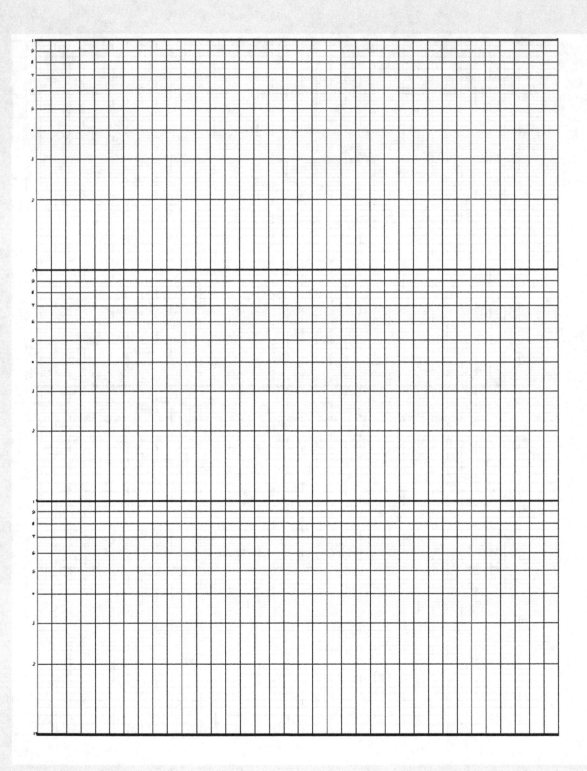

Figure 5-3. 3-Cycle, 34 division Semi-log paper.

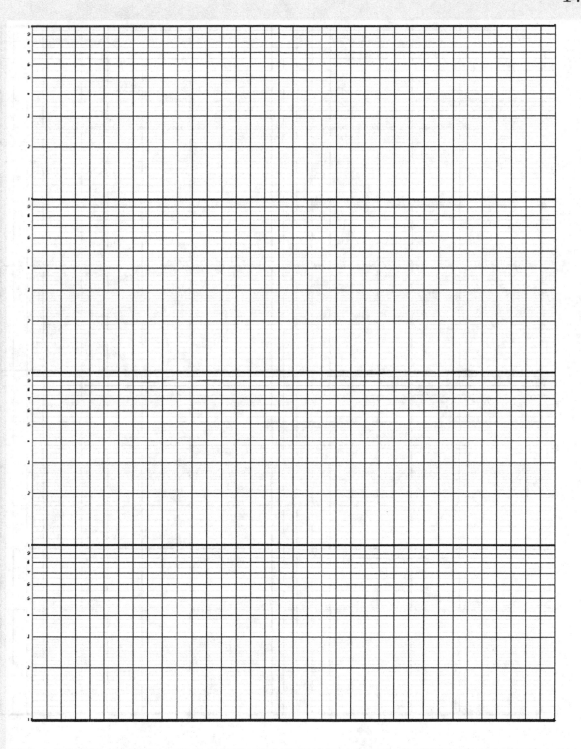

Figure 5-4. 4-Cycle, 34 division Semi-log paper.

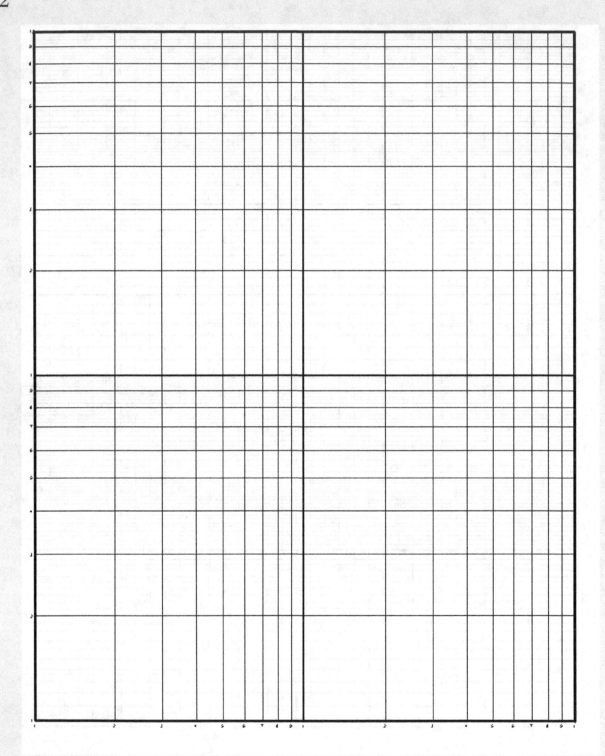

Figure 5-5. 2×2 Cycle log-log paper.

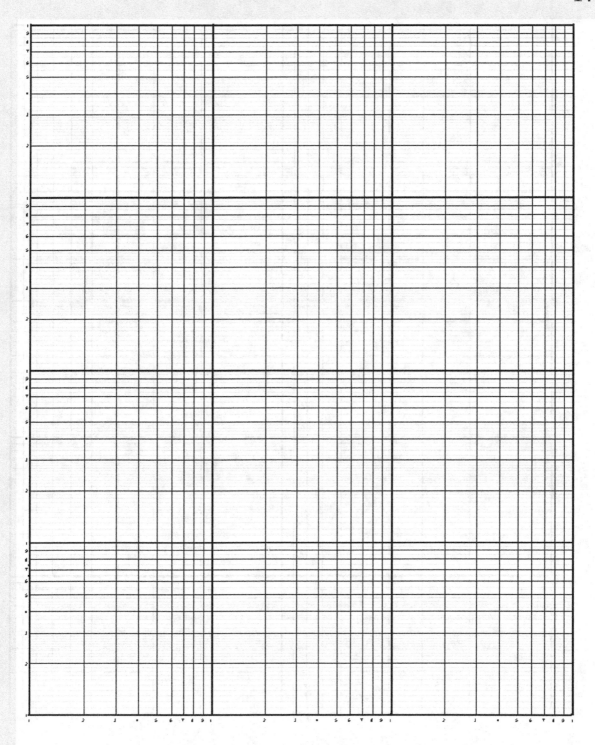

Figure 5-6. 4×3 Cycle log-log paper.

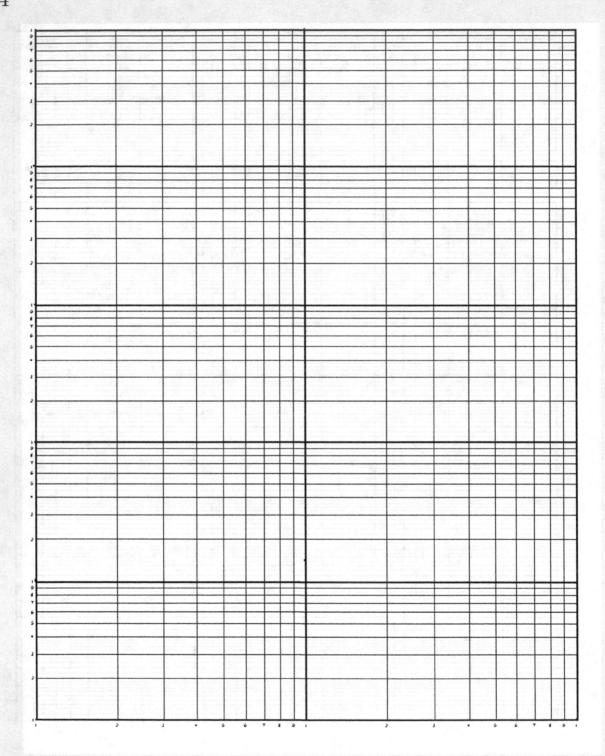

Figure 5-7. 5×2 Cycle log-log paper.

Semi-log paper is classified by the number of logarithmic cycles on the ordinate axis and the number of divisions on the abscissa axis. The figure 5-2 illustrates 2-cycles by 34 divisions, while the figures 5-3 and 5-4 illustrate 3-cycles by 34 divisions and 4-cycles by 34 divisions respectively. Semi-log paper comes in a variety of cycles and divisions.

Example 5-1. (Semi-log paper.)

Verify that the data

x	y
1	0.406006
2	0.0549469
4	0.00100039
6	0.0000184326

can be fitted to an exponential curve $y = \alpha e^{mx}$. Find values for the constants α and m associated with the above data points.

Solution:

We plot the data on semi-log paper and obtain the straight line illustrated. The slope of this line is given by

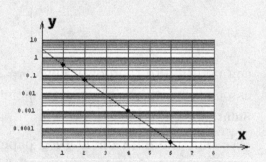

$$m = \frac{\text{Change in } \ln y}{\text{Change in } x}$$

$$m = \frac{\ln(0.0000184326) - \ln(0.406006)}{5} = -2.0.$$

The value of α is given by the y-intercept. This is where the line intersects the y-axis when $x = 0$. We find the values

$$\alpha = 3.0 \quad \text{and} \quad m = -2.0.$$

This gives the exponential curve $y = 3\,e^{-2x}$.

■

Data from the power function $y = \alpha x^m$ can be represented in the form of a straight line by taking the natural logarithm of both sides of the equation to obtain

$$\ln y = \ln \alpha + m \ln x. \tag{5.3}$$

Now let $Y = \ln y$, $X = \ln x$ and $b = \ln \alpha$ so that equation (5.3) takes on the form of the straight line

$$Y = mX + b. \tag{5.4}$$

This shows that the power function becomes a straight line when plotted on log-log paper where both the abscissa and ordinate axes are logarithmic. Some selected examples of log-log paper are given in the figures 5-5, 5-6 and 5-7. Note that log-log paper is classified according to the number of cycles on each axis. The log-log classification is usually denoted by the notation "$n \times m$ cycles". Here n is the number of cycles on the ordinate axis and m is the number of cycles on the abscissa axis. The log-log paper comes in a variety of integer and fractional cycles.

Example 5-2. (Log-log paper.)

   Verify that the data

x	y
1	4.00000
3	0.76980
5	0.35777
7	0.21598

can be fitted to the power curve $y = \alpha x^m$. Find values for the constants α and m associated with the above data points.

Solution:

 We plot the data on Log-log paper and obtain the straight line illustrated. The slope of this line is given by

$$m = \frac{\text{Change in } \ln y}{\text{Change in } \ln x}$$

$$m = \frac{\ln(0.21598) - \ln(4.0)}{\ln(7) - \ln(1)} = -1.5.$$

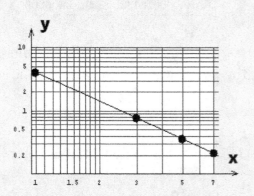

The value of α results when $x = 1$ and so can be read off the log-log graph. We find the values $\alpha = 4.0$ and $m = -1.5$. This gives the power function $y = 4\,x^{-1.5}$.

■

The hyperbolic curve $y = \alpha + \beta/x$ becomes a straight line if one plots $1/x$ vs y since the substitution $X = 1/x$ reduces the hyperbolic curve to the line $y = \alpha + \beta X$. In a similar manner the logarithmic curve $y = \alpha + \beta \ln x$ becomes a straight line if one plots $\ln x$ vs y since the substitution $X = \ln x$ reduces the logarithmic curve to the form $y = \alpha + \beta X$. Many other special curves can be converted to straight lines with the appropriate substitutions. Some additional examples of such conversions are given in the exercises.

The bell shaped curve $y = \alpha \exp[-\beta(x - x_0)^2]$ can be associated with a normal probability distribution and probability graph paper which we discuss in the next section.

Probability Graph Paper

Given a set of numbers it is sometimes convenient to order the numbers and place them within discrete intervals called classes. The number of data points in each class is called the class frequency. A tabulated arrangement of the classes produces a frequency distribution associated with the data set.

For example, assume we are given a data set of n-numbers consisting of m-values, $v_1, v_2, \ldots, v_m$ with $m < n$. Let $\hat{f}_1$ denote the number of times v_1 occurs, $\hat{f}_2$ the number of times v_2 occurs,..., and $\hat{f}_m$ the number of times that v_m occurs, then the numbers $\hat{f}_1, \hat{f}_2, \ldots, \hat{f}_m$ are called the absolute frequencies of occurrence with

$$\hat{f}_1 + \hat{f}_2 + \cdots + \hat{f}_m = n. \tag{5.5}$$

The quantities

$$f_1 = \frac{\hat{f}_1}{n}, \quad f_2 = \frac{\hat{f}_2}{n}, \quad \cdots, \quad f_m = \frac{\hat{f}_m}{n} \tag{5.6}$$

are called the relative frequencies associated with the sample where

$$f_1 + f_2 + \cdots + f_m = 1. \tag{5.7}$$

Define the quantity

$$f(x) = \begin{cases} f_j & \text{when } x = x_j \\ 0 & \text{otherwise} \end{cases} \tag{5.8}$$

as the relative frequency function of the sample. This function is also called the probability distribution function associated with the sample and illustrates how the values in the sample are distributed. Define the quantity

$$F(x) = \sum_{x_j \leq x} f(x_j) \tag{5.9}$$

as the cumulative relative frequency function of the sample or sample distribution function. The figure 5-8 (a)(b) illustrates an example of a frequency function and cumulative frequency function where it is assumed that as the number of data points becomes very large, then the discrete distributions approach the continuous distributions suggested by the figures 5-8 (c)(d).

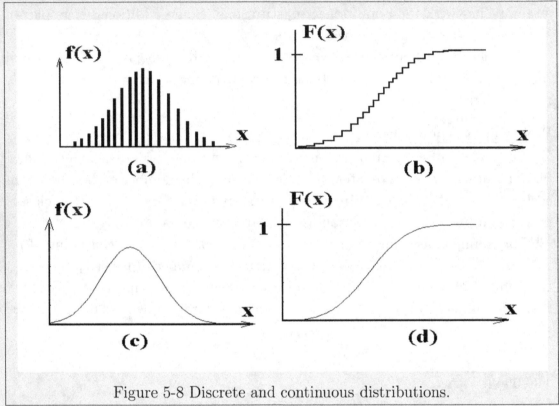

Figure 5-8 Discrete and continuous distributions.

The standard form for the normal distribution with mean 0 and standard deviation of unity is given by

$$f(z) = \frac{1}{\sqrt{2\pi}} e^{-z^2/2} \tag{5.10}$$

with cumulative distribution function

$$F(z) = \frac{1}{\sqrt{2\pi}} \int_{-\infty}^{z} e^{-z^2/2} \, dz \tag{5.11}$$

The change of variable $z = \dfrac{x - \mu}{\sigma}$ gives the cumulative distribution function

$$F\left(\frac{x - \mu}{\sigma}\right) = \phi(x) = \frac{1}{\sigma\sqrt{2\pi}} \int_{-\infty}^{x} e^{-(x-\mu)^2/2\sigma^2} \, dx \tag{5.12}$$

which has the integrand

$$f(x) = \frac{1}{\sigma\sqrt{2\pi}} e^{-(x-\mu)^2/2\sigma^2} \tag{5.13}$$

representing the continuous normal probability density function or distribution (sometimes called the Gaussian distribution), where μ represents the mean and σ represents the standard deviation. The figure 5-9 illustrates the percentage of the total area under this curve between the limits $\mu \pm \sigma$, $\mu \pm 2\sigma$, and $\mu \pm 3\sigma$.

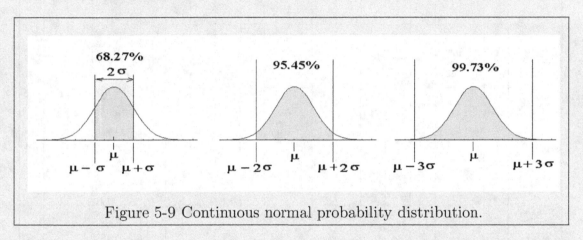

Figure 5-9 Continuous normal probability distribution.

Note that the inverse of the cumulative distribution function gives a straight line, since

$$y = F^{-1}(\phi(x)) = \frac{x}{\sigma} - \frac{\mu}{\sigma}. \tag{5.14}$$

The special graph paper with ordinate axis $y = F^{-1}(\phi(x))$ representing the inverse cumulative frequency and equally spaced abscissa values is called normal probability paper and is illustrated in the figure 5-10. Data which is characterized by a normal probability distribution will have cumulative relative frequency data which plots as a straight line on normal probability paper. The integral of equation (5.13) from $-\infty$ to μ is 0.5 and so an estimate of the mean μ is obtained from the 50.0% value which is the value of x where $y = .50$ on the normal probability graph. An estimate for the standard deviation σ is obtained from the graph in the figure 5-9 where the area between $\mu \pm \sigma$ represents approximately 68.27 percent of the total area under the probability density function. Therefore, one can write as an estimate for the standard deviation σ the relation

$$\sigma \approx \frac{x_{y=84.135\%} - x_{y=15.865\%}}{2}. \tag{5.15}$$

180

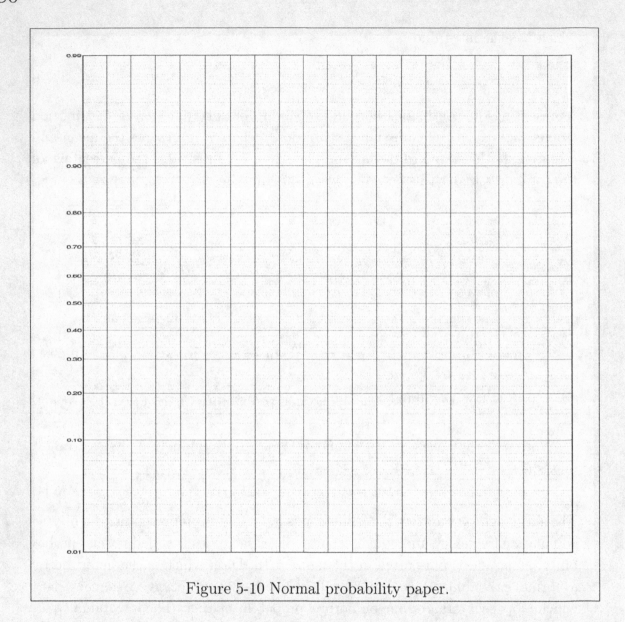

Figure 5-10 Normal probability paper.

Example 5-3.   (Normal distribution test.)
Test cylinders made from concrete are subjected to a tensile force of T lbs/in^2

The following data is the value of T (rounded to the nearest integer) when failure of the test cylinder occurs. Determine if the given data follows a normal distribution and if so estimate the mean and standard deviation.

334	379	359	419	348	402	411	423	360	365	322	327	339
363	330	337	390	395	420	399	370	365	382	322	328	421
320	383	343	335	384	349	375	400	349	410	325	371	332
327	344	396	365	345	351	350	370	347	352	325	374	398
386	358	371	362	370	378	322	404	368	357	324	378	
336	342	342	341	345	352	352	354	355	354	332	380	
415	393	395	385	385	386	371	375	374	376	333	392	
357	358	325	360	364	357	364	370	378	391	331	394	

Solution: We divide the data into class intervals ordered from low to high and form the following table of values.

	x				y
Class Interval	Class Mark	Tally	Frequency	Relative Frequency	Cumulative Relative Frequency
315-321	318	1	1	.01	.01
322-328	325	11111 11111	10	.10	.11
329-335	332	11111 11	7	.07	.18
336-342	339	11111 1	6	.06	.24
343-349	346	11111 111	8	.08	.32
350-356	353	11111 111	8	.08	.40
357-363	360	11111 11111	10	.10	.50
364-370	367	11111 11111	10	.10	.60
371-377	374	11111 111	8	.08	.68
378-384	381	11111 111	8	.08	.76
385-391	388	11111 1	6	.06	.82
392-398	395	11111 1	6	.06	.88
399-405	402	11111	5	.05	.93
406-412	409	11	2	.02	.95
413-419	416	11	2	.02	.97
420-426	423	111	3	.03	1.00

We plot the data points (x, y) on normal probability paper to obtain the straight line illustrated in the figure 5-11.

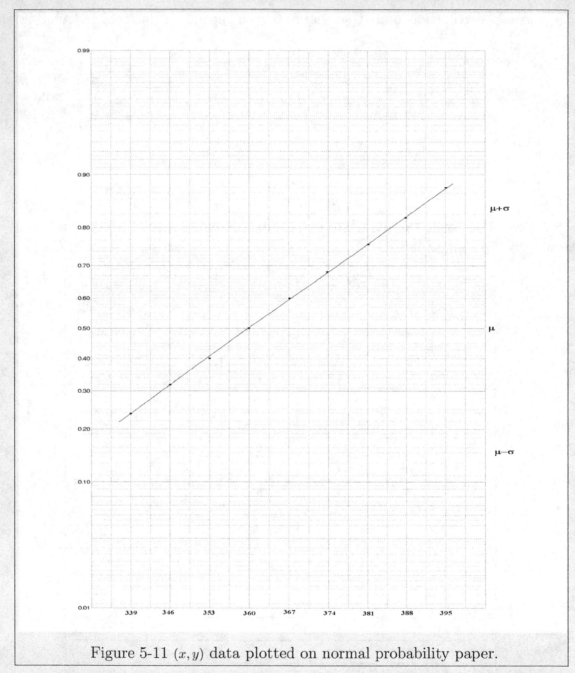

Figure 5-11 (x, y) data plotted on normal probability paper.

The cumulative relative frequency plots as a straight line on normal probability paper and so the data fits a normal distribution. We use linear interpolation to estimate the x-values when $y = 0.84135$ and $y = 0.15865$ and find

$$x_{y=0.84135} = 390.491 \qquad x_{y=0.15865} = 329.865$$

These values produce the approximation

$$\sigma \approx \frac{390.491 - 329.865}{2} = 30.313.$$

The estimate for the mean μ is the value of x corresponding to $y = 0.50$. The mean value for the normal distribution is found to be $\mu = 360$.

■

Least Squares

Many times the data obtained from an experiment has errors from various sources and so the data points do not plot exactly along a straight line. Suppose you are given a set of data points

$$(x_1, y_1), (x_2, y_2), (x_3, y_3), \ldots, (x_i, y_i), \ldots, (x_{n-1}, y_{n-1}), (x_n, y_n)$$

and you plot these points on ordinary graph paper to obtain a figure such as illustrated in the figure 5-12.

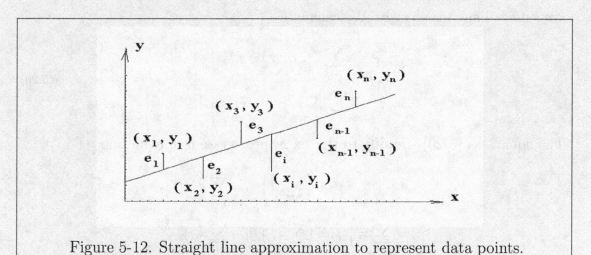

Figure 5-12. Straight line approximation to represent data points.

We assume that the data points can be normally distributed about some straight line and that the errors occur in the y-variable. We want to approximate the given data by using a straight line representation $y = \beta_0 + \beta_1 x$ where β_0 and β_1 are constants. What would be the "best" straight line to represent the given data points? There are many ways to define "best". Define the error e_i associated with the ith data point (x_i, y_i) as

$$e_i = (y \text{ of line at } x_i) - (y \text{ data value at } x_i)$$
$$e_i = \beta_0 + \beta_1 x_i - y_i$$

(5.16)

That is, we define errors

$$
\begin{aligned}
e_1 &= \beta_0 + \beta_1 x_1 - y_1 \\
e_2 &= \beta_0 + \beta_1 x_2 - y_2 \\
e_3 &= \beta_0 + \beta_1 x_3 - y_3 \\
&\quad\vdots \\
e_n &= \beta_0 + \beta_1 x_n - y_n.
\end{aligned}
\tag{5.17}
$$

We can now define "best" straight line as the line $y = \beta_0 + \beta_1 x$ with constants β_0 and β_1 that minimize the sum of squares error E, where

$$
E = E(\beta_0, \beta_1) = \sum_{i=1}^{n} e_i^2 = \sum_{i=1}^{n} (\beta_0 + \beta_1 x_i - y_i)^2.
\tag{5.18}
$$

We know from calculus that the minimum value for E occurs when the conditions

$$
\frac{\partial E}{\partial \beta_0} = 0 \quad \text{and} \quad \frac{\partial E}{\partial \beta_1} = 0
\tag{5.19}
$$

are satisfied simultaneously. We calculate

$$
\begin{aligned}
\frac{\partial E}{\partial \beta_0} &= 2 \sum_{i=1}^{n} (\beta_0 + \beta_1 x_i - y_i)(1) = 0 \\
\frac{\partial E}{\partial \beta_1} &= 2 \sum_{i=1}^{n} (\beta_0 + \beta_1 x_i - y_i)(x_i) = 0.
\end{aligned}
\tag{5.20}
$$

The equations (5.20) simplify to the 2×2 linear system of equations

$$
\begin{aligned}
n\beta_0 + \left(\sum_{i=1}^{n} x_i \right) \beta_1 &= \sum_{i=1}^{n} y_i \\
\left(\sum_{i=1}^{n} x_i \right) \beta_0 + \left(\sum_{i=1}^{n} x_i^2 \right) \beta_1 &= \sum_{i=1}^{n} x_i y_i
\end{aligned}
\tag{5.21}
$$

which can then be solved for the coefficients β_0 and β_1.

Alternatively, we can set all of the equations (5.17) equal to zero, then they can be written in the matrix form

$$
A\overline{\beta} = \overline{\mathbf{y}}
$$

$$
\begin{bmatrix} 1 & x_1 \\ 1 & x_2 \\ 1 & x_3 \\ \vdots & \vdots \\ 1 & x_n \end{bmatrix}
\begin{bmatrix} \beta_0 \\ \beta_1 \end{bmatrix}
=
\begin{bmatrix} y_1 \\ y_2 \\ y_3 \\ \vdots \\ y_n \end{bmatrix}.
\tag{5.22}
$$

By doing this we represent the data as an overdetermined system of equations. That is, we have more equations than there are unknowns. We desired to select the unknowns β_0, β_1 to minimize the sum of squares error associated with the overdetermined system of equations. Observe that if we left multiply both sides of equation (5.22) by the transpose matrix A^T we obtain $A^T A \bar{\beta} = A^T \bar{y}$ or

$$\begin{bmatrix} 1 & 1 & 1 & \dots & 1 \\ x_1 & x_2 & x_3 & \dots & x_n \end{bmatrix} \begin{bmatrix} 1 & x_1 \\ 1 & x_2 \\ 1 & x_3 \\ \vdots & \vdots \\ 1 & x_n \end{bmatrix} \begin{bmatrix} \beta_0 \\ \beta_1 \end{bmatrix} = \begin{bmatrix} 1 & 1 & 1 & \dots & 1 \\ x_1 & x_2 & x_3 & \dots & x_n \end{bmatrix} \begin{bmatrix} y_1 \\ y_2 \\ y_3 \\ \vdots \\ y_n \end{bmatrix}$$

which simplifies to

$$\begin{bmatrix} n & \sum_{i=1}^{n} x_i \\ \sum_{i=1}^{n} x_i & \sum_{i=1}^{n} x_i^2 \end{bmatrix} \begin{bmatrix} \beta_0 \\ \beta_1 \end{bmatrix} = \begin{bmatrix} \sum_{i=1}^{n} y_i \\ \sum_{i=1}^{n} x_i y_i \end{bmatrix} \tag{5.23}$$

which is the matrix form of the equations (5.21).

Example 5-4.   (Least Squares Fit)

Find the best straight line fit to the data points

$$(1,3), \ (2,1), \ (3,5), \ (4,4), \ (5,4), \ (6,7)$$

where "best" is defined in the least squares sense.

Solution: We write the equation of the line as $\beta_0 + \beta_1 x = y$ and substitute the given data values to obtain the overdetermined system of equations

$$A\bar{\beta} = \bar{y}$$

$$\begin{bmatrix} 1 & 1 \\ 1 & 2 \\ 1 & 3 \\ 1 & 4 \\ 1 & 5 \\ 1 & 6 \end{bmatrix} \begin{bmatrix} \beta_0 \\ \beta_1 \end{bmatrix} = \begin{bmatrix} 3 \\ 1 \\ 5 \\ 4 \\ 4 \\ 7 \end{bmatrix} \tag{5.24}$$

Left multiply both the right and left-hand sides of equation (5.24) by A^T to obtain the system of equations

$$\begin{bmatrix} 6 & 21 \\ 21 & 91 \end{bmatrix} \begin{bmatrix} \beta_0 \\ \beta_1 \end{bmatrix} = \begin{bmatrix} 24 \\ 98 \end{bmatrix}$$

We use the methods from chapter 3 and find the solution $\beta_0 = 6/5 = 1.2$ and $\beta_1 = 4/5 = 0.8$. This gives the least squares straight line $y = 1.2 + 0.8x$. The straight line along with the data points are illustrated in the figure 5-13.

186

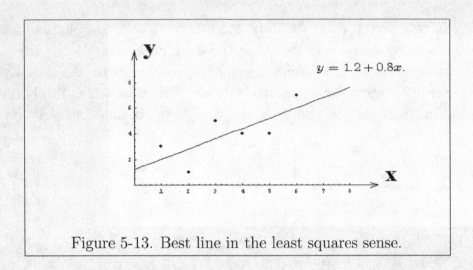

Figure 5-13. Best line in the least squares sense.

Linear Regression

Given a set of (x, y) data points one can assume a curve fit function of the form

$$y = \beta_0 f_0(x) + \beta_1 f_1(x) + \beta_2 f_2(x) + \cdots + \beta_k f_k(x) \qquad (5.25)$$

where $\beta_0, \beta_1, \ldots, \beta_k$ are unknown coefficients and $f_0(x), f_1(x), f_2(x), \ldots, f_k(x)$ represent linearly independent functions, called the basis of the representation. We desire to select the β coefficients such that the sum of squares error

$$E = E(\beta_0, \beta_1, \ldots, \beta_k) = \sum_{i=1}^{n} \left[\beta_0 f_0(x_i) + \beta_1 f_1(x_i) + \beta_2 f_2(x_i) + \cdots + \beta_k f_k(x_i) - y_i \right]^2 \quad (5.26)$$

is a minimum. This requires that we solve the set of simultaneous least square equations

$$\frac{\partial E}{\partial \beta_0} = 0, \quad \frac{\partial E}{\partial \beta_1} = 0, \quad \cdots, \quad \frac{\partial E}{\partial \beta_k} = 0. \qquad (5.27)$$

Another way to obtain the system of equations (5.27) is to first represent the data in the matrix form

$$A\bar{\beta} = \bar{\mathbf{y}}$$

$$\begin{bmatrix} f_0(x_1) & f_1(x_1) & f_2(x_1) & \cdots & f_k(x_1) \\ f_0(x_2) & f_1(x_2) & f_2(x_2) & \cdots & f_k(x_2) \\ \vdots & \vdots & \vdots & \ddots & \vdots \\ f_0(x_n) & f_1(x_n) & f_2(x_n) & \cdots & f_k(x_n) \end{bmatrix} \begin{bmatrix} \beta_0 \\ \beta_1 \\ \beta_2 \\ \vdots \\ \beta_k \end{bmatrix} = \begin{bmatrix} y_1 \\ y_2 \\ \vdots \\ y_n \end{bmatrix} \qquad (5.28)$$

Both sides of the equation (5.28) can be left multiplied by the transpose matrix A^T and the resulting system can be solved for the unknown coefficients. In matrix notation one can write

$$A\bar{\beta} = \bar{\mathbf{y}}$$
$$A^T A \bar{\beta} = A^T \bar{\mathbf{y}} \tag{5.29}$$
$$\bar{\beta} = (A^T A)^{-1} A^T \bar{\mathbf{y}}.$$

The equation (5.29) gives the coefficients which minimizes the sum of square error. In practice, if the number of equations is large, the inverse $(A^T A)^{-1}$ is not calculated. Instead the solution methods from chapter 3 are employed to solve for the β coefficients.

Example 5-5. **(Least Squares Fit)**

Find the best fit parabola to the data points

$$(1,1), \ (2,4), \ (3,9), \ (4,10)$$

where "best" means minimization of the sum of squares error.

Solution: Let $y = \beta_0 + \beta_1 x + \beta_2 x^2$ denote the parabola and substitute the given data points into this equation to obtain the system of equations

$$A\bar{\beta} = \bar{\mathbf{y}}$$
$$\begin{bmatrix} 1 & 1 & 1 \\ 1 & 2 & 4 \\ 1 & 3 & 9 \\ 1 & 4 & 16 \end{bmatrix} \begin{bmatrix} \beta_0 \\ \beta_1 \\ \beta_2 \end{bmatrix} = \begin{bmatrix} 1 \\ 4 \\ 9 \\ 10 \end{bmatrix}$$

where β_0, β_1 and β_2 are to be determined. Left multiply both sides of this equation by A^T to obtain the least squares system of equations

$$\begin{bmatrix} 4 & 10 & 30 \\ 10 & 30 & 100 \\ 30 & 100 & 354 \end{bmatrix} \begin{bmatrix} \beta_0 \\ \beta_1 \\ \beta_2 \end{bmatrix} = \begin{bmatrix} 24 \\ 76 \\ 258 \end{bmatrix}$$

We use the methods of chapter 3 to solve this system of equations and find the values $\beta_0 = -4.5$, $\beta_1 = 5.7$ and $\beta_2 = -0.5$. Therefore, the least squares parabola is represented $y = -4.5 + 5.7\,x - 0.5\,x^2$. A plot of both the data points and the least squares parabola is given in the figure 5-14.

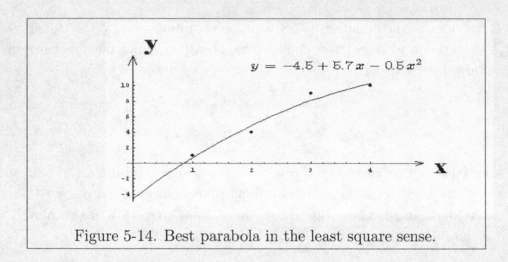

Figure 5-14. Best parabola in the least square sense.

Example 5-6. (Least Square Fit)

Find the best cubic to fit the data points $(1,1)$, $(2,4)$, $(3,9)$, $(4,10)$, $(5,11)$ where "best" means minimization of the sum of squares error.

Solution: Let $y = \beta_0 + \beta_1 x + \beta_2 x^2 + \beta_3 x^3$ denote the cubic curve which is to represent the given data points. We substitute the given data points into this equation to obtain the system of equations

$$A\overline{\beta} = \overline{\mathbf{y}}$$

$$\begin{bmatrix} 1 & 1 & 1 & 1 \\ 1 & 2 & 4 & 8 \\ 1 & 3 & 9 & 27 \\ 1 & 4 & 16 & 64 \\ 1 & 5 & 25 & 125 \end{bmatrix} \begin{bmatrix} \beta_0 \\ \beta_1 \\ \beta_2 \\ \beta_3 \end{bmatrix} = \begin{bmatrix} 1 \\ 4 \\ 9 \\ 10 \\ 11 \end{bmatrix}.$$

Left multiply both sides of this equation by A^T to obtain the least squares system of equations

$$\begin{bmatrix} 5 & 15 & 55 & 225 \\ 15 & 55 & 225 & 979 \\ 55 & 225 & 979 & 4425 \\ 22 & 979 & 4425 & 20515 \end{bmatrix} \begin{bmatrix} \beta_0 \\ \beta_1 \\ \beta_2 \\ \beta_3 \end{bmatrix} = \begin{bmatrix} 35 \\ 131 \\ 533 \\ 2291 \end{bmatrix}$$

We use the methods of chapter 3 and solve this system of equations and find $\beta_0 = -2$, $\beta_1 = 2.09524$, $\beta_2 = 0.928571$ and $\beta_3 = -0.166667$. This gives the least squares cubic equation

$$y = -2 + 2.09524\,x + 0.928571\,x^2 - 0.166667\,x^3.$$

A graph illustrating both the data points and the least square cubic equation is given in the figure 5-15.

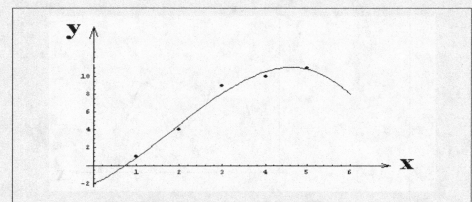

Figure 5-15. Best cubic in the least square sense is given by
$$y = -2 + 2.09524\,x + 0.928571\,x^2 - 0.166667\,x^3.$$

∎

Example 5-7. (Least Squares Fit)

Find the best fit (in the least squares sense) of the trigonometric representation

$$y = \beta_0 + \beta_1 \sin x + \beta_2 \sin 2x$$

to the data points $(1, 2.0)$, $(2, 5.4)$, $(3, 3.7)$, $(4, 0.26)$, $(6, 3.8)$.

Solution: Substitute the data points into the given trigonometric expression to obtain the system of equations

$$A\overline{\beta} = \overline{y}$$

$$\begin{bmatrix} 1 & \sin 1 & \sin 2 \\ 1 & \sin 2 & \sin 4 \\ 1 & \sin 3 & \sin 6 \\ 1 & \sin 4 & \sin 8 \\ 1 & \sin 6 & \sin 12 \end{bmatrix} \begin{bmatrix} \beta_0 \\ \beta_1 \\ \beta_2 \end{bmatrix} = \begin{bmatrix} 2.0 \\ 5.4 \\ 3.7 \\ 0.26 \\ 3.8 \end{bmatrix}.$$

Left multiply both sides of this equation by A^T to obtain the least squares system of equations

$$\begin{bmatrix} 5. & 0.85567 & 0.25865 \\ 0.85567 & 2.20563 & -0.561264 \\ 0.325865 & -0.561264 & 2.74439 \end{bmatrix} \begin{bmatrix} \beta_0 \\ \beta_1 \\ \beta_2 \end{bmatrix} = \begin{bmatrix} 15.16 \\ 5.85674 \\ -5.08372 \end{bmatrix}$$

We use one of the methods from chapter 3 to solve this system of equations and determine the coefficients $\beta_0 = 2.99459$, $\beta_1 = 0.982903$, and $\beta_2 = -2.00696$. This gives the least square fit $y = 2.99459 + 0.982903 \sin x - 2.00696 \sin 2x$. A plot of both the data points and the least square trigonometric curve is given in the figure 5-16.

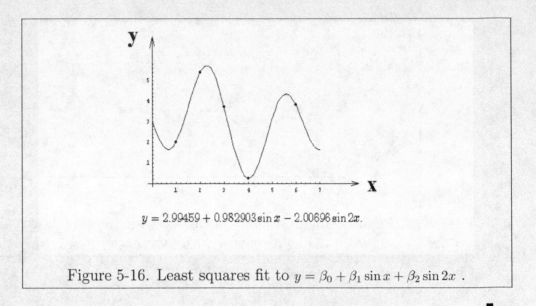

$$y = 2.99459 + 0.982903\sin x - 2.00696\sin 2x.$$

Figure 5-16. Least squares fit to $y = \beta_0 + \beta_1 \sin x + \beta_2 \sin 2x$.

Weighted Least Squares

Suppose that you have a collection of data points where you have great confidence in the accuracy or validity of some of the points. You can introduce weights w_i to place more emphasis on these points during the calculations. For example, to find the best straight line fit $\beta_0 + \beta_1 x$ to a collection of data points (x_i, y_i) for $i = 1, \ldots, n$, one would again form the errors

$$e_1 = \beta_0 + \beta_1 x_1 - y_1$$
$$e_2 = \beta_0 + \beta_1 x_2 - y_2$$
$$\vdots$$
$$e_n = \beta_0 + \beta_1 x_n - y_n$$

and then place weights w_i, $i = 1, \ldots, n$, on the square of these errors and form the weighted sum of squares

$$E(\beta_0, \beta_1) = \sum_{i=1}^{n} w_i (\beta_0 + \beta_1 x_i - y_i)^2.$$

The weighted sum of squares errors is minimized when

$$\frac{\partial E}{\partial \beta_0} = 0 \quad \text{and} \quad \frac{\partial E}{\partial \beta_1} = 0$$

simultaneously. This produces the equations

$$\beta_0 \sum_{i=1}^{n} w_i + \beta_1 \sum_{i=1}^{n} w_i x_i = \sum_{i=1}^{n} w_i y_i$$

$$\beta_0 \sum_{i=1}^{n} w_i x_i + \beta_1 \sum_{i=1}^{n} w_i x_i^2 = \sum_{i=1}^{n} w_i x_i y_i$$

(5.30)

which we must solve for the coefficients β_0 and β_1. Note that the equations (5.30) reduce to the equations (5.21) in the case all the weights reduce to unity so that $w_i = 1$ for $i = 1, \ldots, n$. Weighted least squares for the general case is left for the exercises.

Example 5-8. (Weighted least square straight line)

Find the least squares straight line through the data points

$$(1,3), (2,1), (3,5), (4,4), (5,4), (6,7)$$

if one places weights of $w = 4$ on the points $(2,1), (4,4), (6,7)$ and weights of $w = 1$ on the other points. (See example 5-4 for comparison.)

Solution: We solve the weighted least square equations (5.30) and obtain the straight line $y = -51/106 + (125/106)\,x$. The details are given below.

x	y	w	w*x	w*y	w*x*y	w*x^2
1	3	1	1	3	3	1
2	1	4	8	4	8	16
3	5	1	3	5	15	9
4	4	4	16	16	64	64
5	4	1	5	4	20	25
6	7	4	24	28	168	144
SUM =		15	57	60	278	259

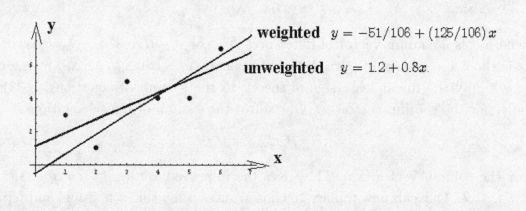

weighted $y = -51/106 + (125/106)\,x$

unweighted $y = 1.2 + 0.8x$.

■

Nonlinear Regression

To fit a function $y = f(x; \beta_1, \beta_2, \ldots, \beta_m)$, which is nonlinear in the parameters $\beta_1, \beta_2, \ldots, \beta_m$, to a set of data points $(x_1, y_1), \ldots, (x_n, y_n)$, where $n > m$, one must find values of the parameters $\beta_1, \beta_2, \ldots, \beta_m$ such that

$$
\begin{aligned}
y_1 &= f(x_1; \beta_1, \ldots, \beta_m) + e_1 \\
y_2 &= f(x_2; \beta_1, \ldots, \beta_m) + e_2 \\
&\vdots \qquad \vdots \\
y_n &= f(x_n; \beta_1, \ldots, \beta_m) + e_n
\end{aligned}
\tag{5.31}
$$

where $e_1, \ldots, e_n$ denote errors. Let $\beta_1^0, \beta_2^0, \ldots, \beta_m^0$ denote an initial guess for the parameter values. Each of the equations (5.31) can then be expanded in a Taylor series about these initial values. We use the notation $\beta^0 = (\beta_1^0, \ldots, \beta_m^0)$ and write the Taylor series expansions as the over determined system of equations

$$
\begin{aligned}
y_1 &= f(x_1; \beta_1^0, \ldots, \beta_m^0) + \frac{\partial f}{\partial \beta_1}\bigg|_{x_1, \beta^0} \Delta\beta_1 + \frac{\partial f}{\partial \beta_2}\bigg|_{x_1, \beta^0} \Delta\beta_2 + \cdots + \frac{\partial f}{\partial \beta_m}\bigg|_{x_1, \beta^0} \Delta\beta_m \\
y_2 &= f(x_2; \beta_1^0, \ldots, \beta_m^0) + \frac{\partial f}{\partial \beta_1}\bigg|_{x_2, \beta^0} \Delta\beta_1 + \frac{\partial f}{\partial \beta_2}\bigg|_{x_2, \beta^0} \Delta\beta_2 + \cdots + \frac{\partial f}{\partial \beta_m}\bigg|_{x_2, \beta^0} \Delta\beta_m \\
&\vdots \qquad \vdots \\
y_n &= f(x_n; \beta_1^0, \ldots, \beta_m^0) + \frac{\partial f}{\partial \beta_1}\bigg|_{x_n, \beta^0} \Delta\beta_1 + \frac{\partial f}{\partial \beta_2}\bigg|_{x_n, \beta^0} \Delta\beta_2 + \cdots + \frac{\partial f}{\partial \beta_m}\bigg|_{x_n, \beta^0} \Delta\beta_m
\end{aligned}
\tag{5.32}
$$

where higher order terms and the errors have been neglected. The system of equations (5.32) can be written in the matrix form

$$
\vec{D} = J\vec{\Delta\beta}
\tag{5.33}
$$

where $\vec{D}$ is a column vector of differences $\vec{D} = \mathrm{col}(y_1 - f(x_1, \beta^0), \ldots, y_n - f(x_n, \beta^0))$, J is the $n \times m$ matrix of partial derivatives and $\vec{\Delta\beta} = \mathrm{col}(\Delta\beta_1, \ldots, \Delta\beta_m)$. One can now apply the linear least square theory to the system of equations (5.33) to solve for the column vector $\vec{\Delta\beta}$. One solves the system of linear equations

$$
\left(J^T J\right) \vec{\Delta\beta} = J^T \vec{D}
\tag{5.34}
$$

for the column vector $\vec{\Delta\beta}$. This gives the improved values $\beta_i^{new} = \beta_i^0 + \Delta\beta_i$ for $i = 1, \ldots, m$. One can now replace the initial guess β_i^0 by the β_i^{new} values and repeat the above steps to get a set of still more improved values. One hopes that this procedure converges to a solution.

Exercises Chapter 5

▶ **1.**

(a) Show that equations of the form $y = \dfrac{x}{a+bx}$, with a, b constants, plots as a straight line if one plots x/y vs x.

(b) Show that equations of the form $y = ax^b + c$, with a, b, c constants, plots as a straight line if one plots $\ln(y-c)$ vs $\ln x$.

(c) Describe what variables to plot in order to obtain a straight line if
$y = ae^{bx+c} + d$, where a, b, c, d are constants.

(d) Consider the parabola $y = a + bx + cx^2$ where a, b, c are constants. Observe that if (x_0, y_0) is a point on the curve, then $y_0 = a + bx_0 + cx_0^2$ so that one can write $y - y_0 = b(x - x_0) + c(x^2 - x_0^2)$ and consequently

$$\frac{y - y_0}{x - x_0} = b + c(x + x_0)$$

What variables should be plotted in order to obtain a straight line?

(e) Follow the example given in part (d) and consider the family of hyperbolic curves $y = \dfrac{x}{a+bx} + c$ where a, b, c are constants. Assume (x_0, y_0) is a point on the curve. What variables should be plotted in order to obtain a straight line?

(f) For curves of the form $y = \dfrac{a}{x+b}$, with a, b constants, what variables should be plotted in order to obtain a straight line?

▶ **2.** **(Least Squares)** Verify that both of the equations (5.27) and (5.29), which determine the least square coefficients, can be represented in the matrix form

$$\begin{bmatrix} \sum_{i=1}^{n} f_0(x_i)f_0(x_i) & \sum_{i=1}^{n} f_1(x_i)f_0(x_i) & \cdots & \sum_{i=1}^{n} f_k(x_i)f_0(x_i) \\ \sum_{i=1}^{n} f_0(x_i)f_1(x_i) & \sum_{i=1}^{n} f_1(x_i)f_1(x_i) & \cdots & \sum_{i=1}^{n} f_k(x_i)f_1(x_i) \\ \vdots & \vdots & \ddots & \vdots \\ \sum_{i=1}^{n} f_0(x_i)f_k(x_i) & \sum_{i=1}^{n} f_1(x_i)f_k(x_i) & \cdots & \sum_{i=1}^{n} f_k(x_i)f_k(x_i) \end{bmatrix} \begin{bmatrix} \beta_0 \\ \beta_1 \\ \vdots \\ \beta_k \end{bmatrix} = \begin{bmatrix} \sum_{i=1}^{n} y_i f_0(x_i) \\ \sum_{i=1}^{n} y_i f_1(x_i) \\ \vdots \\ \sum_{i=1}^{n} y_i f_k(x_i) \end{bmatrix}$$

▶ **3.** **(Least Squares)** If the best polynomial fit to the data $\{x_i, y_i\}$, $i = 1, \ldots, n$, has coefficients $\beta_m, m = 0, 1, 2, \ldots, k$ such that the error

$$E = E(\beta_0, \beta_1, \ldots, \beta_k) = \sum_{i=1}^{n} \left[\beta_0 + \beta_1 x_i + \beta_2 x_i^2 + \cdots + \beta_k x_i^k - y_i \right]^2$$

is minimized, then show that the coefficients must be solutions of the system of equations.

$$\begin{bmatrix} n & \sum_{i=1}^{n} x_i & \sum_{i=1}^{n} x_i^2 & \cdots & \sum_{i=1}^{n} x_i^k \\ \sum_{i=1}^{n} x_i & \sum_{i=1}^{n} x_i^2 & \sum_{i=1}^{n} x_i^3 & \cdots & \sum_{i=1}^{n} x_i^{k+1} \\ \vdots & \vdots & \vdots & \ddots & \vdots \\ \sum_{i=1}^{n} x_i^k & \sum_{i=1}^{n} x_i^{k+1} & \sum_{i=1}^{n} x_i^{k+2} & \cdots & \sum_{i=1}^{n} x_i^{2k} \end{bmatrix} \begin{bmatrix} \beta_0 \\ \beta_1 \\ \vdots \\ \beta_k \end{bmatrix} = \begin{bmatrix} \sum_{i=1}^{n} y_i \\ \sum_{i=1}^{n} x_i y_i \\ \vdots \\ \sum_{i=1}^{n} x_i^k y_i \end{bmatrix}$$

▶ **4.** (**Weighted Least Squares**)

(a) What conditions must be satisfied for the weighted sum of squares error

$$E = E(\beta_0, \beta_1, \ldots, \beta_k) = \sum_{i=1}^{n} w_i \left[\beta_0 f_0(x_i) + \beta_1 f_1(x_i) + \beta_2 f_2(x_i) + \cdots + \beta_k f_k(x_i) - y_i \right]^2$$

to be a minimum.

(b) Write out the system of equations that must be satisfied for the weighted sum of squares to be a minimum.

▶ **5.** Describe in detail what is meant by saying that the best least squares straight line fit to the data $\{(0,1), (1,2), (2,11), (3,8)\}$ is given by $y = 1 + 3x$. Is the previous statement true? Justify your answer.

▶ **6.**

(a) Graph the functions $y = 10^x$, $y = 3e^{-2x}$, $y = 4(3)^x$, on semilog paper.

(b) Graph the functions $y = x^2$, $y = x^3$, $y = x^4$, $y = 4\sqrt{x}$, $y = 3x^{1.2}$ on loglog paper.

▶ **7.** Divide the data into classes and calculate the frequency, relative frequency and cumulative relative frequency for the given data. Does the data follow a normal distribution?

411	423	360	365	322	327	339
363	330	337	390	395	420	399
375	400	349	410	325	371	332
327	344	396	365	345	351	350
363	404	368	357	324	378	365
380	342	342	356	345	352	352
415	393	395	365	385	386	371
357	358	345	360	364	357	364

▶ **8.** Find values of β_1, β_2 so that $y = \sin \beta_1 x + \cos \beta_2 x$ fits the data points
$(0.25, 0.856)$, $(0.50, 0.318)$, $(0.75, -0.262)$, $(1.0, -.510)$.

▶ 9.

Select a curve from figure 5.1 that can be used to represent the given data. Find all necessary parameter values so that your curve is a good fit in some sense. Why did you pick the curve you used?

x	y
0.000	0.000
0.286	0.514
0.571	1.315
0.857	2.461
1.143	5.170
1.714	6.769
2.000	8.575

▶ 10.

x	y
1.000	2.500
1.286	3.436
1.571	4.101
1.857	4.748
2.143	5.226
2.429	5.654
2.714	6.030

Select a curve from figure 5.1 that can be used to represent the given data. Find all necessary parameter values so that your curve is a good fit in some sense. Why did you pick the curve you used?

▶ 11.

Select a curve from figure 5.1 that can be used to represent the given data. Find all necessary parameter values so that your curve is a good fit in some sense. Why did you pick the curve you used?

x	y
-2.00	5.46
-1.71	5.07
-1.43	4.62
-1.14	4.30
-0.86	3.94
-0.57	3.61
-0.28	3.30

▶ 12.

x	y
1.000	3.500
1.286	3.112
1.571	2.792
1.857	2.658
2.143	2.492
2.428	2.372
2.714	2.272

Select a curve from figure 5.1 that can be used to represent the given data. Find all necessary parameter values so that your curve is a good fit in some sense. Why did you pick the curve you used?

► **13.** Copper wire, Gauge No. 00, has a resistance of R ohms per 1000 feet at a temperature of $T°C$ given by the following table of values.

T	20	40	60	80
R	$7.8(10)^{-2}$	$8.4(10)^{-2}$	$9.0(10)^{-2}$	$9.6(10)^{-2}$

Find the least squares representation of resistance versus temperature.

► **14.**

The following (x, y) data values have been sampled from a curve that is known to be composed of the functions $\{e^{-x}, \sin x, \sin 2x\}$

$$Data = \{(0, 3.00), (1, -1.76), (2, 6.00), (3, 1.83), (4, -6.41), (5.5, 3.60), (7, -3.64),$$

$$(9, 4.58), (10.5, -5.94), (11, -1.96), (12, 3.45), (12.5, 0.53)\}$$

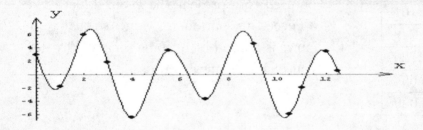

Can you find a reasonable reproduction of the original function?

► **15.** Plot the following curves.
   (i) Plot the curve $y = 3e^{-7x} + 4e^{-x}$ for $-2 \leq x \leq 6$ on semi-log paper.
   (ii) Plot the curve $y = 3e^{-7x}$ for $-2 \leq x \leq 1$ on semi-log paper.
   (iii) Plot the curve $y = 4e^{-x}$ for $-2 \leq x \leq 6$ on semi-log paper.
      Give an analysis of the above graphs. If you had the curve (i) on semi-log paper could you infer the graphs of (ii) and (iii)? Explain how one can find the functions in (ii) and (iii) from the graph (i)?

► **16.** Given the (x, y) data values

$$Data = \{(0, 2), (1, 3), (2, 4), (3, 4), (4, 6), (5, 5)\}$$

(a) Find the weighted least square straight line if a weight of 4 is assign the data points $(1, 3), (2, 4), (4, 6)$ and a weight of 1 is assigned to the other data points.
(b) Find the weighted least square straight line if a weight of 4 is assign the data points $(1, 3), (3, 4), (5, 5)$ and a weight of 1 is assigned to the other data points.

"Words differently arranged have a different meaning and meanings differently arranged have a different effect."

Blaise Pascal (1623-1662)

Chapter 6

Difference Equations and Z-transforms

There are many concepts in science and engineering that can be approached from either a discrete or a continuous viewpoint. For example, consider how you might record the temperature outside at some specific place as a function of time. One technique would be to purchase a chart recorder capable of measuring and plotting the temperature as a function of time. This would give a continuous record of the temperature over some interval of time. Another way to record the temperature would be to measure the temperature, at the specified place, at discrete time intervals. The contrast between these two methods is that one method measures temperature continuously while the other method measures the temperature in a discrete fashion.

In any laboratory experiment, one must make a decision as to how data from the experiment is to be collected. Whether discrete measurements or continuous measurements are recorded depends upon many factors as well as the type of experiment being considered. The techniques used to analyze the data collected depends upon whether the data is continuous or discrete.

The investment of money at compound interest is an example of a physical problem which requires analysis of discrete values. Say, $1,000.00 is to be invested at R percent interest compounded quarterly. How do we determine the discrete values representing the amount of money available at the end of each compound period? To solve this problem, we let P_0 denote the amount of money initially invested, R the percent interest yearly with $\frac{1}{4}\frac{R}{100} = i$ the quarterly interest and P_n the principal due at the end of the nth compound period. We can then set up the equations for the determination of P_n. We have

$$P_0 = \text{Initial amount invested}$$
$$P_1 = P_0 + P_0 i = P_0(1+i)$$
$$P_2 = P_1 + P_1 i = P_1(1+i) = P_0(1+i)^2$$
$$P_3 = P_2 + P_2 i = P_2(1+i) = P_0(1+i)^3$$
$$\vdots$$
$$P_n = P_{n-1} + P_{n-1} i = P_{n-1}(1+i) = P_0(1+i)^n$$

For $i = \frac{1}{4}\frac{R}{100}$ and $P_0 = 1,000.00$, figure 6-1 illustrates a graph of P_n vs time, for a 30 year period, where one year represents four payment periods. In this figure values of R for 4%, 5.5%, 7%, 8.5% and 10% were used in the above calculations.

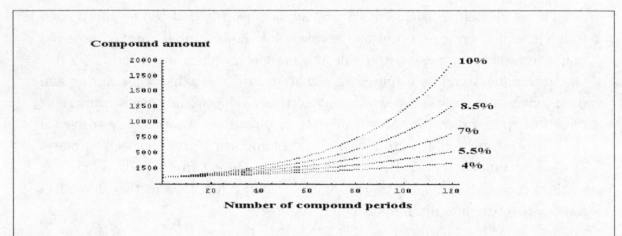

Figure 6-1. Return from $1,000 investment compounded quarterly over 30 year period.

In this chapter we investigate some techniques that can be used in the analysis of discrete phenomena like the compound interest problem just considered.

The study of calculus has demonstrated that derivatives are the mathematical quantities that represent continuous change. We find that if we replace derivatives (continuous change) by differences (discrete change), then linear ordinary differential equations become linear difference equations. We shall investigate these difference equations and find ways to construct solutions to such equations.

In the following discussions, note that the various techniques developed for analyzing discrete systems are very similar to many of the methods used for studying continuous systems.

Differences and Difference Equations

Consider the function $y = f(x)$ illustrated in the figure 6-1 which is evaluated at the equally spaced x–values of $x_0, x_1, x_2, \ldots, x_n, x_{n+1}, \ldots$ where $x_{n+1} = x_n + h$ for $n = 0, 1, 2, \ldots$, where h is the distance between two consecutive points.

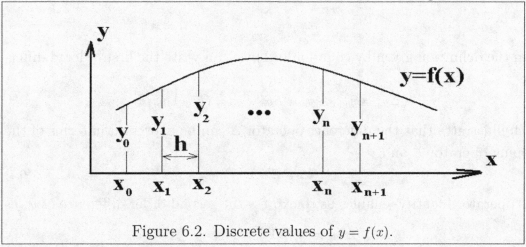

Figure 6.2. Discrete values of $y = f(x)$.

Let $y_n = f(x_n)$ and consider the approximation of the derivative $\frac{dy}{dx}$ at the discrete value x_n. By using the definition of a derivative we may write the approximation as

$$\frac{dy}{dx}\bigg|_{x=x_n} \approx \frac{y_{n+1} - y_n}{h}.$$

This is called a forward difference approximation. By letting $h = 1$ in the above equation, we can define the first forward difference of y_n as

$$\Delta y_n = y_{n+1} - y_n. \tag{6.1}$$

There is no loss in generality in letting $h = 1$, since we can always rescale the x-axis by defining the new variable X defined by the transformation equation $x = x_0 + Xh$, then when $x = x_0, x_0 + h, x_0 + 2h, \ldots, x_0 + nh, \ldots$ the scaled variable X takes on the values $X = 0, 1, 2, \ldots, n, \ldots$.

Define the second forward difference as a difference of the first forward difference. A second difference is denoted by the notation $\Delta^2 y_n$ and

$$\Delta^2 y_n = \Delta(\Delta y_n) = \Delta y_{n+1} - \Delta y_n = (y_{n+2} - y_{n+1}) - (y_{n+1} - y_n)$$

$$\text{or} \quad \Delta^2 y_n = y_{n+2} - 2y_{n+1} + y_n. \tag{6.2}$$

Higher ordered difference are defined in a similar manner. A nth order forward difference is defined as differences of $(n-1)$st forward differences for $n = 2, 3, \ldots$.

Analogous to the differential operator $D = \frac{d}{dx}$, there is a stepping operator E defined as follows:

$$Ey_n = y_{n+1}$$

$$E^2 y_n = y_{n+2}$$

$$\ldots \tag{6.3}$$

$$E^m y_n = y_{n+m}.$$

From the definition given by equation (6.1) we can write the first ordered difference

$$\Delta y_n = y_{n+1} - y_n = Ey_n - y_n = (E-1)y_n$$

which illustrates that the difference operator Δ can be expressed in terms of the stepping operator E and

$$\Delta = E - 1. \tag{6.4}$$

This operator identity, enables us to express the second-order difference of y_n as

$$\Delta^2 y_n = (E-1)^2 y_n$$

$$= (E^2 - 2E + 1)y_n$$

$$= E^2 y_n - 2Ey_n + y_n$$

$$= y_{n+2} - 2y_{n+1} + y_n.$$

Higher order differences such as $\Delta^3 y_n = (E-1)^3 y_n$, $\Delta^4 y_n = (E-1)^4 y_n, \ldots$ and higher ordered differences are quickly calculated by applying the binomial expansion to the operators operating on y_n.

Difference equations are equations which involve differences. For example, the equation

$$L_2(y_n) = \Delta^2 y_n = 0$$

is an example of a second-order difference equation, and

$$L_1(y_n) = \Delta y_n - 3y_n = 0$$

is an example of a first-order difference equation. The symbols $L_1()$, $L_2()$ are operator symbols. Using the operator E, the above equations can be written as

$$L_2(y_n) = \Delta^2 y_n = (E-1)^2 y_n = y_{n+2} - 2y_{n+1} + y_n = 0 \quad \text{and}$$

$$L_1(y_n) = \Delta y_n - 3y_n = (E-1)y_n - 3y_n = y_{n+1} - 4y_n = 0,$$

respectively.

There are many instances where variable quantities are assigned values at uniformly spaced time intervals. We shall be interested in studying these discrete variable quantities by using differences and difference equations. An equation which relates values of a function y and one or more of its differences is called a difference equation. In dealing with difference equations one assumes that the function y and its differences Δy_n, $\Delta^2 y_n, \ldots$, evaluated at x_n, are all defined for every number x in some set of values $\{x_0, \; x_0 + h, \; x_0 + 2h, \ldots, x_0 + nh, \ldots\}$. A difference equation is called linear and of order m if it can be written in the form

$$L(y_n) = a_0(n)y_{n+m} + a_1(n)y_{n+m-1} + \cdots + a_{m-1}(n)y_{n+1} + a_m(n)y_n = g(n), \qquad (6.5)$$

where the coefficients $a_i(n)$, $i = 0, \; 1, \; 2, \ldots, m$, and the right-hand side $g(n)$ are known functions of n. If $g(n) \neq 0$, the difference equation is said to be nonhomogeneous and if $g(n) = 0$, the difference equation is called homogeneous.

The difference equation (6.5) can be written in the operator form

$$L(y_n) = [a_0(n)E^m + a_1(n)E^{m-1} + \cdots + a_{m-1}(n)E + a_m(n)]y_n = g(n),$$

where E is the stepping operator.

A mth-order linear initial value problem associated with a mth-order linear difference equation consists of a linear difference equation of the form given in the equation (6.5) together with a set of m initial values of the type

$$y_0 = \alpha_0, \quad y_1 = \alpha_1, \quad y_2 = \alpha_2, \quad \ldots, \quad y_{m-1} = \alpha_{m-1},$$

where $\alpha_0, \; \alpha_1, \ldots, \; \alpha_{m-1}$ are specified constants.

Difference equations may be solved using techniques which are very similar to the solution methods associated with ordinary differential equations. All the concepts and theorems derived for linear differential equations have analogs in the study of linear difference equations. Instead of presenting each derivation, we list the following summary of these important results. In this summary D.E. can represent either "differential equation" or "difference equation".

1. An nth-order, linear, homogeneous D.E. possesses n independent solutions.

2. The general solution of an nth-order, linear D.E. has n arbitrary constants.

3. The general solution of an nth-order, linear, nonhomogeneous D.E. can be formed by adding the general complementary solution of the homogeneous equation to any particular solution of the nonhomogeneous equation.

4. If two independent solutions of a linear, homogeneous D.E. are known, then any linear combination of these solutions is also a solution.

5. If n-independent solutions to a linear, nth-order, homogeneous D.E. are known, then any linear combination of these solutions produces the general solution.

6. An nth-order linear initial value problem associated with a D.E. possesses a unique solution.

The above analogies between difference equations and differential equations can be anticipated if one writes the forward difference approximation of a derivative in operator form as $Dy = \lim_{h \to 0} \frac{\Delta y}{h}$, where $D = \frac{d}{dx}$ and Δ is the difference operator.

Example 6-1.

Show $\Delta a^k = (a-1)a^k$, for a constant and k an integer.

Solution: Let $y_k = a^k$, then by definition

$$\Delta y_k = y_{k+1} - y_k = a^{k+1} - a^k = (a-1)a^k.$$

∎

Example 6-2.

The function

$$k^{[N]} = k(k-1)(k-2) \cdots [k-(N-2)][k-(N-1)], \qquad k^{[0]} \equiv 1$$

is called a factorial polynomial, see equation (4.72). Here $k^{[N]}$ is a product of N terms.

Show $\Delta k^{[N]} = N k^{[N-1]}$ for N a positive integer and fixed.

Solution: Observe that the factorial polynomials are

$$k^{[0]} = 1, \quad k^{[1]} = k, \quad k^{[2]} = k(k-1), \quad k^{[3]} = k(k-1)(k-2), \quad \cdots$$

Use $y_k = k^{[N]}$ and calculate the forward difference

$$\Delta y_k = y_{k+1} - y_k = (k+1)^{[N]} - k^{[N]}$$

$$= (k+1) \underbrace{(k)(k-1) \cdots [k+1-(N-1)]}_{k^{[N-1]}} - \underbrace{k(k-1)(k-2) \cdots [k-(N-2)]}_{k^{[N-1]}}[k-(N-1)]$$

which simplifies to

$$\Delta y_k = \{(k+1) - [k-(N-1)]\} k^{[N-1]} = N k^{[N-1]}.$$

∎

Example 6-3.

Verify the forward difference relation

$$\Delta(U_k V_k) = U_k \Delta V_k + V_{k+1} \Delta U_k$$

Solution: Let $y_k = U_k V_k$, then we can write

$$\Delta y_k = y_{k+1} - y_k$$
$$= U_{k+1} V_{k+1} - U_k V_k + [U_k V_{k+1} - U_k V_{k+1}]$$
$$= U_k [V_{k+1} - V_k] + V_{k+1} [U_{k+1} - U_k]$$
$$= U_k \Delta V_k + V_{k+1} \Delta U_k.$$

Special Differences

The table 6.1 contains a list of some well known forward differences which are useful in many applications. The verification of these differences is left as an exercise.

Table 6.1 Some common forward differences		
1.	$\Delta a^k = (a-1)a^k$	
2.	$\Delta k^{[N]} = N\, k^{[N-1]} \qquad N$ fixed	$k^{[N]}$ is factorial function See equation (4.72)
3.	$\Delta \sin(\alpha + \beta k) = 2\sin(\beta/2)\cos(\alpha + \beta/2 + \beta k)$	α, β constants
4.	$\Delta \cos(\alpha + \beta k) = -2\sin(\beta/2)\sin(\alpha + \beta/2 + \beta k)$	α, β constants
5.	$\Delta \binom{k}{N} = \binom{k}{N-1} \quad N$ fixed	$\binom{k}{N}$ are binomial coefficients
6.	$\Delta(k!) = k(k!)$	
7.	$\Delta(U_k V_k) = U_k \Delta V_k + V_{k+1} \Delta U_k$	
8.	$\Delta \left(\dfrac{1}{k_{[N]}} \right) = \dfrac{-N}{k_{[N+1]}}, \quad N$ fixed	$k_{[N]}$ is factorial function See equation (4.72)
9.	$\Delta k^2 = 2k + 1$	
10.	$\Delta \log k = \log(1 + 1/k)$	

Finite Integrals

Associated with finite differences are finite integrals. If $\Delta y_k = f_k$, then the function y_k, whose difference is f_k, is called the finite integral of f_k. The inverse of the difference operation Δ is denoted Δ^{-1} and one can write $y_k = \Delta^{-1} f_k$, if $\Delta y_k = f_k$. For example, consider the difference of the factorial function $k^{[n]}$, defined by equation (4.72). If $\Delta k^{[n]} = nk^{[n-1]}$, then $\Delta^{-1} nk^{[n-1]} = k^{[n]}$. Associated with the difference table 6.1 is the finite integral table 6.2. The derivation of the entries is left as an exercise.

Table 6.2 Some selected finite integrals		
1.	$\Delta^{-1} a^k = \dfrac{a^k}{a-1} \quad a \neq 1$	
2.	$\Delta^{-1} k^{[n]} = \dfrac{k^{[n+1]}}{n+1}$	$k^{[n]}$ is factorial function See equation (4.72)
3.	$\Delta^{-1} \sin(\alpha + \beta k) = \dfrac{-1}{2\sin(\beta/2)} \cos(\alpha - \beta/2 + \beta k)$	α, β constants
4.	$\Delta^{-1} \cos(\alpha + \beta k) = \dfrac{1}{2\sin(\beta/2)} \sin(\alpha - \beta/2 + \beta k)$	α, β constants
5.	$\Delta^{-1} \dbinom{k}{n} = \dbinom{k}{n+1} \quad n\,fixed$	$\dbinom{k}{n}$ are binomial coefficients
6.	$\Delta^{-1} (a+bk)^{[n]} = \dfrac{(a+bk)^{[n+1]}}{b(n+1)}$	a, b constants.

Summation of Series

Let $y_{k+1} - y_k = f_k$, then one can substitute $k = 0, 1, 2, \ldots$ to obtain

$$
\begin{aligned}
y_1 - y_0 &= f_0 \\
y_2 - y_1 &= f_1 \\
y_3 - y_2 &= f_2 \\
&\vdots \\
y_n - y_{n-1} &= f_{n-1} \\
y_{n+1} - y_n &= f_n
\end{aligned}
\tag{6.6}
$$

Adding these equations one obtains

$$
\sum_{i=0}^{n} f_i = y_{n+1} - y_0 = \Delta^{-1} f_i \big]_{i=0}^{n+1} = y_i \big]_{i=0}^{n+1} \qquad \text{where} \quad \Delta y_k = f_k.
$$

One can verify that by adding the equations (6.6) from some point $i = m$ to n, one obtains the more general result

$$\sum_{i=m}^{n} f_i = y_{n+1} - y_m = \Delta^{-1} f_i\big]_{i=m}^{n+1} = y_i\big]_{i=m}^{n+1}. \tag{6.7}$$

Example 6-4.

Evaluate the sum

$$S = 1 \cdot 2 + 2 \cdot 3 + 3 \cdot 4 + \cdots + n(n+1)$$

Solution: Let $f_k = k(k+1) = k^2 + k$ and show one can write f_k as the factorial function $f_k = (k+1)^{[2]}$. Therefore,

$$S = \sum_{i=1}^{n} f_i = \sum_{i=1}^{n} (i+1)^{[2]} = \Delta^{-1} f_i\big]_{i=1}^{n+1} = \frac{(i+1)^{[3]}}{3}\Bigg]_{i=1}^{n+1} = \frac{(n+2)^{[3]}}{3} - \frac{2^{[3]}}{3}$$

which simplifies to $S = \dfrac{1}{3} n(n+1)(n+2)$.

■

Difference Equations with Constant Coefficients

Difference equations arise in a variety of situations. The following are some examples of where difference equations arise in applications. In assuming a power series solution to differential equations, the coefficients must satisfy certain recurrence formula which are nothing more than difference equations. In the study of stability of numerical methods there occurs difference equations which must be analyzed. In the computer simulation of various types of real-world processes, difference equations frequently occur. Difference equations also are studied in the areas of probability, statistics, economics, physics, and biology. We begin our investigation of difference equations by studying those with constant coefficients as these are the easiest to solve.

Example 6-5.

Given the difference equation

$$y_{n+1} - y_n - 2y_{n-1} = 0$$

with the initial conditions $y_0 = 1$, $y_1 = 0$. Find values for y_2 through y_{10}.

Solution: In the given difference equation, replace n by $n+1$ in all terms, to obtain

$$y_{n+2} = y_{n+1} + 2y_n,$$

then one can verify

$$n = 0, \qquad y_2 = y_1 + 2y_0 = 2$$

$$n = 1, \qquad y_3 = y_2 + 2y_1 = 2$$

$$n = 2, \qquad y_4 = y_3 + 2y_2 = 6$$

$$n = 3, \qquad y_5 = y_4 + 2y_3 = 10$$

$$n = 4, \qquad y_6 = y_5 + 2y_4 = 22$$

$$n = 5, \qquad y_7 = y_6 + 2y_5 = 42$$

$$n = 6, \qquad y_8 = y_7 + 2y_6 = 86$$

$$n = 7, \qquad y_9 = 7_8 + 2y_7 = 170$$

$$n = 8, \qquad y_{10} = y_9 + 2y_8 = 342.$$

The study of difference equations with constant coefficients closely parallels the development of ordinary differential equations. Our goal is to determine functions $y_n = y(n)$, defined over a set of values of n, which reduce the given difference equation to an identity. Such functions are called solutions of the difference equation. For example, the function $y_n = 3^n$ is a solution of the difference equation $y_{n+1} - 3y_n = 0$ because $3^{n+1} - 3 \cdot 3^n = 0$ for all $n = 0, 1, 2, \ldots$. Recall that for linear differential equations with constant coefficients we assumed a solution $y(x) = \exp(\omega x)$. We did this to obtain the characteristic equation and characteristic roots associated with the differential equation. When $x = n$, we obtain $y(n) = y_n = \exp(\omega n) = \lambda^n$, where $\lambda = \exp(\omega)$ is a constant. This suggests in our study of difference equations with constant coefficients that we should assume a solution of the form $y_n = \lambda^n$, where λ is a constant. Analogous to ordinary linear differential equations with constant coefficients, we find that a linear, nth-order, homogeneous difference equation with constant coefficients has associated with it a characteristic equation with characteristic roots $\lambda_1, \lambda_2, \ldots, \lambda_n$. The characteristic equation is found by assuming a solution $y_n = \lambda^n$, where λ is a constant. The various cases that can arise are illustrated by the following examples.

Example 6-6. (**Characteristic equation with real roots**)

Solve the second-order difference equation

$$y_{k+2} - 3y_{k+1} + 2y_k = 0.$$

Solution: We look for solutions of the form $y_k = \lambda^k$, where λ is a constant. We find $y_{k+1} = \lambda^{k+1}$ and $y_{k+2} = \lambda^{k+2}$. Substituting these values into the difference equation produces the equation

$$(\lambda^2 - 3\lambda + 2)\lambda^k = 0,$$

which tells us the required values for λ in order that $y_k = \lambda^k$ satisfy the difference equation. For a nontrivial solution we require $\lambda \neq 0$. This produces the characteristic equation

$$\lambda^2 - 3\lambda + 2 = 0.$$

A short cut for writing down the characteristic equation is to observe the form of the given difference equation when written in an operator form involving the stepping operator E. One can quickly obtain the characteristic equation from this operator form. For example, the given difference equation can be expressed in the form $(E^2 - 3E + 2)y_k = 0$, where the operator $E^2 - 3E + 2$ shows us the general form of the characteristic equation when E is replaced by λ. The characteristic equation has the roots $\lambda_1 = 2$ and $\lambda_2 = 1$, and hence two linearly independent solutions are

$$y_1(k) = 2^k \quad \text{and} \quad y_2(k) = 1^k = 1.$$

The general solution can be written as a linear combination of this fundamental set and so one can write

$$y(k) = y_k = c_1(2)^k + c_2,$$

where c_1 and c_2 are arbitrary constants. Given a set of initial conditions of the form $y_0 = A$ and $y_1 = B$, where A and B are given constants, we form the equations

$$y_0 = A = c_1 + c_2$$

$$y_1 = B = 2c_1 + c_2,$$

from which the constants c_1 and c_2 can be determined. Solving for these constants we obtain the solution of the initial value problem which satisfies the given initial conditions. The desired solution is unique and found to be

$$y_k = (B - A)(2)^k + (2A - B).$$

Example 6-7. (Characteristic equation with repeated roots)

Find the general solution to the difference equation

$$y_{n+2} - 4y_{n+1} + 4y_n = 0.$$

Solution: Write the difference equation in operator form $(E^2 - 4E + 4)y_n = 0$ and assume a solution of the form $y_n = \lambda^n$. By substituting the assumed solution into the difference equation, we obtain the characteristic equation $\lambda^2 - 4\lambda + 4 = 0$ which has the repeated roots $\lambda = 2,\ 2$. As with ordinary differential equations, one solution is $y_1(n) = 2^n$ and the second independent solution can be obtained by a multiplication of the first solution by the independent variable n. This is analogous to the case of repeated roots for ordinary differential equations with constant coefficients. A second independent solution is therefore $y_2(n) = n2^n$, and the general solution can be expressed as

$$y(n) = y_n = c_0 2^n + c_1 n 2^n,$$

where c_0 and c_1 are arbitrary constants. To verify that $n2^n$ is a second independent solution, the method of variation of parameters is used. We assume that a second solution has the form $y_n = U_n 2^n$, where U_n is an unknown function of n to be determined. Substituting this assumed solution into the difference equation produces the equation

$$2^{n+2}(U_{n+2} - 2U_{n+1} + U_n) = 0$$

which can be written as

$$(E^2 - 2E + 1)U_n = (E - 1)^2 U_n = \Delta^2 U_n = 0. \tag{6.8}$$

It is left as an exercise to verify that the general solution of $\Delta^k U_n = 0$ is given by

$$U_n = c_0 + c_1 n + c_2 n^2 + c_3 n^3 + \cdots + c_{k-1} n^{k-1},$$

and therefore, equation (6.8) has the solution $U_n = c_0 + c_1 n$, which when substituted into the assumed solution gives the result $y(n) = c_0(2)^n + c_1 n(2)^n$ as the general solution.

∎

In general, if the characteristic equation associated with a linear difference equation with constant coefficients has a characteristic root $\lambda = a$ of multiplicity K, then

$$y_1(n) = a^n,\ \ y_2(n) = na^n,\ \ y_3(n) = n^2 a^n, \ldots, y_K(n) = n^{K-1} a^n$$

are K linearly independent solutions of the difference equation. To show this, we shall solve the difference equation

$$(E - a)^K y_n = 0, \tag{6.9}$$

which has the characteristic equation $(\lambda - a)^K = 0$ with $\lambda = a$ as a root of multiplicity K. We use the method of variation of parameters and assume a solution to equation (6.9) of the form $y_n = a^n U_n$. Observe that

$$E y_n = a^{n+1} U_{n+1} \quad \text{and} \quad (E - a) y_n = a^{n+1} \Delta U_n$$

$$E(E - a) y_n = a^{n+2} \Delta U_{n+1} \quad \text{and} \quad (E - a)^2 y_n = a^{n+2} \Delta^2 U_n$$

$$E(E - a)^2 y_n = a^{n+3} \Delta^2 U_{n+1} \quad \text{and} \quad (E - a)^3 y_n = a^{n+3} \Delta^3 U_n.$$

Continuing in this manner, we can show that U_n must satisfy

$$(E - a)^K y_n = a^{n+k} \Delta^K U_n = 0.$$

We want a nonzero solution as so we require that a^{n+K} be different from zero, and so U_n must be chosen such that $\Delta^K U_n = 0$. The general solution of this equation is

$$U_n = c_0 + c_1 n + c_2 n^2 + \cdots + c_{K-1} n^{K-1},$$

and hence the general solution of equation (6.9) is $y_n = a^n U_n$.

One can compare the case of repeated roots for difference and differential equations and readily discern the analogies that exist.

Example 6-8. (**Characteristic equation with complex or imaginary roots**)
Solve the difference equation

$$y_{n+2} - 10 y_{n+1} + 74 y_n = (E^2 - 10E + 74) y_n = 0.$$

Solution: Assume a solution of the form $y_n = \lambda^n$ and obtain the characteristic equation $\lambda^2 - 10\lambda + 74 = 0$ which has the complex roots $\lambda_1 = 5 + 7i$ and $\lambda_2 = 5 - 7i$. Two independent solutions are therefore

$$y_1(n) = (5 + 7i)^n \quad \text{and} \quad y_2(n) = (5 - 7i)^n.$$

To obtain solutions in the form of real quantities we represent the roots λ_1, λ_2 in the polar form $re^{i\theta}$, as illustrated in figure 6-3.

210

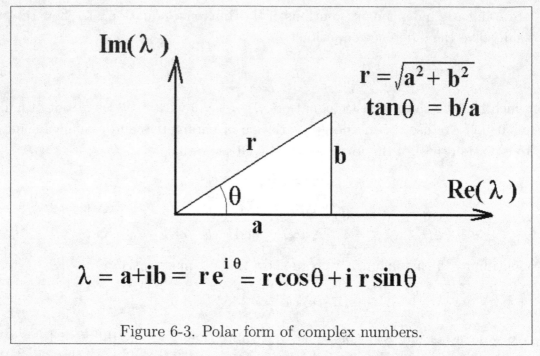

Figure 6-3. Polar form of complex numbers.

In this figure r is called the modulus or length of the complex number λ and θ is called an argument of the complex number λ. Values of 2π can be added to obtain other arguments of λ. A value of θ satisfying $-\pi < \theta \leq \pi$ is called the principal value of the argument of λ. The complex root λ_1 has a modulus and argument of

$$r = \sqrt{5^2 + 7^2} \qquad \text{and} \qquad \theta = \arctan(7/5). \tag{6.10}$$

The polar form of the characteristic roots produce solutions to the difference equations that can then be expressed in the form

$$y_1(n) = r^n e^{in\theta} \qquad \text{and} \qquad y_2(n) = r^n e^{-in\theta}.$$

The Euler's identity $e^{i\theta} = \cos\theta + i\sin\theta$, is used to write these solutions in the form

$$y_1(n) = r^n(\cos n\theta + i\sin n\theta)$$

$$y_2(n) = r^n(\cos n\theta + i\sin n\theta).$$

The solutions $y_1(n)$ and $y_2(n)$ are independent solutions of the given difference equation and hence any linear combination of these solutions is also a solution. We form the linear combinations

$$y_3(n) = \frac{1}{2}[y_1(n) + y_2(n)] \quad \text{and}$$

$$y_4(n) = \frac{1}{2i}[y_1(n) - y_2(n)]$$

to obtain the real solutions
$$y_3(n) = r^n \cos n\theta$$
$$y_4(n) = r^n \sin n\theta.$$

The general solution is any linear combination of these functions and can be expressed
$$y(n) = r^n[c_1 \cos n\theta + c_2 \sin n\theta],$$

where r and θ are defined by equation (6.10) and c_1 and c_2 are arbitrary constants. Therefore, when complex roots arise, these roots are expressed in polar form in order to obtain a real solution and imaginary solution to the given difference equation. If real solutions are desired, then one can take linear combinations of the real solution and imaginary solution in order to construct a general solution.

Nonhomogeneous Difference Equations

Nonhomogeneous difference equations can be solved in a manner analogous to the solution of nonhomogeneous differential equations and we may use the method of undetermined coefficients or the method of variation of parameters to obtain particular solutions.

Example 6-9.   (Undetermined coefficients)
Solve the first order difference equation

$$L(y_n) = y_{n+1} + 2y_n = 3n.$$

Solution: We first solve the homogeneous equation

$$L(y_n) = y_{n+1} + 2y_n = 0$$

Assume a solution $y_n = \lambda^n$ and obtain the characteristic equation $\lambda + 2 = 0$ with characteristic root $\lambda = -2$. The complementary solution is then $y_c(n) = c_1(-2)^n$, where c_1 is an arbitrary constant. Next find any particular solution $y_p(n)$ which produces the right-hand side. Analogous to what has been done with differential equations, we examine the differences of the right-hand side of the given equation. Let $r(n) = 3n$, then the first difference is a constant since $\Delta r(n) = r(n+1) - r(n) = 3$. The basic terms occurring in the right-hand side and the difference of the right-hand side are listed as members of the set $S = \{1, n\}$. If any member of S occurs in the complementary solution, then the set S is modified by multiplying each term of the set S by n. If any member of the new set S also occurs in the complementary solution, then members of the set S are modified again. This is analogous to what one does in the study of ordinary differential equations. Here one can assume a particular solution of the given difference equation which is some linear combination of the functions in S. We therefore assume a particular solution of the form

$$y_p(n) = An + B$$

where A and B are undetermined coefficients. We substitute this assumed particular solution into the difference equation and obtain

$$A(n+1) + B + 2An + 2B = 3n$$

which simplifies to

$$(A + 3B) + 3An = 3n.$$

Comparing terms, we chose $3A = 3$ and $A + 3B = 0$. We solve for A and B and obtain $A = 1$ and $B = -1/3$. Hence, the particular solution becomes

$$y_p(n) = n - \frac{1}{3}.$$

The general solution can be written as the sum of the complementary and particular solutions.

$$y_n = y_c(n) + y_p(n) = c_1(-2)^n + n - \frac{1}{3}.$$

■

Example 6-10. (Variation of parameters)
   Determine a particular solution to the difference equation

$$y_{n+2} + a_1(n)y_{n+1} + a_2(n)y_n = f_n, \tag{6.11}$$

where $a_1(n)$, $a_2(n)$, f_n are given functions of n.
Solution: Assume we know two independent solutions to the linear homogeneous equation

$$L(y_n) = y_{n+2} + a_1(n)y_{n+1} + a_2(n)y_n = 0.$$

We call these solutions u_n and v_n so that by hypothesis $L(u_n) = 0$ and $L(v_n) = 0$. We assume a particular solution to the nonhomogeneous equation (6.11) of the form

$$y_n = \alpha_n u_n + \beta_n v_n, \tag{6.12}$$

where α_n, β_n are to be determined. There are two unknowns and we need two conditions to determine these quantities. As with ordinary differential equations, we assume for our first condition the relation

$$\Delta\alpha_n u_{n+1} + \Delta\beta_n v_{n+1} = 0. \tag{6.13}$$

The second condition is obtained by substituting the assumed solution, given by equation (6.12), into the given difference equation. Starting with the assumed solution given by equation (6.12) we find that

$$y_{n+1} = y_n + \Delta y_n = \alpha_n u_n + \beta_n v_n + \Delta(\alpha_n u_n + \beta_n v_n)$$
$$y_{n+1} = \alpha_n u_n + \beta_n v_n + \alpha_n \Delta u_n + \beta_n \Delta v_n + [(\Delta\alpha_n)u_{n+1} + (\Delta\beta_n)v_{n+1}].$$

This equation simplifies since by assumption equation (6.13) must hold. One can then show that y_{n+1} reduces to

$$y_{n+1} = \alpha_n u_{n+1} + \beta_n v_{n+1}. \tag{6.14}$$

In equation (6.14) replace n by $n+1$ everywhere and establish the result

$$y_{n+2} = \alpha_{n+1} u_{n+2} + \beta_{n+1} v_{n+2}. \tag{6.15}$$

By substituting equations (6.12), (6.14) and (6.15) into the equation (6.11), we obtain the second condition that α_n and β_n must satisfy the equation

$$\alpha_{n+1} u_{n+2} + \beta_{n+1} v_{n+2} + a_1(n)(\alpha_n u_{n+1} + \beta_n v_{n+1}) + a_2(n)(\alpha_n u_n + \beta_n v_n) = f_n.$$

If we rearrange terms in this equation, we find that it can be written in the form

$$(\alpha_{n+1} - \alpha_n)u_{n+2} + (\beta_{n+1} - \beta_n)v_{n+2} + \alpha_n L(u_n) + \beta_n L(v_n) = f_n. \tag{6.16}$$

By hypothesis $L(u_n) = 0$ and $L(v_n) = 0$, thus simplifying the equation (6.16). The equations (6.13) and (6.16) give the two conditions

$$\Delta\alpha_n u_{n+1} + \Delta\beta_n v_{n+1} = 0$$
$$\Delta\alpha_n u_{n+2} + \Delta\beta_n v_{n+2} = f_n$$

from which we may solve for α_n and β_n. This system of equations can be solve by Cramer's rule and we find

$$\Delta\alpha_n = \alpha_{n+1} - \alpha_n = -\frac{f_n v_{n+1}}{C_{n+1}}, \qquad \Delta\beta_n = \beta_{n+1} - \beta_n = \frac{f_n u_{n+1}}{C_{n+1}}, \tag{6.17}$$

where $C_{n+1} = u_{n+1}v_{n+2} - u_{n+2}v_{n+1}$ is called the Casoratian (the analog of the Wronskian for continuous systems). It can be shown that C_{n+1} is never zero if u_n, v_n are linearly independent solutions of equation (6.11). The first order difference equations are a special case of problem 21 of the exercises at the end of this chapter, where it is demonstrated that the solutions can be written in the form

$$\alpha_n = \alpha_0 - \sum_{i=0}^{n-1} \frac{f_i v_{i+1}}{C_{i+1}}$$
$$\beta_n = \beta_0 + \sum_{i=0}^{n-1} \frac{f_i u_{i+1}}{C_{i+1}} \tag{6.18}$$

and the general solution to equation (6.11) can be expressed as

$$y_n = \alpha_0 u_n + \beta_0 v_n - u_n \sum_{i=0}^{n-1} \frac{f_i v_{i+1}}{C_{i+1}} + v_n \sum_{i=0}^{n-1} \frac{f_i u_{i+1}}{C_{i+1}}. \tag{6.19}$$

■

Analogous to the nth-order linear differential equation with constant coefficients

$$L(D) = (D^n + a_1 D^{n-1} + \cdots + a_{n-1}D + a_n)y = r(x) \tag{6.20}$$

with $D = \frac{d}{dx}$ a differential operator, there is the nth-order difference equation

$$L(E) = (E^n + a_1 E^{n-1} + \cdots + a_{n-1}E + a_n)y_k = r(k), \tag{6.21}$$

where E is the stepping operator satisfying $Ey_k = y_{k+1}$. Most theorems and techniques which can be applied to the ordinary differential equation (6.20) have analogous results applicable to the difference equation (6.21).

Z-transforms

Let $f(t)$ represent a sectionally continuous function which is defined for all $t > 0$. If we sample this function at regular intervals of period T we obtain an impulse representation of $f(t)$ which can be thought of as a discrete representation of $f(t)$. We denote this sample function by the symbol $f^*(t)$. The sample function $f^*(t)$ can be defined as

$$f^*(t) = f(0)\delta_\epsilon(t) + f(T)\delta_\epsilon(t - T) + f(2T)\delta_\epsilon(t - 2T) + \cdots$$
$$f^*(t) = \sum_{n=0}^{\infty} f(nT)\delta_\epsilon(t - nT), \tag{6.22}$$

where $\delta_\epsilon(t - nT)$ is the unit impulse function defined by

$$\delta_\epsilon(t - t_0) = \begin{cases} 1/\epsilon & t_0 < t < t_0 + \epsilon \\ 0, & \text{otherwise.} \end{cases}$$

In the limit as ϵ tends towards zero, the impulse function approaches the Dirac delta function $\delta(t - nT)$. The sample function $f^*(t)$ represents a sampling of the function $f(t)$ at the discrete set of points $\{0, T, \ldots, nT, \ldots\}$ as illustrated in figure 6.4. The quantity $\omega = 2\pi/T$ is called the sampling rate and T is referred to as the sampling period. The function $f^*(t)$ is called the sampling function for the function $f(t)$.

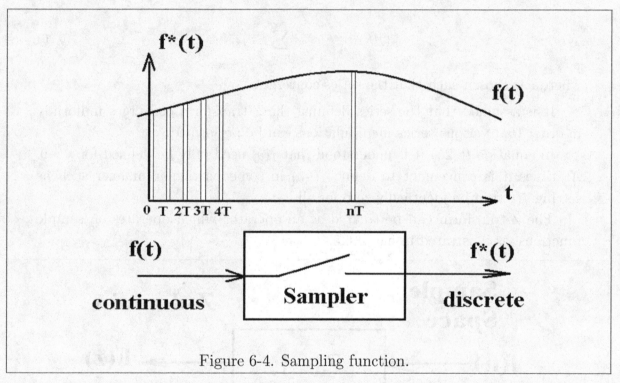

Figure 6-4. Sampling function.

Recall in the limit as ϵ tends toward zero, the Laplace transform of the impulse function $\delta_\epsilon(t - nT)$ is given by

$$\lim_{\epsilon \to 0} \mathcal{L}\{\delta_\epsilon(t - nT)\} = \exp\left(-snT\right). \tag{6.23}$$

Therefore, the Laplace transform of $f^*(t)$ in the limit as ϵ tends toward zero can be written as

$$\lim_{\epsilon \to 0} \mathcal{L}\{f^*(t)\} = \sum_{n=0}^{\infty} f(nT) \exp\left(-snT\right) \tag{6.24}$$

216

We can rescale the t-axis, such that the sampling period is unity, by letting $t = \tau T$. The values $t = 0,\ T,\ 2T, \ldots, nT$, then correspond to the integer values $\tau = 0,\ 1,\ 2, \ldots, n$. This is equivalent to letting $T = 1$ in equations (6.22) through (6.24). With the value $T = 1$, equation (6.24) can then be written as

$$\lim_{\epsilon \to 0} \mathcal{L}\{f^*(t)\} = \sum_{n=0}^{\infty} f(n)e^{-sn}. \tag{6.25}$$

The substitution $z = e^s$ in equation (6.25) produces the Z-transform of the discrete samples $f(0), f(1), \ldots, f(n), \ldots$ and the definition is as follows.

Definition: (Z-transform)   The Z-transform of $f(n)$ is given by

$$\mathcal{Z}\{f(n)\} = F(z) = \sum_{n=0}^{\infty} f(n)z^{-n}, \tag{6.26}$$

where z is chosen such that the series converges.

It is assumed that the series defining the Z-transform converges uniformly in order that various series manipulations can be performed.

In equation (6.25) it is understood that $f(n)$ need only be defined for $n > 0$. At times it is convenient to define $f(-n)$ in some particular manner such as setting $f(-n)$ to be identically zero for all $n < 0$.

The Z-transform can be treated as an operator which operates on sample functions, as illustrated in figure 6.5.

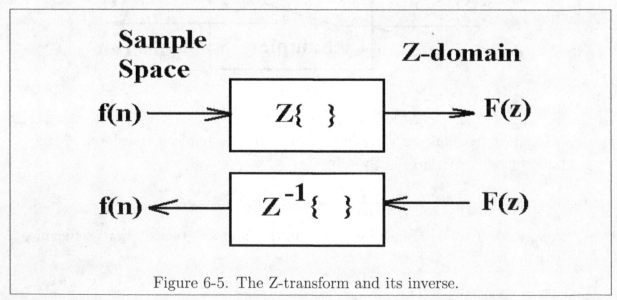

Figure 6-5. The Z-transform and its inverse.

217

Properties of the Z-transform

The Z-transform is a linear operator. To show the linearity of the operator $\mathcal{Z}\{\ \}$, let $\mathcal{Z}\{f(n)\} = F(z)$ and $\mathcal{Z}\{g(n)\} = G(z)$ represent the Z-transform of the sample functions $f(n)$ and $g(n)$. Then we have

$$(i) \quad \mathcal{Z}\{cf(n)\} = c\mathcal{Z}\{f(n)\} = cF(z)$$

for all constants c and

$$(ii) \quad \mathcal{Z}\{f(n) + g(n)\} = \mathcal{Z}\{f(n)\} + \mathcal{Z}\{g(n)\} = F(z) + G(z)$$

which establishes the linearity of the operator.

The inverse Z-transform is denoted by the notation $\mathcal{Z}^{-1}\{\ \}$ and is defined as

$$\mathcal{Z}^{-1}\{F(z)\} = f(n) \quad \text{if and only if} \quad \mathcal{Z}\{f(n)\} = F(z).$$

The inverse Z-transform operator is also a linear operator. In this text the inverse Z-transform of functions in the z-domain are obtained by table look up and by using various properties of the Z-transform which shall be developed shortly. The inverse Z-transform can also be obtained by using methods from complex variable theory.

Each continuous function $f(t)$ has only one possible sampling function $f^*(t)$. However, the converse is not true and there is no unique value of $f(t)$ corresponding to a given sample function $f^*(t)$. Observe that the inverse Z-transform operator $\mathcal{Z}^{-1}\{\ \}$ gives $f^*(t)$ and not $f(t)$.

Analogous to the methods developed for the Laplace transform, we use partial fractions and table look-up for finding the inverse Z-transform. Inverse Z-transforms are obtained by using tables 6.2 and 6.3. An examination of table 6.3 shows that many of the Z-transforms have a z in the numerator and one should make use of this fact when dealing with Z-transforms. For example, it is usually the practice to apply partial fractions to the quantity $z^{-1}F(z)$ and then multiply by z, because this will make the algebra more tractable in the simplification of inverse Z-transforms to recognizable forms. Many of the properties of the Z-transform are analogous to the properties for the Laplace transform.

Example 6-11. (Heaviside unit step function)

Find the Z-transform of the Heaviside unit step function $H(t)$ sampled at the points $t = n$, $n = 0, 1, 2, \ldots$, where

$$H(t) = \begin{cases} 0, & t < 0 \\ 1, & t \geq 0, \end{cases}$$

Solution: With the definition (6.26) we have

$$\mathcal{Z}\{H(n)\} = \sum_{n=0}^{\infty} z^{-n} = 1 + \frac{1}{z} + \frac{1}{z^2} + \cdots = \frac{1}{1 - \frac{1}{z}} = \frac{z}{z-1} = F(z). \qquad (6.27)$$

This result can also be represented as

$$H(n) = \mathcal{Z}^{-1}\left\{\frac{z}{z-1}\right\} \qquad (6.28)$$

and is given in table 6.3.

∎

Like the Laplace transform the Z-transform has many interesting properties which we now develop for later use. Numerous examples are also presented to illustrate the use of these Z-transform properties.

Property 1. (Increment property) If $\mathcal{Z}\{f(n)\} = F(z)$, then

$$\mathcal{Z}\{f(n+1)\} = zF(z) - zf(0) \quad \text{or} \quad f(n+1) = \mathcal{Z}^{-1}\{zF(z) - zf(0)\}. \qquad (6.29)$$

Note the Z-transform of the incremented function is z times the transform of the nonincremented function minus z times the initial value of the nonincremented function.

Proof: From the definition of the Z-transform we have

$$\mathcal{Z}\{f(n+1)\} = \sum_{n=0}^{\infty} f(n+1)z^{-n}$$

$$= z \sum_{n=0}^{\infty} f(n+1)z^{-(n+1)}$$

$$= z \left[\sum_{m=1}^{\infty} f(m)z^{-m} + f(0) - f(0)\right]$$

$$= z \left[\sum_{m=0}^{\infty} f(m)z^{-m} - f(0)\right]$$

$$= z[F(z) - f(0)]$$

Property 2. (Increment property) If $\mathcal{Z}\{f(n)\} = F(z)$, then

$$\mathcal{Z}\{f(n+2)\} = z^2 F(z) - z^2 f(0) - zf(1) \quad \text{or}$$
$$f(n+2) = \mathcal{Z}^{-1}\{z^2 F(z) - z^2 f(0) - zf(1)\} \qquad (6.30)$$

Proof: We know from the property 1 that, $\mathcal{Z}\{f(n+1)\} = zF(z) - zf(0)$. By applying property 1 again, we have the Z-transform of the incremented function $f(n+2)$ given by

$$\mathcal{Z}\{f(n+2)\} = z[zF(z) - f(0)] - zf(1)$$

which is z times the transform of the nonincremented function $f(n+1)$ minus z times the initial value of $f(n+1)$. Continuing in this manner we can develop the properties 3 and 4 given in table 6.2.

Property 3. (Increment property) If $\mathcal{Z}\{f(n)\} = F(z)$, then

$$\begin{aligned}
\mathcal{Z}\{f(n+3)\} &= z[z^2F(z) - z^2f(0) - zf(1)] - zf(2) \\
&= z^3F(z) - z^3f(0) - z^2f(1) - zf(2)
\end{aligned} \tag{6.31}$$

Property 4. (Increment property) If $\mathcal{Z}\{f(n)\} = F(z)$, then

$$\mathcal{Z}\{f(n+m)\} = z^mF(z) - z^mf(0) - z^{m-1}f(1) - \cdots - z^2f(m-2) - zf(m-1) \tag{6.32}$$

Property 5. (Multiplication by n) If $\mathcal{Z}\{f(n)\} = F(z)$, then

$$\mathcal{Z}\{nf(n)\} = -\frac{dF}{dz}(z) \quad \text{or}$$
$$nf(n) = \mathcal{Z}^{-1}\left\{-z\frac{dF}{dz}(z)\right\}. \tag{6.33}$$

This property shows that a multiplication by n in the sample space corresponds to $-z$ times the derivative of $F(z)$ in the z-domain.

Proof: Applying the definition of the Z-transform we may write

$$\mathcal{Z}\{f(n)\} = F(z) = \sum_{n=0}^{\infty} f(n)z^{-n}.$$

When we differentiate this expression with respect to z, we have

$$\frac{dF(z)}{dz} = \sum_{n=0}^{\infty} -nf(n)z^{-n-1} = \frac{-1}{z}\sum_{n=0}^{\infty} nf(n)z^{-n}$$

which implies

$$-z\frac{dF(z)}{dz} = \mathcal{Z}\{nf(n)\}.$$

	$f(n) = Z^{-1}\{F(z)\}$	$Z\{f(n)\} = F(z) = \sum_{n=0}^{\infty} f(n)z^{-n}$
	Table 6.2 Properties of the Z-transform	
1.	$f(n+1)$	$zF(z) - zf(0)$
2.	$f(n+2)$	$z^2F(z) - z^2f(0) - zf(1)$
3.	$f(n+3)$	$z^3F(z) - z^3f(0) - z^2f(1) - zf(2)$
4.	$f(n+m)$	$z^mF(z) - z^mf(0) - z^{m-1}f(1) - \cdots - z^2f(m-2) - zf(m-1)$
5.	$nf(n)$	$-z\frac{dF(z)}{dz}$
6.	$a^nf(n)$	$F(z/a)$
7.	$\sum_{k=0}^{n} f(k)$	$\frac{z}{z-1}F(z)$
8.	$\sum_{k=0}^{n} f(n-k)g(k)$	$F(z)G(z)$
9.	$\frac{f(n)}{n}$	$\int_z^{\infty} \frac{F(z)}{z}\, dz + \lim_{n\to 0}\frac{f(n)}{n}$
10.	$f(n-m)H(n-m)$	$z^{-m}F(z), \quad m \geq 0$
11.	$\frac{f(n)}{n+a}, \quad a > 0$	$z^a \int_z^{\infty} \frac{F(z)}{z^{a+1}}\, dz$
12.	$f(n-1)$	$\frac{F(z)}{z}$
13.	$f(0) = \lim_{z\to\infty} F(z)$	Initial value property
14.	$f(\infty) = \lim_{z\to 1}(z-1)F(z)$	Final value property

Property 6. (Scaling property) If $\mathcal{Z}\{f(n)\} = F(z)$, then

$$\mathcal{Z}\{a^nf(n)\} = F\left(\frac{z}{a}\right) \quad \text{or} \quad a^nf(n) = \mathcal{Z}^{-1}\left\{F\left(\frac{z}{a}\right)\right\}. \tag{6.34}$$

Here multiplication by a^n in the sample space, scales the transform variable z, and z is replaced by z/a.

Proof: By the definition of the Z-transform we have

$$\mathcal{Z}\{f(n)\} = F(z) = \sum_{n=0}^{\infty} f(n)z^{-n},$$

and replacing z by z/a results in

$$F\left(\frac{z}{a}\right) = \sum_{n=0}^{\infty} f(n)\left(\frac{z}{a}\right)^{-n}$$

$$= \sum_{n=0}^{\infty} a^n f(n) z^{-n} = \mathcal{Z}\{a^n f(n)\} \quad \text{or}$$

$$\mathcal{Z}^{-1}\left\{F\left(\frac{z}{a}\right)\right\} = a^n f(n).$$

	Table 6.3 Short table of Z-transforms	
	$f(n) = Z^{-1}\{F(z)\}$	$Z\{f(n)\} = F(z) = \sum_{n=0}^{\infty} f(n)z^{-n}$
1.	$H(n)$	$\dfrac{z}{z-1}$
2.	$H(n-k)$	$\dfrac{1}{z^{k-1}(z-1)}$
3.	a^n	$\dfrac{z}{z-a}$
4.	e^{bn}	$\dfrac{z}{z-e^b}$
5.	$e^{i\omega n}$	$\dfrac{z}{z-e^{i\omega}}$
6.	$\cos\omega n$	$\dfrac{z(z-\cos\omega)}{z^2 - 2z\cos\omega + 1}$
7.	$\sin\omega n$	$\dfrac{z\sin\omega}{z^2 - 2z\cos\omega + 1}$
8.	n	$\dfrac{z}{(z-1)^2}$
9.	$n^{[k]} = n(n-1)\cdots(n-k+1)$	$\dfrac{k!z}{(z-1)^{k+1}}$
10.	$\sinh\omega n$	$\dfrac{z\sinh\omega}{z^2 - 2z\cosh\omega + 1}$
11.	$\cosh\omega n$	$\dfrac{z(z-\cosh\omega)}{z^2 - 2z\cosh\omega + 1}$
12.	$n^k e^{an}$	$\dfrac{\partial^k}{\partial a^k}\left[\dfrac{z}{z-e^a}\right]$
13.	$\dfrac{k(k-1)\cdots(k-n+1)}{n!}$	$\left(1+\dfrac{1}{z}\right)^k$
14.	$\dfrac{1}{n!}$	$\exp\left(\dfrac{1}{z}\right)$

Example 6-12. Find $\mathcal{Z}\{n\}$.

Solution: From example 6-11 we have

$$\mathcal{Z}\{H(n)\} = \frac{z}{z-1} = F(z).$$

Here $n = nH(n)$ is a sample function and by property 5 we may write

$$\mathcal{Z}\{nH(n)\} = \mathcal{Z}\{n\} = -z\frac{d}{dz}\left(\frac{z}{z-1}\right) = \frac{z}{(z-1)^2}. \tag{6.35}$$

■

Example 6-13. Find $\mathcal{Z}\{a^n\}$.

Solution: From the example 6-11 we have shown

$$\mathcal{Z}\{H(n)\} = \frac{z}{z-1}$$

and by property 6 of table 6.2 we may write

$$\mathcal{Z}\{a^n H(n)\} = \mathcal{Z}\{a^n\} = \frac{z/a}{z/a - 1} = \frac{z}{z-a}. \tag{6.36}$$

■

Property 7 (Summation property) If $\mathcal{Z}\{f(n)\} = F(z)$, then

$$\mathcal{Z}\left\{\sum_{k=0}^{n} f(k)\right\} = \frac{z}{z-1}F(z) \quad \text{or} \quad \sum_{k=0}^{n} f(k) = \mathcal{Z}^{-1}\left\{\frac{z}{z-1}F(z)\right\} \tag{6.37}$$

Here multiplication by $z/(z-1)$ in the z-domain corresponds to a summation of $(n+1)$ terms in the sample space.

Proof: Let

$$G(z) = \mathcal{Z}\left\{\sum_{k=0}^{n} f(k)\right\} = \sum_{n=0}^{\infty}\sum_{k=0}^{n} f(k)z^{-n}.$$

Observe that if $f(p) = 0$ for $p < 0$, then we may write

$$\sum_{k=0}^{n} f(k) = \sum_{m=0}^{n} f(n-m) = \sum_{m=0}^{\infty} f(n-m),$$

where in the second summation, we have changed the direction of summation and instead of summing from low to high (as in the first sum) we are summing from high to low. In the third summation, $f(n-m) = 0$ whenever $n - m < 0$ or whenever $m > n$. We may now express the Z-transform $G(z)$ in the form

$$G(z) = \sum_{n=0}^{\infty} \sum_{m=0}^{\infty} f(n-m)z^{-n}.$$

In this double summation, we can interchange the order of the summation and let $n - m = p$. This produces the results

$$G(z) = \sum_{m=0}^{\infty} \sum_{n=0}^{\infty} f(n-m)z^{-n}$$

$$= \sum_{m=0}^{\infty} \sum_{p=-m}^{\infty} f(p)z^{-p}z^{-m}$$

$$= \sum_{m=0}^{\infty} \sum_{p=0}^{\infty} f(p)z^{-p}z^{-m}$$

since $f(-m) = 0$ whenever $m > 0$. The sum on p is the Z-transform $F(z)$ and we may write

$$G(z) = F(z) \sum_{m=0}^{\infty} z^{-m} = \frac{z}{z-1}F(z).$$

Example 6-14.

Let $s(n) = 1 + 2 + 3 + \cdots + n$ denote the sum of the first n integers. Here one can write

$$s(n) = \sum_{k=0}^{n} f(k), \quad \text{where} \quad f(n) = n.$$

We have the result from example 6-12 that

$$\mathcal{Z}\{n\} = \frac{z}{(z-1)^2} = F(z).$$

By the summation property 7 from table 6.2, we may write

$$\mathcal{Z}\{s(n)\} = \mathcal{Z}\left\{\sum_{k=0}^{n} f(k)\right\} = \frac{z}{z-1}F(z) = \frac{z^2}{(z-1)^3} = S(z).$$

To find the inverse Z-transform of $S(z)$, partial fractions are utilized. First, observe the entries in table 6.3 which show that most of the transforms have the variable z in the numerator. Consequently, we write

$$\frac{S(z)}{z} = \frac{z}{(z-1)^3} = \frac{A}{z-1} + \frac{B}{(z-1)^2} + \frac{C}{(z-1)^3}.$$

You may solve for the constants A, B, C by any method you like. A quick way to obtain the constants A, B, C is to observe

$$\frac{S(z)}{z} = \frac{z - 1 + 1}{(z-1)^3} = \frac{1}{(z-1)^2} + \frac{1}{(z-1)^3}$$

which shows that $A = 0$, $B = 1$, $C = 1$. Hence we may write

$$S(z) = \frac{z}{(z-1)^2} + \frac{z}{(z-1)^3},$$

where each term has the form of the entries in table 6.3. Taking the inverse transform we have

$$s(n) = \mathcal{Z}^{-1}\{S(z)\} = \mathcal{Z}^{-1}\left\{\frac{z}{(z-1)^2}\right\} + \mathcal{Z}^{-1}\left\{\frac{z}{(z-1)^3}\right\}.$$

The inverse Z-transforms can be found from table 6.3 and produce the result

$$s(n) = n^{[1]} + \frac{n^{[2]}}{2!} = n + \frac{n(n-1)}{2} = \frac{n(n+1)}{2}.$$

This gives the well known result regarding the sum of the first n integers

$$s(n) = 1 + 2 + 3 + \cdots + n = \frac{n(n+1)}{2}.$$

■

Example 6-15.

Let $\mathcal{Z}\{f(n)\} = F(z)$ and show $\mathcal{Z}\{f(n-1)\} = \frac{F(z)}{z}$. Here division by z in the transform domain corresponds to a translation in the sample space.

Solution: Assume $f(m) = 0$ for $m < 0$, then we may write

$$\mathcal{Z}\{f(n-1)\} = \sum_{n=0}^{\infty} f(n-1)z^{-n}$$

$$= f(-1) + \frac{f(0)}{z} + \frac{f(1)}{z^2} + \frac{f(2)}{z^3} + \cdots$$

$$= \frac{1}{z}\sum_{n=0}^{\infty} f(n)z^{-n} = \frac{F(z)}{z}$$

■

Example 6-16. Find $\mathcal{Z}\{n^{[k]}\}$, where

$$n^{[k]} = n(n-1)(n-2)\cdots(n-k+1)$$

is the kth factorial polynomial.

Solution: From the previous examples we have established the following Z-transforms:

$$n^{[0]} = 1 \quad \text{and} \quad \mathcal{Z}\{n^{[0]}\} = \frac{z}{z-1}$$

$$n^{[1]} = n \quad \text{and} \quad \mathcal{Z}\{n^{[1]}\} = \frac{z}{(z-1)^2}.$$

In general, the factorial polynomials satisfy

$$n^{[k]} = n(n-1)^{[k-1]}.$$

For example, $n^{[2]} = n(n-1) = n(n-1)^{[1]}$ and $n^{[3]} = n(n-1)(n-2) = n(n-1)^{[2]}$. We may use the result from example 6-15 that $\mathcal{Z}\{f(n-1)\} = F(z)/z$, and apply the property 5 to this result to obtain

$$\mathcal{Z}\{nf(n-1)\} = -z\frac{d}{dz}\left[\frac{F(z)}{z}\right].$$

This result can be applied in a recursive manner to develop the following Z-transforms:

$$\mathcal{Z}\{n^{[2]}\} = \mathcal{Z}\{n(n-1)^{[1]}\}$$
$$= -z\frac{d}{dz}(z-1)^{-2}$$
$$= \frac{2!z}{(z-1)^3}.$$

Similarly for the factorial polynomial $n^{[3]}$ we have

$$\mathcal{Z}\{n^{[3]}\} = \mathcal{Z}\{n(n-1)^{[2]}\}$$
$$= -z\frac{d}{dz}[2!(z-1)^{-3}]$$
$$= \frac{3!z}{(z-1)^4}.$$

Continuing in this manner, we have for the factorial polynomial $n^{[4]}$ the following Z-transform:

$$\mathcal{Z}\{n^{[4]}\} = \mathcal{Z}\{n(n-1)^{[3]}\}$$
$$= -z\frac{d}{dz}[3!(z-1)^{-4}]$$
$$= \frac{4!z}{(z-1)^5}.$$

By the principle of mathematical induction, we can establish the more general result:

$$\mathcal{Z}\{n^{[k]}\} = \frac{k!z}{(z-1)^{k+1}}.$$

■

Property 8. (Convolution property)

Let $\mathcal{Z}\{f(n)\} = F(z)$ and $\mathcal{Z}\{g(n)\} = G(z)$, then

$$\mathcal{Z}\left\{\sum_{k=0}^{n} f(n-k)g(k)\right\} = F(z)G(z) \quad \text{or}$$

$$\sum_{k=0}^{n} f(n-k)g(k) = \mathcal{Z}^{-1}\{F(z)G(z)\}$$

(6.38)

Proof: Let

$$H(z) = \mathcal{Z}\left\{\sum_{k=0}^{n} f(n-k)g(k)\right\}$$

$$= \sum_{n=0}^{\infty}\sum_{k=0}^{n} f(n-k)g(k)z^{-n}.$$

For $f(n-k) = 0$, when the argument $n - k < 0,$, we may write this result as

$$H(z) = \sum_{n=0}^{\infty}\sum_{k=0}^{\infty} f(n-k)g(k)z^{-n}.$$

We next interchange the order of summation and while holding k fixed we let $n = p + k$ so that p ranges from $-k$ to ∞. This produces

$$H(z) = \sum_{k=0}^{\infty}\sum_{p=-k}^{\infty} f(p)g(k)z^{-p}z^{-k}$$

$$= \sum_{p=0}^{\infty} f(p)z^{-p}\sum_{k=0}^{\infty} g(k)z^{-k}$$

$$= F(z)G(z)$$

which establishes the result.

Properties 13 and 14. (Initial and final value properties)

Let $f(0)$ denote the initial value of the sample function $f(n)$ and let $F(z)$ denote the Z-transform of $f(n)$. The initial value property can then be expressed as

$$\lim_{z \to \infty} F(z) = f(0).$$

(6.39)

Let $f(\infty) = \lim_{n\to\infty} f(n)$ denote the final value of the sample function $f(n)$ if this limit exists. Then for $F(z) = \mathcal{Z}\{f(n)\}$ we have the final value property

$$\lim_{z\to 1}(z-1)F(z) = f(\infty). \tag{6.40}$$

Proof of initial value property Using the definition of the Z-transform we may write

$$\lim_{z\to\infty} F(z) = \lim_{z\to\infty} \sum_{n=0}^{\infty} f(n)z^{-n}$$
$$= \lim_{z\to\infty}\left[f(0) + \frac{f(1)}{z} + \frac{f(2)}{z^2} + \cdots + \frac{f(n)}{z^n} + \cdots\right]$$
$$= f(0)$$

which establishes the initial value property.

Proof of final value property: Examine the Z-transform of the difference $f(n+1) - f(n)$ and show

$$\mathcal{Z}\{f(n+1) - f(n)\} = (z-1)F(z) - zf(0).$$

We take the limit of this result as z tends toward unity and find

$$\lim_{z\to 1}\mathcal{Z}\{f(n+1) - f(n)\} = \lim_{\substack{k\to\infty\\z\to 1}}\sum_{n=0}^{k}[f(n+1) - f(n)]z^{-n}.$$

When this result is expanded, there results, as z approaches unity, the sum

$$\lim_{k\to\infty}\{[f(1) - f(0)] + [f(2) - f(1)] + \cdots + [f(k+1) - f(k)]\}.$$

In this sum, all the terms add to zero except the first and the last, to give the result

$$\lim_{k\to\infty}[f(k+1) - f(0)] = f(\infty) - f(0)$$
$$= \lim_{z\to 1}(z-1)F(z) - f(0).$$

Adding the initial value to both sides of this last result produces the final value property.

Example 6-17.

Properties 7 and 9 can be combined to find the sum of certain infinite series. Let

$$g(n) = \sum_{k=0}^{n} f(k) \quad \text{and let} \quad \mathcal{Z}\{f(n)\} = F(z),$$

so that by property 7 we have

$$\mathcal{Z}\{g(n)\} = \mathcal{Z}\left\{\sum_{k=0}^{n} f(k)\right\} = G(z) = \frac{z}{z-1}F(z).$$

To obtain the final value we use the final value property, and write

$$g(\infty) = \sum_{k=0}^{\infty} f(k) = \lim_{z \to 1}(z-1)G(z)$$

$$= \lim_{z \to 1} zF(z) = F(1).$$

This result can be utilized to sum certain special series. For example, to find the sum

$$\sum_{k=0}^{\infty} \frac{x^k}{k!} = 1 + x + \frac{x^2}{2!} + \frac{x^3}{3!} + \cdots$$

we let $h(n) = 1/n!$, then $\mathcal{Z}\{1/n!\} = e^{1/z}$. From table 6.2 we find the Z-transform $\mathcal{Z}\{x^n/n!\} = e^{x/z} = F(z)$. By using the summation property, we let $g(n) = \sum_{k=0}^{n} x^k/k!$ and then

$$\mathcal{Z}\{g(n)\} = \frac{z}{z-1}e^{x/z} = G(z).$$

The final value property produces the well-known result

$$\lim_{z \to 1}(z-1)G(z) = F(1) = e^x = \sum_{k=0}^{\infty} \frac{x^k}{k!}.$$

■

Example 6-18. (Difference equation)

Solve the difference equation initial value problem

$$y_{n+1} - 2y_n = 4, \qquad y_n = 0 \quad \text{for} \quad n < 0,$$

where $y_n = y(n)$ represents the sample function.

Solution: Let $Y = Y(z) = \mathcal{Z}\{y_n\}$ and then take the Z-transform of the difference equation to obtain

$$\mathcal{Z}\{y_{n+1}\} - 2\mathcal{Z}\{y_n\} = 4\mathcal{Z}\{H(n)\},$$

where $H(n)$ is the sampling function associated with the unit step function. We utilize the properties in table 6.2, to obtain the algebraic equation

$$zY - zy_0 - 2Y = \frac{4z}{z-1},$$

where $y_0 = y(0)$ is the initial value of the sampling function. From the given difference equation, with $n = -1$, we find $y_0 = 4 + 2y_{-1} = 4$, since y_{-1} is selected to be zero. We solve the algebraic equation for the transform function $Y = Y(z)$ and find

$$Y = Y(z) = \frac{8z}{z-2} - \frac{4z}{z-1}.$$

By using table 6.3, we may take the inverse Z-transform of this result and obtain the solution

$$\mathcal{Z}^{-1}\{Y(z)\} = y_n = 8(2)^n - 4H(n).$$

∎

Example 6-19. (Determinant of tridiagonal matrix)

Consider the $n \times n$ tridiagonal matrix with constant elements α, β, γ,

$$A_n = \begin{bmatrix} \beta & \gamma & & & & & \\ \alpha & \beta & \gamma & & & & \\ & \alpha & \beta & \gamma & & & \\ & & \cdot & \cdot & \cdot & & \\ & & & \cdot & \cdot & \cdot & \\ & & & & \alpha & \beta & \gamma \\ & & & & & \alpha & \beta \end{bmatrix}$$

where the element β is along the main diagonal and the elements α and γ are below and above the main diagonal and zeros everywhere else. It is assumed that the product $\alpha\gamma$ is positive. Tridiagonal matrices occur in a variety of applied situations in numerical analysis. Let us find the number D_n which represents the determinant of A_n in the following cases: $\beta^2 > 4\alpha\gamma$, $\beta^2 = 4\alpha\gamma$, and $\beta^2 < 4\alpha\gamma$.

Solution: Examine the Sturm sequence of subdeterminants formed from the matrix A_m. We define $D_0 = 1$, $D_1 = \beta$ and calculate

$$D_m = \beta D_{m-1} - \alpha\gamma D_{m-2}, \quad m = 2, 3, \ldots, n, \tag{6.41}$$

where D_m is the determinant of A_m. For example, consider the case where $n = 3$ illustrated in figure 6-6.

$$A_3 = \begin{bmatrix} \beta & \gamma & \\ \alpha & \beta & \gamma \\ & \alpha & \beta \end{bmatrix}$$

Figure 6-6. Subdeterminants in the case $n = 3$.

Here we have $D_1 = \beta$, $D_2 = \beta^2 - \alpha\gamma$, and $D_3 = \beta^3 - \alpha\gamma\beta - \alpha\gamma\beta = \beta D_2 - \alpha\gamma D_1$ as the subdeterminants. We solve the difference equation (6.41) by Z-transforms in the general case of an arbitrary n. Let $\tilde{D} = \tilde{D}(z) = \mathcal{Z}\{D_n\}$ denote the Z-transform of D_n. In equation (6.41), we replace m by $m + 2$ to obtain

$$D_{m+2} - \beta D_{m+1} + \alpha\gamma D_m = 0.$$

Taking the Z-transform of this equation we find

$$z^2\tilde{D} - z^2 D_0 - zD_1 - \beta[z\tilde{D} - zD_0] + \alpha\gamma\tilde{D} = 0. \tag{6.42}$$

The transformed equation is now an algebraic equation, and we may solve for the transformed function $\tilde{D}$, which we find has the form

$$\tilde{D} = \frac{z^2}{z^2 - \beta z + \alpha\gamma}. \tag{6.43}$$

We write equation (6.43) in the form

$$\tilde{D} = \frac{\frac{z^2}{\alpha\gamma}}{\frac{z^2}{\alpha\gamma} - \frac{\beta}{\sqrt{\alpha\gamma}}\frac{z}{\sqrt{\alpha\gamma}} + 1} = F\left(\frac{z}{\sqrt{\alpha\gamma}}\right), \tag{6.44}$$

where

$$F(z) = \frac{z^2}{z^2 - \frac{\beta}{\sqrt{\alpha\gamma}}z + 1} \tag{6.45}$$

because this scaled form is easier to analyze. An examination of the Z-transform tables produces the following cases.

Case 1 ($\beta^2 > 4\alpha\gamma$) By making the substitution $\beta = 2\sqrt{\alpha\gamma}\cosh\omega$, equation (6.45) can be written in the form:

$$F(z) = \frac{z^2 - z\cosh\omega}{z^2 - 2z\cosh\omega + 1} + \frac{z\cosh\omega}{z^2 - 2z\cosh\omega + 1}$$

which has the inverse Z-transform

$$f(n) = \cosh\omega n + \frac{\cosh\omega}{\sinh\omega}\sinh\omega n = \frac{\sinh[(n+1)\omega]}{\sinh\omega}.$$

Employing the scaling property 6, we can find the determinant D_n and write

$$D_n = \mathcal{Z}^{-1}\left\{F\left(\frac{z}{\sqrt{\alpha\gamma}}\right)\right\} = (\alpha\gamma)^{\frac{n}{2}}\frac{\sinh[(n+1)\omega]}{\sinh\omega},$$

where $\beta = 2\sqrt{\alpha\gamma}\cosh\omega$ and $\beta^2 > 4\alpha\gamma$.

Case 2 ($\beta^2 = 4\alpha\gamma$) In the case $\beta^2 = 4\alpha\gamma$, equation (6.45) reduces to

$$F(z) = \frac{z^2}{z^2 - 2z + 1} = \frac{z^2}{(z-1)^2},$$

and from the tables of Z-transforms, we find the inverse transform is

$$f(n) = \mathcal{Z}^{-1}\{F(z)\} = n + 1.$$

Consequently, by the scaling property 6, the determinant D_n is represented by the relation

$$D_n = \mathcal{Z}^{-1}\left\{ F\left(\frac{z}{\sqrt{\alpha\gamma}} \right) \right\} = (n+1)\left(\frac{\beta}{2} \right)^n, \qquad \text{where} \quad \beta^2 = 4\alpha\gamma.$$

Case 3 ($\beta^2 < 4\alpha\gamma$) In the case $\beta^2 < 4\alpha\gamma$, we let $\beta = 2\sqrt{\alpha\gamma}\cos\omega$, then equation (6.45) reduces to

$$F(z) = \frac{z^2 - z\cos\omega + z\cos\omega}{z^2 - 2z\cos\omega + 1}$$

which has the inverse Z- transform

$$f(n) = \cos\omega n + \frac{\cos\omega n}{\sin\omega n}\sin\omega n = \frac{\sin[(n+1)\omega]}{\sin\omega}.$$

By the scaling property 6, we find that the determinant D_n can be represented in the form

$$D_n = \mathcal{Z}^{-1}\left\{ F\left(\frac{z}{\sqrt{\alpha\gamma}} \right) \right\} = (\alpha\gamma)^{\frac{n}{2}} \frac{\sin[(n+1)\omega]}{\sin\omega},$$

where $\beta = 2\sqrt{\alpha\gamma}\cos\omega$.

$\blacksquare$

Example 6-20. **(Eigenvalues)**

Find the eigenvalues of the tridiagonal matrix A_n of the previous example.

Solution: Let $C_n = \det(\lambda I_n - A_n)$, where I_n is the $n \times n$ identity matrix. The eigenvalues λ, are the roots of the characteristic equation $C_n = 0$. Proceeding as in the previous example, we find the Sturm sequence $C_0 = 1$, $C_1 = \lambda - \beta$ and, in general, we find C_m is a solution of the difference equation

$$C_{m+2} - (\lambda - \beta)C_{m+1} + \alpha\gamma C_m = 0. \tag{6.46}$$

Let $2s = \lambda - \beta$, and let $\tilde{C} = \mathcal{Z}\{C_m\}$ denote the Z-transform of C_m. Taking the Z-transform of equation (6.46) produces the algebraic equation

$$z^2\tilde{C} - z^2 - 2sz - 2s[z\tilde{C} - z] + \alpha\gamma\tilde{C} = 0,$$

which enables us to solve for the Z-transform of C_m, and we find

$$\widetilde{C} = \frac{z^2}{z^2 - 2sz + \alpha\gamma}. \tag{6.47}$$

We write this Z-transform in the scaled form

$$\widetilde{C} = \frac{\frac{z^2}{\alpha\gamma}}{\frac{z^2}{\alpha\gamma} - \frac{2s}{\alpha\gamma}z + 1} = F\left(\frac{z}{\sqrt{\alpha\gamma}}\right), \tag{6.48}$$

where

$$F(z) = \frac{z^2}{z^2 - \frac{2sz}{\sqrt{\alpha\gamma}} + 1}, \tag{6.49}$$

because the scaled equation is easier to analyze. Making the substitution $\cos\omega = s/\sqrt{\alpha\gamma}$ we may express equation (6.49) in the form

$$F(z) = \frac{(z^2 - z\cos\omega) + z\cos\omega}{z^2 - 2z\cos\omega + 1}. \tag{6.50}$$

The inverse Z-transform of this equation can be found in table 6.3 and

$$f(n) = \mathcal{Z}^{-1}\{F(z)\} = \cos\omega n + \frac{\cos\omega}{\sin\omega}\sin\omega n = \frac{\sin[(n+1)\omega]}{\sin\omega}$$

and, consequently, by the scaling property 6, we find that

$$C_n(\lambda) = (\alpha\gamma)^{\frac{n}{2}}\frac{\sin[(n+1)\omega]}{\sin\omega}, \tag{6.51}$$

where $\lambda - \beta = 2\sqrt{\alpha\gamma}\cos\omega$. From equation (6.51), we see $C_n(\lambda) = 0$ for

$$(n+1)\omega = m\pi \qquad m = 1, 2, 3, \ldots, n.$$

To emphasize that there is more than one nonzero solution for ω, (why can't $\omega = 0$?), the notation is changed, and we write

$$\omega = \omega_m = \frac{m\pi}{n+1}, \quad m = 1, 2, 3, \ldots, n.$$

Hence the eigenvalues of the $n \times n$ tridiagonal matrix are given by

$$\lambda = \lambda_m = \beta + 2\sqrt{\alpha\gamma}\cos\left(\frac{m\pi}{n+1}\right), \quad m = 1, 2, 3, \ldots, n. \tag{6.52}$$

Exercises Chapter 6

▶ **1.** Verify each of the following forward differences:

    (a) $\Delta 4^k = 3 \cdot 4^k$

    (b) $\Delta k^{[3]} = 3\, k^{[2]}, \quad k^{[3]} = k(k-1)(k-2)$

    (c) $\Delta \sin(\alpha + \beta k) = 2 \sin(\beta/2) \cos(\alpha + \beta/2 + \beta k)$

    (d) $\Delta \cos(\alpha + \beta k) = 2 \sin(\beta/2) \sin(\alpha + \beta/2 + \beta k)$

    (e) $\Delta \binom{k}{3} = \binom{k}{2}, \quad \binom{k}{3} = \dfrac{k!}{3!(3-k)!}$

    (f) $\Delta(k!) = k\,(k!)$

    (g) $\Delta(w_k y_k) = w_k \Delta y_k + y_{k+1} \Delta w_k$

    (h) $\Delta \left(\dfrac{1}{(k)_{[3]}} \right) = \dfrac{-3}{(k)_{[4]}}, \quad$ where $\quad (k)_{[3]} = k(k+1)(k+2)$

▶ **2.** Solve the following difference equations:

    (a) $y_{n+1} - 4\,y_n = 0$ (e) $y_{n+1} - y_n = n$

    (b) $y_{n+1} - a\,y_n = 0$ (f) $y_{n+1} - y_n = 5$

    (c) $y_{n+1} - a\,y_n = b, \quad a \neq 1$ (g) $y_{n+1} - y_n = 6n + 3$

    (d) $y_{n+1} - a\,y_n = n, \quad a \neq 1$ (h) $y_{n+1} - 2\,y_n = 5$

▶ **3.** Find Δy_n and $\Delta^2 y_n$ for:

    (a) $y_n = n^2$ (d) $y_n = a^n$

    (b) $y_n = n!$ (e) $y_n = n2^n$

    (c) $y_n = n(n-1)$ (f) $y_n = an^2 + bn + c$

▶ **4.** Solve the difference equations:

(a) $y_{n+2} - 5y_{n+1} + 6y_n = 0$ (b) $y_{n+2} + 2y_{n+1} + 4y_n = 0$ (c) $y_{n+2} - 6y_{n+1} + 9y_n = 0$

▶ **5.**

If $y_n = c_1(2)^n + c_2(4)^n$, calculate y_{n+1} and y_{n+2}. Now eliminate the constants c_1 and c_2 to find the difference equation satisfied by y_n.

► **6.**

Consider the factorial polynomial $n^{[m]} = n(n-1)(n-2)\cdots(n-m+1)$, with $n^{[0]} = 1$, $n^{[1]} = n,\ldots$ Show that

    (a) $\Delta n^{[2]} = 2n^{[1]}$

    (b) $\Delta n^{[3]} = 3n^{[2]}$

    (c) $\Delta n^{[4]} = 4n^{[3]}$

    (d) Take a guess at the following:

        (i) $\Delta n^{[m]}$ (ii) $\Delta\left(\dfrac{n^{[m]}}{m!}\right)$

► **7.**

Use the factorial polynomials from the previous problem and show

$$y(n) = n^2 + 4n + 7 = An^{[2]} + Bn^{[1]} + C$$

for some constants A, B, C. This suggests that polynomials can be expressed in terms of factorial polynomials if one desires.

    (a) Show that $C = y(0) = 7$

    (b) Show that $B = y(1) - y(0) = 5 = \Delta y_0$

    (c) Show that $2!A = y(2) - 2y(1) + y(0) = 14 = \Delta^2 y_0$

    (d) Generalize your results above for polynomials $f(n)$ of the form:

      $f(n) = n^m + \alpha_0 n^{m-1} + \alpha_1 n^{m-2} + \cdots + \alpha_{n-2} n^2 + \alpha_{n-1} n + \alpha_n$

      Assume that

      $f(n) = A_0 n^{[m]} + A_1 n^{[m-1]} + \cdots + A_{m-2} n^{[2]} + A_{m-1} n^{[1]} + A_m$

      and show that

      $f(0) = A_m$

      $\Delta f(n)|_{n=0} = A_{m-1}$

      $\Delta^2 f(n)|_{n=0} = 2!A_{m-2}$

      etc.

► **8.**   Represent the following polynomials in terms of factorial polynomials

      (a) $f(n) = n^4 + 3n^3 + 2n^2 + 1$

      (b) $f(n) = n^4 - 1$

      (c) $f(n) = n^3 - 1$

      (d) $f(n) = n^3 + n^2$

      (e) $f(n) = n(n-1)(n-2)(n+3)$

▶ **9.**

In the study of queues, Poisson processes are described by the difference equation

$$\lambda P_n = \frac{(n+1)}{\tau} P_{n+1}, \qquad n = 0,\, 1,\, 2,\ldots, M-1,$$

where λ and τ are constants. Here λ denotes an arrival rate, τ denotes an average service time, and P_n denotes a probability.

    (a)   Find the solution of the above difference equation

    (b)   Find P_0, if $\displaystyle\sum_{j=0}^{\infty} P_j = 1$ (*i.e.*, $M \to \infty$)

▶ **10.**

If $U_n = c_1 + c_2 n + c_3 n^2 + \cdots + c_K n^{K-1}$, show that $\Delta^K U_n = 0$. Hint: Use problem number 7 if you desire.

▶ **11.**   Solve the difference equations:

    (a) $y_{k+2} - 6y_{k+1} + 8y_k = 3^k$ (c) $(\Delta^2 - 4\Delta + 3)y_k = k^2$

    (b) $y_{k+2} - 6y_{k+1} + 8y_k = k^{[2]}$ (d) $y_{k+2} - 4y_{k+1} + 4y_k = k - 2 + 8(2)^k$

▶ **12.**

In reproduction problems in genetics, there arises the Fibonacci sequence defined by the difference equation

$$U_{n+2} = U_{n+1} + U_n, \quad \text{where} \quad U_0 = 1 \text{ and } U_1 = 1.$$

(a)   Find U_n.

(b)   Show that $\displaystyle\lim_{k \to \infty} \frac{U_{k+1}}{U_k} = \frac{1}{2}(1 + \sqrt{5})$

▶ **13.**

Find the general solution to $y_{n+1} - y_n - 2y_{n-1} = 0$ satisfying the conditions $y_0 = 1$, $y_1 = 0$. (See example 6-5 and compare your solution with the values given.)

▶ **14.**   If $y_{n+1} - 2y_n = 1$, with $y_0 = 0$, then find y_{46}.

▶ **15.**

Show that if $\cos\theta = \alpha$, then

$$\cos n\theta = \frac{1}{2}\left[\left(\alpha + i\sqrt{1-\alpha^2}\right)^n + \left(\alpha - i\sqrt{1-\alpha^2}\right)^n\right], \quad i^2 = -1, \, n = 0, \, 1, \, 2,\ldots.$$

Hint: If $y_n = \cos n\theta$, then $y_{n+2} - 2\alpha y_{n+1} + y_n = 0$.

▶ **16.** (Gambler's ruin problem)

A gambling game is such that you can win 1 dollar with probability ν and lose 1 dollar with probability $(1-\nu)$ at each turn. Let P_n denote the probability of financial ruin if you currently have n dollars. Let N denote the total money possessed by both you and your opponent.

(a) Show $P_n = \nu P_{n+1} + (1-\nu)P_{n-1}$. Notice this game is equivalent to a drunk staggering on a table top which is divided into N equal segments. The left edge of the table is marked 0 and the right-edge of the table is marked N and the segments in between are labeled consecutively $1, \, 2, \, 3,\ldots,n,\ldots,N-1$. At time $t = 0$, the drunk is at position n, and at each successive time interval, he staggers (random walks) either toward N with probability ν or toward 0 with probability $(1-\nu)$. Here P_n denotes the probability a drunk, starting at n, falls off the table at 0 .

(b) Solve the difference equation and find P_n.

Hint: For $n = 0$, $P_0 = 1$, and for $n = N$, $P_N = 0$.

(c) What is the solution in the case $\nu = 1/2$?

▶ **17.** (Difference approximation to $e = 2.71828\ldots$)

(a) Solve the initial value problem $\dfrac{dy}{dx} = y, \quad y(0) = 1$.

(b) In part(a) approximate $\frac{dy}{dx}$ by $\frac{y(x+h)-y(x)}{h}$ and let

$y_n = y(x_n), \quad y_{n+1} = y(x_n + h), \quad$ and $\quad x_0 = 0$ to obtain the difference equation

$$y_{n+1} = (1+h)y_n, \quad y_0 = 1.$$

Show that the solution of this difference equation is $y_k = (1+h)^k$.

(c) From part(a) show $y_k = y(x_k) = e^{kh}$, where $x_k = kh$ and hence show

$$y_k = e^{kh} \approx [(1+h)^{1/h}]^{kh} \quad \text{or} \quad e \approx (1+h)^{1/h}.$$

Fill in the following table using a calculator

h	$e \approx (1+h)^{1/h}$
1.000	
0.500	
0.100	
0.010	
0.001	
0.0001	

▶ **18.** For A and B constants, find the unique solution of

$$y_{n+1} = Ay_n + B, \quad n = 0, 1, 2, \ldots$$

(a) In the case where $A \neq 1$.   (b) In the case where $A = 1$.

▶ **19.** **(Interest)**

(a) (Simple interest) Simple interest earned at an annual rate r (100r is annual rate in percent) is defined $A_{n+1} = A_n + rA_0$, $n = 0, 1, 2, \ldots$, where A_0 is the initial deposit and A_n is the amount after the nth year. Find A_n.

(b) (Compound interest) Compound interest earned at an interest rate i per conversion period is defined by $A_{n+1} = A_n + iA_n$, $n = 0, 1, 2, \ldots$, where A_0 is the initial deposit and A_n is the amount after the nth conversion period. Find A_n. (Note: if the annual rate is 8%, compounded quarterly, then the interest per conversion period $i = .08/4 = 0.02$).

▶ **20.** **(Amortization)** Amortization is the periodic payment of an amount which reduces a debt. (Debt equals principal plus interest). The periodic payment is broken into two parts. One part reduces the principal and the other part pays the interest. Let D be the debt to be repaid subject to compound interest charges at a rate of i per conversion period. We let A denote the amount of the periodic payment to be made at the end of each conversion period and let P_n denote the principal remaining after the nth payment. Before the $(n+1)$st payment is made, D increases by an amount equal to the interest due on the principal P_n. After the $(n+1)$st payment of A, the new debt is P_{n+1}. Here we have:

$$P_0 = D, \qquad P_{n+1} = P_n + iP_n - A, \quad \text{for} \quad n = 0,\ 1,\ 2,\dots$$

(a) Show $P_n = (1+i)^n D - A \left[\dfrac{(1+i)^n - 1}{i} \right]$

(b) To determine the periodic payment A, to amortize a debt D in exactly n payments, we let $P_n = 0$ in part(a) and then solve for A. Show that the periodic payments are given by $A = \dfrac{Di}{1 - (1+i)^{-n}}$.

▶ **21.** For the first order homogeneous equation

$$y_{k+1} + P(k)y_k = 0$$

show that

$$y_1 = -P(0)y_0$$

$$y_2 = (-1)^2 P(1)P(0)y_0$$

$$y_3 = (-1)^3 P(2)P(1)P(0)y_0$$

$$\vdots$$

$$y_k = (-1)^k \left[\prod_{i=0}^{k-1} P(k) \right] y_0$$

▶ **22.**

(a) If $\Delta y_k = y_{k+1} - y_k = g(k)$, show that

$$\sum_{k=0}^{n-1} (y_{k+1} - y_k) = \sum_{k=0}^{n-1} g(k) \quad \text{implies} \quad y_n = y_0 + \sum_{i=0}^{n-1} g(i).$$

(b) Verify the solutions given by equations (6.18) and (6.19).

▶ **23.** For the first-order nonhomogeneous equation

$$y_{k+1} + P(k)y_k = g(k) \tag{23a}$$

let

$$\pi(k) = (-1)^k \prod_{i=0}^{k-1} P(i)$$

and write

$$\frac{y_{k+1}}{\pi(k+1)} + \frac{P(k)y_k}{\pi(k+1)} = \frac{g(k)}{\pi(k+1)} \qquad (23b).$$

(a) Show that

$$\frac{P(k)}{\pi(k+1)} = \frac{-1}{\pi(k)}$$

and consequently equation (23b) can be expressed in the form

$$\Delta\left[\frac{y_k}{\pi(k)}\right] = \frac{g(k)}{\pi(k+1)}.$$

(b) Show that the solution of equation (23a) can be written

$$y_k = \pi(k)\sum_{i=0}^{k-1}\frac{g(i)}{\pi(i+1)} + c\pi(k),$$

where c is a constant.

▶ **24.** Use the results from the previous problem to solve the initial value problems

(a) $y_{k+1} - \left(\dfrac{k+1}{k+2}\right)y_k = 0, \qquad y_0 = 1$

(b) $y_{k+1} - (k+1)y_k = 2^k \qquad y_0 = 1.$

▶ **25.** Find the general solution to the given difference equations

(a) $y_{n+3} - 7y_{n+2} + 16y_{n+1} - 12y_n = 0$

(b) $y_{n+3} - 2y_{n+2} - y_{n+1} + 2y_n = 0$

(c) $y_{n+2} + y_n = 0$

(d) $y_{n+4} - y_n = 0$

▶ **26.** Solve the given difference equations

(a) $y_{n+2} - 7y_{n+1} + 12y_n = 2(5)^n$

(b) $y_{n+2} - 5y_{n+1} + 6y_n = -2(2)^n$

▶ **27.** Find the general solution to the given difference equations

  (a) $(E^2 + E + 1)^2 y_n = 0$

  (b) $(E - 2)^3 y_n = 0$

  (c) $(E - 1)^2 (E - 2)^2 (E - 3)^2 y_n = 0$

  (d) $y_{n+2} - 6y_{n+1} + 9y_n = 6n^{[1]} - 12n^{[2]} + 4n^{[3]}$

  (e) $y_{n+2} - 6y_{n+1} + 9y_n = 3^n$

  (f) $y_{n+2} - 2y_{n+1} - 3y_n = \sin \dfrac{n\pi}{2}$

▶ **28.** Show that

$$\mathcal{Z}\{e^{bn}\} = \frac{z}{z - e^b}.$$

▶ **29.** In problem 28 , let $b = i\omega$ ($i^2 = -1$), and use the Euler identity $e^{i\theta} = \cos\theta + i\sin\theta$ to establish the results

$$\mathcal{Z}\{\cos\omega n\} = \frac{z(z - \cos\omega)}{z^2 - 2z\cos\omega + 1} \quad \text{and}$$

$$\mathcal{Z}\{\sin\omega n\} = \frac{z\sin\omega}{z^2 - 2z\cos\omega + 1}.$$

▶ **30.** In problem 29, replace ω by $i\omega$ and find

$$\mathcal{Z}\{\cosh\omega n\} \quad \text{and} \quad \mathcal{Z}\{\sinh\omega n\}$$

  Hint: $\sin ix = i\sinh x$ and $\cos ix = \cosh x$.

▶ **31.** Show that $\mathcal{Z}\left\{\dfrac{1}{n!}\right\} = \exp\left(\dfrac{1}{z}\right).$

▶ **32.** If $\mathcal{Z}\{e^{an}\} = \dfrac{z}{z - e^a} = \displaystyle\sum_{n=0}^{\infty} e^{an} z^{-n}$, then show by differentiating both sides of this equation with respect to a and using induction that

$$\mathcal{Z}\{n^k e^{an}\} = \frac{\partial^k}{\partial a^k}\left[\frac{z}{z - e^a}\right].$$

▶ **33.** Show that $\mathcal{Z}\{H(n - k)\} = \dfrac{1}{z^{k-1}(z - 1)}$

▶ **34.** Show that $\mathcal{Z}\left\{\dfrac{k^{[n]}}{n!}\right\} = \left(1 + \dfrac{1}{z}\right)^k$, where k is a nonnegative integer.

  Hint: $\displaystyle\sum_{n=0}^{\infty} \frac{k^{[n]}}{n!} = 1 + \frac{k}{z} + \frac{k(k - 1)}{2!z^2} + \cdots$ Examine this series for $k = 0, \ 1, \ 2, \ldots$ and use induction.

▶ **35.** Let $\mathcal{Z}\{f(n)\} = F(z)$ and show that $\mathcal{Z}\left\{\frac{f(n)}{n}\right\} = \int_z^\infty \frac{F(z)}{z}\, dz$ provided that the limit $\lim_{n \to 0} \frac{f(n)}{n}$ exists.

▶ **36.** Show that $\mathcal{Z}\{f(n-m)H(n-m)\} = z^{-m}F(z), \quad m > 0$.

▶ **37.** Solve the difference equation by using Z-transform methods

$$y_{n+2} - 5y_{n+1} + 6y_n = n$$

such that $y_0 = 2$ and $y_1 = 5$.

▶ **38.** Find the sum of the series by using properties of the Z-transform.

(a) $\displaystyle\sum_{k=0}^{\infty} a^k \sin(kx)$

(b) $\displaystyle\sum_{k=0}^{\infty} a^k \cos kx$

(c) $\displaystyle\sum_{k=0}^{\infty} \frac{(-1)^k x^{k+1}}{k+1} = x - \frac{x^2}{2} + \frac{x^3}{3} + \cdots$

▶ **39.** Express the given polynomials in terms of factorial polynomials $n^{[k]}$ and find their Z-transform.

$\qquad\qquad (a)\quad n^2 \qquad\qquad (b)\quad n^3 \qquad (c)\quad n^4$

▶ **40.** Using Z-transforms verify the following summations:

(a) $1 + 2^2 + 3^2 + \cdots + n^2 = \dfrac{n(n+1)(2n+1)}{6}$

(b) $1 + 2^3 + 3^3 + \cdots + n^3 = \left[\dfrac{n(n+1)}{2}\right]^2$

(c) $\displaystyle\sum_{m=0}^{n} m(m+1) = \dfrac{n(n+1)(n+2)}{3}$

(d) $\displaystyle\sum_{k=0}^{n} (2k+1) = (n+1)^2$

▶ **41.** Use Z-transform, to show that the solution of the difference equation

$$y_{n+1} + 4y_n = n, \quad y_0 = A$$

is given by $\quad y_n = f(n) = A(-4)^n + \dfrac{1}{25}(-4)^n - \dfrac{1}{25}H(n) + \dfrac{n}{5}.$

▶ **42.** Use Z-transforms to solve the difference equation

$$y_{n+1} - 3y_n = n^2, \quad y_0 = 0$$

▶ **43.** Use Z-transforms to solve the difference equation

$$y_{n+2} - 3y_{n+1} + 2y_n = 2$$

and obtain the solution satisfying $y_0 = 0$ and $y_1 = 1$.

▶ **44.** Use Z-transforms to find $\displaystyle\sum_{k=1}^{\infty} \frac{2k-1}{2^{k-1}}$.

▶ **45.** Show that $\mathcal{Z}\{e^{-an}f(n)\} = F(e^a z)$.

▶ **46.** Find $\mathcal{Z}\{H(n-1) - H(n-2)\}$.

▶ **47.** Show that $\mathcal{Z}\{a^n H(n-1)\} = \dfrac{a}{z-a}$.

▶ **48.** Show that $\mathcal{Z}\{n^2 f(n)\} = z^2 \dfrac{d^2 F(z)}{dz^2} + z\dfrac{dF(z)}{dz}$

▶ **49.** Show that $\mathcal{Z}\{n^3 f(n)\} = -z^3 \dfrac{d^3 F(z)}{dz^3} - 3z^2 \dfrac{d^2 F(z)}{dz^2} - z\dfrac{dF(z)}{dz}$

▶ **50.** Show that $\mathcal{Z}\{\dfrac{a^n - b^n}{a-b}\} = \dfrac{z}{(z-a)(z-b)}$, where a,b are constants.

▶ **51.** Show that if $\mathcal{Z}\{f(n)\} = F(z)$, then $\mathcal{Z}\{c^{-an}f(n)\} = F(c^\alpha z)$

▶ **52.** Show that $\mathcal{Z}\{\dfrac{a^n}{n!}\} = e^{a/z}$

▶ **53.** Use Z-transforms to solve $y_{n+2} + 8y_{n+1} + 4y_n = 0$ and obtain the solution satisfying $y_0 = 2$ and $y_1 = -8$.

▶ **54.** Show $\displaystyle\sum_{m=0}^{n} m(m+2)(m+4) = \dfrac{6n^{[4]} + 96n^{[3]} + 396n^{[2]} + 360n^{[1]}}{4!}$.

▶ **55.** Let y_n be a solution of the first-order difference equation $y_{n+1} - y_n = U_n$, where U_n is a given function. Show

$$\sum_{n=1}^{m} U_n = y_{m+1} - y_1.$$

(a) If $y_n = (\cos\theta)^n \dfrac{\sin(n-1)\theta}{\sin\theta}$ satisfies $y_{n+1} - y_n = (\cos\theta)^n \cos n\theta$, then find

$$S_m = \sum_{n=1}^{m} (\cos\theta)^n \cos n\theta.$$

(b) Find $S_m = \displaystyle\sum_{n=1}^{m} a^n \sin n\theta$

(c) Find $S_m = \displaystyle\sum_{n=1}^{m} a^n \qquad a \neq 1.$

▶ **56.** Show $S_n = \dfrac{1}{2} + \displaystyle\sum_{m=1}^{n} \cos m\,u = \dfrac{\sin(n + \frac{1}{2})u}{2\sin\frac{u}{2}}, \quad S_0 = \dfrac{1}{2}$

Hint: Find ΔS_n and assume solution $S_n = A\sin(\alpha + \beta n)$.

▶ **57.** (Finite Fourier transforms) Let

$$F(k) = \sum_{n=0}^{N-1} f(n) \exp(-\frac{i2\pi nk}{N}), \quad k = 0, 1, \ldots, N-1 \qquad (57a)$$

where N is an even integer and $i^2 = -1$. The numbers $F(k)$ are called the discrete Fourier transform of the numbers $f(n)$.

(a) Show that for ℓ, n integers $0 \leq n, \ell \leq N-1$ we have

$$\sum_{k=0}^{N-1} \exp\left(\frac{i2\pi k\ell}{N}\right) \exp\left(-\frac{i2\pi kn}{N}\right) = \sum_{k=0}^{N-1} \exp\left(\frac{i2\pi k(\ell-n)}{N}\right) = \begin{cases} 0, & \ell \neq n \\ N, & \ell = n \end{cases}$$

Hint: Let $S_N = 1 + e^{i\omega} + e^{i2\omega} + \cdots$

(b) Multiply equation (57a) by $\exp(\frac{i2\pi k\ell}{N})$ and sum on k from 0 to $N-1$ and show that

$$f(n) = \frac{1}{N} \sum_{k=0}^{N-1} F(k) \exp\left(\frac{i2\pi kn}{N}\right), \quad n = 0, 1, \ldots, N-1. \qquad (57b)$$

Here the numbers $f(n)$ are called the inverse transform of the numbers $F(k)$ defined by the equation (57a).

▶ **58.** Show the eigenvalues of the $N \times N$ tridiagonal matrix

$$A = \begin{bmatrix} 2-2r & r & & & & \\ r & 2-2r & r & & & \\ & r & 2-2r & r & & \\ & & & \ddots & & \\ & & & r & 2-2r & r \\ & & & & r & 1-2r \end{bmatrix}$$

are given by $2 - 4r\sin^2\left(n\pi/2(N+1)\right)$ for $n = 1, 2, \ldots, N$.

▶ **59.** Show the eigenvalues of the $N \times N$ tridiagonal matrix

$$A = \begin{bmatrix} 2+2r & -r & & & & \\ -r & 2+2r & -r & & & \\ & -r & 2+2r & -r & & \\ & & & \ddots & & \\ & & & -r & 2+2r & -r \\ & & & & -r & 1+2r \end{bmatrix}$$

are given by $2 + 4r\sin^2\left(n\pi/2(N+1)\right)$ for $n = 1, 2, \ldots, N$.

▶ **60.**

(a) Show that if λ_i is an eigenvalue of A, then $1/\lambda_i$ is an eigenvalue of A^{-1}.

(b) Show that if ξ_i is an eigenvalue of B and λ_i is an eigenvalue of A, then ξ_i/λ_i is an eigenvalue of $A^{-1}B$.

▶ **61.**

Consider the tridiagonal matrix A_n given in example 6-18.

    (i) Find the determinant of A_n in the case where $\beta = -2\sqrt{\alpha\gamma}$.

    (ii) Find the determinant of A_n in the case where $\beta < -2\sqrt{\alpha\gamma}$.

▶ **62.** Let T_n denote the tridiagonal matrix $T_n = \begin{bmatrix} \alpha & \beta & & & \\ -\beta & \alpha & \beta & & \\ & & \ddots & & \\ & & -\beta & \alpha & \beta \\ & & & -\beta & \alpha \end{bmatrix}$

(a) Find the determinant of T_n in the case n is even.

(b) Find the determinant of T_n in the case n is odd.

(c) Find the eigenvalues of T_n in the case n is even.

(d) Find the eigenvalues of T_n in the case n is odd.

"We have a habit in writing articles published in scientific journals to make the work as finished as possible, to cover up all the tracks, to not worry about the blind alleys or describe how you had the wrong idea first, and so on. So there isn't any place to publish, in a dignified manner, what you actually did in order to get to do the work. "

Richard Philips Feynman (1918 - 1988)

Chapter 7

Numerical Differentiation and Integration

A given set of $(n+1)$ data points (x_i, y_i), $i = 0, 1, 2, \ldots, n$ is assumed to represent some function $y = y(x)$. The data can come from some experiment or statistical study, where $y = y(x)$ is unknown, or the data can be generated from a known function $y = y(x)$. We assume the data points are equally spaced along the x-axis so that $x_{i+1} - x_i = h$ is a constant for $i = 0, 1, 2, \ldots, n-1$. In this chapter we develop ways to approximate the derivatives of $y = y(x)$ given only the data points. We also develop ways to integrate the function $y = y(x)$ based solely upon the data points given.

Numerical Approximation for Derivative

To approximate the derivative function $y'(x)$, evaluated at one of the given data points (x_i, y_i), say at $x = x_m$, $x_0 < x_m < x_n$, we assume that the function $y(x)$ has a Taylor series expansion about the point x_m given by either of the forms

$$y(x_m + h) = y(x_m) + y'(x_m)h + y''(x_m)\frac{h^2}{2!} + y'''(x_n)\frac{h^3}{3!} + \cdots \qquad (7.1)$$

or

$$y(x_m - h) = y(x_m) - y'(x_m)h + y''(x_m)\frac{h^2}{2!} - y'''(x_m)\frac{h^3}{3!} + \cdots \qquad (7.2)$$

By solving the equation (7.1) for the first derivative one obtains the forward derivative approximation

$$y'(x_m) = \frac{y(x_m + h) - y(x_m)}{h} + \mathcal{O}(h). \qquad (7.3)$$

Solving the equation (7.2) for the first derivative gives the backward derivative approximation

$$y'(x_m) = \frac{y(x_m) - y(x_m - h)}{h} + \mathcal{O}(h). \qquad (7.4)$$

Subtracting the equation (7.2) from the equation (7.1) gives

$$y(x_m + h) - y(x_m - h) = 2y'(x_m) + 2y'''(x_m)\frac{h^3}{3!} + \cdots \tag{7.5}$$

from which one can obtain the central derivative approximation

$$y'(x_m) = \frac{y(x_m + h) - y(x_m - h)}{2h} + \mathcal{O}(h^2) \tag{7.6}$$

which is more accurate than the results from equations (7.3) or (7.4). By using Taylor series expansions one can develop a variety of derivative approximations.

One can derive a derivative approximation for any order derivative. Consider an approximation for the jth derivative

$$\frac{d^j y}{dx^j}\bigg|_{x=x_i} = y^{(j)}(x_i), \tag{7.7}$$

where j a positive integer. The derivative can be approximated by assuming the derivative can be represented in the form

$$\begin{aligned} y^{(j)}(x_i) = \frac{1}{h^j} [\beta_m y(x_i - mh) + \beta_{m-1}y(x_i - (m-1)h) + \cdots + \beta_1 y(x_i - h) \\ \alpha_0 y(x_i) + \cdots + \alpha_{n-1}y(x_i + (n-1)h) + \alpha_n y(x_i + nh)] + \mathcal{O}(h^N) \end{aligned} \tag{7.8}$$

involving $(m + n + 1)$ data points, where $\beta_m, \beta_{m-1}, \ldots, \beta_1, \alpha_0, \alpha_1, \ldots, \alpha_n$ and N are constants to be determined. Let $y_{i+j} = y(x_i + jh)$ for the index j ranging over the values $j = -m, -(m-1), \ldots, (n-1), n$ and expand these terms in a Taylor series which are then substitute into the equation (7.8). One can then collect like terms and force the right-hand side of equation (7.8) to equal the left-hand side of equation (7.8) by setting certain coefficients equal to either zero or one. This will produce a system of equations where the coefficients $\beta_m, \ldots, \beta_1, \alpha_0, \ldots, \alpha_n$ and the order N of the error term can be determined .

Example 7-1. (Derivative formula)
Derive a formula for the first derivative of the form

$$y'(x_m) = \frac{1}{h}[\alpha_0 y_m + \alpha_1 y_{m+1} + \alpha_2 y_{m+2}] + \mathcal{O}(h^N) \tag{7.9}$$

where $\alpha_0, \alpha_1, \alpha_2$ and N are constants to be determined.
Solution: Substitute the Taylor series expansions

$$\begin{aligned} y_{m+1} = y(x_m + h) &= y(x_m) + y'(x_m)h + y''(x_m)\frac{h^2}{2!} + \cdots \\ y_{m+2} = y(x_m + 2h) &= y(x_m) + y'(x_m)(2h) + y''(x_m)\frac{(2h)^2}{2!} + \cdots \end{aligned} \tag{7.10}$$

into the assumed form for the derivative to obtain

$$y'_m = \frac{1}{h}\left[\alpha_0 y_m + \alpha_1\left(y_m + y'_m h + y''_m \frac{h^2}{2} + \cdots\right) + \alpha_2\left(y_m + y'_m(2h) + y''_m \frac{(2h)^2}{2} + \cdots\right)\right].$$

We collect like terms and write the above equation in the form

$$y'_m = \frac{1}{h}\left[(\alpha_0 + \alpha_1 + \alpha_2)y_m + (\alpha_1 h + \alpha_2(2h))y'_m + (\alpha_1 \frac{h^2}{2} + \alpha_2(2h^2))y''_m + \mathcal{O}(h^3)\right] \quad (7.11)$$

In order that the right-hand side of equation (7.11) reduce to y'_m we require the unknown coefficients to satisfy the equations

$$\begin{array}{rrrl} \alpha_0 & + \quad \alpha_1 & + \alpha_2 & = 0 \\ & \alpha_1 & + 2\alpha_2 & = 1 \\ & (1/2)\,\alpha_1 & + 2\alpha_2 & = 0. \end{array} \quad (7.12)$$

We solve this system of equations and find $\alpha_0 = -3/2$, $\alpha_1 = 2$ and $\alpha_2 = -1/2$. This gives the derivative formula

$$y'(x_m) = \frac{-3y_m + 4y_{m+1} - y_{m+2}}{2h} + \mathcal{O}(h^2) \quad (7.13)$$

where the $1/h$ factor has simplified the error term in equation (7.11). By including more terms in the expansions above one can determine the exact form for the error term.

∎

Example 7-2.   (Derivative formula)
Derive an approximation formula for the second derivative of the form

$$y''(x_m) = \frac{1}{h^2}[\beta_1 y_{m-1} + \alpha_0 y_m + \alpha_1 y_{m+1}] + \mathcal{O}(h^N) \quad (7.14)$$

where $\beta_1, \alpha_0, \alpha_1$ and N are constants to be determined.
Solution: Substitute the Taylor series expansions given by equations (7.1) and (7.2) into the equation (7.14) and then combine like terms to obtain

$$\begin{aligned} y''(x_m) = \frac{1}{h^2}\Big[&(\beta_1 + \alpha_0 + \alpha_1)y_{m-1} + (\alpha_0 h + \alpha_1(2h))y'_m \\ &+(\alpha_0 h^2/2 + \alpha_1(2h^2))y''_m + (\alpha_1 - \beta_1)y'''_m \frac{h^3}{3!} + (\alpha_1 + \beta_1)y_m^{(iv)}\frac{h^4}{4!} + \cdots\Big]. \end{aligned} \quad (7.15)$$

In order for the right-hand side of this equation to reduce to y''_m we require the coefficients to satisfy the conditions

$$\begin{array}{rrrl} \beta_1 & + \quad \alpha_0 & + \alpha_1 & = 0 \\ & \alpha_0 & + 2\alpha_1 & = 0 \\ & (1/2)\,\alpha_0 & + 2\alpha_1 & = 1 \end{array}$$

We solve this system of equations and find $\beta_1 = 1$, $\alpha_0 = -2$ and $\alpha_1 = 1$. Observe that these values for $\beta_1, \alpha_0, \alpha_1$ make the y_m''' coefficient zero and so one obtains the second derivative approximation

$$y''(x_m) = \frac{y_{m-1} - 2y_m + y_{m+1}}{h^2} + \mathcal{O}(h^2). \tag{7.16}$$

■

Derivative approximations of a function can also be derived by differentiating a polynomial approximation of the function. For example, one can use a polynomial approximations such as the Newton forward, Newton backward or Stirling polynomial approximations for $y(x)$ and then one can differentiate the polynomial approximation and use that as an approximation for the derivative. For the first derivative one obtains the approximation

$$\frac{dy}{dx}\bigg|_{x=x_0} \approx \frac{dP_n(s)}{ds}\frac{ds}{dx}\bigg|_{x=x_0}, \quad \text{where} \quad s = \frac{x - x_0}{h} \tag{7.17}$$

Approximations for higher derivatives can be obtain by taking higher order derivatives of the approximating polynomials. This gives the approximation

$$\frac{d^m y}{dx^m} \approx \frac{1}{h^m}\frac{d^m P_n(s)}{ds^m} \tag{7.18}$$

for $m = 1, 2, 3, \ldots$. Differentiation is a roughening process and so one should expect to obtain large errors when using collocation polynomials to approximate a derivative. The error term associated with a derivative of an interpolating polynomial is obtained by differentiating the error term of the interpolating polynomial.

Example 7-3. (**Derivative formula**)
Obtain approximations for the derivatives $y'(x_0), y''(x_0)$ and $y'''(x_0)$ by differentiation of the Stirling polynomial approximation which we obtain from the lozenge diagram of figure 4-1

$$y(x) \approx P_n(x) = y_0 + \binom{s}{1}\frac{\Delta y_0 + \Delta y_{-1}}{2} + \frac{\binom{s+1}{2} + \binom{s}{2}}{2}\Delta^2 y_{-1}$$
$$+ \binom{s+1}{3}\frac{\Delta^3 y_{-1} + \Delta^3 y_{-2}}{2} + \frac{\binom{s+2}{4} + \binom{s+1}{4}}{2}\Delta^4 y_{-2} + \cdots \tag{7.19}$$

where $s = \dfrac{x - x_0}{h}$ takes on integer values at $x_0, x_1, \ldots$.
Solution: We use chain rule differentiation to differentiate the approximating

polynomial and then use these derivatives to approximate the derivatives of $y(x)$. Expanding the equation (7.19) we find

$$y(x) \approx y_0 + s\left(\frac{\Delta y_0 + \Delta y_{-1}}{2}\right) + \frac{1}{2}s^2\Delta^2 y_{-1} + \frac{1}{6}(s^3 - s)\left(\frac{\Delta^3 y_{-1} + \Delta^3 y_{-2}}{2}\right)$$
$$+ \frac{1}{24}(s^4 - s^2)\Delta^4 y_{-2} + \cdots$$

with derivatives

$$y'(x) \approx \frac{1}{h}\left[\left(\frac{\Delta y_0 + \Delta y_{-1}}{2}\right) + s\Delta^2 y_{-1} + \frac{1}{6}(3s^2 - 1)\left(\frac{\Delta^3 y_{-1} + \Delta^3 y_{-2}}{2}\right)\right.$$
$$\left. + \frac{1}{24}(4s^3 - 2s)\Delta^4 y_{-2} + \cdots\right]$$

$$y''(x) \approx \frac{1}{h^2}\left[\Delta^2 y_{-1} + s\left(\frac{\Delta^3 y_{-1} + \Delta^3 y_{-2}}{2}\right) + \frac{1}{12}(6s^2 - 1)\Delta^4 y_{-2} + \cdots\right]$$

$$y'''(x) \approx \frac{1}{h^3}\left[\left(\frac{\Delta^3 y_{-1} + \Delta^3 y_{-2}}{2}\right) + s\Delta^4 y_{-2} + \cdots\right]$$

At the point $x = x_0$ we have $s = 0$ and so we obtain the approximations

$$y'(x_0) \approx \frac{1}{h}\left[\left(\frac{\Delta y_0 + \Delta y_{-1}}{2}\right) - \frac{1}{6}\left(\frac{\Delta^3 y_{-1} + \Delta^3 y_{-2}}{2}\right)\right]$$

$$y''(x_0) \approx \frac{1}{h^2}\left[\Delta^2 y_{-1} - \frac{1}{12}\Delta^4 y_{-2}\right]$$

$$y'''(x_0) \approx \frac{1}{h^3}\left[\left(\frac{\Delta^3 y_{-1} + \Delta^3 y_{-2}}{2}\right)\right]$$

These same results can be obtain by differentiating the equation (4.95) considered earlier. Note also that derivative approximations can be obtained from the appropriate values of a difference table. Alternatively, the differences can be expressed in terms of ordinate values and so the above derivative formulas can also be expressed in terms of ordinate values.

■

Error Terms for Derivative Approximations

To derive error terms associated with numerical differentiation or integration we will need the following results.

(1.) If $F(x)$ is a continuous function over the interval $a \leq x \leq b$, then there exists at least one point c such that $a \leq c \leq b$ and

$$\alpha F(a) + \beta F(b) = (\alpha + \beta)F(c) \tag{7.20}$$

for positive constants α and β.

(2.) A generalization of the above result is the following. For $F(x)$ continuous over the interval $a < x < b$, with points x_i satisfying $a \le x_i \le b$ for $i = 1, \ldots, n$ then one can write

$$F(x_1) + F(x_2) + \cdots + F(x_n) = nF(\xi) \tag{7.21}$$

for some value ξ lying in the interval $[a, b]$.

The result (7.20) follows from the inequalities, that if $F(a) \le F(b)$, then for positive weights α and β

$$(\alpha + \beta)F(a) \le \alpha F(a) + \beta F(b) \le (\alpha + \beta)F(b)$$

or

$$F(a) \le \frac{\alpha F(a) + \beta F(b)}{\alpha + \beta} \le F(b).$$

Hence, if $F(x)$ is continuous over the interval $[a, b]$, then there exists at least one point c such that

$$\frac{\alpha F(a) + \beta F(b)}{\alpha + \beta} = F(c).$$

The result (7.21) is obtained by similar arguments.

The error associated with an ith derivative approximation evaluated at a point x_0 is defined

$$Error = y^{(i)}(x_0) - y^{(i)}(x_0)_{approx} \tag{7.22}$$

Most error terms can be obtained by truncation of appropriate Taylor series expansions. For example, to find an error term associated with the forward derivative approximation $y_0' = \dfrac{y_1 - y_0}{h}$ we truncate the Taylor series expansion and write

$$y_1 = y(x_0 + h) = y(x_0) + y'(x_0)h + y''(\xi)\frac{h^2}{2!}, \qquad x_0 < \xi < x_0 + h \tag{7.23}$$

The error term is then found from the relation

$$y'(x_0) - \left(\frac{y(x_0 + h) - y(x_0)}{h} \right) = Error \tag{7.24}$$

Substituting the Taylor series expansion for $y(x_0 + h)$ gives

$$Error = y_0' - \frac{1}{h}\left[y_0 + y_0'h + y(\xi)''\frac{h^2}{2!} - y_0 \right] = -\frac{h}{2}y''(\xi). \tag{7.25}$$

Sometimes it is necessary to use Taylor series expansions on one or more terms in a derivative approximation. For example, to find an error term associated with the central difference approximation for the second derivative $y_0'' = \dfrac{y_{-1} - 2y_0 + y_1}{h^2}$ we use truncated Taylor series expansions from equations (7.1) and (7.2) to obtain

$$Error = y_0'' - \left(\frac{y_{-1} - 2y_0 + y_1}{h^2} \right)$$

$$Error = y_0'' - \frac{1}{h^2}\left[y_0 - y_0'h + y_0''\frac{h^2}{2} - y_0'''\frac{h^3}{6} + y_0^{(iv)}(\xi_1)\frac{h^4}{24} - 2y_0 \right. \tag{7.26}$$

$$\left. + y_0 + y_0'h + y_0''\frac{h^2}{2} + y_0'''\frac{h^3}{6} + y_0^{(iv)}(\xi_2)\frac{h^4}{24} \right]$$

which simplifies to

$$Error = -\frac{h^2}{12} y_0^{(iv)}(\zeta) \tag{7.27}$$

To derive the result given by equation (7.27) we have made the assumption that the derivative $y_0^{(iv)}(x)$ is a continuous function so that

$$\frac{h^4}{24} y_0^{(iv)}(\xi_1) + \frac{h^4}{24} y_0^{(iv)}(\xi_2) = \left(\frac{h^4}{24} + \frac{h^4}{24} \right) y_0^{(iv)}(\zeta)$$

which is a special case of the result (7.20) previously cited. Note also we had to go to fourth order terms in the expansions because the third order terms added to zero.

Method of Undetermined Coefficients

One can assume a derivative formula for $f'(x)$ involving undetermined coefficients and then select the coefficients so that the assumed derivative representation is exact when the function $f(x)$ is a polynomial. For example, with equal spacing where $x_{i+1} = x_i + h$, one can assume a derivative formula

$$f'(x_i) = \beta_0 f(x_i) + \beta_1 f(x_{i+1}) + \beta_2 f(x_{i+2}) \tag{7.28}$$

where $\beta_0, \beta_1, \beta_2$ are undetermined coefficients, and then require that this formula be exact for the cases $f(x) = 1$, $f(x) = x - x_i$ and $f(x) = (x - x_i)^2$. In this way one obtains three equations from which the three unknowns $\beta_0, \beta_1, \beta_2$ can be determined. We have:

$$\text{For } f(x) = 1, \text{ the equation (7.28) becomes} \quad 0 = \beta_0 + \beta_1 + \beta_2$$

$$\text{For } f(x) = x - x_i, \text{ the equation (7.28) becomes} \quad 1 = \beta_1 h + \beta_2 2h \tag{7.29}$$

$$\text{For } f(x) = (x - x_i)^2, \text{ the equation (7.28) becomes} \quad 0 = \beta_1 h^2 + \beta_2 4h^2$$

We solve the equations (7.29) and find $\beta_0 = -3/2h$, $\beta_1 = 4/2h$, and $\beta_2 = -1/2h$. This gives the first derivative formula

$$f'(x_i) = \frac{1}{2h}\left[-3f(x_i) + 4f(x_{i+1}) - f(x_{i+2})\right] \tag{7.30}$$

The error term associated with this formula can be obtained from equation (7.22) together with appropriate Taylor series expansions. The method of undetermined coefficients is applicable for determining both derivative and integration formulas.

Numerical Integration

In this section we develop integration formulas and associated error terms which can be used for evaluating integrals of the form

$$I_1 = \int_a^b f(x)\,dx \qquad \text{or} \qquad I_2 = \int_a^b w(x)f(x)\,dx \tag{7.31}$$

where $w(x)$ is called a weight function. Integration formulas are also referred to as quadrature formulas. The term quadrature coming from the ancient practice of constructing squares with area equivalent to that of a given plane surface. The integrands $f(x)$ or $w(x)f(x)$ in equations (7.31) are assumed to be continuous with known values over the interval $a \le x \le b$. The interval $[a,b]$, over which the integral is desired, is divided by $(n+1)$ points into sections with $a = x_0 < x_1 < x_2 < \ldots < x_n = b$. This is called partitioning the interval into n-panels. These panels can be of equal lengths or unequal lengths as illustrated in the figure 7-1.

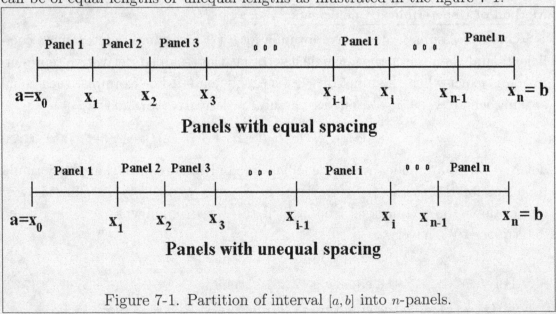

Figure 7-1. Partition of interval $[a,b]$ into n-panels.

By developing integration formulas for the area under the curve associated with one or more panels, one can repeat the integration formula until the area associated with all panels is calculated. We begin by developing a one-panel formula.

Assume that the interval $[a, b]$ is partitioned with equal spacing with

$$a = x_0, \quad b = x_n, \quad h = \frac{b-a}{n}, \quad x_j = x_0 + jh, \quad \text{for} \quad j = 0, 1, 2, \ldots, n. \tag{7.32}$$

The area under the curve $y = f(x)$ between x_{i-1} and x_i is approximated by constructing a straight line interpolating polynomial through the points (x_{i-1}, y_{i-1}) and (x_i, y_i) and then integrating this interpolation function. We use $y_{i-1} = f(x_{i-1})$ and $y_i = f(x_i)$ and obtain from the lozenge diagram of figure 4-1, with appropriate notation change, the straight line

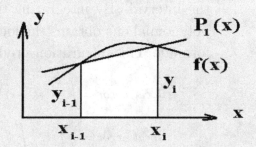

$$P_1(x) = y_{i-1} + s\Delta y_{i-1} + s(s-1)\frac{h^2}{2}f''(\xi(x)), \quad \text{where} \quad s = \frac{x - x_{i-1}}{h} \tag{7.33}$$

and $\Delta y_{i-1} = y_i - y_{i-1}$. The area associated with one-panel is then approximated by

$$\int_{x_{i-1}}^{x_i} f(x)\, dx \approx \int_{x_{i-1}}^{x_i} P_1(x)\, dx = \int_{x_{i-1}}^{x_i} \left[y_{i-1} + s\Delta y_{i-1} + s(s-1)\frac{h^2}{2}f''(\xi(x)) \right] dx. \tag{7.34}$$

We use the change of variable for s given by equation (7.33) and integrate the first two terms of equation (7.34) to obtain

$$\int_{x_{i-1}}^{x_i} f(x)\, dx = h \left[sy_{i-1} + \frac{s^2}{2}\Delta y_{i-1} \right]_0^1 = \frac{h}{2}[y_{i-1} + y_i] \tag{7.35}$$

which is known as the trapezoidal rule since the area of a trapezoid is the average height times the base. Alternatively the trapezoidal rule can be derived by integrating the Lagrange interpolating polynomial

$$P_1(x) = \frac{x - x_i}{x_{i-1} - x_i}f(x_{i-1}) + \frac{x - x_{i-1}}{x_i - x_{i-1}}f(x_i)$$

from x_{i-1} to x_i.

The last integral in equation (7.34) represents an integral of the error term accompanying the approximation polynomial. In order to evaluate this last integral we use the mean value theorem

$$\int_{x_m}^{x_n} f(x)g(x)\,dx = f(\zeta) \int_{x_m}^{x_n} g(x)\,dx, \quad x_m < \zeta < x_n, \tag{7.36}$$

where it is assumed that the functions f, g are continuous in the interval $[x_m, x_n]$ and $g(x)$ remains of one sign over the interval. i.e. Either $g(x) \geq 0$ or $g(x) \leq 0$ over the interval. By integrating the error term of the straight line approximating polynomial one obtains the local error term associated with the trapezoidal rule formula. This integration produces

$$local\ error = \int_{x_{i-1}}^{x_i} (s^2 - s) \frac{h^2}{2} f''(\xi(x))\,dx, \quad s = \frac{x - x_{i-1}}{h}, \quad h\,ds = dx$$

$$local\ error = \frac{h^3}{2} f''(\zeta) \int_0^1 (s^2 - s)\,ds = \frac{h^3}{2} f''(\zeta) \left[\frac{s^3}{3} - \frac{s^2}{2} \right]_0^1 \tag{7.37}$$

$$local\ error = -\frac{h^3}{12} f''(\zeta).$$

The one-panel trapezoidal formula with local error term can be written in either of the forms

$$\int_{x_{i-1}}^{x_i} f(x)\,dx = \frac{h}{2} [y_{i-1} + y_i] - \frac{h^3}{12} f''(\zeta)$$

$$or \quad \int_{x_{i-1}}^{x_i} f(x)\,dx = \frac{h}{2} [f(x_{i-1}) + f(x_i)] - \frac{h^3}{12} f''(\zeta) \tag{7.38}$$

By partitioning an interval $[x_0, x_n]$ into $n+1$ points one can write

$$\int_{x_0}^{x_n} f(x)\,dx = \sum_{j=1}^{n} \int_{x_{j-1}}^{x_j} f(x)\,dx. \tag{7.39}$$

Now one can apply the trapezoidal rule to each of the n-panels. The sum that results gives a representation of the integral over the interval $[x_0, x_n]$. This representation is called the extended trapezoidal rule or composite trapezoidal rule and can be represented for equal or unequal panel spacing. For unequal panel spacing the extended trapezoidal rule is written

$$\int_{x_0}^{x_n} f(x)\,dx = \frac{h_1}{2}(y_0 + y_1) + \frac{h_2}{2}(y_1 + y_2) + \cdots + \frac{h_n}{2}(y_{n-1} + y_n) + global\ error \tag{7.40}$$

where h_i is the length of the ith panel, $i = 1, 2, \ldots, n$ and the global error is the sum of the local errors associated with each panel. One finds that for unequal panel spacing the global error can be written

$$global\ error = -\frac{h_1^3}{12}f''(\zeta_1) - \frac{h_2^3}{12}f''(\zeta_2) - \cdots - \frac{h_n^3}{12}f''(\zeta_n). \tag{7.41}$$

For the case of equal panel spacing the extended trapezoidal rule simplifies to

$$\int_{x_0}^{x_n} f(x)\,dx = \frac{h}{2}\left[y_0 + 2y_1 + 2y_2 + \cdots 2y_{n-1} + y_n\right] + global\ error \tag{7.42}$$

and the result (7.21) simplifies the global error to the form

$$global\ error = -\frac{h^3}{12}nf''(\zeta), \qquad a < \zeta < b. \tag{7.43}$$

We use the result $n = \dfrac{b-a}{h}$ from the equation (7.32) to simplify the global error (7.43) to the form

$$global\ error = -\frac{(b-a)}{12}h^2 f''(\zeta) \tag{7.44}$$

The extended trapezoidal rule for equal panel spacing can be written in the form

$$\int_{x_0}^{x_n} f(x)\,dx = \frac{h}{2}\left[y_0 + 2y_1 + 2y_2 + \cdots 2y_{n-1} + y_n\right] - \frac{(b-a)}{12}h^2 f''(\zeta), \quad a < \zeta < b \tag{7.45}$$

or by letting $y_i = f(x_i)$ the trapezoidal rule can be expressed

$$\int_{x_0}^{x_n} f(x)\,dx = \frac{h}{2}\left[f(x_0) + 2f(x_1) + 2f(x_2) + \cdots 2f(x_{n-1}) + f(x_n)\right] - \frac{(b-a)}{12}h^2 f''(\zeta). \tag{7.46}$$

A two-panel integration formula results when an interpolating polynomial is constructed to pass through the points $(x_{i-1}, y_{i-1}), (x_i, y_i)$ and (x_{i+1}, y_{i+1}) and then the interpolating polynomial is integrated over the two panels. This requires the construction of a second degree polynomial or parabola through the given points. An integration of the interpolating polynomial plus error term gives an approximation to the area under the curve $f(x)$ which is associated with two panels.

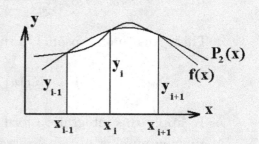

We assume equal spacing and obtain from the lozenge diagram of figure 4-1, with appropriate notation change, the parabola

$$P_2(x) = y_{i-1} + \binom{s}{1}\Delta y_{i-1} + \binom{s}{2}\Delta^2 y_{i-1} + \binom{s}{3}h^3 f'''(\xi(x)). \tag{7.47}$$

A two panel integration formula results when this interpolating polynomial is integrated from x_{i-1} to x_{i+1}. We find

$$\int_{x_{i-1}}^{x_{i+1}} f(x)\,dx = \int_{x_{i-1}}^{x_{i+1}} \left[y_{i-1} + s\Delta y_{i-1} + \frac{s(s-1)}{2}\Delta^2 y_{i-1} + \frac{s(s-1)(s-2)}{6}h^3 f'''(\xi(x)) \right] dx$$

where $s = \dfrac{x - x_{i-1}}{h}$ is the scaled variable. An integration of the first three terms in the above integral gives the two panel formula

$$\int_{x_{i-1}}^{x_{i+1}} f(x)\,dx = \frac{h}{3}\left[y_{i-1} + 4y_i + y_{i+1} \right]$$

$$\text{or} \qquad \int_{x_{i-1}}^{x_{i+1}} f(x)\,dx = \frac{h}{3}\left[f(x_{i-1}) + 4f(x_i) + f(x_{i+1}) \right] \tag{7.48}$$

which is known as Simpson's 1/3 rule. We find that the integral of the error term gives zero and so we select the next term of the interpolating polynomial as the term to integrate. This gives the local error term associated with the Simpson's 1/3 rule as

$$local\ error = \int_{x_{i-1}}^{x_{i+1}} \binom{s}{4} h^4 f^{(iv)}(\xi(x))\,dx$$

$$local\ error = \frac{h^4}{24}\int_{x_{i-1}}^{x_{i+1}} (s^4 - 6s^3 + 11s^2 - 6s)f^{(iv)}(\xi(x))\,dx, \qquad s = \frac{x - x_{i-1}}{h}$$

The mean value theorem is used to simplify this integral to the form

$$local\ error = \frac{h^5}{24} f^{(iv)}(\zeta) \int_0^2 (s^4 - 6s^3 + 11s^2 - 6s)\,ds = -\frac{h^5}{90} f^{(iv)}(\zeta). \tag{7.49}$$

This gives the Simpson 1/3 rule with local error term

$$\int_{x_{i-1}}^{x_{i+1}} f(x)\,dx = \frac{h}{3}\left[y_{i-1} + 4y_i + y_{i+1} \right] - \frac{h^5}{90} f^{(iv)}(\zeta). \tag{7.50}$$

To apply this integration formula over the extended interval $[a, b]$, there must be an even number of panels associated with the extended interval.

The composite Simpson's 1/3 rule results when the area of all groups of panel-pairs are added. This formula has the form

$$\int_{x_0}^{x_{2n}} f(x)\,dx = \frac{h}{3}\left(y_0 + 4y_1 + 2y_2 + 4y_3 + \cdots + 2y_{2n-2} + 4y_{2n-1} + y_{2n}\right) + global\ error \quad (7.51)$$

where $1, 4, 2, 4, 2, 4, \ldots, 2, 4, 1$ is the pattern for the coefficients of the ordinates, with $a = x_0$ and $b = x_{2n}$. Addition of the error terms from each group of 2-panels gives

$$global\ error = -\frac{h^5}{90}f^{(iv)}(\zeta_1) - \frac{h^5}{90}f^{(iv)}(\zeta_2) - \cdots - \frac{h^5}{90}f^{(iv)}(\zeta_n)$$

$$global\ error = -\frac{h^5}{90}nf^{(iv)}(\zeta) = -\frac{(b-a)}{90}h^4f^{(iv)}(\zeta) \quad (7.52)$$

where the sum has been simplified by using the results from equation (7.21).

Newton-Cotes Formula

The integration of a Newton forward interpolating polynomial associated with equal spacing of n-panels results in a n-panel formula of the form

$$\int_{x_0}^{x_n} f(x)\,dx = c_0 h\left[w_0 f(x_0) + w_1 f(x_1) + \cdots + w_n f(x_n)\right] + Error \quad (7.53)$$

where $c_0, w_0, w_1, \ldots, w_n$ are constants. The resulting integration formulas are called Newton-Cotes closed formulas because the end points x_0 and x_n are used. In contrast Newton-Cotes open formulas do not use the end points. One can divide the Newton-Cotes closed formulas into two types. One type considers all panel groupings with an even number of ordinates and the other type considers panel groupings having and odd number of ordinates. The tables 7.1 and table 7.2 give selected Newton-Cotes formula.

Table 7.1 Newton-Cotes Formulas (Even number of ordinates)									
$\int_{x_0}^{x_n} f(x)\,dx = c_0 h\left[w_0 f(x_0) + w_1 f(x_1) + \cdots + w_n f(x_n)\right] + Error$									
n	c_0	w_0	w_1	w_2	w_3	w_4	w_5	w_6	w_7	Error
1	1/2	1	1							$-\frac{1}{12}h^3 f''(\xi)$
3	3/8	1	3	3	1					$-\frac{3}{80}h^5 f^{(iv)}(\xi)$
5	5/288	19	75	50	50	75	19			$-\frac{275}{12096}h^7 f^{(vi)}(\xi)$
7	7/17280	751	3577	1323	2989	2989	1323	3577	751	$-\frac{8183}{518400}h^9 f^{(viii)}(\xi)$

										Table 7.2 Newton-Cotes Formulas (Odd number of ordinates)

$$\int_{x_0}^{x_n} f(x)\, dx = c_0 h\left[w_0 f(x_0) + w_1 f(x_1) + \cdots + w_n f(x_n)\right] + Error$$

n	c_0	w_0	w_1	w_2	w_3	w_4	w_5	w_6	w_7	w_8	Error
2	1/3	1	4	1							$-\frac{1}{90}h^5 f^{(iv)}(\xi)$
4	2/45	7	32	12	32	7					$-\frac{8}{945}h^7 f^{(vi)}(\xi)$
6	1/140	41	216	27	272	27	216	41			$-\frac{9}{1400}h^9 f^{(viii)}(\xi)$
8	4/14175	989	5888	-928	10496	-4540	10496	-928	5888	989	$-\frac{2368}{467775}h^{11} f^{(x)}(\xi)$

In the tables 7.1 and 7.2 the quantity n represents the number of panels. Note that in going from n an odd number, to the next higher even number, the order of the error term jumps by a factor of two. Therefore, the Newton-Cotes formulas with an odd number of ordinates, or even number of panels, is preferred.

Integration formulas which do not include the end points x_0 and x_n are called Newton-Cotes open formulas and have the form

$$\int_{x_0}^{x_n} f(x)\, dx = c_0 h\left[w_1 f(x_1) + w_2 f(x_2) + \cdots + w_{n-1} f(x_{n-1})\right] + Error \qquad (7.54)$$

for $n = 2, 3, 4, \ldots$. The formulas are derived by constructing an interpolating polynomial through the interior points and then integrating over the whole interval.

Example 7-4. (Newton-Cotes open formula)

Construct a Newton-Cotes open formula associated with the integration over 5-panels as illustrated in the figure.

Solution:

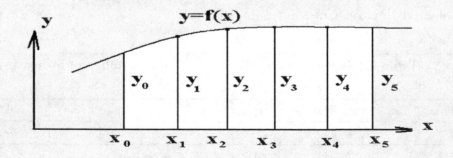

From the lozenge diagram of figure 4-1 we construct an interpolating polynomial $P_3(x)$ through the interior points $(x_1, y_1), (x_2, y_2), (x_3, y_3), (x_4, y_4)$ and then approximate the area under the curve $y = f(x)$ by integrating the polynomial over the five panels. We find

$$P_3(x) = y_1 + \binom{s-1}{1}\Delta y_1 + \binom{s-1}{2}\Delta^2 y_1 + \binom{s-1}{3}\Delta^3 y_1 + \binom{s-1}{4}h^4 f^{(iv)}(\xi) \qquad (7.55)$$

and then integrate this polynomial from x_0 to x_5 to obtain

$$\int_{x_0}^{x_5} f(x)\,dx = h\int_{s=0}^{5}\left[y_1 + \binom{s-1}{1}\Delta y_1 + \binom{s-1}{2}\Delta^2 y_1 + \binom{s-1}{3}\Delta^3 y_1\right]ds$$

with error term

$$local\ error = h^5 f^{(iv)}(\zeta)\int_{s=0}^{5}\binom{s-1}{4}ds.$$

One can evaluate the above integrals and simplify them to obtain the Newton-Cotes open formula

$$\int_{x_0}^{x_5} f(x)\,dx = \frac{5}{24}h[11y_1 + y_2 + y_3 + 11y_4] + \frac{95}{144}h^5 f^{(iv)}(\zeta). \qquad (7.56)$$

■

The table 7.3 depicts some selected Newton-Cotes open formulas. The most popular of these formulas is the midpoint rule

$$\int_{x_0}^{x_2} f(x)\,dx = 2hf(x_1) + \frac{h^3}{3}f''(\zeta), \qquad x_0 < \zeta < x_2 \qquad (7.57)$$

which is sometimes expressed in the form

$$\int_{a}^{b} f(x)\,dx = (b-a)f\left(\frac{a+b}{2}\right) + \frac{(b-a)^3}{24}f''(\zeta), \qquad a < \zeta < b. \qquad (7.58)$$

Table 7.3 Newton-Cotes Open Formulas						
$\int_{x_0}^{x_n} f(x)\,dx = c_0 h\left[w_1 f(x_1) + w_2 f(x_2) + \cdots + w_{n-1} f(x_{n-1})\right] + Error$						
n	c_0	w_1	w_2	w_3	w_4	w_5	*Error*
2	2	1					$\frac{1}{3}h^3 f''(\xi)$
3	3/2	1	1				$\frac{3}{4}h^3 f''(\xi)$
4	4/3	2	-1	2			$\frac{14}{45}h^5 f^{(iv)}(\xi)$
5	5/24	11	1	1	11		$\frac{95}{144}h^5 f^{(iv)}(\xi)$
6	3/10	11	-14	26	-14	11	$\frac{41}{140}h^7 f^{vi}(\xi)$

The Newton-Cotes open formulas are not as accurate as the Newton-Cotes closed formulas. They are sometimes used to evaluate integrals of functions which have a singularity at an endpoint. The higher order Newton-Cotes formulas are seldom used because they tend to oscillate and propagate round off errors and can have convergence problems for some types of functions. The lower order Newton-Cotes formulas are easy to use with computer programming and very accurate results can be obtained with proper utilization of the panel lengths. One important application of the trapezoidal rule involves Romberg's method of integration.

Romberg Integration

The simplest Newton-Cotes formula is the trapezoidal rule. Using the trapezoidal rule along with interval halving can produce very accurate answers. To numerically evaluate an integral $I = \int_a^b f(x)\,dx$ we partition the interval $[a, b]$ into n-panels of equal width h and write

$$I = \int_a^b f(x)\,dx = \sum_{j=1}^n \int_{x_{j-1}}^{x_j} f(x)\,dx = \sum_{j=1}^n \int_{x_{j-1}}^{x_j} P(x)\,dx + Error = V_1(h) + Error \quad (7.59)$$

where $V_1(h)$ represents a value obtained by integrating an approximating polynomial $P(x)$ in place of the function $f(x)$.

The Richardson extrapolation applies whenever the error term associated with an approximation formula can be written in the form

$$Error = c_2 h^2 + c_4 h^4 + c_6 h^6 + \cdots + c_{2m} h^{2m} + \mathcal{O}(h^{2m+2}) \quad (7.60)$$

with coefficients $c_2, c_4, \ldots, c_{2m}$ which are constants. The trapezoidal rule satisfies this condition and so one can write

$$I = V_1(h) + c_2 h^2 + c_4 h^4 + c_6 h^6 + \cdots + c_{2m} h^{2m} + \mathcal{O}(h^{2m+2}) \quad (7.61)$$

Now if we halve the step size h, we expect the numerical approximation obtained for the integral would be more accurate. Therefore, we replace h by $h/2$ in equation (7.61) and obtain

$$I = V_1(h/2) + c_2 \frac{h^2}{2^2} + c_4 \frac{h^4}{2^4} + \cdots + c_{2m} \frac{h^{2m}}{2^{2m}} + \mathcal{O}(h^{2m+2}) \quad (7.62)$$

Now eliminate c_2 from the equations (7.61) and (7.62) and solve for I to obtain

$$I = \frac{4V_1(h/2) - V_1(h)}{4 - 1} + D_4 h^4 + D_6 h^6 + \cdots \quad (7.63)$$

where $D_4, D_6, \ldots$ are new coefficients. Let

$$V_2(h) = \frac{4V_1(h/2) - V_1(h)}{4 - 1} \tag{7.64}$$

denote an improved value for the integral I. If we halve the step size again one can obtain $\mathcal{O}(h^6)$ error terms. That is, if

$$I = V_2(h) + D_4 h^4 + D_6 h^6 + \cdots \tag{7.65}$$

then halving the step size gives

$$I = V_2(h/2) + D_4 \frac{h^4}{2^4} + D_6 \frac{h^6}{2^6} + \cdots \tag{7.66}$$

so by eliminating the $D_4 h^4$ term from the equations (7.65) and (7.66) and solving for I we find the still more improved value

$$I = \frac{4^2 V_2(h/2) - V_2(h)}{4^2 - 1} + E_6 h^6 + \cdots \tag{7.67}$$

where

$$V_3(h) = \frac{4^2 V_2(h/2) - V_2(h)}{4^2 - 1} \tag{7.68}$$

is an improved value for the integral I. Continue to halve the step size, then one can verify the relation

$$V_{j+1}(h) = \frac{4^j V_j(h/2) - V_j(h)}{4^j - 1} \tag{7.69}$$

is an improved value for the integral I with error $\mathcal{O}(h^{2j+2})$ for $j = 1, 2, \ldots$. The equation (7.69) is known as the Richardson extrapolation formula.

The Romberg integration method makes use of the fact that the trapezoidal rule can be improved by halving the step size and then the results obtained can be improved using the Richardson extrapolation formula. The table 7.4 illustrates the Romberg integration scheme associated with the halving of the panel size.

In table 7.4, one starts off with a large step size h_1 and approximates the area by the trapezoidal rule to obtain the value $V_{1,1}$. The step size is then halved and the area $V_{1,2}$ is calculated. The column three in the table 7.4 are the areas calculated by the trapezoidal rule for the given step size each time it is halved. Note that certain shortcuts can be taken in calculating the values in column three. The first approximation uses a step size $h_1 = b - a$ and we calculate by the trapezoidal rule

$$V_{1,1} = \frac{h_1}{2}\left[f(a) + f(b)\right].$$

	Error	$\mathcal{O}(h^2)$	$\mathcal{O}(h^4)$	$\mathcal{O}(h^6)$	$\mathcal{O}(h^8)$	$\mathcal{O}(h^{10})$	$\mathcal{O}(h^{12})$
n	Step Size	$j = 1 \rightarrow$	$j = 2 \rightarrow$	$j = 3 \rightarrow$	$j = 4 \rightarrow$	$j = 5 \rightarrow$	$j = 6 \rightarrow$
1	h_1	$V_{1,1}$					
2	$h_2 = h_1/2$	$V_{1,2}$	$V_{2,2}$				
3	$h_3 = h_2/2$	$V_{1,3}$	$V_{2,3}$	$V_{3,3}$			
4	$h_4 = h_3/2$	$V_{1,4}$	$V_{2,4}$	$V_{3,4}$	$V_{4,4}$		
5	$h_5 = h_4/2$	$V_{1,5}$	$V_{2,5}$	$V_{3,5}$	$V_{4,5}$	$V_{5,5}$	
6	$h_6 = h_5/2$	$V_{1,6}$	$V_{2,6}$	$V_{3,6}$	$V_{4,6}$	$V_{5,6}$	$V_{6,6}$
$\vdots$	$\vdots$	$\vdots$	$\vdots$	$\vdots$	$\vdots$	$\vdots$	$\vdots$

Table 7.4 Romberg Integration with trapezoidal rule for the evaluation of $\int_a^b f(x)\,dx$

One then halves the step size to obtain $h_2 = h_1/2$ and calculates by the trapezoidal rule the improved value

$$V_{1,2} = \frac{h_2}{2}[f(a) + f(a+h_2)] + \frac{h_2}{2}[f(a+h_2) + f(b)].$$

The step size is halved again with $h_3 = h_2/2$ and by the trapezoidal rule we calculate

$$V_{1,3} = \frac{h_3}{2}[f(a) + f(b)] + h_3 \sum_{i=1}^{3} f(a + ih_3).$$

After n-halving operations the area is calculated by the trapezoidal rule from the relation

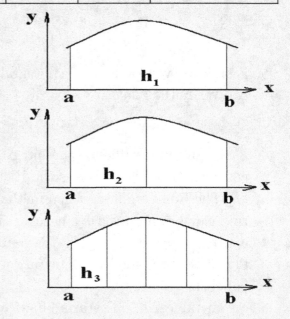

$$V_{1,m} = \frac{h_m}{2}[f(a) + f(b)] + h_m \sum_{i=1}^{m} f(a + ih_m)$$

where $h_m = (b-a)/2^{m-1}$ for $m = 1, 2, 3, \ldots$. A shortcut in calculating the area $V_{1,m+1}$

is to observed that one can make use of the previous calculations and write

$$V_{1,m+1} = \frac{1}{2}V_{1,m} + h_{m+1}\sum_{i=1}^{2^{m-1}} f(a + (2i-1)h_{m+1}) \qquad \text{where} \qquad h_{m+1} = \frac{b-a}{2^m} \qquad (7.70)$$

which can be readily verified.

In table 7.4 the fourth column represents the improved values which are calculated using the Richardson extrapolation formula (7.69) with $j = 1$. For example

$$V_{2,2} = \frac{4V_{1,2} - V_{1,1}}{3}, \quad V_{2,3} = \frac{4V_{1,3} - V_{1,2}}{3}, \dots$$

The values in column five of table 7.4 are improvements of the column four values which are calculated using the Richardson extrapolation formula (7.69) with $j = 2$. For example,

$$V_{3,3} = \frac{4^2 V_{2,3} - V_{2,2}}{4^2 - 1}, \quad V_{3,4} = \frac{4^2 V_{2,4} - V_{2,3}}{4^2 - 1}, \dots$$

The columns $6, 7, \dots$ in table 7.4 are improvements over the values in the preceding column using the Richardson extrapolation formula with the proper value of the index j in equation (7.69).

Note the Richardson extrapolation formula can be viewed as a weighted average of the previous calculations. One can write

$$Improved\ value = \frac{w_1(More\ accurate) + w_2(Less\ accurate)}{w_1 + w_2}$$

where the weight $w_2 = -1$ and the weight $w_1 = 4^j$ gets larger as the accuracy of the calculations improve.

Example 7-5.   (Romberg Integration)

Use Romberg integration to evaluate the integral $I = \int_0^1 e^{-x^2}\,dx$.

Solution: As a first approximation to the integral I we use the trapezoidal rule with $h_1 = 1$, $a = 0$ and $b = 1$ with $f(x) = e^{-x^2}$ to obtain

$$V_{1,1} = \frac{h_1}{2}[f(a) + f(b)] = \frac{1}{2}[e^0 + e^{-1}] = 0.68394$$

Now halve the interval with $h_2 = h_1/2 = 1/2$ and calculate

$$V_{1,2} = \frac{1}{2}V_{1,1} + h_2 f(a + h_2) = \frac{1}{2}V_{1,1} + \frac{1}{2}e^{-1/4} = 0.73137$$

We halve the interval again and using $h_3 = h_2/2 = 1/4$ we calculate the improved value

$$V_{1,3} = \frac{1}{2}V_{1,2} + h_3 \left[f(a + h_3) + f(a + 3h_3) \right]$$

$$V_{1,3} = \frac{1}{2}V_{1,2} + \frac{1}{4} \left[e^{-1/16} + e^{-9/16} \right] = 0.742984$$

Another halving of the interval gives $h_4 = h_3/2 = 1/8$ with

$$V_{1,4} = \frac{1}{2}V_{1,3} + h_4 \left[f(a + h_4) + f(a + 3h_4) + f(a + 5h_4) + f(a + 7h_4) \right]$$

$$V_{1,4} = \frac{1}{2}V_{1,3} + \frac{1}{8} \left[e^{-1/64} + e^{-9/64} + e^{-25/64} + e^{-49/64} \right] = 0.745866$$

The table 7.5 is constructed using the Richardson extrapolation formula for the improved values.

	Table 7.5 Romberg Integration with trapezoidal rule for the evaluation of $\int_0^1 e^{-x^2}\,dx$				
	Error	$\mathcal{O}(h^2)$	$\mathcal{O}(h^4)$	$\mathcal{O}(h^6)$	$\mathcal{O}(h^8)$
n	Step Size	$j = 1 \rightarrow$	$j = 2 \rightarrow$	$j = 3 \rightarrow$	$j = 4 \rightarrow$
1	$h_1 = 1$	0.68394			
2	$h_2 = 1/2$	0.73137	0.74718		
3	$h_3 = 1/4$	0.742984	0.746855	0.746834	
4	$h_4 = 1/8$	0.745866	0.746826	0.746824	0.746824

One can see that the final answer is accurate to the digits displayed.

Adaptive Quadrature

The Romberg method of integration requires the calculation of many equally spaced ordinate values in addition to the many applications of the Richardson extrapolation formula. In an effort to cut down on the number of calculations necessary for the evaluation of an integral, the adaptive quadrature method was devised. Before evaluating an integral

$$I = \int_a^b f(x)\,dx \tag{7.71}$$

one should sketch a graph of the integrand $f(x)$ over the interval $[a, b]$ to estimate what kind of step size is necessary for the calculation. If the function $f(x)$

is a fairly flat curve, then one can use a large step size h. However, if the function tends to oscillate rapidly, then one will be forced to use a small step size h to maintain accuracy. The adaptive quadrature takes the accuracy of the calculations into account in calculating the integral. To evaluate the integral (7.71) with known error $\epsilon > 0$ the adaptive quadrature method employs the Simpson 1/3 rule and interval halving in order to achieve the specified error tolerance.

Begin by calculating the area A_1 associated with halving the interval $[a, b]$ to create two panels. We use $h_1 = \frac{b-a}{2}$ and calculate by Simpson's 1/3 rule the approximate area

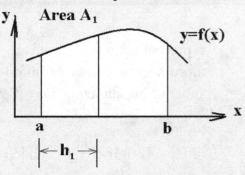

$$A_1 = \frac{h_1}{3}\left[f(a) + 4f(a + h_1) + f(b)\right] \qquad (7.72)$$

One can then write

$$I = A_1 + Error_1 \qquad (7.73)$$

where

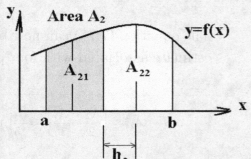

$$Error_1 = -\frac{1}{90}f^{(iv)}(\xi_1)h_1^5, \qquad a < \xi_1 < b.$$

Now halve the step size and use $h_2 = h_1/2$ as the step size associated with four panels and calculate by Simpson's 1/3 rule the area estimate

$$A_2 = A_{21} + A_{22}$$

where

$$A_{21} = \frac{h_2}{3}\left[f(a) + 4f(a + h_2) + f(a + 2h_2)\right]$$
$$A_{22} = \frac{h_2}{3}\left[f(a + 2h_2) + 4f(a + 3h_2) + f(a + 4h_2)\right] \qquad (7.74)$$

with respective local errors

$$-\frac{1}{90}f^{(iv)}(\xi_2)h_2^5 \quad \text{and} \quad -\frac{1}{90}f^{(iv)}(\xi_3)h_2^5.$$

By the composite Simpson's rule we can write

$$I = A_2 + Error_2 \quad \text{where} \quad Error_2 = -\frac{2}{90}f^{(iv)}(\xi_4)h_2^5, \qquad a < \xi_4 < b \qquad (7.75)$$

is a sum of the local errors.

Adaptive quadrature makes the assumption that $\xi_1 = \xi_4 = \xi$ so the errors can be compared by examining the equations

$$I - A_1 = -\frac{1}{90} f^{(iv)}(\xi) h_1^5 = Error_1$$

$$I - A_2 = -\frac{2}{90} f^{(iv)}(\xi) \frac{h_1^5}{2^5} = Error_2$$

(7.76)

representing a less accurate area estimate associated with two panels and a more accurate area estimate associated with four panels. One can obtain by subtracting the equations (7.76) either of the relations

$$A_2 - A_1 = \frac{h_1^5}{90} f^{(iv)}(\xi) \left(\frac{1}{2^4} - 1\right) \quad \text{or} \quad \frac{h_1^5}{90} f^{(iv)}(\xi) = \frac{2^4}{1 - 2^4}(A_2 - A_1).$$

(7.77)

The equation (7.77) can now be used to produce a bound on the improved area estimate associated with four panels, since

$$|I - A_2| \leq \left|\frac{A_2 - A_1}{2^4 - 1}\right| \leq \frac{1}{15}|A_2 - A_1| = \frac{1}{15}\left|\begin{pmatrix} More \\ accurate \\ area \end{pmatrix} - \begin{pmatrix} Less \\ accurate \\ area \end{pmatrix}\right|$$

(7.78)

Note that if $\frac{1}{15}|A_2 - A_1| < \epsilon$ we have achieved the desired error bound. In general, this will not be achieved after just one interval halving. Therefore, we repeat what we have just accomplished on each half of the area with the new error criteria that the error associated with each half be less than $\epsilon/2$. If it becomes necessary to halve an interval again the new error criteria becomes $\epsilon/4$. We continue to use the equation (7.78) applied to each half section until the error criteria is satisfied.

Example 7-6. (Adaptive quadrature)

Use adaptive quadrature to evaluate the integral $I = \int_0^1 e^{-x^2} \, dx$ accurate to within an error tolerance of $\epsilon = 1.0 \times 10^{-5}$.

Solution: We use $h_1 = 1/2$ and calculate by Simpson's 1/3 rule

$$A_1 = \frac{h_1}{3}\left[e^{-(0)^2} + 4e^{-h_1^2} + e^{-(2h_1)^2}\right] = 0.747180$$

Now halve the interval using $h_2 = 1/4$ and calculate the improved estimate for the area using Simpson's 1/3 rule on each half. One finds

$$A_{21} = \frac{h_2}{3}\left[e^{-(0)^2} + 4e^{-h_2^2} + e^{-(2h_2)^2}\right] = 0.461371$$

$$A_{22} = \frac{h_2}{3}\left[e^{-2h_2)^2} + 4e^{-(3h_2)^2} + e^{-(4h_2)^2}\right] = 0.285484$$

This gives the improved area estimate

$$A_2 = A_{21} + A_{22} = 0.746855$$

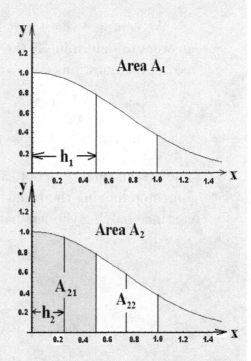

By comparing the 4-panel approximation with the 2-panel approximation we find

$$\frac{1}{15}\left|\left(\begin{array}{c} More \\ accurate \\ area \end{array}\right) - \left(\begin{array}{c} Less \\ accurate \\ area \end{array}\right)\right| = \frac{1}{15}|A_2 - A_1| = 2.16699 \times 10^{-5}$$

Now ask the question, "Is this difference less than the given ϵ?" The answer is "No", and so the intervals must be halved again. Let $h_3 = h_2/2 = 1/8$ and use Simpson's 1/3 rule to calculate the areas

$$A_{31} = \frac{h_3}{3}\left[e^{-(0)^2} + 4e^{-(h_3)^2} + e^{-(2h_3)^2}\right] = 0.244892$$

$$A_{32} = \frac{h_3}{3}\left[e^{-(2h_3)^2} + 4e^{-(3h_3)^2} + e^{-(4h_3)^2}\right] = 0.216395$$

$$A_{33} = \frac{h_3}{3}\left[e^{-(4h_3)^2} + 4e^{-(5h_3)^2} + e^{-(6h_3)^2}\right] = 0.168963$$

$$A_{34} = \frac{h_3}{3}\left[e^{-(6h_3)^2} + 4e^{-(7h_3)^2} + e^{-(8h_3)^2}\right] = 0.116578$$

which gives

$$A_3 = A_{31} + A_{32} + A_{33} + A_{34} = 0.746826.$$

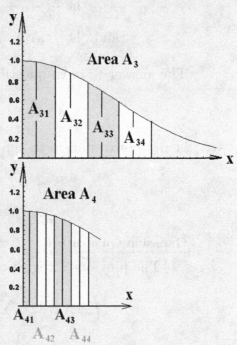

We compare the 4-panel improved areas with the old 2-panel less accurate answers to determine if the errors associated with each section are satisfied by asking the questions

$$\text{``Is} \quad \frac{1}{15}\,|A_{31} + A_{32} - A_{21}| = 5.64774 \times 10^{-6} < \frac{\epsilon}{2} = 5.0 \times 10^{-6} \quad ?\text{''}$$

$$\text{and ``Is} \quad \frac{1}{15}\,|A_{33} + A_{34} - A_{22}| = 3.697124 \times 10^{-6} < \frac{\epsilon}{2} = 5.0 \times 10^{-6} \quad ?\text{''}$$

The first inequality is not satisfied and the second inequality is satisfied. Therefore, stop halving the right-half and use the calculated areas $A_{33} + A_{34}$. We must continue halving the left-half. Let $h_4 = h_3/2 = 1/16$ and calculate

$$A_{41} = \frac{h_4}{3}\left[e^{-(0)^2} + 4e^{-h_4^2} + e^{-(2h_4)^2}\right] = 0.124352$$

$$A_{42} = \frac{h_4}{3}\left[e^{-(2h_4)^2} + 4e^{-(3h_4)^2} + e^{-(4h_4)^2}\right] = 0.120536$$

$$A_{43} = \frac{h_4}{3}\left[e^{-(4h_4)^2} + 4e^{-(5h_4)^2} + e^{-(6h_4)^2}\right] = 0.113251$$

$$A_{44} = \frac{h_4}{3}\left[e^{-(6h_4)^2} + 4e^{-(7h_4)^2} + e^{-(8h_4)^2}\right] = 0.103142$$

Now compare the 4-panel more accurate results with the less accurate 2-panel results previously calculated and ask the questions

$$\text{``Is} \quad \frac{1}{15}\,|A_{41} + A_{42} - A_{31}| = 2.33331 \times 10^{-7} < \frac{\epsilon}{4} = 2.5 \times 10^{-6} \ ?\text{''}$$

$$\text{and ``Is} \quad \frac{1}{15}\,|A_{43} + A_{44} - A_{32}| = 1.02156 \times 10^{-7} < \frac{\epsilon}{4} = 2.5 \times 10^{-6} \ ?\text{''}$$

The answer to both questions is "Yes" and therefore we can write

$$A = A_{33} + A_{34} + A_{41} + A_{42} + A_{43} + A_{44} = 0.746821$$

is an approximation to the integral I which satisfies

$$|I - A| = \left|\int_0^1 e^{-x^2}\,dx - A\right| < \epsilon = 1.0 \times 10^{-5}.$$

$\blacksquare$

Gaussian Quadrature

The previous integration formulas can be expressed in the form

$$\int_{x_0}^{x_n} f(x)\,dx = \sum_{i=0}^{n} \alpha_i f(x_i) = \alpha_0 f(x_0) + \alpha_1 f(x_1) + \cdots + \alpha_n f(x_n) \tag{7.79}$$

where the points $x_0, x_1, \ldots, x_n$, called nodes, are equally spaced over the interval $a = x_0 < x_1 < \cdots < x_n = b$ with step size $h = (b-a)/n$ and α_i, $i = 0, 1, \ldots, n$ are constants called weights associated with the integration formula. For example, the Newton-Cotes open and closed formulas were derived by integrating interpolating polynomials of degree m over some interval. These type of integration formulas then give exact values in the special case $f(x)$ is a polynomial of degree k where $k \leq m$.

In contrast to equally spaced intervals being used to calculate area under a curve the Gaussian quadrature formulas use unequal spacing for determining area under a curve. Each Gaussian quadrature formula is associated with an orthogonal set of functions over some interval. (See pages 149-155.) These quadrature formulas have the general form

$$I = \int_{x_0}^{x_n} w(x)f(x)\,dx = \sum_{i=1}^{m} \beta_i f(x_i) = \beta_1 f(x_1) + \beta_2 f(x_2) + \cdots + \beta_m f(x_m) \qquad (7.80)$$

where $w(x)$ is a weighting function associated with the orthogonal set of functions and $f(x)$ is continuous over the interval (x_0, x_n). In the equation (7.80) both the nodes $x_1, \ldots, x_m$ and the weights $\beta_1, \ldots, \beta_m$, are treated as unknowns which are to be selected such that the integration formula (7.80) gives exact results for polynomials of largest possible degree. Consequently, the Gaussian integration formulas have $2m$, $(\beta_1, x_1, \beta_2, x_2, \ldots, \beta_m, x_m)$, unknowns to be determined.

We start by developing selected special Gaussian quadrature formulas and conclude by developing a general technique for determining the unknowns $x_1, x_2, \ldots, x_m$ and $\beta_1, \beta_2, \ldots, \beta_m$ associated with orthogonal polynomials in general. For illustrative purposes we select from the large class of Gaussian quadrature formulas only the special integration formulas known by the names of (i) Gauss-Legendre (ii) Gauss-Laguerre and (iii) Gauss-Chebyshev.

Gauss-Legendre Integration

The Legendre polynomials $P_n(x)$ are given by the Rodriques' formula

$$P_n(x) = \frac{1}{2^n n!} \frac{d^n}{dx^n}(x^2-1)^n \qquad (7.81)$$

One can verify the first couple of Legendre polynomials are given by

$$\begin{aligned} P_0(x) &= 1 & P_2(x) &= \frac{1}{2}(3x^2-1) \\ P_1(x) &= x & P_3(x) &= \frac{1}{2}(5x^3-3x) \end{aligned} \qquad (7.82)$$

270

and then generate higher order Legendre polynomials by use of the recurrence relation

$$P_{n+1}(x) = \frac{(2n+1)xP_n(x) - nP_{n-1}(x)}{n+1} \tag{7.83}$$

for $n = 1, 2, 3, \ldots$. The Legendre polynomials $P_n(x)$ for $n = 0, 1, 2, 3, 4$ are illustrated in the figure 7-2.

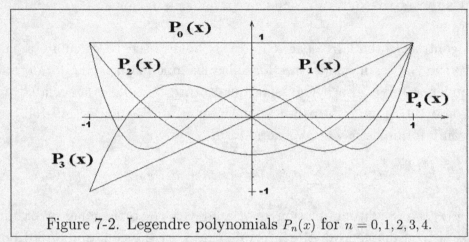

Figure 7-2. Legendre polynomials $P_n(x)$ for $n = 0, 1, 2, 3, 4$.

The set of functions $\{P_n(x)\}$, for $n = 0, 1, 2, \ldots$ and $-1 \le x \le 1$ are orthogonal over the interval $(-1, 1)$ and have the inner product property

$$(P_n, P_m) = \int_{-1}^{1} P_n(x)P_m(x)\, dx = \begin{cases} 0, & \text{if } m \ne n \\ \| P_n \|^2, & \text{if } m = n \end{cases} \tag{7.84}$$

where m, n are integers and

$$\| P_n \|^2 = (P_n, P_n) = \int_{-1}^{1} P_n^2(x)\, dx = \frac{2}{2n+1} \tag{7.85}$$

for $n = 0, 1, 2, \ldots$ is called the norm squared. Gauss-Legendre integration formulas have the form

$$\int_{-1}^{1} f(x)\, dx = \sum_{i=1}^{m} \beta_i f(x_i) \tag{7.86}$$

Example 7-7. (Gauss-Legendre formula $m = 1$)
In the case $m = 1$ the Gauss-Legendre integration formula (7.86) reduces to

$$\int_{-1}^{1} f(x)\, dx = \beta_1 f(x_1) \tag{7.87}$$

where β_1 and x_1 are unknowns to be determined. We require that the equation (7.87) be exact for $f(x) = 1$ and $f(x) = x$. Note that the integration formula will then be exact for $f(x)$ some linear combination of these functions. For $f(x) = 1$ the equation (7.87) becomes

$$\int_{-1}^{1} dx = 2 = \beta_1(1) \quad \Rightarrow \beta_1 = 2 \tag{7.88}$$

and for $f(x) = x$ the equation (7.87) becomes

$$\int_{-1}^{1} x\, dx = 0 = \beta_1 x_1 \quad \Rightarrow x_1 = 0 \tag{7.89}$$

This gives the midpoint rule

$$\int_{-1}^{1} f(x)\, dx = 2f(0) \tag{7.90}$$

∎

Example 7-8. **(Gauss-Legendre formula $m = 2$)**
In the case $m = 2$ the Gauss-Legendre formula (7.86) becomes

$$\int_{-1}^{1} f(x)\, dx = \beta_1 f(x_1) + \beta_2 f(x_2) \tag{7.91}$$

where $\beta_1, \beta_2, x_1, x_2$ are unknowns to be determined. We require that the equation (7.91) give exact results if $f(x)$ is one of the functions $1, x, x^2$ or x^3. It therefore produces an exact result whenever $f(x)$ is a linear combination of these functions. For $f(x) = 1$ the equation (7.91) becomes

$$\int_{-1}^{1} 1\, dx = 2 = \beta_1(1) + \beta_2(1) \tag{7.92}$$

For $f(x) = x$ the equation (7.91) becomes

$$\int_{-1}^{1} x\, dx = 0 = \beta_1 x_1 + \beta_2 x_2 \tag{7.93}$$

For $f(x) = x^2$, the equation (7.91) becomes

$$\int_{-1}^{1} x^2\, dx = \frac{2}{3} = \beta_1 x_1^2 + \beta_2 x_2^2 \tag{7.94}$$

For $f(x) = x^3$, the equation (7.91) becomes

$$\int_{-1}^{1} x^3\, dx = 0 = \beta_1 x_1^3 + \beta_2 x_2^3 \tag{7.95}$$

We must now solve the equations (7.92), (7.93), (7.94), and (7.95) for the unknowns β_1, β_2, x_1 and x_2. To solve these equations we form an equivalent set of equations which involve the Legendre polynomials. Multiply equation (7.92) by $-1/2$ and equation (7.94) by $3/2$ and add the results to obtain

$$\beta_1 P_2(x_1) + \beta_2 P_2(x_2) = 0.$$

Now multiply the equation (7.93) by $-1/2$ and equation (7.95) by $3/2$ and add the results to obtain

$$\beta_1 x_1 P_2(x_1) + \beta_2 x_2 P_2(x_2) = 0.$$

We can now replace the equations (7.92), (7.93), (7.94), (7.95) by the equivalent system

$$\begin{aligned}
\beta_1 + \beta_2 &= 2 \\
\beta_1 x_1 + \beta_2 x_2 &= 0 \\
\beta_1 P_2(x_1) + \beta_2 P_2(x_2) &= 0 \\
\beta_1 x_1 P_2(x_1) + \beta_2 x_2 P_2(x_2) &= 0
\end{aligned}$$ (7.96)

Note the last two equations of the system (7.96) are now easily solved if we select x_1, x_2 as the roots of the polynomial

$$P_2(x) = \frac{3}{2}x^2 - \frac{1}{2} = 0$$

which are $x_1 = -\dfrac{1}{\sqrt{3}} = -0.57735026919$ and $x_2 = +\dfrac{1}{\sqrt{3}} = 0.57735026919$. Substituting these values into the first two equations of the system (7.96) one finds the values $\beta_1 = 1$ and $\beta_2 = 1$. This gives the Gauss-Legendre integration formula

$$\int_{-1}^{1} f(x)\, dx = f(-\frac{1}{\sqrt{3}}) + f(\frac{1}{\sqrt{3}})$$ (7.97)

∎

Example 7-9. **(Gauss-Legendre formula $m = 3$)**
In the case $m = 3$ the Gauss-Legendre formula (7.86) becomes

$$\int_{-1}^{1} f(x)\, dx = \beta_1 f(x_1) + \beta_2 f(x_2) + \beta_3 f(x_3)$$ (7.98)

where $\beta_1, \beta_2, \beta_3, x_1, x_2, x_3$ are unknowns to be determined. We require that the integration formula (7.98) give exact results whenever $f(x)$ is one of the functions

$\{1, x, x^2, x^3, x^4, x^5\}$. It will then give exact results whenever $f(x)$ is some linear combination of these functions. The equation (7.98) becomes

$$\text{for } f(x) = 1 \qquad 2 = \beta_1 + \beta_2 + \beta_3 \tag{7.99}$$

$$\text{for } f(x) = x \qquad 0 = \beta_1 x_1 + \beta_2 x_2 + \beta_3 x_3 \tag{7.100}$$

$$\text{for } f(x) = x^2 \qquad \frac{2}{3} = \beta_1 x_1^2 + \beta_2 x_2^2 + \beta_3 x_3^2 \tag{7.101}$$

$$\text{for } f(x) = x^3 \qquad 0 = \beta_1 x_1^3 + \beta_2 x_2^3 + \beta_3 x_3^3 \tag{7.102}$$

$$\text{for } f(x) = x^4 \qquad \frac{2}{5} = \beta_1 x_1^4 + \beta_2 x_2^4 + \beta_3 x_3^4 \tag{7.103}$$

$$\text{for } f(x) = x^5 \qquad 0 = \beta_1 x_1^5 + \beta_2 x_2^5 + \beta_3 x_3^5 \tag{7.104}$$

We form an equivalent set of equations which are easier to solve. Multiply equation (7.100) by $-3/2$ and equation (7.102) by $5/2$ and add the results to obtain

$$\beta_1 P_3(x_1) + \beta_2 P_3(x_2) + \beta_3 P_3(x_3) = 0.$$

Now multiply equation (7.101) by $-3/2$ and equation (7.103) by $5/2$ and add the results to obtain

$$\beta_1 x_1 P_3(x_1) + \beta_2 x_2 P_3(x_2) + \beta_3 x_3 P_3(x_3) = 0.$$

Finally, multiply equation (7.102) by $-3/2$ and equation (7.104) by $5/2$ and add the results to obtain

$$\beta_1 x_1^2 P_3(x_1) + \beta_2 x_2^2 P_3(x_2) + \beta_3 x_3^2 P_3(x_3) = 0.$$

The equations (7.99) through (7.104) can now be replaced by the equivalent system of equations

$$\begin{aligned}
\beta_1 + \beta_2 + \beta_3 &= 2 \\
\beta_1 x_1 + \beta_2 x_2 + \beta_3 x_3 &= 0 \\
\beta_1 x_1^2 + \beta_2 x_2^2 + \beta_3 x_3^2 &= 2/3 \\
\beta_1 P_3(x_1) + \beta_2 P_3(x_2) + \beta_3 P_3(x_3) &= 0 \\
\beta_1 x_1 P_3(x_1) + \beta_2 x_2 P_3(x_2) + \beta_3 x_3 P_3(x_3) &= 0 \\
\beta_1 x_1^2 P_3(x_1) + \beta_2 x_2^2 P_3(x_2) + \beta_3 x_3^2 P_3(x_3) &= 0
\end{aligned} \tag{7.105}$$

The last three equations are satisfied if we select x_1, x_2, x_3 as the roots of

$$P_3(x) = \frac{1}{2}(5x^3 - 3x) = 0.$$

These roots are easily determined and one finds that $x_1 = -\sqrt{3/5} = -0.77459666924$, $x_2 = 0$, and $x_3 = \sqrt{3/5} = 0.77459666924$. One can now substitute these values into the first three equations of the system (7.105) and find the values $\beta_1 = 5/9$, $\beta_2 = 8/9$ and $\beta_3 = 5/9$. This gives the Gauss-Legendre integration formula

$$\int_{-1}^{1} f(x)\, dx = \frac{5}{9} f(-\sqrt{3/5}) + \frac{8}{9} f(0) + \frac{5}{9} f(\sqrt{3/5}) \tag{7.106}$$

Higher order Gauss-Legendre integration formulas can be derived in a similar fashion. The table 7.6 lists the nodes and weights associated with selected Gauss-Legendre integration formulas. A more extensive table of parameter values can be found in the reference Abramowitz and Stegun listed in the bibliography.

An integral of the form

$$I = \int_{a}^{b} F(x)\, dx \tag{7.107}$$

can be reduced to a Gauss-Legendre integral by using the transformation of coordinates

$$x = \frac{1}{2}\left[a + b + t(b - a)\right], \qquad -1 \le t \le 1 \tag{7.108}$$

with $dx = \frac{1}{2}(b - a)\, dt$. The equation (7.107) can then be written in the form

$$I = \frac{b - a}{2} \int_{-1}^{1} F\left(\frac{a + b + t(b - a)}{2}\right) dt \tag{7.109}$$

Table 7.6 Gauss-Legendre integration formula

$$\int_{-1}^{1} f(x)\,dx = \beta_1 f(x_1) + \beta_2 f(x_2) + \cdots + \beta_m f(x_m)$$

m	Nodes x_i	Weights β_i
1	$x_1 = 0.0000000000$	$\beta_1 = 2$
2	$x_1 = -0.57735026921$	$\beta_1 = 1$
	$x_2 = 0.57735026921$	$\beta_2 = 1$
3	$x_1 = -0.7745966692$	$\beta_1 = 5/9$
	$x_2 = 0.0000000000$	$\beta_2 = 8/9$
	$x_3 = 0.7745966692$	$\beta_3 = 5/9$
4	$x_1 = -0.8611363116$	$\beta_1 = 0.3478548451$
	$x_2 = -0.3399810436$	$\beta_2 = 0.6521451548$
	$x_3 = 0.3399810436$	$\beta_3 = 0.6521451548$
	$x_4 = 0.8611363116$	$\beta_4 = 0.3478548451$
5	$x_1 = -0.9061798459$	$\beta_1 = 0.2369268851$
	$x_2 = -0.5384693101$	$\beta_2 = 0.4786286705$
	$x_3 = 0.0000000000$	$\beta_3 = 0.5688888889$
	$x_4 = 0.5384693101$	$\beta_4 = 0.4786286705$
	$x_5 = 0.9061798459$	$\beta_5 = 0.2369268851$
6	$x_1 = -0.9324695142$	$\beta_1 = 0.1713244924$
	$x_2 = -0.6612093865$	$\beta_2 = 0.3607615730$
	$x_3 = -0.2386191861$	$\beta_3 = 0.4679139345$
	$x_4 = 0.2386191861$	$\beta_4 = 0.4679139345$
	$x_5 = 0.6612093865$	$\beta_5 = 0.3607615730$
	$x_6 = 0.9324695142$	$\beta_6 = 0.1713244924$

Example 7-10. (Gauss-Legendre integration)

Use the Gauss-Legendre formula with $m = 2$ and $m = 3$ to evaluate the integral

$$I = \int_0^\pi \sin x\,dx$$

Solution: Let $x = \frac{\pi}{2}(t+1)$ for $-1 \le t \le 1$ and write the given integral as

$$I = \frac{\pi}{2}\int_{-1}^{1} \sin[\frac{\pi}{2}(t+1)]\,dt$$

Using the Gauss-Legendre formula (7.97) we find

$$I = \frac{\pi}{2}\left(\sin[\frac{\pi}{2}(-\frac{1}{\sqrt{3}}+1)] + \sin[\frac{\pi}{2}(\frac{1}{\sqrt{3}}+1)]\right) = 1.93582$$

which has an error of -0.06418. The Gauss-Legendre integration formula (7.106) gives the result

$$I = \frac{\pi}{2}\left(\frac{5}{9}\sin[\frac{\pi}{2}(-\sqrt{3/5}+1)] + \frac{8}{9}\sin[\frac{\pi}{2}(0+1)] + \frac{5}{9}\sin[\frac{\pi}{2}(\sqrt{3/5}+1)]\right) = 2.00139$$

which has an error of 0.00139.

∎

The higher order Gauss-Legendre quadrature formulas are very accurate when integrating functions that can be well approximated by polynomials.

Gauss-Laguerre Integration

The Laguerre polynomials $L_n(x)$ are given by the Rodriques' formula

$$L_n(x) = e^x \frac{d^n}{dx^n}(x^n e^{-x}) \tag{7.110}$$

One can verify the first couple of Laguerre polynomials are given by

$$
\begin{aligned}
L_0(x) &= 1 & L_2(x) &= x^2 - 4x + 2 \\
L_1(x) &= 1 - x & L_3(x) &= -x^3 + 9x^2 - 18x + 6
\end{aligned}
\tag{7.111}
$$

and if higher order Laguerre polynomials are desired they can be generated from the recurrence relation

$$L_{n+1}(x) = (2n+1-x)L_n(x) - n^2 L_{n-1}(x) \tag{7.112}$$

for $n = 1, 2, 3\ldots$.

The Laguerre polynomials $L_n(x)$ for $0 \le x < 5$ and $n = 0, 1, 2, 3, 4$ are illustrated in the figure 7-3.

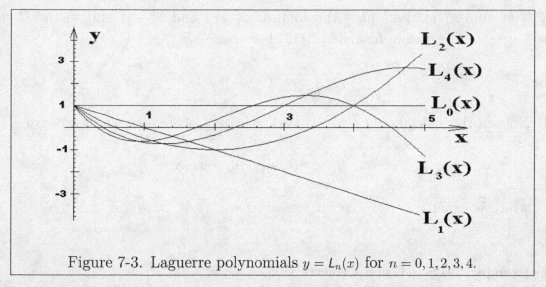

Figure 7-3. Laguerre polynomials $y = L_n(x)$ for $n = 0, 1, 2, 3, 4$.

The Laguerre polynomials $\{L_n(x)\}$, are orthogonal over the interval $(0, \infty)$ with respect to the weight function $w(x) = e^{-x}$. The inner product of two Laguerre polynomials is found to satisfy

$$(L_n, L_m) = \int_0^\infty e^{-x} L_n(x) L_m(x)\, dx = \begin{cases} 0, & \text{for } m \neq n \\ \| L_n \|^2, & \text{for } m = n \end{cases} \tag{7.113}$$

where the norm squared is given by

$$\| L_n \|^2 = (L_n, L_n) = \int_0^\infty e^{-x} L_n^2(x)\, dx = (n!)^2 \tag{7.114}$$

for $n = 0, 1, 2, 3, \ldots$.

The Gauss Laguerre integration formulas are of the form

$$\int_0^\infty e^{-x} f(x)\, dx = \sum_{i=1}^{m} \beta_i f(x_i) \tag{7.115}$$

where x_i and β_i, $i = 1, \ldots, m$ are special constants to be determined.

Example 7-11.   (Gauss-Laguerre integration $m = 1$)
In the case $m = 1$, the Gauss-Laguerre integration formula given by equation (7.115) reduces to

$$\int_0^\infty e^{-x} f(x)\, dx = \beta_1 f(x_1) \tag{7.116}$$

where β_1 and x_1 are constants to be determined. We require that equation (7.116) be exact when $f(x) = 1$ and $f(x) = x$.

For $f(x) = 1$, equation (7.116) reduces to $1 = \beta_1(1)$

For $f(x) = x$, equation (7.116) reduces to $1 = \beta_1 x_1$

The above equations have the solution $\beta_1 = 1$ and $x_1 = 1$ and so the Gauss-Laguerre integration formula (7.115) becomes

$$\int_0^\infty e^{-x} f(x)\, dx = f(1) \tag{7.117}$$

Note in the special case $f(x) = c_0 + c_1 x$, where c_0, c_1 are constants, we have

$$\int_0^\infty e^{-x}(c_0 + c_1 x)\, dx = c_0 + c_1$$

which is exact.

∎

Example 7-12.   (Gauss-Laguerre integration $m = 2$)
In the case $m = 2$, the Gauss-Laguerre integration formula (7.115) is written

$$\int_0^\infty e^{-x} f(x)\, dx = \beta_1 f(x_1) + \beta_2 f(x_2) \tag{7.118}$$

where $\beta_1, \beta_2, x_1, x_2$ are constants to be determined. We require that the integration formula (7.118) be exact when $f(x)$ is one of the functions $\{1, x, x^2, x^3\}$ or some linear combination of these functions. We construct the following equations for determining the unknowns.

$$\text{Substituting } f(x) = 1 \text{ into equation (7.118) gives} \quad 1 = \beta_1 + \beta_2 \tag{7.119}$$

$$\text{Substituting } f(x) = x \text{ into equation (7.118) gives} \quad 1 = \beta_1 x_1 + \beta_2 x_2 \tag{7.120}$$

$$\text{Substituting } f(x) = x^2 \text{ into equation (7.118) gives} \quad 2 = \beta_1 x_1^2 + \beta_2 x_2^2 \tag{7.121}$$

$$\text{Substituting } f(x) = x^3 \text{ into equation (7.118) gives} \quad 6 = \beta_1 x_1^3 + \beta_2 x_2^3 \tag{7.122}$$

To solve the system of equations (7.119), (7.120),   (7.121), (7.122) multiply equation (7.120) by -4 and equation (7.119) by 2 and then add the resulting equations to obtain

$$\beta_1 L_2(x_1) + \beta_2 L_2(x_2) = 0.$$

Then multiply equation (7.121) by -4 and equation (7.120) by 2 and then add the resulting equations to equation (7.122) to obtain

$$\beta_1 x_1 L_2(x_1) + \beta_2 x_2 L_2(x_2) = 0.$$

The system of equation (7.119), (7.120), (7.121), (7.122) can now be replaced by the equivalent system

$$\beta_1 + \beta_2 = 1$$
$$\beta_1 x_1 + \beta_2 x_2 = 1$$
$$\beta_1 L_2(x_1) + \beta_2 L_2(x_2) = 0$$
$$\beta_1 x_1 L_2(x_1) + \beta_2 x_2 L_2(x_2) = 0$$

(7.123)

Let x_1, x_2 denote the roots of the second degree Laguerre polynomial

$$L_2(x) = x^2 - 4x + 2 = 0 \tag{7.124}$$

which can be denoted by

$$x_1 = 2 - \sqrt{2} = 0.5857864376, \qquad x_2 = 2 + \sqrt{2} = 3.4142135624 \tag{7.125}$$

These values substituted into the first two equations of the system (7.123) produces the results

$$\beta_1 = \frac{1}{4}(2 + \sqrt{2}) = 0.8535533906, \qquad \beta_2 = \frac{1}{4}(2 - \sqrt{2}) = 0.1464466094 \tag{7.126}$$

These are the values to be substituted into the Gauss-Laguerre integration formula (7.118).

■

Example 7-13. **(Gauss-Laguerre integration $m = 3$)**
Consider the Gauss-Laguerre integration formula

$$\int_0^\infty e^{-x} f(x)\, dx = \beta_1 f(x_1) + \beta_2 f(x_2) + \beta_3 f(x_3) \tag{7.127}$$

where the weights $\beta_1, \beta_2, \beta_3$ and the abscissa spacing x_1, x_2, x_3 are to be determined. We require that equation (7.127) be exact whenever $f(x)$ is one of the functions $\{1, x, x^2, x^3, x^4, x^5\}$ or some linear combination of these functions. This produces the equations

$$\begin{aligned}
\text{For } f(x) = 1, \text{ we find} \quad & 1 = \beta_1 + \beta_2 + \beta_3 \\
\text{For } f(x) = x, \text{ we find} \quad & 1 = \beta_1 x_1 + \beta_2 x_2 + \beta_3 x_3 \\
\text{For } f(x) = x^2, \text{ we find} \quad & 2 = \beta_1 x_1^2 + \beta_2 x_2^2 + \beta_3 x_3^2 \\
\text{For } f(x) = x^3, \text{ we find} \quad & 6 = \beta_1 x_1^3 + \beta_2 x_2^3 + \beta_3 x_3^3 \\
\text{For } f(x) = x^4, \text{ we find} \quad & 24 = \beta_1 x_1^4 + \beta_2 x_2^4 + \beta_3 x_3^4 \\
\text{For } f(x) = x^5, \text{ we find} \quad & 120 = \beta_1 x_1^5 + \beta_2 x_2^5 + \beta_3 x_3^5
\end{aligned} \tag{7.128}$$

Table 7.7 Gauss-Laguerre integration formula		
$\int_0^\infty e^{-x} f(x)\,dx = \beta_1 f(x_1) + \beta_2 f(x_2) + \cdots + \beta_m f(x_m)$		
m	Nodes x_i	Weights β_i
1	$x_1 = 1.0000000000$	$\beta_1 = 1.0000000000$
2	$x_1 = 0.5857864376$	$\beta_1 = 0.8535533906$
	$x_2 = 3.4142135624$	$\beta_2 = 0.1464466094$
3	$x_1 = 0.4157745568$	$\beta_1 = 0.7110930099$
	$x_2 = 2.2942803603$	$\beta_2 = 0.2785177336$
	$x_3 = 6.2899450829$	$\beta_3 = 0.0103892565$
4	$x_1 = 0.3225476896$	$\beta_1 = 0.60315410434$
	$x_2 = 1.745611012$	$\beta_2 = 0.3574186924$
	$x_3 = 4.5366202969$	$\beta_3 = 0.03888790851$
	$x_4 = 9.3950709123$	$\beta_4 = 0.000539294706$
5	$x_1 = 0.2635603197$	$\beta_1 = 0.52175561058$
	$x_2 = 1.4134030591$	$\beta_2 = 0.39866681108$
	$x_3 = 3.5964257710$	$\beta_3 = 0.07594244968$
	$x_4 = 7.0858100058$	$\beta_4 = 0.003611758699$
	$x_5 = 12.6408008442$	$\beta_5 = 0.000023369973$
6	$x_1 = 0.22284660418$	$\beta_1 = 0.45896467395$
	$x_2 = 1.1889321017$	$\beta_2 = 0.417000830772$
	$x_3 = 2.9927363261$	$\beta_3 = 0.113373382074$
	$x_4 = 5.7751435691$	$\beta_4 = 0.0103991974531$
	$x_5 = 9.8374674184$	$\beta_5 = 0.000261017203$
	$x_6 = 15.982873981$	$\beta_6 = 0.000000898547$

One can now take linear combinations of the above equations which will create the Laguerre polynomial $L_3(x)$ and consequently the system of equations (7.128) can be replaced by the equivalent system of equations

$$\begin{aligned}
\beta_1 + \beta_2 + \beta_3 &= 1 \\
\beta_1 x_1 + \beta_2 x_2 + \beta_3 x_3 &= 1 \\
\beta_1 x_1^2 + \beta_2 x_2^2 + \beta_3 x_3^2 &= 2 \\
\beta_1 L_3(x_1) + \beta_2 L_3(x_2) + \beta_3 L_3(x_3) &= 0 \\
\beta_1 x_1 L_3(x_1) + \beta_2 x_2 L_3(x_2) + \beta_3 x_3 L_3(x_3) &= 0 \\
\beta_1 x_1^2 L_3(x_1) + \beta_2 x_2^2 L_3(x_2) + \beta_3 x_3^2 L_3(x_3) &= 0
\end{aligned}$$

(7.129)

The last three equations from the system (7.129) are satisfied if we require that x_1, x_2, x_3 are the roots of the Laguerre equation

$$L_3(x) = -x^3 + 9x^2 - 18x + 6 = 0. \tag{7.130}$$

These roots are given by

$$x_1 = 0.4157745567, \qquad x_2 = 2.2942803602, \qquad x_3 = 6.2899450829 \tag{7.131}$$

and substituting these values into the first three equations of the system (7.129) one can verify the values

$$\beta_1 = 0.7110930099, \qquad \beta_2 = 0.2785177336, \qquad \beta_3 = 0.0103892565. \tag{7.132}$$

∎

The table 7.7 lists the weights and abscissa values (nodes) for selected Gauss-Laguerre formula.

Gauss-Chebyshev Integration

The Chebyshev[†] polynomials are defined over the interval $-1 \leq x \leq 1$ and can be represented in the form

$$T_n(x) = \cos(n \arccos x) \tag{7.133}$$

The Chebyshev polynomials of order zero through four are obtained by letting $\alpha = \arccos x$, so that

$$
\begin{aligned}
T_0(x) &= \cos 0 = 1 & T_2(x) &= \cos 2\alpha = 2\cos^2 \alpha - 1 = 2x^2 - 1 \\
T_1(x) &= \cos \alpha = x & T_3(x) &= \cos 3\alpha = 4\cos^3 \alpha - 3\cos \alpha = 4x^3 - 3x
\end{aligned}
\tag{7.134}
$$

and higher order Chebyshev polynomials can be constructed using the recursion formula

$$T_{n+1}(x) = 2nT_n(x) - T_{n-1}(x) \tag{7.135}$$

for $n = 1, 2, 3, \ldots$.

The Chebyshev polynomials $\{T_n(x)\}$, for $-1 \leq x \leq 1$ and $n = 0, 1, 2, 3, 4$ are illustrated in the figure 7-4.

[†] Chebyshev is a translation of the Russian Чебышев. Some other translations used in the literature are Tchebicheff and Tschebyscheff.

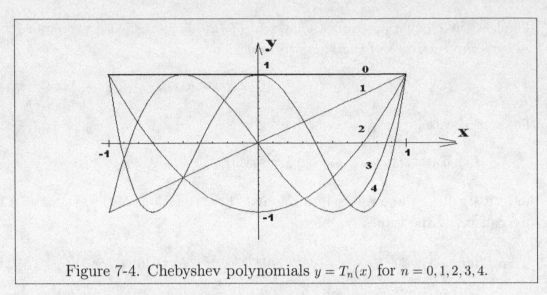

Figure 7-4. Chebyshev polynomials $y = T_n(x)$ for $n = 0, 1, 2, 3, 4$.

These polynomials are orthogonal over the interval $-1 \leq x \leq 1$ with respect to the weight function $1/\sqrt{1 - x^2}$. The weighted inner product satisfies

$$(T_n, T_m) = \int_{-1}^{1} \frac{T_n(x)T_m(x)}{\sqrt{1 - x^2}} \, dx = \begin{cases} 0 & \text{if } m \neq n \\ \| T_n \|^2 & \text{if } m = n \end{cases} \tag{7.136}$$

where the norm squared is

$$\| T_n \|^2 = (T_n, T_n) = \int_{-1}^{1} \frac{T_n^2(x)}{\sqrt{1 - x^2}} \, dx = \begin{cases} \frac{\pi}{2} & n \neq 0 \\ \pi & n = 0 \end{cases} \tag{7.137}$$

The Gauss-Chebyshev integration formulas are of the form

$$\int_{-1}^{1} \frac{f(x)}{\sqrt{1 - x^2}} \, dx = \sum_{i=1}^{m} \beta_i f(x_i) \tag{7.138}$$

where β_i and x_i, for $i = 1, \ldots, m$, are unknowns to be determined.

Example 7-14. **(Gauss-Chebyshev integration $m = 1$)**
We require the Gauss-Chebyshev integration formula

$$\int_{-1}^{1} \frac{f(x)}{\sqrt{1 - x^2}} \, dx = \beta_1 f(x_1)$$

be exact for $f(x) = 1$ and $f(x) = x$. This requires that

$$\int_{-1}^{1} \frac{1}{\sqrt{1 - x^2}} \, dx = \pi = \beta_1(1)$$

$$\text{and} \qquad \int_{-1}^{1} \frac{x}{\sqrt{1 - x^2}} \, dx = 0 = \beta_1 x_1$$

which gives $\beta_1 = 1/\pi$ and $x_1 = 0$. This gives the Gauss-Chebyshev integration formula

$$\int_{-1}^{1} \frac{f(x)}{\sqrt{1-x^2}}\, dx = \frac{1}{\pi} f(0).$$

∎

Example 7-15. **(Gauss-Chebyshev integration $m = 2$)**
We require the Gauss-Chebyshev integration formula

$$\int_{-1}^{1} \frac{f(x)}{\sqrt{1-x^2}}\, dx = \beta_1 f(x_1) + \beta_2 f(x_2)$$

be exact if $f(x)$ is one of the functions $\{1, x, x^2, x^3\}$. This requires that

$$\begin{aligned}
\int_{-1}^{1} \frac{1}{\sqrt{1-x^2}}\, dx &= \pi = \beta_1 + \beta_2 \\
\int_{-1}^{1} \frac{x}{\sqrt{1-x^2}}\, dx &= 0 = \beta_1 x_1 + \beta_2 x_2 \\
\int_{-1}^{1} \frac{x^2}{\sqrt{1-x^2}}\, dx &= \pi/2 = \beta_1 x_1^2 + \beta_2 x_2^2 \\
\int_{-1}^{1} \frac{x^3}{\sqrt{1-x^2}}\, dx &= 0 = \beta_1 x_1^3 + \beta_2 x_2^3
\end{aligned} \qquad (7.139)$$

By taking linear combinations of the equations in the system (7.139) one can produce the equivalent system of equations

$$\begin{aligned}
\beta_1 + \beta_2 &= \pi \\
\beta_1 x_1 + \beta_2 x_2 &= 0 \\
\beta_1 T_2(x_1) + \beta_2 T_2(x_2) &= 0 \\
\beta_1 x_1 T_2(x_1) + \beta_2 T_2(x_2) &= 0
\end{aligned} \qquad (7.140)$$

The last two equations in the system (7.140) are satisfied if x_1, x_2 are the roots of the equation

$$T_2(x) = 2x^2 - 1 = 0$$

which gives

$$x_1 = -\cos[\pi/4] = -\frac{1}{\sqrt{2}} \qquad \text{and} \qquad x_2 = \cos[\pi/4] = \frac{1}{\sqrt{2}}.$$

These values for x_1, x_2 are then substituted into the first two equations of the system (7.140) to obtain $\beta_1 = \beta_2 = \pi/2$. This gives the Gauss-Chebyshev formula

$$\int_{-1}^{1} \frac{f(x)}{\sqrt{1-x^2}}\, dx = \frac{\pi}{2} \left[f(-1/\sqrt{2}) + f(1/\sqrt{2}) \right] \qquad (7.141)$$

∎

Example 7-16. (**Gauss-Chebyshev integration** $m = 3$)

We require the Gauss-Chebyshev integration formula

$$\int_{-1}^{1} \frac{f(x)}{\sqrt{1-x^2}} \, dx = \beta_1 f(x_1) + \beta_2 f(x_2) + \beta_3 f(x_3) \tag{7.142}$$

to be exact if $f(x)$ is one of the functions $\{1, x, x^2, x^3, x^4, x^5\}$ or a linear combination of these functions. This requires

$$\begin{aligned}
\int_{-1}^{1} \frac{1}{\sqrt{1-x^2}} \, dx &= \pi = \beta_1 + \beta_2 + \beta_3 \\
\int_{-1}^{1} \frac{x}{\sqrt{1-x^2}} \, dx &= 0 = \beta_1 x_1 + \beta_2 x_2 + \beta_3 x_3 \\
\int_{-1}^{1} \frac{x^2}{\sqrt{1-x^2}} \, dx &= \pi/2 = \beta_1 x_1^2 + \beta_2 x_2^2 + \beta_3 x_3^2 \\
\int_{-1}^{1} \frac{x^3}{\sqrt{1-x^2}} \, dx &= 0 = \beta_1 x_1^3 + \beta_2 x_2^3 + \beta_3 x_3^3 \\
\int_{-1}^{1} \frac{x^4}{\sqrt{1-x^2}} \, dx &= 3\pi/8 = \beta_1 x_1^4 + \beta_2 x_2^4 + \beta_3 x_3^4 \\
\int_{-1}^{1} \frac{x^5}{\sqrt{1-x^2}} \, dx &= 0 = \beta_1 x_1^5 + \beta_2 x_2^5 + \beta_3 x_3^5
\end{aligned} \tag{7.143}$$

The system of equations (7.143) can be replaced by the equivalent system of equations

$$\begin{aligned}
\beta_1 + \beta_2 + \beta_3 &= \pi \\
\beta_1 x_1 + \beta_2 x_2 + \beta_3 x_3 &= 0 \\
\beta_1 x_1^2 + \beta_2 x_2^2 + \beta_3 x_3^2 &= \pi/2 \\
\beta_1 T_3(x_1) + \beta_2 T_3(x_2) + \beta_3 T_3(x_3) &= 0 \\
\beta_1 x_1 T_3(x_1) + \beta_2 x_2 T_3(x_2) + \beta_3 x_3 T_3(x_3) &= 0 \\
\beta_1 x_1^2 T_3(x_1) + \beta_2 x_2^2 T_3(x_2) + \beta_3 x_3^2 T_3(x_3) &= 0
\end{aligned} \tag{7.144}$$

The last three equations in the system (7.144) are satisfied if x_1, x_2, x_3 are selected as the roots of the equation

$$T_3(x) = 4x^3 - 3x = 0 \tag{7.145}$$

which are given by

$$x_i = \cos[(2i - 1)\pi/6] \quad \text{for} \quad i = 1, 2, 3 \tag{7.146}$$

and then from the first three equations of the system (7.144) we find the weights $\beta_1 = \beta_2 = \beta_3 = \pi/3$. These values produce the Gauss-Chebyshev integration formula

$$\int_{-1}^{1} \frac{f(x)}{\sqrt{1-x^2}}\, dx = \frac{\pi}{3}\left[f(\cos \pi/6) + f(\cos 3\pi/6) + f(\cos 5\pi/6)\right].$$

∎

The table 7.8 gives the weights and abscissa values for the Gauss-Chebyshev integration formulas.

Table 7.8 Gauss-Chebyshev integration formula		
$\int_{-1}^{1} \dfrac{f(x)}{\sqrt{1-x^2}}\, dx = \beta_1 f(x_1) + \beta_2 f(x_2) + \cdots + \beta_m f(x_m)$		
m	Nodes x_i	Weights β_i
n	$x_i = \cos[(2i-1)\pi/2n]$, for $i = 1, 2, \ldots, n$	$\beta_i = \pi/n$ for $i = 1, 2, \ldots, n$
n is a positive integer.		

Example 7-17.   (Gauss-Chebyshev integration)
Use the Gauss-Chebyshev integration formula with $m = 4$ to evaluate the integral

$$\int_{-1}^{1} \frac{\cos x}{\sqrt{1-x^2}}\, dx$$

Solution:
$$\int_{-1}^{1} \frac{\cos x}{\sqrt{1-x^2}}\, dx = \frac{\pi}{4}\left(\cos\left[\cos \pi/8\right] + \cos\left[\cos 3\pi/8\right] + \cos\left[\cos 5\pi/8\right] + \cos\left[\cos 7\pi/8\right]\right) = 2.403938839$$

which has an error of $5.92389 \times (10)^{-7}$.

∎

General Gaussian Quadrature Integration Formulas

Given a set of polynomial functions defined by $\{\phi_n(x)\}$, for $n = 0, 1, 2, 3, \ldots$ defined over an interval $a \leq x \leq b$ where for each integer m, the function $\phi_m(x)$ is a polynomial of degree m. Further assume that these functions are orthogonal over the interval (a, b) with respect to a weight function $w(x)$ such that the weighted inner product satisfies

$$(\phi_n, \phi_m) = \int_a^b w(x)\phi_n(x)\phi_m(x)\, dx = \begin{cases} 0, & m \neq n \\ \|\phi_n\|^2 & m = n \end{cases} \tag{7.147}$$

where the norm squared is defined

$$\|\phi_n\|^2 = (\phi_n, \phi_n) = \int_a^b w(x)\phi_n^2(x)\, dx \tag{7.148}$$

and is assumed to be a nonzero quantity. The quantity $w(x)$ is called a positive valued weight function. Note that any polynomial $p_n(x)$ of degree n can be expanded in a series of these orthogonal polynomials. The series expansion has the form

$$p_n(x) = \sum_{i=0}^{n} c_i \phi_i(x) = c_0 \phi_0(x) + c_1 \phi_1(x) + \cdots + c_n \phi_n(x) \qquad (7.149)$$

where c_i, $i = 0, 1, 2, \ldots, n$ are constants. The constants can be determined by employing the inner product property of the orthogonal polynomials in the following way. Multiply the equation (7.149) by $w(x)\phi_m(x)$ and then integrate all terms from a to b. This produces the result

$$\int_a^b w(x)p_n(x)\phi_m(x)\,dx = c_0(\phi_0, \phi_m) + c_1(\phi_1, \phi_m) + \cdots + c_m(\phi_m, \phi_m) + \cdots + c_n(\phi_n, \phi_m)$$

where all inner products on the right-hand side of the resulting equation are zero except for the term involving $\phi_m(x)$. This gives the formula

$$c_m = \frac{1}{\parallel \phi_m \parallel^2} \int_a^b w(x)P_n(x)\phi_m(x)\,dx \quad \text{for} \quad m = 0, 1, 2, \ldots, n \qquad (7.150)$$

from which the coefficients of the series expansion (7.149) can be calculated.

The Gaussian integration formula associated with the set of orthogonal polynomials $\{\phi_n(x)\}$ is of the form

$$\int_a^b w(x)f(x)\,dx = \sum_{i=1}^{m} \beta_i f(x_i) \qquad (7.151)$$

where for $i = 1, \ldots, m$, the weights β_i and the nodes x_i are constants to be determined. We require that the equation (7.151) be exact for all polynomials of degree $2m - 1$ or less. This is the same requirement as in all of the previous examples. That is, we require the integration equation (7.151) to be exact if $f(x)$ is one of the functions

$$\{1, x, x^2, x^3, \ldots, x^{2m-2}, x^{2m-1}\}.$$

This implies integrations using (7.151) will give exact results for $f(x)$ any linear combination of these functions. These requirements will produce $2m$ equations with $2m$ unknowns from which the weight factors $\beta_1, \ldots, \beta_m$ and abscissa values $x_1, \ldots, x_m$ can be determined.

Alternatively, there is another way we can construct a polynomial of degree $2m-1$ to be substituted into the equation (7.151) for determining the unknowns. If the abscissa $x_1, \ldots, x_m$ are known, then we can construct a Lagrange interpolating polynomial of degree $m-1$ with error term and construct the special polynomial of degree $2m-1$

$$f(x) = \sum_{i=1}^{m} \ell_i(x) f(x_i) + \frac{\pi(x)}{m!} f^{(m)}(\xi(x)) \tag{7.152}$$

where

$$\ell_k(x) = \frac{(x-x_1)(x-x_2)\cdots(x-x_{k-1})(x-x_{k+1})\cdots(x-x_m)}{(x_k-x_1)(x_k-x_2)\cdots(x_k-x_{k-1})(x_k-x_{k+1})\cdots(x_k-x_m)}$$

$$\ell_k(x) = \frac{\pi(x)}{(x-x_k)\pi'(x_k)} \qquad \text{for } k = 1, \ldots, m \tag{7.153}$$

with

$$\pi(x) = (x-x_1)(x-x_2)\cdots(x-x_m). \tag{7.154}$$

The first term in equation (7.152) is a polynomial of degree $m-1$ and the product $\pi(x)$ is of degree m so that if equation (7.152) is to be a polynomial of degree $2m-1$, then the term $f^{(m)}(\xi(x))$ must be a polynomial of degree $m-1$. We can now substitute the special polynomial (7.152) into the Gaussian integration formula (7.151) and require the results to be exact. Substituting equation (7.152) into the equation (7.151) gives

$$\int_a^b w(x) \left[\sum_{i=1}^{m} \ell_i(x) f(x_i) + \frac{\pi(x)}{m!} f^{(m)}(\xi(x)) \right] dx = \sum_{i=1}^{m} \beta_i f(x_i) \tag{7.155}$$

which simplifies to

$$\sum_{i=1}^{m} f(x_i) \int_a^b w(x) \ell_i(x)\, dx + \int_a^b \frac{w(x)\pi(x)}{m!} f^{(m)}(\xi(x))\, dx = \sum_{i=1}^{m} \beta_i f(x_i) \tag{7.156}$$

Comparing terms in the equation (7.156) we see that if it is to give exact results, then we must require

$$\beta_i = \int_a^b w(x) \ell_i(x)\, dx \tag{7.157}$$

and the error term must satisfy

$$Error = \int_a^b \frac{w(x)\pi(x)}{m!} f^{(m)}(\xi(x))\, dx = 0. \tag{7.158}$$

To show the error term is zero we employ the orthogonal functions $\{\phi_n(x)\}$ and the orthogonality property given by equation (7.147). One can expand the product function $\pi(x)$ in terms of the orthogonal functions and write

$$\pi(x) = a_0\phi_0(x) + a_1\phi_1(x) + \cdots + a_{m-1}\phi_{m-1}(x) + a_m\phi_m(x). \tag{7.159}$$

Similarly, the derivative function $f^{(m)}(\xi(x))/m!$ can be expanded in a series of the orthogonal functions to obtain

$$\frac{f^{(m)}(\xi(x))}{m!} = b_0\phi_0(x) + b_1\phi_1(x) + \cdots + b_{m-1}\phi_{m-1}(x) \tag{7.160}$$

The error term can then be written

$$Error = \int_a^b w(x) \sum_{i=0}^m a_i\phi_i(x) \sum_{j=0}^{m-1} b_j\phi_j(x) = \sum_{i=0}^m \sum_{j=0}^{m-1} a_ib_j \int_a^b w(x)\phi_i(x)\phi_j(x)\,dx \tag{7.161}$$

We make use of the orthogonality property of the orthogonal functions and find that the double summation given by equation (7.161) reduces to

$$Error = a_0 b_0 \parallel \phi_0 \parallel^2 + a_1 b_1 \parallel \phi_1 \parallel^2 + \cdots + a_{m-1}b_{m-1} \parallel \phi_{m-1} \parallel^2 . \tag{7.162}$$

Now the norm squared terms are different from zero and all the b_i coefficients can't be zero simultaneously and so the only way we can force the equation (7.162) to be zero is to require that $a_0 = a_1 = \cdots = a_{m-1} = 0$. This requirement reduces the equation (7.159) to the form

$$\pi(x) = (x - x_1)(x - x_2)\cdots(x - x_m) = a_m\phi_m(x) \tag{7.163}$$

which tells us that the abscissa $x_1, x_2, \ldots, x_m$ must be the roots of the mth degree polynomial $\phi_m(x)$. This gives the Gaussian integration formula

$$\int_a^b w(x)f(x)\,dx = \beta_1 f(x_1) + \beta_2 f(x_2) + \cdots + \beta_m f(x_m) \tag{7.164}$$

where the nodes $x_1, x_2, \ldots, x_m$ are the roots of $\phi_m(x) = 0$ and the weighting coefficients are given by

$$\beta_i = \int_a^b w(x)\ell_i(x)\,dx \qquad i = 1, 2, \ldots, m \tag{7.165}$$

where $\ell_i(x)$ are the Lagrange interpolating polynomials given by equation (7.153) which also employ the above nodes.

Error Term

Whenever $f(x)$ is not a polynomial of degree $2m - 1$ or less, the error term associated with the Gaussian integration formula

$$\int_a^b w(x)f(x)\,dx = \sum_{i=1}^m \beta_i f(x_i)$$

can be constructed as follows. Let $f(x)$ be approximated by a Hermite interpolating polynomial $P(x)$ where we have shown that

$$f(x) - P(x) = \frac{f^{(2m)}(\xi(x))}{(2m)!}[\pi(x)]^2 \qquad (7.166)$$

By construction $P(x_i) = f(x_i)$ and so we can write

$$\int_a^b w(x)P(x)\,dx = \sum_{i=1}^m \beta_i P(x_i) = \sum_{i=1}^m \beta_i f(x_i) = \int_a^b w(x)f(x)\,dx$$

Therefore,

$$Error = \int_a^b w(x)f(x)\,dx - \sum_{i=1}^m \beta_i f(x_i)$$

$$Error = \int_a^b w(x)f(x)\,dx - \int_a^b w(x)P(x)\,dx$$

$$Error = \int_a^b w(x)[f(x) - P(x)]\,dx$$

$$Error = \int_a^b w(x)\frac{f^{(2m)}(\xi(x))}{(2m)!}[\pi(x)]^2\,dx$$

By the mean value theorem this can be written

$$Error = \frac{f^{(2m)}(\zeta)}{(2m)!}\int_a^b w(x)[\pi(x)]^2\,dx \qquad (7.167)$$

There are many specialized Gaussian type integration formulas which are associated with orthogonal functions. The table 7.9 lists some of the many special functions used to construct integration formulas.

Table 7.9 Orthogonal Functions used in Gaussian Integration			
Name	Interval (a, b)	weight function	Nodes x_i are roots of
Legendre	$(-1, 1)$	1	$P_n(x)$
Laguerre	$(0, \infty)$	e^{-x}	$L_n(x)$
Chebyshev(first kind)	$(-1, 1)$	$1/\sqrt{1 - x^2}$	$T_n(x)$
Hermite	$(-\infty, \infty)$	e^{-x^2}	$H_n(x)$
Chebyshev(second kind)	$(-1, 1)$	$\sqrt{1 - x^2}$	$U_n(x)$

Improper Integrals

The majority of integration formulas developed assume that the integrand $f(x)$ is continuous over the interval of integration. In the cases where $f(x)$ is discontinuous, with say an infinite discontinuity at an interior point or at one of the end points of the integration interval, then special care must be taken in evaluating the integral. Integration problems where one or both limits of integration become infinite, again pose problems and one must take special steps to evaluate these types of integrals.

If the integrand $f(x)$ has only simple jump discontinuities at points c_i for $i = 1, \ldots, m$, where $a < c_1 < c_2 < \cdots < c_m < b$, then integrals of the form $\int_a^b f(x)\, dx$ must be written as a sum of integrals over the separate subintervals where the integrand is continuous. One would write

$$\int_a^b f(x)\, dx = \int_a^{c_1} f(x)\, dx + \sum_{i=1}^{m-1} \int_{c_i}^{c_{i+1}} f(x)\, dx + \int_{c_m}^b f(x)\, dx.$$

If the integrand $f(x)$ has an infinite discontinuity at an endpoint or interior point, then one must use a limiting process to determine if the integral exists. These types of integrals are called improper integrals of the second kind. If the discontinuity occurs at the lower limit of integration, one would write

$$\int_a^b f(x)\, dx = \lim_{\epsilon \to 0} \int_{a-\epsilon}^b f(x)\, dx$$

and if the discontinuity occurs at the upper limit of integration, one would write

$$\int_a^b f(x)\, dx = \lim_{\epsilon \to 0} \int_a^{b-\epsilon} f(x)\, dx.$$

If an infinite discontinuity occurs at some point c between the limits of integration, one would write

$$\int_a^b f(x)\, dx = \lim_{\epsilon \to 0} \int_a^{c-\epsilon} f(x)\, dx + \lim_{\epsilon \to 0} \int_{c+\epsilon}^b f(x)\, dx.$$

Integrals with one or both limits of integration infinite, say for example integrals of the form $I_1 = \int_0^\infty f(x)\, dx$ or $I_2 = \int_{-\infty}^\infty f(x)\, dx$ can also be treated using a limiting process. These types of integrals are called improper integrals of the first kind. One would write

$$I_1 = \lim_{T \to \infty} \int_0^T f(x)\, dx, \qquad I_2 = \lim_{T \to \infty} \int_{-T}^T f(x)\, dx.$$

Sometimes if $f(x)$ becomes infinite at an end point, the transformations $x = 1/t^\alpha$ or $x = \frac{1}{\beta}\ln t$ can simplify the integral to a more tractable form.

The Newton-Cotes formula result when the integrand is replaced by some sort of polynomial approximation. The integration problem is then reduced to a problem of making certain function evaluations at equally spaced points. The open Newton-Cotes integration formulas are particularly useful in those cases where the integrand cannot be evaluated at an endpoint.

One generally evaluates an improper integral using a step size h, then another estimate of the integral is obtained using a step size of $h/2$. One can then compare the two results for an estimate of the error occurring in the calculation. One can continue to half the step size to obtain better estimates. The Richardson extrapolation can then be used to obtain still more improved estimates for the value of the integral.

To evaluate an improper integral of the first kind

$$I = \int_0^\infty f(x)\,dx, \tag{7.168}$$

to within an error tolerance ϵ, one would first write $I = \int_0^b f(x)\,dx + \int_b^\infty f(x)\,dx$. Given an error tolerance ϵ one tries to find a value of b such that $|\int_b^\infty f(x)\,dx| < \epsilon/2$ and then calculate the integral $\int_0^b f(x)\,dx$ with accuracy $\epsilon/2$ so that the total error is less than ϵ. The comparison test for improper integrals is usually employed. One tries to find a function $g(x)$ such that $0 \le |f(x)| \le g(x)$ for $x \ge b$, then if $\lim_{T\to\infty}\int_b^T g(x)\,dx$ converges, then the integral $\int_b^\infty f(x)\,dx$ will also converge.

Improper integrals of the form $I = \int_{-\infty}^\infty f(x)\,dx$ are reduced to integrals of the form given by equation (7.168) by writing $I = \int_{-\infty}^0 f(x)\,dx + \int_0^\infty f(x)\,dx$. One then employs the change of variable $x = -u$ to the first integral.

Example 7-18. (Gauss-Chebyshev integration)

The integral $I = \int_0^1 \frac{1}{\sqrt{x}}\,dx$ is a singular integral because the integrand is not defined at $x = 0$. Make the change of variable $x = \frac{1}{2}(1+t)$ with $dx = \frac{1}{2}dt$ and show that $I = \int_{-1}^1 \frac{1}{\sqrt{2}}\frac{1}{\sqrt{1+t}}\,dt$. This integral can now be put into the form of a Gauss-Chebyshev integral by multiplying the numerator and denominator of the integrand by $\sqrt{1-t^2}$ to obtain $I = \int_{-1}^1 \frac{f(t)}{\sqrt{1-t^2}}\,dt$ where $f(t) = \frac{1}{\sqrt{2}}\sqrt{1-t}$. Define $I_n = \frac{\pi}{n}\sum_{i=1}^n f(\cos[(2i-1)\pi/(2n)])$, then one can verify the following values.

$I_2 = 2.05234, \quad I_5 = 2.00825, \quad I_{10} = 2.00206, \quad I_{20} = 2.00051, \quad I_{50} = 2.00008,$

$I_{100} = 2.00002, \quad I_{150} = 2.0000091, \quad I_{200} = 2.0000051, \quad I_{300} = 2.0$

Exercises Chapter 7

▶ **1.** Assume a constant step size h and show how to derive the following forward difference approximations

(a) $\quad y'(x_i) = \dfrac{y_{i+1} - y_i}{h} + \mathcal{O}(h)$

(b) $\quad y''(x_i) = \dfrac{y_{i+2} - 2y_{i+1} + y_i}{h^2} + \mathcal{O}(h)$

(c) $\quad y'''(x_i) = \dfrac{y_{i+3} - 3y_{i+2} + 3y_{i+1} - y_i}{h^3} + \mathcal{O}(h)$

(d) $\quad y^{(iv)}(x_i) = \dfrac{y_{i+4} - 4y_{i+3} + 6y_{i+2} - 4y_{i+1} + y_i}{h^4} + \mathcal{O}(h)$

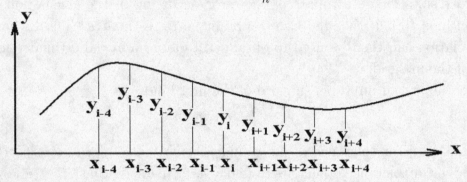

▶ **2.** Assume a constant step size h and show how to derive the following backward difference approximations

(a) $\quad y'(x_i) = \dfrac{y_i - y_{i-1}}{h} + \mathcal{O}(h)$

(b) $\quad y''(x_i) = \dfrac{y_i - 2y_{i-1} + y_{i-2}}{h^2} + \mathcal{O}(h)$

(c) $\quad y'''(x_i) = \dfrac{y_i - 3y_{i-1} + 3y_{i-2} + y_{i-3}}{h^3} + \mathcal{O}(h)$

(d) $\quad y^{(iv)}(x_i) = \dfrac{y_i - 4y_{i-1} + 6y_{i-2} - 4y_{i-3} + y_{i-4}}{h^4} + \mathcal{O}(h)$

▶ **3.** Assume a constant step size h and show how to derive the following forward difference approximations

(a) $\quad y'(x_i) = \dfrac{-3y_i + 4y_{i+1} - y_{i+2}}{2h} + \mathcal{O}(h^2)$

(b) $\quad y''(x_i) = \dfrac{2y_i - 5y_{i+1} + 4y_{i+2} - y_{i+3}}{h^2} + \mathcal{O}(h^2)$

(c) $\quad y'''(x_i) = \dfrac{-5y_i + 18y_{i+1} - 24y_{i+2} + 14y_{i+3} - 3y_{i+4}}{2h^3} + \mathcal{O}(h^2)$

(d) $\quad y^{(iv)}(x_i) = \dfrac{3y_i - 14y_{i+1} + 26y_{i+2} - 24y_{i+3} + 11y_{i+4} - 2y_{i+5}}{h^4} + \mathcal{O}(h^2)$

▶ **4.** Assume a constant step size h and show how to derive the following backward difference approximations

$$(a) \qquad y'(x_i) = \frac{3y_i - 4y_{i-1} + y_{i-2}}{2h} + \mathcal{O}(h^2)$$

$$(b) \qquad y''(x_i) = \frac{2y_i - 5y_{i-1} + 4y_{i-2} - y_{i-3}}{h^2} + \mathcal{O}(h^2)$$

$$(c) \qquad y'''(x_i) = \frac{5y_i - 18y_{i-1} + 24y_{i-2} - 14y_{i-3} + 3y_{i-4}}{2h^3} + \mathcal{O}(h^2)$$

$$(d) \qquad y^{(iv)}(x_i) = \frac{3y_i - 14y_{i-1} + 26y_{i-2} - 24y_{i-3} + 11y_{i-4} - 2y_{i-5}}{h^4} + \mathcal{O}(h^2)$$

▶ **5.** Assume a constant step size h and show how to derive the following central difference approximations

$$(a) \qquad y'(x_i) = \frac{y_{i+1} - y_{i-1}}{2h} + \mathcal{O}(h^2)$$

$$(b) \qquad y''(x_i) = \frac{y_{i+1} - 2y_i + y_{i-1}}{h^2} + \mathcal{O}(h^2)$$

$$(c) \qquad y'''(x_i) = \frac{y_{i+2} - 2y_{i+1} + 2y_{i-1} - y_{i-2}}{2h^3} + \mathcal{O}(h^2)$$

$$(d) \qquad y^{(iv)}(x_i) = \frac{y_{i+2} - 4y_{i+1} + 6y_i - 4y_{i-1} + y_{i-2}}{h^4} + \mathcal{O}(h^2)$$

▶ **6.** Assume a constant step size h and show how to derive the following central difference approximations

$$(a) \qquad y'(x_i) = \frac{-y_{i+2} + 8y_{i+1} - 8y_{i-1} + y_{i-2}}{12h} + \mathcal{O}(h^4)$$

$$(b) \qquad y''(x_i) = \frac{-y_{i+2} + 16y_{i+1} - 30y_i + 16y_{i-1} - y_{i-2}}{12h^2} + \mathcal{O}(h^4)$$

$$(c) \qquad y'''(x_i) = \frac{-y_{i+3} + 8y_{i+2} - 13y_{i+1} + 13y_{i-1} - 8y_{i-2} + y_{i-3}}{8h^3} + \mathcal{O}(h^4)$$

$$(d) \qquad y^{(iv)}(x_i) = \frac{-y_{i+3} + 12y_{i+2} - 39y_{i+1} + 56y_i - 39y_{i-1} + 12y_{i-2} - y_{i-3}}{6h^4} + \mathcal{O}(h^4)$$

▶ **7.** Use the trapezoidal rule with $h = 1/4$ to estimate the integral $I = \int_0^2 \frac{4}{1 + x^2}\, dx$ and state the accuracy of your answer.

▶ **8.** Use Simpson's 1/3 rule with $h = 1/4$ to estimate the integral $I = \int_0^2 \frac{4}{1 + x^2}\, dx$ and state the accuracy of your answer.

▶ **9.**   Use the trapezoidal rule to evaluate the integral $I = \int_0^\pi x \sin x \, dx$ and state the accuracy of your answer.

▶ **10.**   Use Simpson's 1/3 rule to estimate the integral $I = \int_0^\pi x \cos x \, dx$ and state the accuracy of your answer.

▶ **11.**   Use Romberg integration to evaluate the integral $I = \int_0^\pi x \sin x \, dx$ and state the accuracy of your answer.

▶ **12.**   Use Romberg integration to evaluate the integral $I = \int_0^\pi x \cos x \, dx$ and state the accuracy of your answer.

▶ **13.**   Use adaptive quadrature to evaluate the integral $I = \int_0^1 x e^{-x^2} \, dx$ accurate to within an error tolerance of $\epsilon = 1.0 \times 10^{-6}$.

▶ **14.**   Use adaptive quadrature to evaluate the integral $I = \int_0^2 \frac{1}{1+x^2} \, dx$ accurate to within an error tolerance of $\epsilon = 1.0 \times 10^{-6}$.

▶ **15.**   Use Gauss-Legendre quadrature with $n = 1, 2, 3, 4$ to evaluate the integral $I = \int_0^1 x^2 e^{-x} \, dx$ and compare your result with the exact value of the integral. What is the percent error in each case?
Hint: You must first transform the integral to the form $\int_{-1}^1 f(x) \, dx$.

▶ **16.**   Use Gauss-Legendre quadrature for the cases $n = 1, 2, 3, 4$ to evaluate the integral $I = \int_0^1 (x^2 + x + 1) \, dx$ and compare your result with the exact value of the integral. What is the percent error in each case?
Hint: You must first transform the integral to the form $\int_{-1}^1 f(x) \, dx$.

▶ **17.**   Use Gauss-Laguerre integration for the cases $m = 1, 2, 3$ and 4 to evaluate the integral $I = \int_0^\infty e^{-x} \sin x \, dx$ and state the accuracy of your answer in each case.

▶ **18.**   Use Gauss-Laguerre integration for the cases $m = 1, 2, 3$ and 4 to evaluate the integral $I = \int_0^\infty e^{-x} \cos x \, dx$ and state the accuracy of your answer in each case.

▶ **19.**   Use Gauss-Chebyshev integration for the cases $m = 1, 2, 3$ and 4 to evaluate the integral $I = \int_{-1}^1 \frac{e^{-x}}{\sqrt{1 - x^2}} \, dx$.

▶ **20.** Use Gauss-Chebyshev integration for the cases $m = 1, 2, 3$ and 4 to evaluate the integral $I = \int_{-1}^{1} \dfrac{\sin^2 x}{\sqrt{1 - x^2}}\, dx$.

▶ **21.** Discuss possible ways to approximate the derivative $\dfrac{df(x)}{dx}$ and the integral $I = \int_{0}^{x} f(x)\, dx$ if $f(x)$ is given by the tabular values

x	0	1	2	3	4	5
$f(x)$	0.0000	0.8532	1.3863	3.2958	5.5452	8.0472

(a) Approximate $\dfrac{df(x)}{dx}\Big|_{x=2.5} = f'(2.5)$.

(b) Approximate $\int_{0}^{5} f(x)\, dx$

▶ **22.**

(a) Verify that a composite integration rule for the midpoint formula given by equation (7.57) can be expressed

$$\int_{x_0}^{x_{2n}} f(x)\, dx = 2h \sum_{j=1}^{n} f(x_{2j-1}) + Error$$

(b) Verify that the error term for the above composite rule can be expressed

$$Error = \frac{x_{2n} - x_0}{6} h^2 f''(\zeta)$$

▶ **23. Simpson's 3/8 rule**

(a) Derive the 3-panel formula

$$\int_{x_{i-1}}^{x_{i+2}} f(x)\, dx = \frac{3}{8} h \left(y_{i-1} + 3y_i + 3y_{i+1} + y_{i+2} \right) - \frac{3}{80} h^5 f^{(iv)}(\zeta), \qquad x_{i-1} < \zeta < x_{i+2}$$

which is known as Simpson's 3/8 rule.

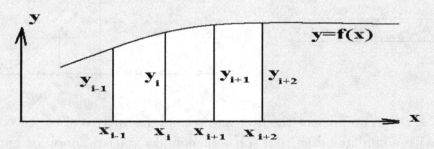

(Let $i = 1$ for simplicity in deriving the formula.)

(b) Compare Simpson's 3/8 rule with that of the Simpson's 1/3 rule. Which is the more accurate integration formula?

(c) Derive a composite or extended formula where Simpson's 3/8 rule can be used to calculate the integral $\int_a^b f(x)\,dx$. What restrictions must you place upon your formula?

(d) Use Simpson's 3/8 rule to evaluate the integral $I = \int_0^\pi \sin x\,dx$ using an equal step size of $h = \pi/6$. How accurate is your answer?

(e) Use Simpson's 1/3 rule to evaluate the integral $I = \int_0^\pi \sin x\,dx$ using an equal step size of $h = \pi/6$. How accurate is your answer?

▶ 24.   **Newton Cotes Open Formulas**

Derive the following Newton-Cotes open formula

$$\int_{x_0}^{x_6} f(x)\,dx = \frac{6}{20}h\left[11y_1 - 14y_2 + 26y_3 - 14y_4 + 11y_5\right] + \frac{41}{140}h^7 f^{(vi)}(\xi)$$

for the integration of $y = f(x)$ over 6-panels.

▶ 25.   **Computer program- Standard Normal Distribution**

Elementary statistic books have tables which evaluate the integral

$$\phi(z) = \int_0^z f(x)\,dx \text{ where } f(x) = \frac{1}{\sqrt{2\pi}}e^{-x^2/2}.$$

The value of this integral represents the probability that a random variable having a standard normal distribution takes on a value between 0 and z.

(a) Write a computer program which evaluates the above integral for the values of z from 0.00 to 4.50 in steps of 0.01.

(a) Write a computer program which creates a table of values

z	$\phi(z)$	$\phi'(z)$	$\phi''(z)$

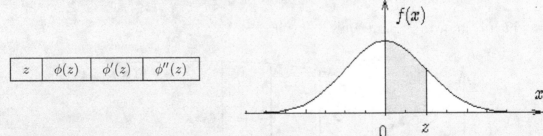

where $\phi'(z), \phi''(z)$ represent the first and second derivatives of $\phi(z)$. Compare your table with a similar table from some elementary statistic book or handbook.

► 26.   **Computer Problem**

The chi-square distribution with degrees of freedom $n = 1, 2, 3, \ldots$ is given by the probability density function

$$f(x) = \frac{1}{2^{n/2}\Gamma(n/2)} x^{(n/2)-1} \exp(-x/2)$$

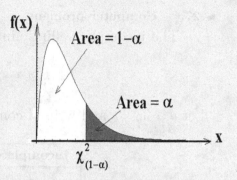

illustrated in the accompanying graph.

Write a computer program to reproduce the accompanying chi-square table of critical values. The problem is to determine values of $p = \chi^2_{(1-\alpha)}$ corresponding to the areas of $1 - \alpha = 0.995,\ 0.990,\ 0.975,\ 0.950$ and 0.900. That is find p such that

$$\int_0^p f(x)\, dx = 1 - \alpha$$

for $\alpha = 0.005,\ 0.010,\ 0.025,\ 0.050,$ and 0.100.

Hint: For the case $n = 1$ use integration by parts. The Gamma function satisfies $\Gamma(n+1) = n\Gamma(n)$ and $\Gamma(1/2) = \sqrt{\pi}$.

Critical Values for the chi-square distribution with ν degrees of freedom $\displaystyle\int_0^{\chi^2_{(1-\alpha)}} \frac{1}{2^{\nu/2}\Gamma(\nu/2)} x^{(\nu/2)-1} \exp(-x/2)\, dx$					
α	0.005	0.010	0.025	0.050	0.100
ν	$\chi^2_{0.995}$	$\chi^2_{0.990}$	$\chi^2_{0.975}$	$\chi^2_{0.950}$	$\chi^2_{0.900}$
1	7.8794	6.63479	5.02389	3.84146	2.70554
2	10.5966	9.21033	7.37776	5.99145	4.60517
3	12.8381	11.3447	9.34840	7.81473	6.25138
4	14.8602	13.2765	11.1433	9.48773	7.77944
5	16.7496	15.0863	12.8325	11.0705	9.23636
6	18.5476	16.8119	14.4494	12.5916	10.6446
7	20.2777	18.4753	16.0128	14.0671	12.0170
8	21.9550	20.0902	17.5345	15.5073	13.3616
9	23.5893	21.6659	19.0228	16.9190	14.6837
10	25.1882	23.2092	20.4831	18.3070	15.9872

▶ **27.**   **Computer problem**

The incomplete elliptic integral of the first kind is defined

$$F(k, \phi) = \int_0^\phi \frac{d\phi}{\sqrt{1 - k^2 \sin^2 \phi}}, \qquad 0 < k < 1.$$

Let $\theta = \sin^{-1} k$ and write a computer program to create a table of values

Incomplete elliptic integral $F(\sin\theta, \phi)$					
ϕ ╲ θ	5°	10°	15°	$\cdots$	90°
1°					
2°					
$\vdots$					
90°					▄▄

where ϕ is taken in steps of 1 degree and θ is taken in steps of 5 degrees.
Hint: Be sure to convert from degrees to radians.

▶ **28.**

(a) Integrals of the form $I = \int_0^1 \frac{f(x)}{\sqrt{1 - x^2}} \, dx$ have a singularity at $x = 1$. Show that the transformation $x = \sin u$ removes the singularity and reduces the integral to the form $I = \int_0^{\pi/2} f(\sin u) \sin u \, du$.

(b) Evaluate the integral $I = \int_0^1 \frac{e^x}{\sqrt{1 - x^2}} \, dx$ and determine the accuracy of your answer.

▶ **29.**   **(Double integral)**

(a) Write a subroutine that evaluates the integral

$$h(x) = \int_{f_1(x)}^{f_2(x)} g(x, y) \, dy$$

for a fixed value of x.

(b) Write a subroutine to evaluate the integral

$$\int_a^b h(x) \, dx$$

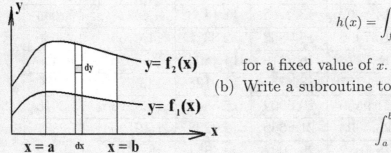

(c) Combine your results in parts (a) and (b) and describe how you would evaluate the double integral

$$I = \int_a^b \int_{f_1(x)}^{f_2(x)} g(x,y)\,dydx$$

Discuss the error associated with your method.

(d) Test your method on the double integrals

$$(i) \quad I_1 = \int_{-2}^4 \int_{-x^2+x+1}^{x^2+x+3} y\,dydx \qquad (ii) \quad I_2 = \int_0^2 \int_{-\sqrt{4-x^2}}^{\sqrt{4-x^2}} dydx$$

▶ **30.** (**Triple integral**)

(a) Write a subroutine which evaluates the integral
$$p(x,y) = \int_{g_1(x,y)}^{g_2(x,y)} h(x,y,z)\,dz \text{ for fixed values of } x \text{ and } y.$$

(b) Write a subroutine which evaluates the integral
$$q(x) = \int_{f_1(x)}^{f_2(x)} p(x,y)\,dy \text{ for a fixed value of } x.$$

(c) Write a subroutine which evaluates the integral
$$\int_a^b q(x)\,dx \text{ where } a \text{ and } b \text{ are constants.}$$

(d) Combine your results in parts (a),(b),(c) and describe how you would evaluate the triple integral

$$r = \int_a^b \int_{f_1(x)}^{f_2(x)} \int_{g_1(x,y)}^{g_2(x,y)} h(x,y,z)\,dz\,dy\,dx$$

(e) Test your method on the triple integrals

$$(i) \quad I_1 = \int_0^1 \int_x^{2-x^2} \int_0^{x+y} (x^2+z^2)\,dz\,dy\,dx \qquad (ii) \quad I_2 = \int_0^2 \int_0^x \int_0^{x+y} z\,dz\,dy\,dx$$

▶ **31.** (**Computer problem**) Write a computer program to calculate the Gamma function
$$\Gamma(\alpha) = \int_0^\infty t^{\alpha-1}e^{-t}\,dt, \qquad \alpha > 0$$

and create a table of values for $\Gamma(x)$ for x going from 1.00 to 2.0 in steps of 0.01.

(a) Show that $\Gamma(x+1) = x\Gamma(x)$

(b) Show that $\Gamma(1) = 1$

(c) Show that $\Gamma(n+1) = n!$ for n a positive integer.

300

▶ **32.** (**Computer problem**) A black-body radiator is one that emits the maximum possible intensity of radiation for a given temperature. Let q denote the rate at which radiant energy of wavelength λ leaves a unit of surface area. According to Planck's law

$$q = q(\lambda, T) = \frac{2\pi h c^2}{\lambda^5 (e^{hc/k\lambda T} - 1)} \qquad \left(\frac{\text{ergs}}{\text{cm}^2 \sec A^\circ}\right)$$

where $c = 2.99793(10)^{10}$ (cm/sec) is the speed of light, $h = 6.6256(10)^{-27}$ (erg sec) is Planck's constant, $k = 1.38054(10)^{-16}$ (erg/°K) is the Boltzmann constant, T is the absolute temperature (°K) and λ is the wavelength (cm). Note the conversion factors $1\,°A = 1$ Angstrom unit $= 10^{-8}$ cm and 1 calorie $= 4.184(10)^7$ ergs. A graph of $q = q(\lambda, T)$ is illustrated.

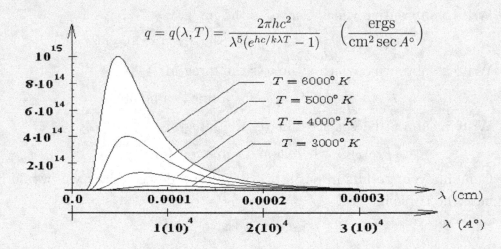

The rate Q at which radiation is being emitted per unit of time per unit of area between wavelengths λ_1 and λ_2 is the corresponding area under the q curve given by $Q = \int_{\lambda_1}^{\lambda_2} q(\lambda, T)\,d\lambda$. The total rate of emission is given by the integral $Q_{total} = \int_0^\infty q(\lambda, T)\,d\lambda = \sigma T^4 = E$, where $\sigma = 2\pi^5 k^4 / 15 c^2 h^3 \approx 5.67(10)^{-8}$ W/m² (°K)⁴ is called the Stefan-Boltzmann constant.   Write a computer program to numerically calculate, for a fixed value of the temperature T, the quantities

(a) The integral $Q = \int_{\lambda_1}^{\lambda_2} q(\lambda, T)\,d\lambda$ and the value of Q_{total}.

(b) The fraction of the total emission $Q_{frac} = Q/Q_{total}$

(c) Stefan's law states that Q_{total} is proportional to the fourth power of the temperature T. Verify numerically Stefan's law.

(d) Wien's displacement law states that at each temperature T where q has a maximum value, the value of the wavelength corresponding to this maximum, call it λ_{max}, is such that $\lambda_{max}T$ has a constant value. Can you numerically verify this law? (The constant value found experimentally is $2.9(10)^{-3}\,\text{m K}$ for a black-body radiator.)

(e) Assume $\lambda_{max} \approx 5(10)^{-7}\text{m}$ for the black-body temperature of the Sun. Use Wien's law to find the approximate temperature of the Sun.

▶ **33.** Use Romberg integration to evaluate the integrals

$$(a) \quad I_1 = \int_0^\pi \frac{1}{1 + \frac{1}{2}\cos x}\,dx \qquad (b) \quad I_2 = \int_0^\pi \ln(5 + 3\cos x)\,dx$$

▶ **34.** Use adaptive quadrature to evaluate the integrals

$$(a) \quad I_1 = \int_0^\pi \frac{1}{1 + \frac{1}{2}\cos x}\,dx \qquad (b) \quad I_2 = \int_0^\pi \ln(5 + 3\cos x)\,dx$$

▶ **35.**

Use the method of undetermined coefficients to derive a one panel integration formula involving both ordinates and derivatives. Your integration formula is to have the form

$$\int_{x_0}^{x_1} f(x)\,dx = Af(x_0) + Bf(x_1) + Cf'(x_0) + Df'(x_1)$$

where A, B, C, D are constants to be determined.

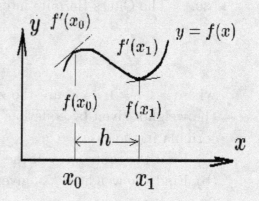

▶ **36.**

(a) Show that the change of variable $x = \left(\frac{b+a}{2}\right) + \left(\frac{b-a}{2}\right)\xi$ for $-1 \le \xi \le 1$, transforms the following integrals to a form where the Gauss-Legendre formulas will be applicable

$$(i) \quad I_1 = \int_a^b f(x)\sqrt{b-x}\,dx \qquad (ii) \quad I_2 = \int_a^b \frac{f(x)}{\sqrt{b-x}}\,dx$$

(b) Evaluate the integrals

$$(i) \quad I_1 = \int_0^2 e^{-x^2}\sqrt{2-x}\,dx \qquad (ii) \quad I_2 = \int_0^2 \frac{e^{-x^2}}{\sqrt{2-x}}\,dx$$

▶ **37.**

(a) Show that the change of variable $x = \left(\frac{b+a}{2}\right) + \left(\frac{b-a}{2}\right)\xi$ for $-1 \le \xi \le 1$, transforms the following integral to a form where the Gauss-Chebyshev formula will be applicable

$$I = \int_a^b \frac{f(x)}{\sqrt{x-a}\sqrt{b-x}}\, dx$$

(b) Evaluate the integrals

$$(i) \quad I_1 = \int_0^2 \frac{e^{-x^2}}{\sqrt{x}\sqrt{2-x}}\, dx \qquad (ii) \quad I_2 = \int_0^2 \frac{\sin x}{\sqrt{x}\sqrt{2-x}}\, dx$$

▶ **38.** Derive the Bode's rule for integration over four panels

$$\int_{x_0}^{x_1} f(x)\, dx = \frac{2h}{45}\left[7f(x_0) + 32f(x_1) + 12f(x_2) + 32f(x_3) + 7f(x_4)\right] - \frac{8}{945}h^7 f^{(vi)}(\xi)$$

▶ **39.** The Gauss-Hermite integration formula has the form

$$\int_{-\infty}^{\infty} e^{-x^2} f(x)\, dx = \sum_{i=1}^n \beta_i f(x_i)$$

where $x_i,$, $i = 1,\ldots,n$, are the zero's of the nth Hermite polynomial and β_i are the weights given by equation (7.165). Develop the Gauss-Hermite integration formula in the case $n = 2$.

(a) Find the roots x_1, x_2 of the Hermite polynomial $H_2(x) = 4x^2 - 2$.

(b) Find the weights β_1, β_2 given by equation (7.165) where $w(x) = e^{-x^2}$.

$$\beta_i = \int_{-\infty}^{\infty} e^{-x^2} \ell_i(x)\, dx, \qquad i = 1,2$$

where $\ell_i(x)$ are the Lagrange interpolating polynomials associated with the roots x_1, x_2.

Hint: Let $I_n = \int_{-\infty}^{\infty} e^{-x^2} x^n\, dx, \quad I_0 = \sqrt{\pi}, \quad I_1 = 0, \quad I_{n+2} = \frac{(n+1)}{2} I_n.$

(c) Use the Gauss-Hermite formula with $n = 2$ to evaluate the integrals

$$(i) \quad I_1 = \int_{-\infty}^{\infty} e^{-x^2} x^2\, dx \qquad (ii) \quad I_2 = \int_{-\infty}^{\infty} \frac{e^{-x^2}}{1+x^4}\, dx$$

"On two occasions I have been asked [by members of Parliament], 'Pray, Mr. Babbage, if you put into the machine wrong figures, will the right answers come out?' I am not able rightly to apprehend the kind of confusion of ideas that could provoke such a question. "

Charles Babbage (1792-1871)

Chapter 8
Ordinary Differential Equations

Many ordinary differential equations encountered do not have easily obtainable closed form solutions, and we must seek other methods by which solutions can be constructed. Numerical methods provide an alternative way of constructing solutions to these sometimes difficult problems. In this chapter we present an introduction to some numerical methods which can be applied to a wide variety of ordinary differential equations. These methods can be programmed into a digital computer or even programmed into some hand-held calculators. Many of the numerical techniques introduced in this chapter are readily available in the form of subroutine packages available from the internet.

We consider the problem of developing numerical methods to solve a first order initial value problem of the form

$$\frac{dy}{dx} = f(x,y), \qquad y(x_0) = y_0 \tag{8.1}$$

and then consider how to generalize these methods to solve systems of ordinary differential equations having the form

$$\frac{dy_1}{dx} = f_1(x, y_1, y_2, \ldots, y_m), \qquad y_1(x_0) = y_{10}$$

$$\frac{dy_2}{dx} = f_2(x, y_1, y_2, \ldots, y_m), \qquad y_2(x_0) = y_{20}$$

$$\vdots$$

$$\frac{dy_m}{dx} = f_m(x, y_1, y_2, \ldots, y_m), \qquad y_m(x_0) = y_{m0}$$

$$\tag{8.2}$$

Coupled systems of ordinary differential equations are sometimes written in the vector form

$$\frac{d\vec{y}}{dx} = \vec{f}(x, \vec{y}), \qquad \vec{y}(x_0) = \vec{y}_0 \tag{8.3}$$

where $\vec{y}$, $\vec{y}(x_0)$ and $\vec{f}(x, \vec{y})$ are column vectors given by $\vec{y} = \text{col}(y_1, y_2, y_3, \ldots, y_m)$, $\vec{y}(x_0) = \text{col}(y_{10}, y_{20}, \ldots, y_{m0})$ and $\vec{f}(x, \vec{y}) = \text{col}(f_1, f_2, \ldots, f_m)$.

We start with developing numerical methods for obtaining solutions to the first order initial value problem (8.1) over an interval $x_0 \leq x \leq x_n$. Many of the techniques developed for this first order equation can, with modifications, also be applied to solve a first order system of differential equations.

Higher Order Equations

By defining new variables, higher order differential equations can be reduced to a first order system of differential equations. As an example, consider the problem of converting a nth order linear homogeneous differential equation

$$\frac{d^n y}{dx^n} + a_1 \frac{d^{n-1} y}{dx^{n-1}} + a_2 \frac{d^{n-2} y}{dx^{n-2}} + \cdots + a_{n-1} \frac{dy}{dx} + a_n y = 0 \tag{8.4}$$

to a vector representation. To convert this equation to vector form we define new variables. Define the vector quantities

$$\vec{y} = \text{col}(y_1, y_2, y_3, \cdots, y_n) = \text{col}\left(y, \frac{dy}{dx}, \frac{d^2 y}{dx^2}, \cdots, \frac{d^{n-1} y}{dx^{n-1}} \right)$$

$$\vec{f}(x, \vec{y}) = A\vec{y},$$

$$\text{where} \quad A = \begin{bmatrix} 0 & 1 & 0 & 0 & \cdots & 0 & 0 \\ 0 & 0 & 1 & 0 & \cdots & 0 & 0 \\ 0 & 0 & 0 & 1 & \cdots & 0 & 0 \\ 0 & 0 & 0 & 0 & \cdots & 0 & 0 \\ \vdots & \vdots & \vdots & \vdots & \ddots & \vdots & \vdots \\ 0 & 0 & 0 & 0 & \cdots & 0 & 1 \\ -a_n & -a_{n-1} & -a_{n-2} & -a_{n-3} & \cdots & -a_2 & -a_1 \end{bmatrix}. \tag{8.5}$$

Observe that the linear nth order differential equation (8.4) can now be represented in the form of equation (8.3). In this way higher order linear ordinary differential equations can be represented as a first order vector system of differential equations.

Numerical Solution

In our study of the scalar initial value problem (8.1) it is assumed that $f(x, y)$ and its partial derivative f_y both exist and are continuous in a rectangular region about a point (x_0, y_0). If these conditions are satisfied, then theoretically there exists a unique solution of the initial value problem (8.1) which is a continuous curve $y = y(x)$, which passes through the point (x_0, y_0) and satisfies the differential equation. In contrast to the solution being represented by a continuous function $y = y(x)$, the numerical solution to the initial value problem (8.1) is represented by a set of data points (x_i, y_i) for $i = 0, 1, 2, \ldots, n$ where y_i is an approximation to the true solution $y(x_i)$. We shall investigate various methods for constructing the data points (x_i, y_i), for $i = 1, 2, \ldots, n$ which approximate the true solution. This data set is then called a numerical solution to the given initial value problem. The given rule or technique used to obtain the numerical solution is called a numerical method or algorithm. There are many numerical methods for solving ordinary differential equations. In this chapter we will consider only a select few of the more popular methods. The numerical methods considered can be classified as either single-step methods or multi-step methods. We begin our introduction to numerical methods for ordinary differential equations by considering single-step methods.

Single Step Methods

From calculus a function $y = y(x)$, which possesses derivatives of all orders, can be expanded in a Taylor series about a point $x = x_0$. The Taylor series expansion has the form

$$y(x_0 + h) = y(x_0) + y'(x_0)h + y''(x_0)\frac{h^2}{2!} + \ldots + y^{(n)}(x_0)\frac{h^n}{n!} + R_n, \qquad (8.6)$$

where R_n is a remainder term. If the $(n+1)$st derivative of y is bounded such that $|y^{(n+1)}(x)| < K$ for $x \in (x_0, x_0 + h)$, then we can say that the remainder term satisfies

$$R_n = y^{(n+1)}(\xi)\frac{h^{n+1}}{(n+1)!} = O(h^{n+1}) \qquad (8.7)$$

with ξ lying somewhere between x_0 and $x_0 + h$. We can use the Taylor series expansion to derive many numerical methods for solving initial value problems.

Example 8-1. Euler's Method (First-order Taylor series method)
Consider the specific initial value problem to solve

$$y'(x) = \frac{dy}{dx} = f(x, y) = x + y, \quad y(0) = 1, \quad 0 \le x \le 1. \qquad (8.8)$$

306

We use the first and second term of the Taylor series expansion to approximate the value of the solution y at a nearby point $x_0 + h$. We calculate the slope of the curve at the initial point (x_0, y_0) directly from the given differential equation and find $y'(x_0) = f(x_0, y_0) = x_0 + y_0 = 1$. We then select a step size h, and use the value $y_1 = y(x_1) = y(x_0 + h)$ as an approximate value for the true solution. This gives $y_1 = y(x_0 + h) = y(x_0) + y'(x_0)h + O(h^2)$, where the error of the approximation is of order h^2. Letting $h = 0.1$ and substituting in the values for x_0 and y_0 we find $y_1 = 1.1$ at $x_1 = 0.1$. If we repeat this step-by-step process with (x_1, y_1) as the new initial point, we obtain the algorithm illustrated in figure 8-1, which is called Euler's method or a first-order Taylor series method. Notice the algorithm in figure 8-1 is a single-step method. That is, if we know the approximate value y_m of the solution curve $y(x)$ at $x = x_m$, then from the point (x_m, y_m) we can take a single-step to the next approximating value (x_{m+1}, y_{m+1}), where

$$x_{m+1} = x_m + h \qquad \text{and} \qquad y_{m+1} = y_m + y'_m h + O(h^2). \qquad (8.9)$$

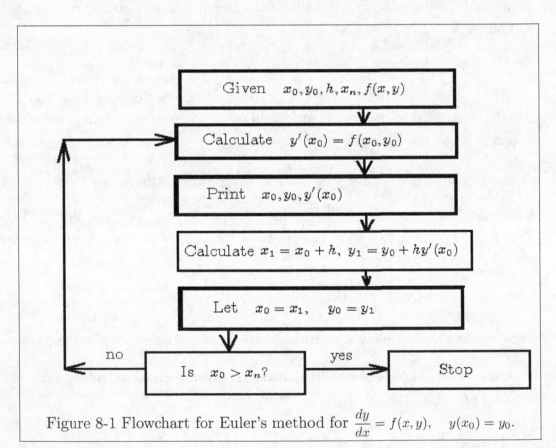

Figure 8-1 Flowchart for Euler's method for $\dfrac{dy}{dx} = f(x, y), \quad y(x_0) = y_0$.

Applying the Euler method single step algorithm illustrated in figure 8-1 to the initial value problem (8.8) and using a step size of $h = 0.1$, we obtain the numerical values in table 8.1. The fourth column in this table gives the exact solution for comparison purposes. Analysis of the error term associated with the Euler method is considered in the exercises at the end of this chapter.

Table 8.1

Numerical Results for Euler's method applied to

$$\frac{dy}{dx} = x + y, \quad y(0) = 1, \quad h = 0.1$$

x	$y(x)$	$y'(x)$	$y(x) = 2e^x - x - 1$	% error
0.0	1.000	1.000	1.000	0.00
0.1	1.100	1.200	1.110	0.93
0.2	1.220	1.420	1.243	1.84
0.3	1.362	1.662	1.400	2.69
0.4	1.528	1.928	1.584	3.51
0.5	1.721	2.221	1.797	4.25
0.6	1.943	2.543	2.044	4.95
0.7	2.197	2.897	2.328	5.61
0.8	2.487	3.287	2.651	6.19
0.9	2.816	3.716	3.019	6.73
1.0	3.187	4.187	3.437	7.26

Taylor Series Method

Other numerical methods for solving differential equations can be developed from the Taylor series equation (8.6). If we retain the first $(m + 1)$ terms in the Taylor series expansion given by equation (8.6), one can write

$$y(x_0 + h) = y(x_0) + hT_m(x_0, y_0, h) + R_m \tag{8.10}$$

with
$$T_m = T_m(x_0, y_0, h) = y'(x_0) + y''(x_0)\frac{h}{2!} + \ldots + y^{(m)}(x_0)\frac{h^{m-1}}{m!} \tag{8.11}$$

and $R_m = y^{(m+1)}(\xi)\dfrac{h^{m+1}}{(m+1)!} = O(h^{m+1})$ representing the error term of the approximation. Equation (8.10) represents an mth-order Taylor series approximation to the value $y_1 = y(x_0 + h)$. In order to use the above formula, it is necessary for us to obtain the various derivative terms $y^{(k)}(x_0)$, $k = 1, 2, 3, \ldots, m$. These derivatives may be computed by differentiating the given differential equation (8.1) using the chain rule for differentiation. The first couple of derivatives are

$$y'(x_0) = f \Big|_{(x_0, y_0)} \qquad\qquad y'''(x_0) = f_{xx} + 2f_{xy}f + f_x f_y + f_{yy}f^2 + f_y^2 f \Big|_{(x_0, y_0)}$$

$$y''(x_0) = f_x + f_y f \Big|_{(x_0, y_0)} \qquad\qquad y^{(iv)}(x_0) = \frac{d}{dx}y'''(x) \Big|_{(x_0, y_0)} \tag{8.12}$$

Higher derivatives become more difficult to calculate if $f = f(x, y)$ is a complicated expression. In equation (8.12) the subscripts denote partial derivatives. For example, $f_x = \frac{\partial f}{\partial x}$, $f_{xx} = \frac{\partial^2 f}{\partial x^2}$, $f_{xy} = \frac{\partial^2 f}{\partial x \partial y}$, etc. Of course the larger the value of m in the Taylor series expansion (8.10) (8.11), the more work is involved in calculating these higher derivatives. In most applications the Taylor series method should be used only when the derivatives of $f = f(x, y)$ are easily obtainable.

Example 8-2. (Second-order Taylor series method)

A second-order Taylor series algorithm for approximating solutions to the initial value problem (8.1) is given in figure 8-2. Applying the second-order Taylor series algorithm to the initial value problem (8.8) we obtain the results in table 8.2. Compare the results of table 8.1 with the entries in table 8.2 to see the difference in the errors between a first- and second-order Taylor series method.

Table 8.2

Numerical Results for second-order Taylor series method applied to

$$\frac{dy}{dx} = x + y, \quad y(0) = 1, \quad h = 0.1$$

x	$y(x)$	$y'(x)$	$y''(x)$	$y(x) = 2e^x - x - 1$	% error
0.0	1.000	1.000	2.000	1.000	0.000
0.1	1.100	1.210	2.210	1.110	0.031
0.2	1.242	1.442	2.442	1.243	0.061
0.3	1.398	1.698	2.698	1.400	0.089
0.4	1.582	1.982	2.982	1.584	0.117
0.5	1.795	2.295	3.295	1.797	0.142
0.6	2.041	2.641	3.641	2.044	0.165
0.7	2.323	3.023	4.023	2.328	0.187
0.8	2.646	3.446	4.446	2.651	0.298
0.9	3.012	3.912	4.912	3.019	0.227
1.0	3.428	4.428	5.428	3.437	0.244

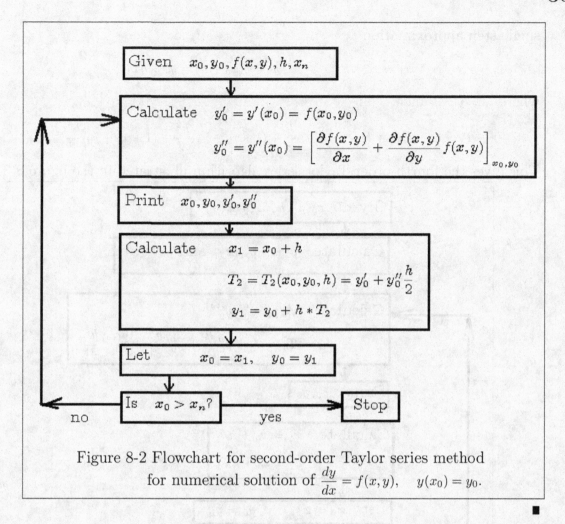

Figure 8-2 Flowchart for second-order Taylor series method
for numerical solution of $\dfrac{dy}{dx} = f(x,y)$, $y(x_0) = y_0$.

Example 8-3. (Fourth-order Taylor series method)

Set up a fourth-order Taylor series algorithm to solve the initial value problem

$$y'(x) = \frac{dy}{dx} = f(x,y) = x + y, \quad y(0) = 1, \quad 0 \le x \le 1.$$

Solution: We differentiate the given differential equation to obtain derivatives through order four. There results the following equations:

$$y'(x) = x + y \qquad\qquad y'''(x) = 1 + y' = 1 + x + y$$
$$y''(x) = 1 + y' = 1 + x + y \qquad y^{(iv)}(x) = 1 + y' = 1 + x + y$$

Substituting the above derivatives into the Taylor series gives us the fourth-order

310

single step approximation

$$x_1 = x_0 + h \qquad\qquad y_1 = y(x_0 + h) = y_0 + hT_4,$$

where $y_0 = y(x_0)$ and

$$T_4 = (x_0 + y_0) + (1 + x_0 + y_0)\frac{h}{2!} + (1 + x_0 + y_0)\frac{h^2}{3!} + (1 + x_0 + y_0)\frac{h^3}{4!}.$$

This gives the fourth-order Taylor series algorithm illustrated in the figure 8-3.

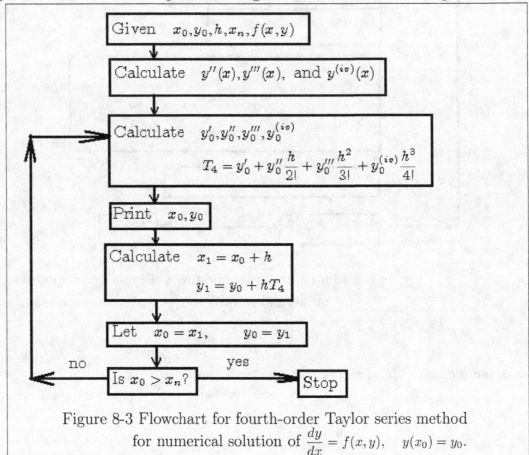

Figure 8-3 Flowchart for fourth-order Taylor series method
for numerical solution of $\frac{dy}{dx} = f(x,y), \quad y(x_0) = y_0$.

Runge-Kutta Methods

There are various types of Runge-Kutta algorithms for the numerical solution
of the initial value problem

$$y' = f(x,y), \quad y(x_0) = y_0, \quad x_0 \le x \le x_n. \tag{8.13}$$

These methods are single step methods with step size h. A p-stage Runge-Kutta algorithm to solve the initial value problem (8.13) is a stepping procedure from a point (x_i, y_i) to the next point (x_{i+1}, y_{i+1}) given by

$$x_{i+1} = x_i + h$$
$$y_{i+1} = y_i + \omega_1 k_1 + \omega_2 k_2 + \omega_3 k_3 + \ldots + \omega_p k_p,$$
(8.14)

where $\omega_1, \omega_2, \ldots, \omega_p$ are called weighting constants and

$$k_1 = hf(x_i, y_i)$$
$$k_2 = hf(x_i + c_2 h, y_i + a_{21} k_1)$$
$$k_3 = hf(x_i + c_3 h, y_i + a_{31} k_1 + a_{32} k_2)$$
$$\ldots$$
$$k_p = hf(x_i + c_p h, y_i + a_{p1} k_1 + a_{p2} k_2 + \ldots + a_{p,p-1} k_{p-1}),$$
(8.15)

are scaled slope calculations at specified points. In the equation (8.15) the quantities $c_2, c_3, \ldots, c_p, a_{21}, a_{31}, a_{32}, \ldots$ are given constants. The k_i, $i = 1, \ldots, p$ values require p function evaluations for the slope $f(x, y)$ to be evaluated at specified (x, y) points. In general, a p-stage Runge-Kutta method is a stepping method, with step size h, which requires p function evaluations for each step taken.

There is an array notation for representing Runge-Kutta methods which was introduced by J.C. Butcher around 1965. The array is called a Butcher array and has the form

$$
\begin{array}{c|cccc}
c_1 & a_{11} & a_{12} & \cdots & a_{1p} \\
c_2 & a_{21} & a_{22} & \cdots & a_{2p} \\
\vdots & \vdots & \vdots & \ddots & \vdots \\
c_p & a_{p1} & a_{p2} & \cdots & a_{pp} \\
\hline
 & \omega_1 & \omega_2 & \cdots & \omega_p
\end{array}
\qquad
\begin{array}{c|c}
\vec{c} & \mathbf{A} \\
\hline
 & \vec{\omega}
\end{array}
$$
(8.16)

where $\vec{c} = (c_1, c_2, \ldots, c_p)^T$ and $\vec{\omega} = (\omega_1, \omega_2, \ldots, \omega_p)$ are vectors and $\mathbf{A} = (a_{ij})$ for $i = 1, \ldots, p$ and $j = 1, \ldots, p$ is a $p \times p$ matrix array. The Runge-Kutta method given by equations (8.14) and (8.15) is denoted by the Butcher array

$$
\begin{array}{c|ccccccc}
0 & 0 \\
c_2 & a_{21} & 0 \\
c_3 & a_{31} & a_{32} & 0 \\
c_4 & a_{41} & a_{42} & a_{43} & 0 \\
\vdots & \vdots & \vdots & \vdots & \vdots & \ddots \\
c_p & a_{p1} & a_{p2} & a_{p3} & \cdots & a_{p,p-1} & 0 \\
\hline
 & \omega_1 & \omega_2 & \omega_3 & \cdots & \omega_{p-1} & \omega_p
\end{array}
\tag{8.17}
$$

where the matrix **A** is lower triangular. Whenever the matrix **A** is strictly lower triangular, the Runge-Kutta method is called an explicit stepping method. If the matrix **A** is not lower triangular, then the Runge-Kutta method is called an implicit stepping method.

The p-stage Runge-Kutta method requires that the weighting constants $\omega_1,\ \omega_2, \ldots, \omega_p$ and the constants

$$
\begin{array}{ll}
c_2 & a_{21} \\
c_3 & a_{31}\ a_{32} \\
c_4 & a_{41}\ a_{42}\ a_{43} \\
 & \cdots \\
c_p & a_{p1}\ a_{p2}\ a_{p3} \cdots a_{p,p-1},
\end{array}
$$

are chosen such that y_{n+1} of equation (8.14) agrees with a Taylor series expansion through some order. It is known that if equation (8.14) agrees with a Taylor series expansion of order m, for $m = 2, 3$ or 4, then $p = m$ and for $m > 4$ it is known that $p > m$. It is also known that for consistency it is required that

$$
c_i = \sum_{j=1}^{i-1} a_{ij}, \quad \text{for} \quad i = 2, 3, \ldots, p.
\tag{8.18}
$$

Note that p-function evaluations are required if one uses a p-stage Runge-Kutta method. The Runge-Kutta methods for $p = 2, 3, 4$ have the same order of accuracy as the Taylor series methods of order $m = 2, 3, 4$ and they have the advantage that higher ordered derivative terms do not have to be calculated. In order to illustrate the general character of Runge-Kutta methods we begin by developing second-order Runge-Kutta methods in detail.

Second-order Runge-Kutta Methods

A second-order Runge-Kutta method designed to solve the equation (8.13) assumes that for $n = 0, 1, 2, \ldots$

$$
\begin{aligned}
x_{n+1} &= x_n + h \\
y_{n+1} &= y_n + \omega_1 k_1 + \omega_2 k_2 + \mathcal{O}(h^3),
\end{aligned}
\tag{8.19}
$$

where

$$
\begin{aligned}
k_1 &= h f(x_n, y_n) \\
k_2 &= h f(x_n + c_2 h, y_n + a_{21} k_1).
\end{aligned}
\tag{8.20}
$$

The problem is to find constants ω_1, ω_2, c_2, a_{21} such that y_{n+1} agrees with the Taylor series expansion of $y(x_n + h)$ through the second-order terms. Recall that a Taylor series expansion through the second-order terms appears as

$$
y(x_n + h) = y(x_n) + y'(x_n)h + y''(x_n)\frac{h^2}{2!} + O(h^3).
$$

Substituting for the first and second derivatives obtained from equation (8.13) gives

$$
y_{n+1} = y_n + f(x_n, y_n)h + \frac{h^2}{2!}\left[\frac{\partial f}{\partial x}(x_n, y_n) + f(x_n, y_n)\frac{\partial f}{\partial y}(x_n, y_n)\right] + O(h^3).
\tag{8.21}
$$

By using equations (8.19) and (8.20), we write the second-order Runge-Kutta algorithm in the expanded form as

$$
y_{n+1} = y_n + \omega_1 h f(x_n, y_n) + \omega_2 h f(x_n + c_2 h, y_n + a_{21} k_1),
\tag{8.22}
$$

where the constants ω_1, ω_2, c_2, and a_{21} are to be selected such that equations (8.22) and (8.21) are the same through second-order terms involving h^2. Expanding the last term in equation (8.22) in a Taylor series about the point (x_n, y_n) gives the expansion

$$
f(x_n + c_2 h, y_n + a_{21} k_1) = f(x_n, y_n) + \frac{\partial f}{\partial x}(x_n, y_n)c_2 h + \frac{\partial f}{\partial y}(x_n, y_n)a_{21}h f(x_n, y_n) + \cdots
$$

plus higher ordered terms which have been neglected. When this expansion is substituted into equation (8.22) and terms are rearranged, the result may be expressed as

$$
y_{n+1} = y_n + (\omega_1 + \omega_2)h f(x_n, y_n) + \omega_2 c_2 h^2 f_x(x_n, y_n) + \omega_2 a_{21} h^2 f(x_n, y_n) f_y(x_n, y_n),
\tag{8.23}
$$

where higher order terms have been neglected. Observe that equation (8.23) is identical with equation (8.21), if we choose the constants to satisfy the relations

$$\omega_1 + \omega_2 = 1, \qquad \omega_2 c_2 = \frac{1}{2}, \qquad \omega_2 a_{21} = \frac{1}{2}. \tag{8.24}$$

To accomplish this we can arbitrarily assign a value to one of the constants ω_1, ω_2, c_2 or a_{21}, and then use equations (8.24) to determine the values of the remaining constants. The arbitrary constant selected is referred to as a parameter. Letting ω_2 be a parameter gives the equations

$$\omega_1 = 1 - \omega_2, \qquad \text{and} \qquad a_{21} = c_2 = \frac{1}{2\omega_2}. \tag{8.25}$$

This process of arbitrarily assigning a value to the parameter and then calculating the remaining constants produces a variety of second-order Runge-Kutta methods.

The **Runge-Kutta modified Euler method**[†] results when $\omega_2 = 1$ and the other constants have the values $\omega_1 = 0, c_2 = a_{21} = 1/2$. The resulting Runge-Kutta stepping equations are

$$
\begin{array}{c|cc}
0 & 0 & 0 \\
1/2 & 1/2 & 0 \\
\hline
 & 0 & 1
\end{array}
\qquad
\begin{aligned}
x_{n+1} &= x_n + h \\
y_{n+1} &= y_n + k_2 \\
k_1 &= hf(x_n, y_n) \\
k_2 &= hf(x_n + \frac{1}{2}h, y_n + \frac{1}{2}k_1)
\end{aligned}
\tag{8.26}
$$

or in expanded form

$$y_{n+1} = y_n + hf(x_n + \frac{1}{2}h, y_n + \frac{1}{2}hf(x_n, y_n)) + \mathcal{O}(h^3) \tag{8.27}$$

The **Runge-Kutta improved Euler method**[†] or **Huen's method**[†] results if one selects the parameter value $\omega_2 = 1/2$ and the other constants then have the values $\omega_1 = \omega_2 = 1/2$ and $c_2 = a_{21} = 1$. The resulting Runge-Kutta stepping equations are

$$
\begin{array}{c|cc}
0 & 0 & 0 \\
1 & 1 & 0 \\
\hline
 & 1/2 & 1/2
\end{array}
\qquad
\begin{aligned}
x_{n+1} &= x_n + h \\
y_{n+1} &= y_n + \frac{1}{2}k_1 + \frac{1}{2}k_2 \\
k_1 &= hf(x_n, y_n) \\
k_2 &= hf(x_n + h, y_n + k_1)
\end{aligned}
\tag{8.28}
$$

[†] The reference R.J. Lopez points out that there is no uniformity in the literature associated with the above Runge-Kutta names.

or in expanded form

$$y_{n+1} = y_n + \frac{h}{2} \left[f(x_n, y_n) + f(x_n + h, y_n + hf(x_n, y_n)) \right] + \mathcal{O}(h^3). \tag{8.29}$$

The **Runge-Kutta minimum error bound method** results selecting the parameter value $\omega_2 = 3/4$ so that the other constants then have the values $\omega_1 = 1/4$, $c_2 = a_{21} = 2/3$. The resulting Runge-Kutta stepping equations are

0	0	0
2/3	2/3	0
	1/4	3/4

$$\begin{aligned} x_{n+1} &= x_n + h \\ y_{n+1} &= y_n + \frac{1}{4}k_1 + \frac{3}{4}k_2 \\ k_1 &= hf(x_n, y_n) \\ k_2 &= hf(x_n + \frac{2}{3}h, y_n + \frac{2}{3}k_1) \end{aligned} \tag{8.30}$$

or in expanded form

$$y_{n+1} = y_n + \frac{h}{4} \left[f(x_n, y_n) + 3f(x_n + \frac{2}{3}h, y_n + \frac{2}{3}hf(x_n, y_n)) \right] + \mathcal{O}(h^3) \tag{8.31}$$

Third-order Runge-Kutta methods

To develop a 3-stage Runge-Kutta stepping method of the form

$$\begin{aligned} y_{n+1} &= y_n + \omega_1 k_1 + \omega_2 k_2 + \omega_3 k_3 + \mathcal{O}(h^4) \\ k_1 &= hf(x_n, y_n) \\ k_2 &= hf(x_n + c_2 h, y_n + a_{21} k_1) \\ k_3 &= hf(x_n + c_3 h, y_n + a_{31} k_1 + a_{32} k_2) \end{aligned} \tag{8.32}$$

one must expand the equation (8.32) for $y_{n+1} - y_n$ by expanding the functions $f(x_n + c_2 h, y_n + a_{21} k_1)$ and $f(x_n + c_3 h, y_n + a_{31} k_1 + a_{32} k_2)$ in a Taylor series about (x_n, y_n) through h^3 terms. The expanded form of $y_{n+1} - y_n$ from equation (8.32) is then compared with the Taylor series expansion

$$y(x_n + h) - y(x_n) = y_{n+1} - y_n = hy_n' + \frac{h^2}{2!}y_n'' + \frac{h^3}{3!}y_n''' + \cdots \tag{8.33}$$

where one uses the differential equation $y' = f(x, y)$ and calculates the derivatives

$$y_n' = f(x_n, y_n)$$
$$y_n'' = \left. \frac{\partial f}{\partial x} + \frac{\partial f}{\partial y} f \right|_{(x_n, y_n)}$$
$$y_n''' = \left. \frac{\partial^2 f}{\partial x^2} + \frac{\partial^2 f}{\partial x \partial y} f + \frac{\partial f}{\partial y} \left(\frac{\partial f}{\partial x} + \frac{\partial f}{\partial y} f \right) + f \left(\frac{\partial^2 f}{\partial y \partial x} + \frac{\partial^2 f}{\partial y^2} f \right) \right|_{(x_n, y_n)}$$

Comparing the expanded forms of equations (8.32) and (8.33) one finds the following equations must be satisfied.

$$\omega_1 + \omega_2 + \omega_3 = 1$$

$$c_2\omega_2 + c_3\omega_3 = 1/2$$

$$c_2^2\omega_2 + c_3^2\omega_3 = 1/3$$

$$c_2 = a_{21}$$

$$c_3 = a_{31} + a_{32}$$

$$c_2 a_{32}\omega_3 = 1/6$$

(8.34)

The comparison produces six equations with eight unknowns to be determined. One can select two of the unknowns as parameters which are then used to determine the remaining quantities. The non-uniqueness of the solutions to the system of equations (8.34) produces a variety of Runge-Kutta 3-stage methods.

The **Runge-Kutta-Huen** method results using the values

$$\omega_1 = 1/4, \; \omega_2 = 0, \; \omega_3 = 3/4, \; c_2 = a_{21} = 1/3, \; c_3 = 2/3, \; a_{31} = 0, \; a_{32} = 2/3.$$

These values give the Runge-Kutta 3-stage stepping procedure

0	0	0	0
1/3	1/3	0	0
2/3	0	2/3	0
	1/4	0	3/4

$$x_{n+1} = x_n + h$$
$$y_{n+1} = y_n + \frac{1}{4}k_1 + \frac{3}{4}k_3 + \mathcal{O}(h^4)$$
$$k_1 = hf(x_n, y_n)$$
$$k_2 = hf(x_n + \frac{1}{3}h, y_n + \frac{1}{3}k_1)$$
$$k_3 = hf(x_n + \frac{2}{3}h, y_n + \frac{2}{3}k_2)$$

(8.35)

The **Runge-Kutta classic** form results using the values

$$\omega_1 = 1/6, \; \omega_2 = 4/6, \; \omega_3 = 1/6, \; c_2 = a_{21} = 1/2, \; c_3 = 1, \; a_{31} = -1, \; a_{32} = 2.$$

The resulting Runge-Kutta 3-stage stepping algorithm is

0	0	0	0
1/2	1/2	0	0
1	-1	2	0
	1/6	4/6	1/6

$$x_{n+1} = x_n + h$$
$$y_{n+1} = y_n + \frac{1}{6}(k_1 + 4k_2 + k_3) + \mathcal{O}(h^4)$$
$$k_1 = hf(x_n, y_n)$$
$$k_2 = hf(x_n + \frac{1}{2}h, y_n + \frac{1}{2}k_1)$$
$$k_3 = hf(x_n + h, y_n - k + 2k_2)$$

(8.36)

Fourth-order Runge-Kutta methods

A 4-stage Runge-Kutta method has the form

$$x_{n+1} = x_n + h$$
$$y_{n+1} = y_n + \omega_1 k_1 + \omega_2 k_2 + \omega_3 k_3 + \omega_4 k_4 + \mathcal{O}(h^5) \tag{8.37}$$

where

$$k_1 = hf(x_n, y_n)$$
$$k_2 = hf(x_n + c_2 h, y_n + a_{21} k_1)$$
$$k_3 = hf(x_n + c_3 h, y_n + a_{31} k_1 + a_{32} k_2) \tag{8.38}$$
$$k_4 = hf(x_n + c_4 h, y_n + a_{41} k_1 + a_{42} k_2 + a_{43} k_3)$$

One must expand these functions through h^4 terms and compare the results with the Taylor series expansion through h^4 terms. Modern computer packages are suggested to handle the messy algebra. Alternatively, a fourth-order derivation can be found in the A. Ralston and P. Rabinowitz reference. There results from the comparing of terms in the above expansions the following eleven equations

$$\omega_1 + \omega_2 + \omega_3 + \omega_4 = 1$$
$$c_2 \omega_2 + c_3 \omega_3 + c_4 \omega_4 = 1/2$$
$$c_2^2 \omega_2 + c_3^2 \omega_3 + c_4^2 \omega_4 = 1/3$$
$$c_2^3 \omega_2 + c_3^3 \omega_3 + c_4^3 \omega_4 = 1/$$
$$c_2 a_{32} + (c_2 a_{42} + c_3 a_{43}) \omega_4 = 1/6$$
$$c_2^2 a_{32} \omega_3 + (c_2^2 a_{42} + c_3^2 a_{43}) \omega_4 = 1/12 \tag{8.39}$$
$$c_2 c_3 a_{32} \omega_3 + (c_2 a_{42} + c_3 a_{43}) c_4 \omega_4 = 1/8$$
$$c_2 a_{32} a_{43} \omega_4 = 1/24$$
$$c_2 = a_{21}$$
$$c_3 = a_{31} + a_{32}$$
$$c_4 = a_{41} + a_{42} + a_{43}$$

with thirteen unknowns. Here again two quantities can be selected as parameters and the remaining quantities can be determined from the above system of equations (8.39). The non-uniqueness of solutions to the system of equations (8.39) results in a variety of fourth-order Runge-Kutta methods.

The **Runge-Kutta classic form** uses the values

$$\omega_1 = 1/6, \ \omega_2 = 2/6, \ \omega_3 = 2/6, \ \omega_4 = 1/6$$
$$c_2 = a_{21} = 1/2, \ c_3 = a_{32} = 1/2, \ a_{31} = 0, \ c_4 = a_{43} = 1, \ a_{41} = a_{42} = 0 \tag{8.40}$$

and results in the stepping equations

0	0	0	0	0
1/2	1/2	0	0	0
1/2	0	1/2	0	
1	0	0	1	0
	1/6	2/6	2/6	1/6

$$x_{n+1} = x_n + h$$
$$y_{n+1} = y_n + \frac{1}{6}(k_1 + 2k_2 + 2k_3 + k_4)$$
$$k_1 = hf(x_n, y_n)$$
$$k_2 = hf(x_n + \frac{1}{2}h, y_n + \frac{1}{2}k_1)$$
$$k_3 = hf(x_n + \frac{1}{2}h, y_n + \frac{1}{2}k_2)$$
$$k_4 = hf(x_n + h, y_n + k_3)$$

$$(8.41)$$

Two other fourth-order Runge Kutta methods which are sometimes found in the literature have the forms

0	0	0	0	0
1/3	1/3	0	0	0
2/3	−1/3	1	0	0
1	1	−1	1	0
	1/8	3/8	3/8	1/8

$$x_{n+1} = x_n + h$$
$$y_{n+1} = y_n + \frac{1}{8}(k_1 + 3k_2 + 3k_3 + k_4)$$
$$k_1 = hf(x_n, y_n)$$
$$k_2 = hf(x_n + \frac{1}{3}h, y_n + \frac{1}{3}k_1)$$
$$k_3 = hf(x_n + \frac{2}{3}h, y_n - \frac{1}{3}k_1 + k_2)$$
$$k_4 = hf(x_n + h, y_n + k_1 - k_2 + k_3)$$

$$(8.42)$$

and

0	0	0	0	0
1/2	1/2	0	0	0
1/2	1/4	1/4	0	0
1	0	−1	2	0
	1/6	0	4/6	1/6

$$x_{n+1} = x_n + h$$
$$y_{n+1} = y_n + \frac{1}{6}(k_1 + 4k_3 + k_4)$$
$$k_1 = hf(x_n, y_n)$$
$$k_2 = hf(x_n + \frac{1}{2}h, y_n + \frac{1}{2}k_1)$$
$$k_3 = hf(x_n + \frac{1}{2}h, y_n + \frac{1}{4}k_1 + \frac{1}{4}k_2)$$
$$k_4 = hf(x_n + h, y_n - k_2 + 2k_3)$$

$$(8.43)$$

Error terms associated with Runge-Kutta methods do not have simple forms and are not easily constructed. If one expresses a Runge-Kutta method in the form

$$x_{n+1} = x_n + h, \qquad y_{n+1} = y_n + h\phi(x_n, y_n, h) \qquad (8.44)$$

then the local error at x_{n+1} associated with a step size h can be defined

$$local\ error = E_\ell = y(x_{n+1}) - y_{n+1} \qquad (8.45)$$

where $y(x)$ is the true solution of the initial value problem. The exact form of this difference is complicated and depends upon the order of the Runge-Kutta method.

Example 8-4. (Fourth-order Runge-Kutta)

Set up an algorithm to solve the initial value problem $\frac{dy}{dx} = f(x,y)$, $y(x_0) = y_0$, $x_0 < x < x_n$ using a fourth-order Runge-Kutta method with step size h.

Solution: The fourth-order Runge-Kutta classic form is selected and its use is illustrated in the flowchart of figure 8-4.

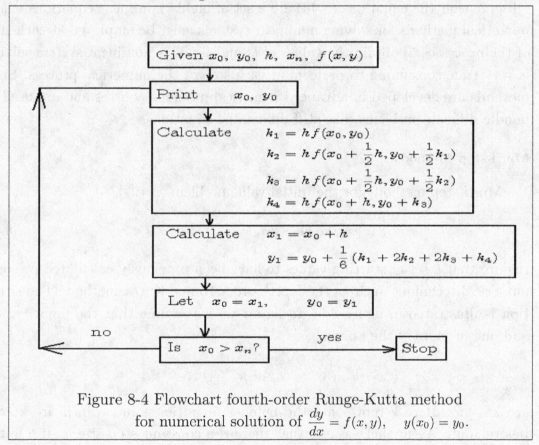

Figure 8-4 Flowchart fourth-order Runge-Kutta method
for numerical solution of $\dfrac{dy}{dx} = f(x,y), \quad y(x_0) = y_0.$

Implicit Runge-Kutta Methods

Implicit Runge-Kutta methods result when the matrix $\mathbf{A}$ of the Butcher array is not strictly lower triangular. An example of a fourth-order, 2-stage, implicit Runge-Kutta method is the following

$$
\begin{array}{c|cc}
1/2 + \sqrt{3}/6 & 1/4 & 1/4 + \sqrt{3}/6 \\
1/2 - \sqrt{3}/6 & 1/4 - \sqrt{3}/6 & 1/4 \\
\hline
 & 1/2 & 1/2
\end{array}
\qquad
\begin{aligned}
x_{n+1} &= x_n + h \\
y_{n+1} &= y_n + \omega_1 k_1 + \omega_2 k_2 \\
k_1 &= h f(x_n + c_1 h, y_n + a_{11} k_1 + a_{12} k_2) \\
k_2 &= h f(x_n + c_2, y_n + a_{21} k_1 + a_{22} k_2)
\end{aligned}
\qquad (8.46)
$$

which is derived in the same way that an explicit Runge-Kutta method is derived. Observe that the equations (8.46) are a set of highly nonlinear equations where numerical methods for solving nonlinear systems must be employed at each step of the processes to obtain the values of k_1 and k_2. This nonlinear system solving is very time consuming to perform at each step of the numerical process. Such methods are developed in advanced numerical methods courses and are used to handle difficult problems, like stiff differential equations.

Multi-step Methods

Multi-step methods for the initial value problem

$$
y'(x) = f(x, y), \quad y(x_0) = y_0, \quad x_0 \le x \le x_1 \qquad (8.47)
$$

require that a set of starting values to have been previously calculated by some numerical technique such as the fourth-order Runge-Kutta method. The situation is illustrated in figure 8-5. In figure 8.5 we assume that the point (x_n, y_n) and one or more of the points

$$
(x_{n-1}, y_{n-1}), \ (x_{n-2}, y_{n-2}), \ldots, (x_{n+1-k}, y_{n+1-k})
$$

are known. Here k represents the number of initial points which are known beforehand. We further assume that there is a constant step size h. If k initial points are assumed known, we can calculate the slope of the solution curve at each of these points because we have the differential equation (8.47) which defines these slopes. We can therefore calculate

$$y'(x_m) = y'_m = f(x_m, y_m) \quad \text{for} \quad m = n, n-1, n-2, \ldots, n+1-k.$$

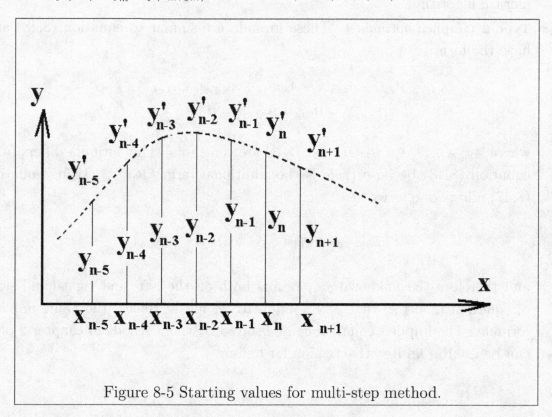

Figure 8-5 Starting values for multi-step method.

Our problem is to determine an approximation to the solution curve of equation (8.47) which passes through the point (x_{n+1}, y_{n+1}). We know $x_{n+1} = x_n + h$ and our problem is to estimate y_{n+1}. Further, we want to make use of our knowledge of the starting points and their slopes in the hope of better approximating y_{n+1}. The methods developed are called multi-step methods because the algorithms incorporate some or all of the k starting values and their slopes. In general, multi-step methods are of two types:

Type 1 (Explicit formula) These are formula where y_{n+1} is estimated by taking some linear combination of the k starting ordinate values and their slopes. These formulas have the general form:

$$\begin{aligned}
y_{n+1} = \ & A_1 y_n + A_2 y_{n-1} + A_3 y_{n-2} + \ldots + A_k y_{n+1-k} \\
& + h \left[B_1 y'_n + B_2 y'_{n-1} + B_3 y'_{n-2} + \ldots + B_k y'_{n+1-k} \right],
\end{aligned} \tag{8.48}$$

where A_1, $A_2, \ldots, A_k$, B_1, $B_2, \ldots, B_k$ are constants which are chosen so that equation (8.48) produces exact results in certain special circumstances. The values assigned to the constants depends upon the conditions imposed upon the numerical algorithm.

Type 2 (Implicit formula) These formula are similar to equation (8.48) and have the form:

$$
\begin{aligned}
y_{n+1} = & \ A_1 y_n + A_2 y_{n-1} + A_3 y_{n-2} + \ldots + A_k y_{n+1-k} \\
& + h \left[B_0 y'_{n+1} + B_1 y'_n + B_2 y'_{n-1} + B_3 y'_{n-2} + \ldots + B_k y'_{n+1-k} \right],
\end{aligned}
\tag{8.49}
$$

where A_1, $A_2, \ldots, A_k$, B_0, B_1, $B_2, \ldots, B_k$ are constants. This formula differs from equation (8.48) because there is the additional term $h B_0 y'_{n+1}$. Using equation (8.47) allows one to write

$$
h B_0 y'_{n+1} = h B_0 y'(x_{n+1}) = h B_0 f(x_{n+1}, y_{n+1}),
$$

and therefore the unknown y_{n+1} occurs both on the left- and right-hand sides of equation (8.49) so that y_{n+1} is implicitly defined. Hence, the name implicit formula. The implicit equation can be solved by the methods in chapter 2 or it can be used in an iterative fashion for finding y_{n+1}.

Open and Closed Adams Formula

Assume that some starting values have been calculated by the Runge-Kutta method and we are at (x_n, y_n) and want to advance to the point (x_{n+1}, y_{n+1}). We can use our previous integration techniques and integrate the differential equation

$$
\frac{dy}{dx} = f(x, y), \qquad y(x_n) = y_n
\tag{8.50}
$$

to obtain

$$
\int_{y_n}^{y_{n+1}} dy = \int_{x_n}^{x_{n+1}} f(x, y) \, dx
$$

$$
y_{n+1} - y_n = \int_{x_n}^{x_{n+1}} f(x, y) \, dx
\tag{8.51}
$$

Our assumption is that a set of starting values is known and so we can replace the function $f = f(x, y)$ in equation (8.51) by an approximate polynomial interpolation function and then perform the integration. The resulting equations are known as Adams formula for advancing the differential equation one more step.

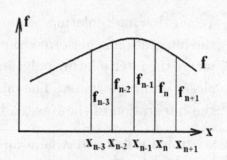

We use the Newton backward interpolating polynomial

$$f = f_n + \binom{s}{1}\Delta f_{n-1} + \binom{s+1}{2}\Delta^2 f_{n-2} + \binom{s+2}{3}\Delta^3 f_{n-3} + \binom{s+3}{4}\Delta^4 f_{n-4} + \cdots \quad (8.52)$$

where $s = \frac{x - x_n}{h}$ is the scaled variable, and substitute this interpolating polynomial into the equation (8.51) and then perform the integration. Various Adams formula result depending upon where the interpolating polynomial is truncated. These formula are called open Adams formula. The table 8.3 shows the various forms that result due to the integration of the Newton backward interpolating polynomial (8.52).

Table 8.3 Open Adams formula for integration of $\frac{dy}{dx} = f(x, y), \quad y(x_n) = y_n$							
$y_{n+1} = y_n + c_0 h \left[a_0 f_n + a_1 f_{n-1} + a_2 f_{n-2} + \cdots + a_m f_{n-m} \right] + Error \quad$ where $\quad f_k = f(x_k, y_k)$							
m	c_0	a_0	a_1	a_2	a_3	a_4	a_5	Error
0	1	1	0					$\frac{1}{2}h^2 y''(\xi)$
1	1/2	3	-1					$\frac{5}{12}h^3 y'''(\xi)$
2	1/12	23	-16	5				$\frac{3}{8}h^4 y^{(iv)}(\xi)$
3	1/24	55	-59	37	-9			$\frac{251}{720}h^5 y^{(v)}(\xi)$
4	1/720	1901	-2774	2616	-1274	251		$\frac{95}{288}h^6 y^{(vi)}(\xi)$
5	1/1440	4277	-7923	9982	-7298	2877	-4751	$\frac{19087}{60480}h^7 y^{(vii)}(\xi)$

Note that these formula are of the explicit type.

In place of the Newton backward interpolating polynomial starting at f_n, given by equation (8.52), one can use the Newton backward interpolating polynomial starting at f_{n+1} which is given by

$$f = f_{n+1} + \binom{s-1}{1}\Delta f_n + \binom{s}{2}\Delta^2 f_{n-1} + \binom{s+1}{3}\Delta^3 f_{n-2} + \binom{s+2}{4}\Delta^4 f_{n-3} \quad (8.53)$$

When this interpolating polynomial is substituted into the equation (8.51) and the integration is performed one obtains various Adams formula depending upon where the interpolating polynomial is truncated. The resulting formula are called closed Adams formula. The table 8.4 shows the various forms that result due to the integration of the Newton backward interpolating polynomial (8.53).

Table 8.4 Closed Adams formula for integration of $\dfrac{dy}{dx} = f(x,y), \quad y(x_n) = y_n$

$$y_{n+1} = y_n + c_0 h \left[b_{-1} f_{n+1} + b_0 f_n + b_1 f_{n-1} + b_2 f_{n-2} + \cdots + b_m f_{n-m} \right] + Error \quad \text{where} \quad f_k = f(x_k, y_k)$$

m	c_0	b_{-1}	b_0	b_1	b_2	b_3	b_4	Error
-1	1	1	0					$-\frac{1}{2}h^2 y''(\xi)$
0	1/2	1	1					$-\frac{1}{12}h^3 y'''(\xi)$
1	1/12	5	8	-1				$-\frac{1}{24}h^4 y^{(iv)}(\xi)$
2	1/24	9	19	-5	1			$-\frac{19}{720}h^5 y^{(v)}(\xi)$
3	1/720	251	646	-264	106	-19		$-\frac{3}{160}h^6 y^{(vi)}(\xi)$
4	1/1440	475	1427	-798	482	-173	27	$-\frac{863}{60480}h^7 y^{(vii)}(\xi)$

Note that these formula are of the implicit type.

Predictor Corrector Methods

One way of using the implicit form (8.49) is in an iterative fashion. We first approximate y_{n+1} by using a formula like (8.48). The value of y_{n+1} obtained is called a predictor and is used as input into the y'_{n+1} term of equation (8.49) in order to calculate a more improved value of y_{n+1}. When used in this way, equations (8.48) and (8.49) are termed a predictor-corrector set of equations sometimes called a predictor-corrector pair. The predictor formula (8.48) estimates y_{n+1}, which is then used in equation (8.49) to give a more correct value to y_{n+1}. Predictor-corrector formulas have the following advantages over Runge-Kutta schemes: (a) Greater speed of computation because only two formulas are needed per step. (b) An error estimate can be more easily obtained.

The simplest example of a predictor corrector multi-step method can be obtained by modifying the Euler method described earlier. Recall that the single-step Euler method of equation (8.47) gives the predicted values

$$x_{n+1} = x_n + h, \quad \text{and} \quad y_{n+1,p} = y_n + y'_n h = y_n + f(x_n, y_n)h, \quad n = 0,\ 1,\ 2,\ldots$$

By using this predicted value of y_{n+1}, we may then estimate the slope at (x_{n+1}, y_{n+1}) by the relation $y'_{n+1} = y'(x_{n+1}) = f(x_{n+1}, y_{n+1,p})$. In using Euler's method to predict y_{n+1}, we have used the slope y_n' at (x_n, y_n) and have assumed that this slope remained constant for all x in the interval $x_n \le x \le x_{n+1}$. Only in special cases does the slope remain constant throughout the interval. We therefore try to better estimate y_{n+1} by using the average slope of $\frac{1}{2}(y_{n+1}' + y_n')$ over the interval from x_n to x_{n+1}. By using the average slope we hope to better approximate y_{n+1} by using the corrector algorithm

$$y_{n+1} = y_n + \frac{h}{2}\left(y_{n+1}' + y_n'\right).$$

This gives the implicit modified Euler predictor-corrector method:

$$x_{n+1} = x_n + h$$

Predictor formula: $\qquad y_{n+1,p} = y_n + y_n'h = y_n + f(x_n, y_n)h$
$$\tag{8.54}$$

Corrector formula: $\qquad y_{n+1,c} = y_n + \frac{h}{2}\left(y_{n+1}' + y_n'\right)$

$$y_{n+1,c} = y_n + \frac{h}{2}\left(f(x_{n+1}, y_{n+1,p}) + f(x_n, y_n)\right) \tag{8.55}$$

These equations have the forms specified by equations (8.48), (8.49) and can also be found as the second entry of table 8.4. The figure 8-6 illustrates a flowchart of how to use the modified Euler predictor-corrector formulas in the solution of the initial value problem (8.47).

Example 8-5. (Predictor-Corrector Method)

Apply the modified Euler algorithm of figure 8.6 to solve the initial value problem

$$y' = f(x, y) = x + y, \qquad y(0) = 1, \qquad h = 0.1, \qquad 0 \le x \le 1.$$

Solution: The results of the Euler modified algorithm, applied to the given problem using the flowchart of figure 8-6, are given in table 8.5. These results can be compared with the results of the Euler method given in table 8.1, and one can see the improvement in the error. The error term associated with this method and other predictor-corrector method error terms are developed later in the chapter.

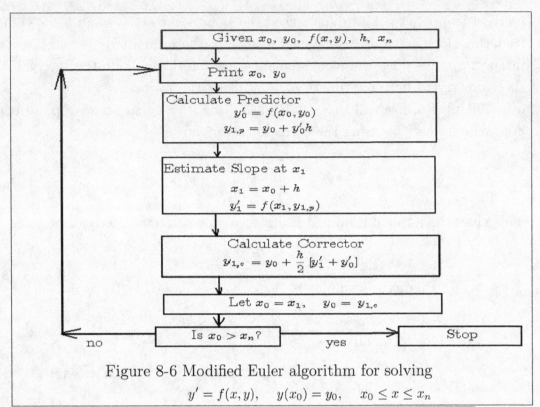

Figure 8-6 Modified Euler algorithm for solving

$$y' = f(x,y), \quad y(x_0) = y_0, \quad x_0 \le x \le x_n$$

Table 8.5

Numerical Results for modified Euler method applied to

$$\frac{dy}{dx} = x + y, \quad y(0) = 1, \quad h = 0.1$$

x	y	$y(x) = 2e^x - x - 1$	% error
.00	1.0000	1.0000	.000
.10	1.1100	1.1103	.031
.20	1.2421	1.2428	.061
.30	1.3985	1.3997	.089
.40	1.5818	1.5836	.117
.50	1.7949	1.7974	.142
.60	2.0409	2.0442	.165
.70	2.3231	2.3275	.187
.80	2.6456	2.6511	.208
.90	3.0124	3.0192	.227
1.00	3.4282	3.4366	.244

There are many predictor-corrector type formulas and they can be derived in various ways. As an example, a four panel formula for integrating the differential equation (8.50) from x_{n-3} to x_{n+1} is obtained from

$$\int_{x_{n-3}}^{x_{n+1}} dy = \int_{x_{n-3}}^{x_{n+1}} f(x,y)\,dx$$

where the right-hand side of this integral is evaluated using the four panel Newton-Cotes open formula from table 7.3 (with appropriate notation changes). There results the predictor formula

$$y_{n+1} = y_{n-3} + \frac{4}{3}h\left[2y'_{n-2} - y'_{n-1} + 2y'_n\right] + \frac{28}{90}h^5 y^{(v)}(\xi_1) \tag{8.56}$$

which is an explicit formula. A corrector formula having an error of the same order is obtained from the integration of the differential equation

$$\int_{x_{n-1}}^{x_{n+1}} dy = \int_{x_{n-1}}^{x_{n+1}} f(x,y)\,dx$$

where the right-hand side is integrated using the Simpson's 1/3-rule from the table 7.2. One obtains the corrector formula

$$y_{n+1} = y_{n-1} + \frac{h}{3}\left[y'_{n+1} + 4y'_n + y'_{n-1}\right] - \frac{1}{90}h^5 y^{(v)}(\xi_2) \tag{8.57}$$

which is an implicit formula. The equations (8.56) and (8.57) produce the Milne predictor-corrector equations.

Milne Predictor-Corrector Pair

$$x_{n+1} = x_n + h$$

Predictor:

$$y_{n+1,p} = y_{n-3} + \frac{4}{3}h\left[2f(x_{n-2}, y_{n-2}) - f(x_{n-1}, y_{n-1}) + 2f(x_n, y_n)\right] \tag{8.58}$$

Corrector:

$$y_{n+1,c} = y_{n-1} + \frac{h}{3}\left[f(x_{n+1}, y_{n+1,p}) + 4f(x_n, y_n) + f(x_{n-1}, y_{n-1})\right]$$

Observe that four starting values are required for the Milne method. Note also that one can define the iterative scheme

$$y_{n+1}^{(k+1)} = y_{n-1} + \frac{h}{3}\left[f(x_{n+1}, y_{n+1}^{(k)}) + 4f(x_n, y_n) + f(x_{n-1}, y_{n-1})\right]$$

for $k = 1, 2, 3, \ldots$ in order to solve for y_{n+1}. The starting value for the iterative scheme is given by $y_{n+1}^{(1)} = y_{n+1,c}$. The use of an iterative scheme to solve the implicit equation is not often done because it is time consuming to perform the iterations at each step of the numerical method. Sometimes the iterations are purposely done to help maintain stability. Stability associated with numerical methods and differential equations is a subject we consider later in the chapter.

As another example of a predictor-corrector pair consider the integration of equation (8.50)

$$\int_{x_n}^{x_{n+1}} dy = \int_{x_n}^{x_{n+1}} f(x, y) \, dx \tag{8.59}$$

where the right-hand side is integrated using an open Adams formula from the table 8.3 with error $\mathcal{O}(h^5)$ to obtain

$$y_{n+1} = y_n + \frac{h}{24} \left[55 y_n' - 59 y_{n-1}' + 37 y_{n-2}' - 9 y_{n-3}' \right] + \frac{251}{720} h^5 y^{(v)}(\xi_1) \tag{8.60}$$

sometime referred to as the fourth order Adams-Bashforth method. Now a corrector formula with the same order of error is obtained from equation (8.59) using a closed Adams formula from table 8.4 to obtain

$$y_{n+1} = y_n + \frac{h}{24} \left[9 y_{n+1}' + 19 y_n' - 5 y_{n-1}' + y_{n-2}' \right] - \frac{19}{720} h^5 y^{(v)}(\xi_2) \tag{8.61}$$

The equations (8.58) and (8.61) are combined to produce the Adams-Moulton predictor correction pair.

Adams-Moulton Predictor-Corrector Method

$$x_{n+1} = x_n + h$$

Predictor:

$$y_{n+1,p} = y_n + \frac{h}{24} \left[55 f(x_n, y_n) - 59 f(x_{n-1}, y_{n-1}) + 37 f(x_{n-2}, y_{n-2}) - 9 f(x_{n-3}, y_{n-3}) \right]$$

Corrector: $\hspace{11cm}$ (8.62)

$$y_{n+1,c} = y_n + \frac{h}{24} \left[9 f(x_{n+1}, y_{n+1,p}) + 19 f(x_n, y_n) - 5 f(x_{n-1}, y_{n-1}) + f(x_{n-2}, y_{n-2}) \right]$$

The Adams-Moulton method also requires four starting values.

Method of Undetermined Coefficients

The method of undetermined coefficients can be used to determine the constant coefficients associated with numerical methods of either the explicit type, equation (8.48), or the implicit type, equation (8.49). As an example, consider the problem of constructing a numerical method of the implicit type having the form

$$y_{n+1} = A_1 y_n + h[B_0 y'_{n+1} + B_1 y'_n] \qquad (8.63)$$

which is a special case of equation (8.49). In this numerical method we have three undetermined coefficients, namely A_1, B_0, B_1. We use special functions for $y(x)$ in order to determine values for the unknown constants A_1, B_0, B_1. The special functions are selected such that equation (8.63) is exact with zero error for these special functions. In order to keep the algebra tractable we choose polynomials of the form $(x - x_n)^m$ as our choice for the special functions $y(x)$ since the resulting equations are easy to work with and the algebra is tractable. Other choices for the function $y(x)$ are possible but are not considered here. By substituting each of the functions from the set $S = \{1, x - x_n, (x - x_n)^2\}$ into the equation (8.63) one obtains the following three equations

$$\text{For } y(x) = 1 \text{ one finds} \qquad 1 = A_1(1) + h[B_0(0) + B_1(0)]$$

$$\text{For } y(x) = x - x_n \text{ there results} \qquad h = A_1(0) + h[B_0(1) + B_1(1)]$$

$$\text{For } y(x) = (x - x_n)^2 \text{ there results} \qquad h^2 = A_1(0) + h[B_0(2h) + B_1(0)].$$

The above three equations are then solved for the unknown coefficients and one finds the values $A_1 = 1$, $B_0 = 1/2$ and $B_1 = 1/2$. For the conditions specified, the assumed form of equation (8.63) reduces to the modified Euler formula

$$y_{n+1} = y_n + \frac{h}{2}[y'_{n+1} + y'_n]. \qquad (8.64)$$

We now know that the modified Euler formula produces exact results in the case where $y(x)$ is a member of the set S because we calculated the coefficients A_1, B_0, B_1 in order to satisfy this requirement. Therefore, the equation (8.64) gives exact results for all polynomials through degree two, since second-degree polynomials are just linear combinations of the elements from the set S. What happens when $y(x)$ is different from one of these polynomials? In this case there is an error associated with the modified Euler method. In order to estimate this error we can argue as follows. We desire to have an error term associated with the modified Euler formula (8.64) which has the same form as the Taylor series error term. We therefore assume that an error term for equation (8.64) would have the form

$$y_{n+1} = y_n + \frac{h}{2}[y'_{n+1} + y'_n] + ch^m y^{(m)}(\xi), \qquad (8.65)$$

where m and c are unknown constants and $x_n < \xi < x_{n+1}$. We want equation (8.65) to hold for all functions $y(x)$ and further we know if $y(x)$ is a polynomial of degree two the error term is zero and hence m is greater than two. The smallest integer value of m greater than two is $m = 3$ and for this value of m we can determine the constant c by letting $y(x) = (x - x_n)^3$. Substitution of the cubic reduces equation (8.65) to the form

$$h^3 = 0 + \frac{h}{2}[3h^2 + 0] + ch^3(6)$$

from which we determine $c = -1/12$. Here we find the error term is $O(h^3)$ and can be expressed as

$$-\frac{h^3}{12} y'''(\xi).$$

The method of undetermined coefficients enables us to derive a large variety of numerical methods of the explicit or implicit type for solving first-order ordinary differential equations.

Example 8-6.   (Undetermined Coefficients)

Develop an algorithm of the explicit type which has the form

$$y_{n+1} = A_2 y_{n-1} + h[B_1 y'_n + B_2 y'_{n-1}]$$

and determine the error term.

Solution: We substitute each of the functions from the set $S = \{1, x - x_n, (x - x_n)^2\}$ into the assumed form and obtain the following three equations

$$\text{For } y(x) = 1 \text{ there results} \qquad 1 = A_2 + h[B_1(0) + B_2(0)]$$

$$\text{For } y(x) = x - x_n \text{ there results} \qquad h = A_2(-h) + h[B_1(1) + B_2(1)]$$

$$\text{For } y(x) = (x - x_n)^2 \text{ there results} \qquad h^2 = A_2 h^2 + h[B_1(0) + B_2(-2h)].$$

We solve these three equations for the constants A_2, B_1, B_2 and find

$$A_2 = 1, \quad B_1 = 2, \quad B_2 = 0.$$

We have chosen the constants such that the above algorithm is exact for polynomials of degree two, and hence we assume an error term of $O(h^3)$ and write

$$y_{n+1} = y_{n-1} + 2h y'_n + ch^3 y'''(\xi).$$

Now let $y(x) = (x - x_n)^3$ and substitute into the above algorithm and show $c = 1/3$. The desired algorithm has the form

$$y_{n+1} = y_{n-1} + 2hy_n' + \frac{1}{3}h^3 y'''(\xi)$$

and we determine that the error term for this algorithm is larger than that for the modified Euler method.

∎

Error Term

An alternate method for deriving explicit methods of the type given by equation (8.48), or implicit methods of the type given by equation (8.49), is to use Taylor series expansions. The unknown coefficients can then be determined by requiring that certain coefficients be zero.

For example, one can define the error associated with an explicit multi-step method as

$$E(x_n, h) = \ y(x_n + h) - A_1 y(x_n) - A_2 y(x_n - h) - \ldots - A_k y(x + n - (k-1)h)$$

$$- h\left[B_1 y'(x_n) + B_2 y'(x_n - h) + \ldots + B_k y'(x_n - (k-1)h)\right].$$

In this expression for the error we expand the terms $y(x_n \pm mh)$ and $y'(x_n \pm mh)$, for $m = 1, 2, \ldots$, in a Taylor series about the point $x = x_n$. We then combine like terms and write the error $E = E(x_n, h)$ as

$$E = \ [1 - A_1 - A_2 - A_3 - \ldots - A_k] y(x_n)$$

$$+ [1 + A_2 + 2A_3 + \ldots + (k-1)A_k - B_1 - B_2 - B_3 - \ldots - B_k] y'(x_n)h$$

$$+ \left[1 - A_2 - 2^2 A_3 - \ldots - (k-1)^2 A_k + 2B_2 + 2(2)B_3 + \ldots + 2(k-1)B_k\right] y''(x_n)\frac{h^2}{2!}$$

$$+ \left[1 + A_2 + 2^3 A_3 + \ldots + (k-1)^3 A_k + 3B_2 - 3(2)^2 B_3 - \ldots - 3(k-1)^2 B_k\right] y'''(x_n)\frac{h^3}{3!}$$

$$+ \left[1 - A_2 - 2^4 A_3 - \ldots - (k-1)^4 A_k + 4B_2 + 4(2)^3 B_3 + \ldots + 4(k-1)^3 B_k\right] y^{(iv)}(x_n)\frac{h^4}{4!}$$

$$+ \ldots$$

Now if there are $N + M$ unknowns

$$A_1, A_2, \ldots, A_N, B_1, B_2, \ldots, B_M,$$

then we want $N + M$ equations from which we may solve for these unknowns. By setting to zero the first $N + M$ coefficients in the error term, we obtain $N + M$ linear equations and consequently the error term is reduced to the form

$$E(x_n, h) = C y^{(N+M)}(x_n) \frac{h^{(N+M)}}{(N+M)!},$$

where C is some constant.

Similarly, we can define an error term associated with the implicit method given by equation (8.49). We again use the Taylor series expansion, collect like terms, and require that a certain number of the lowest ordered coefficients are zero. The expression for the error is then obtained from the remaining terms.

Example 8-7. Use of Taylor series expansions

Determine an implicit multi-step method of the form

$$y_{n+1} = A_2 y_{n-1} + h[B_0 y'_{n+1} + B_1 y'_n + B_2 y'_{n-1}]$$

and find the error term associated with this method.

Solution: Define the error

$$E(x_n, h) = y(x_n + h) - A_2 y(x_n - h) - h[B_0 y'_{n+1} + B_1 y'_n + B_2 y'_{n-1}]$$

as the term which results when the exact solution is substituted into the implicit method. We now use the Taylor series expansions

$$y(x_n \pm h) = y(x_n) \pm y'(x_n)h + y''(x_n)\frac{h^2}{2!} \pm y'''(x_n)\frac{h^3}{3!} + \ldots$$

$$y'(x_n \pm h) = y'(x_n) \pm y''(x_n)h + y'''(x_n)\frac{h^2}{2!} \pm y^{(iv)}(x_n)\frac{h^3}{3!} + \ldots$$

and express the error in the form

$$
\begin{aligned}
E(x_n, h) = \ & [1 - A_2]\, y(x_n) \\
& + [1 + A_2 - B_0 - B_1 - B_2]\, y'(x_n)h \\
& + [1 - A_2 - 2B_0 + 2B_2]\, y''(x_n)\frac{h^2}{2!} \\
& + [1 + A_2 - 3B_0 - 3B_2]\, y'''(x_n)\frac{h^3}{3!} \\
& + [1 - A_2 - 4B_0 + 4B_2]\, y^{(iv)}(x_n)\frac{h^4}{4!} \\
& + [1 + A_2 - 5B_0 - 5B_2]\, y^{(v)}(x_n)\frac{h^5}{5!} \\
& + \cdots
\end{aligned}
$$

Now since there are four unknowns, we require that the error satisfy

$$
\begin{aligned}
1 - A_2 &= 0 \\
1 + A_2 - B_0 - B_1 - B_2 &= 0 \\
1 - A_2 - 2B_0 + 2B_2 &= 0 \\
1 + A_2 - 3B_0 - 3B_2 &= 0.
\end{aligned}
$$

Solving this system of equations we find

$$A_2 = 1, \quad B_0 = \frac{1}{3}, \quad B_1 = \frac{4}{3}, \quad B_2 = \frac{1}{3}.$$

These values make the h^4 coefficient zero, and so we use the next term in the series expansion to obtain the error expression

$$E(x_n, h) = -\frac{4}{3} y^{(v)} \frac{h^5}{5!} = -\frac{1}{90} y^{(v)} h^5.$$

Therefore, the desired implicit method can be written as

$$y_{n+1} = y_{n-1} + \frac{h}{3}[y'_{n+1} + 4y'_n + y'_{n-1}] - \frac{1}{90} y^{(v)}(\xi) h^5, \qquad x_n < \xi < x_{n+1}$$

which is recognized as the Simpson's 1/3 rule with error. Observe that the error term vanishes for all polynomials of degree four or less.

Taylor Series for Systems of Equations

The initial value problem associated with a system of two differential equations in two unknowns can be expressed

$$\frac{dy_1}{dt} = f_1(t, y_1, y_2), \quad y_1(0) = y_{10}$$

$$\frac{dy_2}{dt} = f_2(t, y_1, y_2), \quad y_2(0) = y_{20}$$

where y_{10}, y_{20} are given constant initial values. If the functions f_1, f_2 are analytic in the neighborhood of the initial points y_{10}, y_{20}, at $t = 0$, then solutions in the form of Taylor series expansion can be constructed. These solutions have the form

$$y_1 = y_1(t) = y_{10} + y'_{10}t + y''_{10}\frac{t^2}{2!} + y'''_{10}\frac{t^3}{3!} + \cdots$$

$$y_2 = y_2(t) = y_{20} + y'_{20}t + y''_{20}\frac{t^2}{2!} + y'''_{20}\frac{t^3}{3!} + \cdots$$

where the various derivatives at $t = 0$ must be calculated. That is,

$$y'_{i0} = \frac{dy_i}{dt}\Big|_{t=0}, \quad y''_{i0} = \frac{d^2 y_i}{dt^2}\Big|_{t=0}, \quad y'''_{i0} = \frac{d^3 y_i}{dt^3}\Big|_{t=0}, \quad \cdots$$

for $i = 1, 2$. The above ideas can be extended to a system of n-ordinary differential equations.

Runge-Kutta for Systems of Equations

The previous Runge-Kutta methods developed for a single differential equation can be applied to systems of ordinary differential equations. Consider the following system of ordinary differential equations

$$\frac{dy_1}{dx} = f_1(x, y_1, y_2), \qquad y_1(x_0) = y_{1,0} \tag{8.66}$$

$$\frac{dy_2}{dx} = f_2(x, y_1, y_2), \qquad y_2(x_0) = y_{2,0} \tag{8.67}$$

over the interval $I = \{\, x \mid x_0 \le x \le x_n \,\}$. The system of equations (8.66), (8.67) is called a coupled system because $y_1(x)$ of equation (8.66) occurs in equation (8.67) while simultaneously the solution $y_2(x)$ of equation (8.67) occurs in the differential equation (8.66). Many physical problems can be represented in the form of coupled first-order differential equations in two or more dependent variables.

Example 8-8. **(System of differential equations)**

Consider the example of a linear spring-mass system with damping proportional to the velocity. Such as the system illustrated. Represent the differential equation describing the displacement as an initial value problem associated with a system of coupled first-order equations.

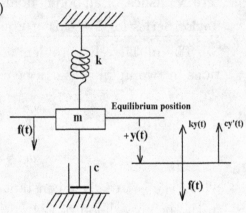

Solution: The equation of motion for the spring-mass system is obtained by summing forces and applying Newton's second law. One finds the displacement $y = y(t)$ is described by the linear differential equation

$$my'' + cy' + ky = f(t), \quad y(0) = A, \quad y'(0) = B$$

we make the substitutions $y = y_1(x)$ and $y' = y_1'(x) = y_2$, then the spring-mass equation can be replaced by the system of first-order equations

$$\frac{dy_1}{dt} = f_1(t, y_1, y_2) = y_2, \qquad\qquad y_1(0) = A$$

$$\frac{dy_2}{dt} = f_2(t, y_1, y_2) = \frac{f(t) - ky_1 - cy_2}{m}, \quad y_2(0) = B$$

which is a system of the form of equations (8.66) and (8.67) with time t replacing x as the independent variable.

Notice all linear second-order initial value problems

$$y'' + p(x)y' + q(x)y = f(x), \qquad y(x_0) = A, \quad y'(x_0) = B \qquad (8.68)$$

can be written as a system of coupled first-order equations. We make the substitutions $y_1 = y$ and $y_2 = y'$ and write the initial value problem as the equivalent system

$$\frac{dy_1}{dx} = f_1(x, y_1, y_2) = y_2, \qquad y_1(x_0) = A$$

$$\frac{dy_2}{dx} = f_2(x, y_1, y_2) = -p(x)y_2 - q(x)y_1 + f(x), \qquad y_2(x_0) = B. \qquad (8.69)$$

∎

Example 8-9. (System of differential equations)

Suppose we desire to solve the system of equations (8.66) and (8.67) simultaneously by using a fourth-order Runge-Kutta classic algorithm. Recall that to solve the initial value problem

$$\frac{dz}{dx} = f(x, z), \qquad z(x_0) = z_0 \qquad (8.70)$$

by the Runge-Kutta method we must first calculate each of the following quantities

$$k_1 = hf(x_n, z_n)$$

$$k_2 = hf\left(x_n + \frac{1}{2}h, z_n + \frac{1}{2}k_1\right)$$

$$k_3 = hf\left(x_n + \frac{1}{2}h, z_n + \frac{1}{2}k_2\right) \qquad (8.71)$$

$$k_4 = hf(x_n + h, z_n + k_3)$$

and then calculate

$$x_{n+1} = x_n + h, \qquad \text{and} \qquad z_{n+1} = z_n + \frac{1}{6}[k_1 + 2k_2 + 2k_3 + k_4] \qquad (8.72)$$

for $n = 0, 1, 2, \ldots$. This algorithm is for a step size h which is chosen by the user.

Consider now equation (8.66) and assume for the time being y_2 is constant in this equation. If we applied to this equation (with y_2 constant) the Runge-Kutta algorithm to solve

$$\frac{dy_1}{dx} = f_1(x, y_1, y_2), \qquad y_1(x_0) = A,$$

it would be necessary to calculate the equations

$$y_{1,n+1} = y_{1,n} + \frac{1}{6}(\alpha_1 + 2(\alpha_2 + \alpha_3) + \alpha_4), \qquad y_{1,0} = A, \qquad (8.73)$$

where

$$\alpha_1 = hf_1(x_n, y_{1,n}, y_2)$$
$$\alpha_2 = hf_1(x_n + \frac{1}{2}h, y_{1,n} + \frac{1}{2}\alpha_1, y_2)$$
$$\alpha_3 = hf_1(x_n + \frac{1}{2}h, y_{1,n} + \frac{1}{2}\alpha_2, y_2) \qquad (8.74)$$
$$\alpha_4 = hf_1(x_n + h, y_{1,n} + \alpha_3, y_2).$$

Here we have replace the Runge-Kutta k_i values with coefficients α_i, $i = 1, 2, 3, 4$ and we have held y_2 constant. In a similar fashion we can apply the Runge-Kutta algorithm to the second differential equation (8.67) and assume this time y_1 is constant. There results the differential equation

$$\frac{dy_2}{dx} = f_2(x, y_1, y_2), \qquad y_2(x_0) = B,$$

which is solved by the Runge-Kutta calculations

$$y_{2,n+1} = y_{2,n} + \frac{1}{6}(\beta_1 + 2(\beta_2 + \beta_3) + \beta_4), \qquad y_{2,0} = B,$$

where

$$\beta_1 = hf_2(x_n, y_1, y_{2,n})$$
$$\beta_2 = hf_2(x_n + \frac{1}{2}h, y_1, y_{2,n} + \frac{1}{2}\beta_1)$$
$$\beta_3 = hf_2(x_n + \frac{1}{2}h, y_1, y_{2,n} + \frac{1}{2}\beta_2) \qquad (8.75)$$
$$\beta_4 = hf_2(x_n + h, y_1, y_{2,n} + \beta_3).$$

Here the Runge-Kutta k_i values have been replaced by β_i $i = 1, 2, 3, 4$ and we have purposely held the dependent variable y_1 constant. Now originally the equations (8.66) and (8.67) were coupled. That is, both y_1 and y_2 occurred in the given equations simultaneously. Therefore we would expect the terms occurring in equations (8.66) must also occur in equations (8.67) and the terms in equations (8.67) must also occur simultaneously in equations (8.66). Substituting these terms yields the following fourth-order Runge-Kutta algorithm for solving the system of first-order differential equations (8.66) and (8.67).

The single-step Runge-Kutta algorithm for obtaining the numerical solution to the system of first-order equations

$$\frac{dy_1}{dx} = f_1(x, y_1, y_2), \qquad y_1(x_0) = A$$

$$\frac{dy_2}{dx} = f_2(x, y_1, y_2), \qquad y_2(x_0) = B$$

can be calculate from the equations

$$y_{1,n+1} = y_{1,n} + \frac{1}{6}[\alpha_1 + 2(\alpha_2 + \alpha_3) + \alpha_4], \qquad y_{1,0} = A$$

$$y_{2,n+1} = y_{2,n} + \frac{1}{6}[\beta_1 + 2(\beta_2 + \beta_3) + \beta_4], \qquad y_{2,0} = B,$$

where

$$\alpha_1 = hf_1(x_n, y_{1,n}, y_{2,n})$$

$$\beta_1 = hf_2(x_n, y_{1,n}, y_{2,n})$$

$$\alpha_2 = f_1\left(x_n + \frac{1}{2}h, y_{1,n} + \frac{1}{2}\alpha_1, y_{2,n} + \frac{1}{2}\beta_1\right)$$

$$\beta_2 = hf_2\left(x_n + \frac{1}{2}h, y_{1,n} + \frac{1}{2}\alpha_1, y_{2,n} + \frac{1}{2}\beta_1\right)$$

$$\alpha_3 = hf_1\left(x_n + \frac{1}{2}h, y_{1,n} + \frac{1}{2}\alpha_2, y_{2,n} + \frac{1}{2}\beta_2\right)$$

$$\beta_3 = f_2\left(x_n + \frac{1}{2}h, y_{1,n} + \frac{1}{2}\alpha_2, y_{2,n} + \frac{1}{2}\beta_2\right)$$

$$\alpha_4 = hf_1(x_n + h, y_{1,n} + \alpha_3, y_{2,n} + \beta_3)$$

$$\beta_4 = hf_2(x_n + h, y_{1,n} + \alpha_3, y_{2,n} + \beta_3),$$

which is obtained from equations (8.74) and (8.75) by replacing the previously held constant terms by the variable terms that occur in the other equations. ∎

The vector differential system given by

$$\frac{d\vec{y}}{dx} = \vec{f}(x, \vec{y}), \qquad \vec{y}(x_0) = \vec{y}_0 \tag{8.76}$$

where $\vec{y}$, $\vec{y}(x_0)$ and $\vec{f}(x, \vec{y})$ are column vectors given by $\vec{y} = \text{col}(y_1, y_2, y_3, \ldots, y_m)$, $\vec{y}(x_0) = \text{col}(y_{1,0}, y_{2,0}, \ldots, y_{m,0})$ and $\vec{f}(x, \vec{y}) = \text{col}(f_1, f_2, \ldots, f_m)$. can be solved by Runge-Kutta methods applied to each equation of the system. This results in the Runge-Kutta method being written in the vector form

$$x_{n+1} = x_n + h$$

$$\vec{y}_{n+1} = \vec{y}_n + \frac{1}{6}\left(\vec{k}_1 + 2\vec{k}_2 + 2\vec{k}_3 + \vec{k}_4\right)$$

where $\quad \vec{k}_1 = h\vec{f}(x_n, \vec{y}_n)$

$$\vec{k}_2 = h\vec{f}(x_n + \frac{h}{2}, \vec{y}_n + \frac{1}{2}\vec{k}_1)$$

$$\vec{k}_3 = h\vec{f}(x_n + \frac{h}{2}, \vec{y}_n + \frac{1}{2}\vec{k}_2)$$

$$\vec{k}_4 = h\vec{f}(x_n + h, \vec{y}_n + \vec{k}_3)$$

$$(8.77)$$

for $n = 0, 1, 2, \ldots$. Note that the example 8-9 is a special case of the above representation when the system of differential equations is a second order system.

In a similar manner one can convert other algorithms for solving first order differential equations so that they will become applicable for use in solving a vector system of first order differential equations. For example, all the predictor-corrector and linear multi-step methods can be modified to handle systems of differential equations. The conversion of an algorithm to handle systems is accomplished in a manner analogous to what we have done for the Runge-Kutta method in the previous examples.

Example 8-10. (System of differential equations)

Consider the system of differential equations

$$\frac{dx}{dt} = f(t, x, y), \qquad x(0) = x_0, \qquad \text{and} \qquad \frac{dy}{dt} = g(t, x, y), \quad y(0) = y_0$$

Assume that we use a Taylor series expansion on each equation and obtain four starting values. In general if $x_{n-3}, y_{n-3}, x_{n-2}, y_{n-2}, x_{n-1}, y_{n-1}, x_n, y_n$ are starting values corresponding to $t_{n-3}, t_{n-2}, t_{n-1}, t_n$, then the system can be advanced using the Milne method to obtain the predictor formulas

$$x_{n+1,p} = x_{n-3} + \frac{4h}{3}\left[2f(t_{n-2}, x_{n-2}, y_{n-2}) - f(t_{n-1}, x_{n-1}, y_{n-1}) + 2f(t_n, x_n, y_n)\right]$$

$$y_{n+1,p} = y_{n-3} + \frac{4h}{3}\left[2g(t_{n-2}, x_{n-2}, y_{n-2}) - g(t_{n-1}, x_{n-1}, y_{n-1}) + 2g(t_n, x_n, y_n)\right]$$

and corresponding corrector formulas

$$x_{n+1,c} = x_{n-1} + \frac{h}{3}\left[f(t_{n+1}, x_{n+1,p}, y_{n+1,p}) + 4f(t_n, x_n, y_n) + f(t_{n-1}, x_{n-1}, y_{n-1})\right]$$

$$y_{n+1,c} = y_{n-1} + \frac{h}{3}\left[g(t_{n+1}, x_{n+1,p}, y_{n+1,p}) + 4g(t_n, x_n, y_n) + g(t_{n-1}, x_{n-1}, y_{n-1})\right]$$

Local and Global Error

Associated with the single-step methods and multi-step methods for approximating solutions to differential equations there are local errors which occur at each step of the calculations. Consider figure 8-6, where in going from the point (x_0, y_0) to the new point (x_1, y_1) there is a local error ϵ_1. Each successive step also has a local error and so we let ϵ_i denote the local error associated with the ith step. Each local error comes from the algorithm used to approximate the next successive value. Usually numerical methods are applied over many steps, and approximate solutions are constructed over an interval $a \leq x \leq b$ which is built up of many subintervals of size $h = \frac{b-a}{n}$ when there is a constant step size. After n applications of the numerical algorithm, we should be concerned with the total error or global error between the approximate and exact solution.

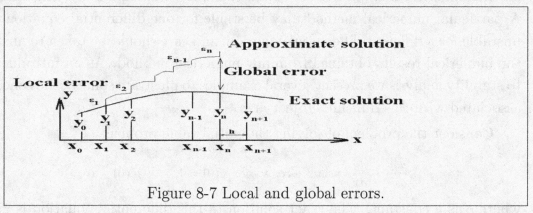

Figure 8-7 Local and global errors.

This global error is given by

$$\text{Global error} = \epsilon_1 + \epsilon_2 + \cdots + \epsilon_n$$

which is the summation of the local errors. If each local error ϵ_i satisfies an inequality of the type $|\epsilon_i| < Mh^m$, where m and M are constants, then we may write

$$|\text{Global error}| < nMh^m.$$

For a constant step size h we have $n = \frac{b-a}{h}$, which enables us to express the global error as

$$|\text{Global error}| < (b-a)Mh^{m-1}. \tag{8.78}$$

This shows that if each local error is $O(h^m)$, then the global error is $O(h^{m-1})$.

Stability

Associated with a numerical method used to solve a differential equation, there is the concept of numerical stability. If an error such as round off error or local truncation error is introduced into the numerical method, then we desire to analyze the difference between the approximate solution y_n at x_n and the exact solution $y(x_n)$ as n becomes large. A numerical method is called stable if any errors that are introduced at a point tend to die out and not affect the errors at future points in a severe way. A numerical method is called unstable if the errors that are introduced into a calculation tend to cause large differences between the approximate and exact solution as n increases. Stability (or instability) is a subject which can become complex and difficult. Stability depends not only on the numerical method but also upon the differential equation being studied. A particular numerical method may be stable for one differential equation and unstable for a different differential equation, and care should be taken to analyze the numerical results obtained from any numerical method. As an introduction to stability analysis we present several examples to illustrate some of the concepts associated with this difficult subject area.

Consider the problem of solving the initial value problem

$$\frac{d^2y}{dx^2} - \frac{dy}{dx} = \sin x + \cos x, \qquad y(0) = 1, \qquad y'(0) = \alpha \tag{8.79}$$

where α is a constant. The exact solution to this differential equation is given by

$$y = y(x) = (1 + \alpha)e^x - \alpha - \sin x \tag{8.80}$$

and represents a family of solutions through the point $(0, 1)$ with slope α. The figure 8-8 illustrates some members of the solution family for selected values of the parameter α.

Consider the initial value problem given by equation (8.79) in the special case where $\alpha = -1$. Observe that the exponential function must occur in all solutions of the family except in the case where $\alpha = -1$. If you use some numerical method to solve the initial value problem (8.79), in the special case where $\alpha = -1$, then you should obtain the wide curve illustrated in the middle of figure 8-8. If any error gets into your numerical method, then you are forced onto one of the other curves of the solution family which must have an exponential solution. The exponential function is forced into your numerical solution by a small error and eventually becomes the dominant term for large values of x.

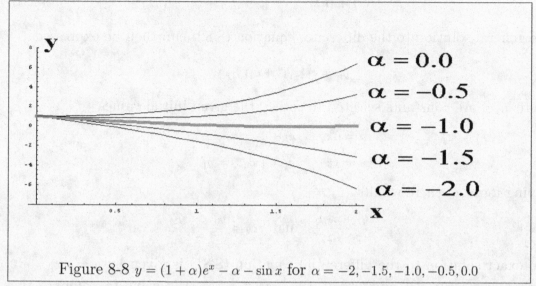

Figure 8-8 $y = (1+\alpha)e^x - \alpha - \sin x$ for $\alpha = -2, -1.5, -1.0, -0.5, 0.0$

Eventually your numerical solution becomes unbounded and an instability is said to have occurred. The exponential function is part of the solution family of the differential equation and you can't get rid of it. In the case where $\alpha = -1$, it is known as a parasitic solution. Even though the exponential function is not part of the solution, in the case $\alpha = -1$, any small error brings it into play.

Parasitic solutions can also arise because of the numerical method selected. Consider the problem to solve the initial value problem

$$\frac{dy}{dx} = \lambda y, \qquad y(0) = 1. \tag{8.81}$$

Evaluate the given differential equation at (x_n, y_n) and replace the derivative by a central difference approximation, with step size h, to obtain the difference equation

$$\frac{y_{n+1} - y_{n-1}}{2h} = \lambda y_n, \qquad n = 1, 2, \ldots \tag{8.82}$$

which you should recognize as Euler's midpoint method. If the initial values y_0 and y_1 are known, then we can assume a solution to the difference equation of the form $y_n = z^n$ where z is a constant to be determined. Substitute the assumed solution into the difference equation and show that z must satisfy the characteristic equation

$$z^2 - 2h\lambda z - 1 = 0. \tag{8.83}$$

The characteristic equation (8.83) has the solutions

$$z_1 = h\lambda + \sqrt{1 + (h\lambda)^2}, \qquad z_2 = h\lambda - \sqrt{1 + (h\lambda)^2} \tag{8.84}$$

The general solution to the difference equation (8.82) can then be expressed

$$y_n = c_1(z_1)^n + c_2(z_2)^n \tag{8.85}$$

where c_1, c_2 are constants selected to satisfy the given initial values

$$n = 0: \qquad c_1 + c_2 = y_0$$

$$n = 1: \qquad c_1 z_1 + c_2 z_2 = y_1$$

Solving for c_1, c_2 there results

$$c_1 = \frac{y_1 - z_2 y_0}{z_1 - z_2} \quad \text{and} \quad c_2 = \frac{y_0 z_1 - y_1}{z_1 - z_2}.$$

The exact solution to the differential equation (8.81) is given by $y = y(x) = e^{\lambda x}$ so that $y_0 = y(0) = 1$ and $y_1 = y(h) = y(x_1) = e^{\lambda h}$. We want to compare the exact solution values from the differential equation with the values obtained from solving the difference equation. The error of the numerical method is then given by

$$E = y(x_n) - y_n = e^{\lambda x_n} - c_1(z_1)^n - c_2(z_2)^n. \tag{8.86}$$

To calculate an approximate value for the error term it would be nice if the quantities z_1 and z_2 were expressed in terms of exponentials like the true solution. Therefore, we assume that z_1 can be expressed in a series of the form

$$z_1 = z_1(h\lambda) = h\lambda + \sqrt{1 + (h\lambda)^2} = e^{h\lambda} \left(\beta_0 + \beta_1(h\lambda) + \beta_2(h\lambda)^2 + \cdots \right) \tag{8.87}$$

where $\beta_0, \beta_1, \ldots$ are constants to be determined. Calculus helps us to determine that $\beta_0 = 1$ and $\beta_1 = 0$ so that we can write

$$z_1 = z_1(h\lambda) = e^{h\lambda} \left[1 + \mathcal{O}(h^2\lambda^2) \right] \tag{8.88}$$

Therefore,

$$z_2 = z_2(h\lambda) = -z_1(-h\lambda) = -e^{-h\lambda} \left[1 + \mathcal{O}(h^2\lambda^2) \right] \tag{8.89}$$

so that the solution to the difference equation can be represented in the approximate form

$$y_n = c_1 e^{x_n\lambda} \left[1 + \mathcal{O}(h^2) \right] + c_2(-1)^n e^{-x_n\lambda} \left[1 + \mathcal{O}(h^2) \right] \tag{8.90}$$

One can use the above approximations and show $c_1 = 1 + \mathcal{O}(h^2)$ so that the error term expression (8.86) reduces to

$$E = e^{\lambda x_n} - \left[e^{\lambda x_n} \left(1 + \mathcal{O}(h^2) \right) + c_2 (-1)^n e^{-\lambda x_n} (1 + \mathcal{O}(h^2)) \right]. \qquad (8.91)$$

The equation (8.91) shows that as $h \to 0$ the term $c_1(z_1)^n$ approaches the true solution and the term $c_2(z_2)^n$ is an extra term which arises because of the numerical method. This extra term is called a parasitic solution. If $\lambda > 0$ the parasitic term decreases as n increases and eventually dies out. However, if $\lambda < 0$, then the parasitic term grows and eventually becomes large which causes a numerical instability.

The above examples illustrate that the terminology "parasitic solution" is used to describe those unwanted terms that arise because of the difference equation resulting from the numerical method or to describe those detrimental solutions of a solution family of the differential equation that cause the numerical solution to become large. Whenever the parasitic terms become large, for large values of x, they cause numerical instability. We desire parasitic solutions which decrease with increasing x values and so die out and not affect the numerical results.

Stability Analysis of Multi-step Methods

Stability or instability of a numerical method is an area of study arising in advanced numerical analysis considerations. The following is a brief introduction into this study area.

Consider the implicit numerical algorithm given by equation (8.49) which we repeat here for convenience

$$
\begin{aligned}
y_{n+1} = \ & A_1 y_n + A_2 y_{n-1} + A_3 y_{n-2} + \ldots + A_k y_{n+1-k} \\
& + h \left[B_0 y'_{n+1} + B_1 y'_n + B_2 y'_{n-1} + B_3 y'_{n-2} + \ldots + B_k y'_{n+1-k} \right],
\end{aligned} \qquad (8.92)
$$

where A_1, $A_2, \ldots, A_k$, B_0, B_1, $B_2, \ldots, B_k$ are constants. Assume that a numerical algorithm of the type (8.92) is applied to solve the differential equation

$$\frac{dy}{dx} = \lambda y, \qquad y(x_n) = y_n.$$

This differential equation is called a test differential equation for the numerical method we have constructed. Substituting this test differential equation into the numerical algorithm (8.92) produces the difference equation

$$(1 - h\lambda B_0) y_{n+1} - (A_1 + h\lambda B_1) y_n - \ldots - (A_k + h\lambda B_k) y_{n+1-k} = 0. \qquad (8.93)$$

We compare the solution of this difference equation with the exact solution to our test problem. The solution of the difference equation is obtained by assuming a solution $y_n = z^n$ to equation (8.93) and then solving for the values of z which satisfy the characteristic equation

$$\rho(z) - h\lambda\sigma(z) = 0 \qquad (8.94)$$

where

$$\begin{aligned}
\rho(z) &= z^k - A_1 z^{k-1} - A_2 z^{k-2} - \ldots - A_k \\
\sigma(z) &= B_0 z^k - B_1 z^{k-1} + \ldots + B_k.
\end{aligned} \qquad (8.95)$$

Observe that for $Re(\lambda) < 0$, if all roots of the characteristic equation lie on or are interior to the unit circle, then the solution y_n of the difference equation will not grow or increase as n gets large. Such solutions are called stable solutions. For $Re(\lambda) < 0$ and at least one of the roots of the characteristic equation outside the unit circle, then the solution of the difference equation will increase in absolute value as n increases. This tells us the corresponding numerical algorithm will be unstable.

We conclude this brief discussion of stability with some important definitions relating to the subject.

Definition: (A-stability)   When a numerical algorithm for solving differential equations is applied to the test problem $\frac{dy}{dx} = \lambda y$, with $\lambda < 0$, there results a difference equation. If all solutions of this difference equation tend toward zero as n increases, then the numerical method is called A-stable.

Definition: (Region of stability)   When a numerical algorithm for solving differential equations is applied to the test problem $\frac{dy}{dx} = \lambda y$, with λ a complex constant, there results a difference equation. The region of stability of the numerical method is the set of all values λ, where $Re(\lambda) < 0$, and where the solutions of the difference equation remain bounded as $n \to \infty$.

Example 8-11.   Stability
Stability can be more easily recognized in the cases where the difference equations can be written in the form

$$y_{n+1} = G y_n \qquad (8.96)$$

where G is a real or complex constant. Equations of the form (8.96) imply

$$y_1 = Gy_0$$

$$y_2 = Gy_1 = G^2 y_0$$

$$y_3 = Gy_2 = G^3 y_0$$

$$\vdots$$

$$y_n = Gy_{n-1} = G^n y_0$$

In order for y_n to remain bounded it is required that $|G| \leq 1$. The quantity G is called an amplification factor. There are many numerical methods where the test differential equation $\dfrac{dy}{dx} = \lambda y$ produces a result where the amplification factor is readily obtained.

The **explicit Euler method** (table 8.3, entry 1) has the numerical method $y_{n+1} = y_n + hf_n$. Applying this algorithm to the differential equation $\frac{dy}{dx} = \lambda y$ results in the difference equation

$$y_{n+1} = y_n + h\lambda y_n = (1 + h\lambda)y_n = Gy_n.$$

Here the amplification factor is $G = 1 + h\lambda$ and for stability it is required that $|1 + h\lambda| \leq 1$ or $-1 \leq 1 + h\lambda \leq 1$. If $h\lambda < 0$, then the right-hand inequality is always satisfied. The left-hand inequality $-1 \leq 1 + h\lambda$ requires that $-h\lambda \leq 2$. This places restrictions upon the step size h and so the method is said to be conditionally stable.

The **implicit Euler method** (table 8.4, entry 1) has the numerical method $y_{n+1} = y_n + hf_{n+1}$. Applying this algorithm to the differential equation $\frac{dy}{dx} = \lambda y$, results in the difference equation $y_{n+1} = y_n + h\lambda y_{n+1}$ or

$$y_{n+1} = \frac{1}{1 - h\lambda} y_n = Gy_n.$$

Here the amplification factor is $G = 1/(1 - h\lambda)$. For stability it is required that $|G| \leq 1$. Note that this condition is satisfied for all $h\lambda < 0$ and consequently, the implicit Euler method is said to be unconditionally stable.

The **modified Euler method** (see the equation (8.64)) gives the algorithm $y_{n+1} = y_n + \frac{h}{2}\left(f(x_n, y_n) + f(x_{n+1}, y_{n+1})\right)$. Applied to the test differential equation $\frac{dy}{dx} = \lambda y$ one obtains $y_{n+1} = y_n + \frac{h}{2}(\lambda y_n + \lambda y_{n+1})$ which simplifies to the form

$$y_{n+1} = \frac{1 + \frac{1}{2}h\lambda}{1 - \frac{1}{2}h\lambda} y_n = Gy_n.$$

346

The stability condition $|G| \leq 1$ requires that $Re\,(h\lambda) < 0$, where $Re\,(h\lambda)$ denotes the real part of $h\lambda$.

The **fourth order Runge-Kutta method** (classic form) requires

$$y_{n+1} = y_n + \frac{1}{6}(k_1 + 2k_2 + 2k_2 + k_4)$$

$$k_1 = hf(x_n, y_n)$$

$$k_2 = hf(x_n + \frac{1}{2}h, y_n + \frac{1}{2}k_1)$$

$$k_3 = hf(x_n + \frac{1}{2}h, y_n + \frac{1}{2}k_2)$$

$$k_4 = hf(x_n + h, y_n + k_3)$$

which applied to the test differential equation $\frac{dy}{dx} = \lambda y$ gives

$$k_1 = h\lambda y_n$$

$$k_2 = h\lambda y_n + \frac{1}{2}(h\lambda)^2 y_n$$

$$k_3 = h\lambda y_n + (h\lambda)^2 y_n + \frac{1}{2}(h\lambda)^3 y_n$$

$$k_4 = h\lambda y_n + (h\lambda)^2 y_n + (h\lambda)^3 y_n + \frac{1}{2}(h\lambda)^4 y_n$$

which gives

$$y_{n+1} = \left(1 + h\lambda + \frac{1}{2}(h\lambda)^2 + \frac{1}{6}(h\lambda)^3 + \frac{1}{24}(h\lambda)^4\right) y_n = G y_n.$$

Using complex variable theory, the condition $|G| \leq 1$ is obtained for $h\lambda$ in the region illustrated where $Im\,(h\lambda)$ denotes the imaginary part of $h\lambda$ and $Re\,(h\lambda)$ denotes the real part of $h\lambda$.

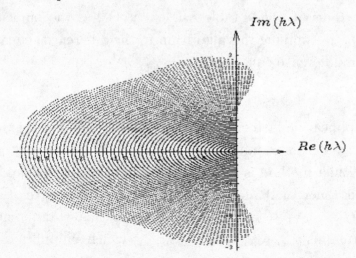

Stiff Differential Equations

The subject of stiff differential equations is another topic discussed in advanced numerical analysis texts. The following is a brief introduction to systems of differential equations referred to as stiff differential equations. Consider the first order linear system

$$\frac{d\vec{y}}{dx} = A\vec{y} + \vec{f}(x) \tag{8.97}$$

where A is a $n \times n$ constant matrix. We assume that A has distinct eigenvalues $\lambda_1, \lambda_2, \ldots, \lambda_n$ with eigenvectors $\vec{z}_1, \vec{z}_2, \ldots, \vec{z}_n$. Further assume each eigenvalue λ_i is such that $\mathrm{Re}(\lambda_i) < 0$. (The case $\mathrm{Re}(\lambda_i) > 0$ leads to exponential solutions where $\vec{y}(x) \to \infty$ as x increases without bound.) The general solution of the equation (8.97) can be represented as a transient solution plus a steady state solution

$$\vec{y} = \vec{y}(x) = \underbrace{\sum_{i=1}^{n} C_i e^{\lambda_i x} \vec{z}_i}_{Transient\ solution} + \underbrace{\vec{S}(x)}_{Steady\ state\ solution} \tag{8.98}$$

where C_i are constants for $i = 1, \ldots, n$. If one of the eigenvalues is very large in absolute value, in comparison with the other eigenvalues, then the corresponding exponential solution dies out quickly and eventually vanishes after x increases a short distance. There are many numerical methods for solving differential equations where the large differences in magnitude of the eigenvalues greatly affects the stability and accuracy of the numerical method and one is forced to select very small step sizes to maintain stability. Whenever the above type of conditions exist, the differential equation or system of differential equations is referred to as a stiff system or stiff equation. Stiff systems are characterized by the presence of transient solutions which have the effect of restricting the step size to be very small.

As an example of a stiff differential equation consider the initial value problem to solve

$$\frac{d^2y}{dx^2} + 101\frac{dy}{dx} + 100y = 0, \qquad y(0) = 1.0, \quad y'(0) = -1.0 \tag{8.99}$$

over the interval $0 \le x \le 4$. This differential equation can be written as a first order system by making the substitutions $y = y_1$ and $\frac{dy}{dt} = y_2$. One then obtains the system

$$\begin{aligned}\frac{dy_1}{dx} &= y_2 & y_1(0) &= 1.0 \\ \frac{dy_2}{dx} &= -101y_2 - 100y_1 & y_2(0) &= -1.0\end{aligned} \tag{8.100}$$

We solve this system using the Runge-Kutta method with double precision arithmetic to produce the results in table 8.6.

Table 8.6
Numerical Results for 4th order Runge-Kutta method applied to
$$\frac{d^2y}{dx^2} + 101\frac{dy}{dx} + 100y = 0 \quad y(0) = 1, \quad y'(0) = -1.0$$
over the interval $0 \le x \le 4$.

Step Size h	Number of Steps to reach $x = 4$	Runge-Kutta $y(4) = y_1(4)$	Exact Solution $y(4)$
0.10000	40	Overflow error	0.0183156
0.05000	80	Overflow error	0.0183156
0.02500	160	0.0187793	0.0183156
0.01000	400	0.0184997	0.0183156
0.00500	800	0.0184074	0.0183156
0.00100	4000	0.0183340	0.0183156
0.00050	8000	0.0183248	0.0183156
0.00010	40000	0.0183174	0.0183156
0.00005	80000	0.0183165	0.0183156
0.00001	400000	0.0183158	0.0183156

An examination of the above table shows that one is forced to use a small step size in order to achieve an accurate solution. The differential equation has the solution family $y = y(x) = c_1 e^{-x} + c_2 e^{-100x}$ where c_1, c_2 are constants. The solution satisfying the given initial conditions is $y = y(x) = e^{-x}$. The term e^{-100x}, even though it does not occur in the desired solution, is the term that is forcing us to use a small step size.

Variable Step Size

Systems of differential equations can have some solutions which are smooth and well behaved while other solutions vary wildly. This type of behavior can cause many numerical methods to fail. In order for a numerical method, with a fixed step size, to maintain accuracy and stability it is sometimes necessary to select a very small step size to handle the ill-behaved solutions within the system of equations being solved. Having to reduce the step size drastically causes the numerical method to become inefficient. It has been found through experience that the variable step size numerical methods usually out perform those methods which employ a fixed step size. At present, the numerical methods of choice for solving a single differential equation or a system of differential equations are those methods which have the following properties.

(i) The method is easily implemented and does not require excessive calculations or manipulation of information each step of the algorithm.

(ii) The numerical method is such that the local error information is known at each step of the algorithm and can be used to help adjust the step size.

(iii) Adjustment of the step size to maintain stability and accuracy of the solution is part of the algorithm.

(iii) The numerical method is rugged enough so that it can be applied to a large number or reasonably behaved systems.

Numerical methods which meet or come close to meeting all of the above requirements and readily available for everyday use are (i) Adaptive step size multi-step methods (ii) The Gragg-Bulirsch-Stoer adaptive step size method and (iii) The Runge-Kutta-Fehlberg method. These methods are developed and analyzed in some advanced numerical textbooks. The following is a brief description of the variable step size Runge-Kutta-Fehlberg method.

Runge-Kutta-Fehlberg method

The Runge-Kutta-Fehlberg method (denoted RKF45) makes two different approximations for the solution at each step of the numerical procedure. It then compares the two approximations to see if they are in agreement. The step size is adjusted based upon whether the two approximations agree or do not agree to some specified degree of accuracy. There are many Runge-Kutta methods because the system of equations resulting from a comparison of the Taylor series expansion and the Runge-Kutta expansions leads to a system of equations with more unknowns than equations. Consequently, certain parameters can be selected arbitrarily and so there results a variety of Runge-Kutta Methods to solve the initial value problem $\frac{dy}{dx} = f(x,y)$, $y(x_0) = y_0$ which have the form

$$x_{n+1} = x_n + h$$

$$y_{n+1} = y_n + \omega_1 k_1 + \omega_2 k_2 + \cdots + \omega_p k_p$$

Fehlberg discovered that the coefficients

$$k_1 = hf(x_n, y_n)$$

$$k_2 = hf(x_n + \frac{1}{4}h, y_n + \frac{1}{4}k_1)$$

$$k_3 = hf(x_n + \frac{3}{8}h, y_n + \frac{3}{32}k_1 + \frac{9}{32}k_2)$$

$$k_4 = hf(x_n + \frac{12}{13}h, y_n + \frac{1932}{2197}k_1 - \frac{7200}{2197}k_2 + \frac{7296}{2197}k_3)$$

$$k_5 = hf(x_n + h, y_n + \frac{439}{216}k_1 - 8k_2 + \frac{3680}{513}k_3 - \frac{845}{4104}k_4)$$

$$k_6 = hf(x_n + \frac{1}{2}h, y_n - \frac{8}{27}k_1 + 2k_2 - \frac{3544}{2565}k_3 + \frac{1859}{4104}k_4 - \frac{11}{40}k_5)$$

can be used simultaneously in a fourth order Runge-Kutta approximation formula

$$y_{n+1} = y_n + \frac{25}{216}k_1 + \frac{1408}{2565}k_3 + \frac{2197}{4101}k_4 - \frac{1}{5}k_5 + \mathcal{O}(h^4) \tag{8.101}$$

and a fifth order Runge-Kutta approximation formula

$$y_{n+1}^* = y_n + \frac{16}{135}k_1 + \frac{6656}{12825}k_3 + \frac{28561}{56430}k_4 - \frac{9}{50}k_5 + \frac{2}{55}k_6 + \mathcal{O}(h^5) \tag{8.102}$$

The two approximations y_{n+1} and y_{n+1}^* can then be compared to adjust the step size. Whenever there is a significant difference in the local truncation errors the step size h is scaled by a factor S. The new scaled step size is $h_{new} = Sh_{old}$. Note that $S > 1$ when everything is going along fine, but if errors start to occur, then $S < 1$ in order to reduce the step size. Further discussion of variable step size methods can be found in more advanced texts on numerical methods and analysis.

Boundary Value Problems

Differential equations of the form

$$\frac{d^2y}{dx^2} = f(x, y, \frac{dy}{dx}), \quad a \leq x \leq b \tag{8.103}$$

and subject to boundary conditions $y(a) = \alpha$ and $y(b) = \beta$, where α, β are given constants, are called boundary value problems for ordinary differential equations.

In the special case the equation (8.103) is a linear differential equation, the boundary value problem can be written in the form

$$\frac{d^2y}{dx^2} + p(x)\frac{dy}{dx} + q(x)y = r(x), \quad a \leq x \leq b \tag{8.104}$$

subject to the boundary conditions $y(a) = \alpha$ and $y(b) = \beta$. In this special case the interval (a, b) can be divided into $N + 1$ parts having a step size given by $h = \Delta x = (b - a)/(N + 1)$, then one can replace the derivatives in equation (8.104) by differences. Then at $x = x_i$ one can write

$$\frac{y_{i+1} - 2y_i + y_{i-1}}{h^2} + p(x_i)\frac{y_{i+1} - y_{i-1}}{2h} + q(x_i)y_i = r(x_i) \tag{8.105}$$

or rearranging terms

$$(1 + \frac{p(x_i)h}{2})y_{i+1} + (-2 + q(x_i)h^2)y_i + (1 - \frac{p(x_i)h}{2})y_{i-1} = h^2 r(x_i). \tag{8.106}$$

Define the quantities

$$F_i = 1 + \frac{p(x_i)h}{2}, \qquad G_i = -2 + q(x_i)h^2, \qquad H_i = 1 - \frac{p(x_i)h}{2} \tag{8.107}$$

and write out the equations (8.106) for $i = 1, 2, \ldots, N$. One finds

$$
\begin{array}{ll}
i = 1 & F_1 y_2 + G_1 y_1 + H_1 y_0 = h^2 r(x_1) \\
i = 2 & F_2 y_3 + G_2 y_2 + H_2 y_1 = h^2 r(x_2) \\
i = 3 & F_3 y_4 + G_3 y_3 + H_3 y_2 = h^2 r(x_3) \\
\vdots & \qquad\qquad\qquad\qquad \vdots \\
i = N-1 & F_{N-1} y_N + G_{N-1} y_{N-1} + H_{N-1} y_{N-2} = h^2 r(x_{N-1}) \\
i = N & F_N y_{N+1} + G_N y_N + H_N y_{N-1} = h^2 r(x_N)
\end{array}
\tag{8.108}
$$

In the system of equations (8.108) note that $y_0 = \alpha$ and $y_{N+1} = \beta$ are the known boundary conditions. The system of equations (8.108) then represents N equations in N unknowns to be solved. This system can also be written in the tridiagonal matrix form

$$
\begin{bmatrix}
G_1 & F_1 & & & & \\
H_2 & G_2 & F_2 & & & \\
 & H_3 & G_3 & F_3 & & \\
 & & \ddots & \ddots & \ddots & \\
 & & & H_{N-1} & G_{N-1} & F_{N-1} \\
 & & & & H_N & G_N
\end{bmatrix}
\begin{bmatrix}
y_1 \\ y_2 \\ y_3 \\ \vdots \\ y_{N-1} \\ y_N
\end{bmatrix}
=
\begin{bmatrix}
h^2 r(x_1) - H_1 \alpha \\
h^2 r(x_2) \\
h^2 r(x_3) \\
\vdots \\
h^2 r(x_{N-1}) \\
h^2 r(x_N) - F_N \beta
\end{bmatrix}
\tag{8.109}
$$

Shooting Methods

Boundary value problems of the form given by equation (8.103) can many times be solved by the shooting method. The shooting method begins by replacing the two point boundary value problem given by equation (8.103), with a related initial value problem. One solves numerically

$$\frac{d^2y}{dx^2} = f(x, y, \frac{dy}{dx}), \quad a \leq x \leq b \tag{8.110}$$

subject to the initial conditions $y(a) = \alpha$ and $y'(a) = \gamma$ where γ represents a guess of the slope at $x = a$. The solution of the initial value problem can be denoted $y(x; \gamma)$ and it can be solved using any of the numerical methods previously considered. The problem then becomes one of selecting the slope γ such that the error

$$Error = |\beta - y(b; \gamma)| = 0. \tag{8.111}$$

The problem now resembles a root finding scheme. That is, find γ if possible, such that the error $|\beta - y(x; \gamma)|$ is zero at $x = b$.

The above shooting method has been devised for Dirichlet boundary conditions where the end point values are specified. Neumann type boundary conditions occur whenever derivative boundary conditions are specified at the end points. If $y'(a) = \alpha$ and $y'(b) = \beta$ are specified derivative boundary conditions one can form an initial value problem to solve equation (8.110) subject to the initial conditions $y(a) = \gamma$ and $y'(a) = \alpha$ where γ represents a guess of the initial value for y at $x = a$. The solution of this initial value problem can be denoted $y(x; \gamma)$ and our objective is solve the equation $Error = |\beta - y'(b; \gamma)| = 0$. This is analogous to the Dirichlet method with the exception that we are now trying to find the initial ordinate γ that gives the correct slope at $x = b$.

Differential equations with singular points

Differential equations of the form $\frac{d^2y}{dx^2} + p(x)\frac{dy}{dx} + q(x)y = 0$ where $p(x)$ and/or $q(x)$ have singularities at a point $x = x_0$, where the singularities are such that $(x - x_0)p(x)$ and $(x - x_0)^2q(x)$ can be expanded in a Taylor series about the point x_0, have Frobenius type solutions of either of the forms

$$y_1(x) = (x - x_0)^r \sum_{n=0}^{\infty} c_n(x - x_0)^n, \qquad y_2(x) = Cy_1(x)\ln(x - x_0) + \sum_{n=0}^{\infty} b_n(x - x_0)^n$$

where r, C are constants and c_n, b_n, $n = 0, 1, 2, \ldots$ are constants. Discussions of the Frobenius method can be found in most ordinary differential equation books.

Exercises Chapter 8

► **1.** (Euler's method)

Construct a table of values and a flowchart of your computations to numerically solve the given differential equations by the Euler method for the step size given. Compare your numerical solution with the exact solution.

(a) $\dfrac{dy}{dx} = -xy$, $\quad y(0) = 1$, $\quad 0 \le x \le 1$, $\quad h = 0.05$

(b) $\dfrac{dy}{dx} = \dfrac{2y}{x+1}$, $\quad y(0) = 1$, $\quad 0 \le x \le 1$, $\quad h = 0.05$

(c) $\dfrac{dz}{dt} = z(1-z)$, $\quad z(0) = 3$, $\quad 0 \le t \le 2$, $\quad h = 0.05$

► **2.** (Taylor series method)

Construct a table of values and a flowchart of your computations to numerically solve the given differential equations by the second-order Taylor series algorithm using the constant step sizes given. Compare your numerical solutions with the exact solution. What happens to the local and global errors each time the step size is halved?

(a) $\dfrac{dy}{dx} = x + xy$, $\quad y(0) = 1$, $\quad 0 \le x \le 1$, $\quad h = 0.1,\ 0.05,\ 0.025$

(b) $\dfrac{dy}{dt} = 0.1y(1-y)$, $\quad y(0) = 2$, $\quad 0 \le t \le 1$, $\quad h = 0.1,\ 0.05,\ 0.025$

(c) $\dfrac{dz}{dt} = \dfrac{z}{t} - t$, $\quad z(1) = 1$, $\quad 1 \le t \le 2$ $\quad h = 0.1,\ 0.05,\ 0.025$

► **3.** (Taylor series method)

Illustrate with a flowchart how each equation can be solved using a third-order Taylor series method. Compare the numerical solution with the exact solution.

(a) $\dfrac{dy}{dx} = x + xy$, $\quad y(0) = 1$, $\quad 0 \le x \le 1$, $\quad h = 0.05$

(b) $\dfrac{dy}{dx} = y \cos x$, $\quad y(0) = 1$, $\quad 0 \le x \le 1$, $\quad h = 0.05$

(c) $\dfrac{dz}{dt} = \dfrac{z}{t} - t$, $\quad z(1) = 1$, $\quad 1 \le t \le 2$, $\quad h = 0.05$

(d) $\dfrac{dz}{dt} = -z \sin x$, $\quad z(0) = 1$, $\quad 0 \le t \le 1$, $\quad h = 0.05$

▶ **4.** **(Runge-Kutta method)**

Numerically solve the initial value problem

$$\frac{dy}{dx} = y + x^2, \quad y(0) = 1, \quad 0 \le x \le 1, \quad h = 0.05$$

by the numerical method specified. Compare your results with the exact solution.

(a) Second-order Runge-Kutta method (improved Euler form).

(b) Third-order Runge-Kutta method (classic form).

(c) Fourth-order Runge-Kutta method (classic form).

▶ **5.** **(Runge-Kutta method)**

Numerically solve the initial value problem $y' = f(x), \quad y(x_0) = y_0$ using a fourth-order Runge-Kutta classic method and show

(a)
$$k_1 = hf(x_0)$$
$$k_2 = k_3 = hf(x_0 + h/2)$$
$$k_4 = hf(x_0 + h)$$

(b) Show the numerical solution can be generated using

$$y(x_0 + h) = y_0 + \int_{x_0}^{x_0+h} f(x)\,dx$$
$$= y_0 + \frac{h}{6}[f(x_0) + 4f(x_0 + h/2) + f(x_0 + h)]$$

(c) Show the replacement of h by $2H$, produces the well-known Simpson's one-third rule for numerically evaluating integrals

$$\int_{x_0}^{x_0+2H} f(x)\,dx = \frac{H}{3}[f(x_0) + 4f(x_0 + H) + f(x_0 + 2H)]$$

▶ **6.** **(Euler's method)**

Consider the initial value problem $y' = f(x, y), \quad y(x_0) = y_0$ and let $y_n = y(x_n)$ denote the exact or true value of the solution at $x = x_n$ and let Y_n denote the approximate value of y_n which is calculated from the Euler algorithm

$$Y_{n+1} = Y_n + hf(x_n, Y_n).$$

Define $e_n = y_n - Y_n$ as the error associated with the approximation Y_n and from the Taylor series equation write

$$y_{n+1} = y_n + h f(x_n, y_n) + \frac{h^2}{2!} y''(\xi_n), \quad x_n \leq \xi_n \leq x_n + h$$

(a) Show that

$$e_{n+1} = e_n + h[f(x_n, y_n) - f(x_n, Y_n)] + \frac{h^2}{2!} y''(\xi_n)$$

(b) Use the mean value theorem

$$f(x_n, y_n) - f(x_n, Y_n) = \frac{\partial f}{\partial y}(x_n, \eta_n)(y_n - Y_n)$$

to show that

$$e_{n+1} = e_n \left[1 + h \frac{\partial f}{\partial y}(x_n, \eta_n) \right] + \frac{h^2}{2!} y''(\xi_n)$$

(c) If $|f_y(x, y)| < K$, for all x, y in a rectangular region R about the point (x_n, η_n), then show that

$$e_{n+1} \leq (1 + hK)e_n + \frac{h^2}{2} y''(\xi_n) \quad \text{for } n = 1, 2, 3, \ldots$$

(d) For $y_0 = Y_0 = y(x_0)$ we have $e_0 = 0$. Show that

$$e_1 \leq \frac{1}{2} h^2 y''(\xi_0)$$

$$e_2 \leq \frac{1}{2} h^2 [(1 + hK)y''(\xi_0) + y''(\xi_1)]$$

$$\cdots$$

$$e_n \leq \frac{1}{2} h^2 [(1 + hK)^{n-1} y''(\xi_0) + (1 + hK)^{n-2} y''(\xi_1) + \cdots + y''(\xi_{n-1})].$$

This shows that the error at each step of the Euler method permeates to each later step and further this error is magnified by the factor $(1 + hK)$. What happens as $h \to 0$?

(e) If we assume $f_y(x, y) < 0$ for all x, y in R and satisfies $|h f_y| < \alpha$, what limits exist for α such that the Euler method is a stable method?

▶ **7.** **(Taylor series)** Construct an error term for the modified Euler method

$$y_{n+1} = y_n + \frac{h}{2}(y'_{n+1} + y'_n)$$

by performing the following steps. (1.) Use the Taylor series expansion and show that

$$y(x_n + h) = y_{n+1} = y_n + y_n'h + y_n''\frac{h^2}{2} + y'''(\xi)\frac{h^3}{3!}.$$

(2.) Differentiate this expression with respect to h and use the result to show that

$$y_{n+1} - y_n - \frac{h}{2}(y_{n+1}' + y_n') = -\frac{h^3}{12}y'''(\xi)$$

▶ **8.** (**Milne method**) For the initial value problem $y' = x + y$, $y(0) = 1$, the following starting values have been calculated using the fourth-order Runge-Kutta method.

x	y
0.00	1.000000
0.05	1.052542
0.10	1.110342
0.15	1.173668

Continue the solution using the Milne method for $0.2 \le x \le 1.0$ with $h = 0.05$.

(a) Sketch a flowchart of your computational algorithm.

(b) Prepare a table of values similar to table 8.2.

▶ **9.** (**Adams-Moulton**) Use the Adams-Moulton method and solve the initial value problem given in problem 8 above.

(a) Sketch a flowchart of your computational algorithm.

(b) Prepare a table of values similar to table 8.2.

▶ **10.** (**Stability**) Use the second order Runge-Kutta improved Euler form, with step size $h = 0.05$, to numerically solve the initial value problem

$$\frac{dy}{dx} = 3y + \cos x - 3\sin x, \qquad y(0) = 0$$

over the region $0 \le x \le 3$.

(a) Compare your results with the exact solution $y = \sin x$, for $0 \le x \le 3$.

(b) What causes the instability of the numerical method?

(c) Can a different numerical method be selected to solve this problem?

▶ **11.** (**System of Differential Equations**) Consider the coupled system of differential equations in two unknowns

$$\frac{du}{dx} = F(x, u, v), \quad u(x_0) = A, \qquad \frac{dv}{dx} = G(x, u, v), \quad v(x_0) = B$$

where u, v are to be determined over the interval $x_0 \le x \le x_n$ with A, B, x_0, x_n constants. Set up a flowchart to illustrate how you would solve such a system numerically using the numerical method specified.

(*a*) Second-order Runge-Kutta method   (*d*) Modified Euler method

(*b*) Third-order Runge-Kutta method   (*e*) Milne method

(*c*) Fourth-order Runge-Kutta method.   (*f*) Adams-Moulton method

▶ **12.** (**Undetermined Coefficients**) Use the method of undetermined coefficients to derive the predictor corrector formula and associated error terms for the numerical method specified.

(a) Milne method   (b)   Adams-Moulton method

▶ **13.** (**Undetermined Coefficients**) The method of undetermined coefficients can be used to derive a variety of numerical methods. Consider the initial value problem associated with the second-order differential equation

$$y'' = f(x, y, y'), \quad y(x_0) = A, \quad y'(x_0) = B.$$

Use the method of undetermined coefficients to construct a single-step numerical algorithm of the form

$$y_{n+1} = \alpha_0 y_n + \alpha_1 y_{n-1} + h^2 [\beta_0 y''_{n+1} + \beta_1 y''_n + \beta_2 y''_{n-1}],$$

where α_0, α_1, β_0, β_1 and β_2 are constants to be determined. These constants are chosen such that the desired algorithm gives exact results whenever $y(x)$ is a polynomial from the set $S = \{\, 1,\ (x - x_n),\ (x - x_n)^2,\ (x - x_n)^3,\ (x - x_n)^4 \,\}$.

(a) Show that the following algebraic equations result when $y(x)$ has the specified polynomial value.

$y(x)$	equation
1	$1 = \alpha_0 + \alpha_1$
$(x - x_n)$	$h = -\alpha_1 h$
$(x - x_n)^2$	$h^2 = \alpha_1 h^2 + h^2[2\beta_0 + 2\beta_1 + 2\beta_2]$
$(x - x_n)^3$	$h^3 = -\alpha_1 h^3 + h^3[6\beta_0 - 6\beta_2]$
$(x - x_n)^4$	$h^4 = \alpha_1 h^4 + h^4[12\beta_0 + 12\beta_2]$

(b) Solve the system of equations in part (a) and obtain the Noumerov recurrence formula

$$y_{n+1} = 2y_n - y_{n-1} + \frac{h^2}{12}[y''_{n+1} + 10y''_n + y''_{n-1}].$$

(c) Show that in the special case $y'' = f(x,y,y') = F(x) - g(x)y$, the Noumerov recurrence formula reduces to

$$\left(1 + \frac{h^2}{12}g_{n+1}\right)y_{n+1} = \left(2 - \frac{10h^2}{12}g_n\right)y_n - \left(1 + \frac{h^2}{12}g_{n-1}\right)y_{n-1}$$

$$+ \frac{h^2}{12}[F_{n+1} + 10F_n + F_{n-1}],$$

where $g_n = g(x_n)$ and $F_n = F(x_n)$. Note that two starting values are needed to use this algorithm. The initial condition is one value and the next value can be obtained by using a Taylor series expansion of step size h. These two values are used to start the Noumerov algorithm.

(d) Show that the change of variables $u = y \exp\left(-\frac{1}{2}\int_{x_0}^x p(\xi)\,d\xi\right)$ can be used to reduce differential equations of the form $\frac{d^2u}{dx^2} + p(x)\frac{du}{dx} + q(x)u = 0$ to the new form $\frac{d^2y}{dx^2} + g(x)y = 0$ where $g(x) = q(x) - \frac{1}{2}\frac{dp(x)}{dx} - \frac{1}{4}p^2(x)$.

▶ **14.** Solve the system of ordinary differential equations

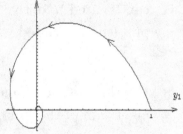

$$\frac{dy_1}{dt} = -y_1 - 2y_2, \qquad y_1(0) = 1.0$$

$$\frac{dy_2}{dt} = 2y_1 - y_2, \qquad y_2(0) = 0.0$$

for $0 \le t \le 10$ and then plot a graph of $y_2(t)$ vs $y_1(t)$.

▶ **15.** **(Crystal Growth)** In the study of crystal growth, single atoms (monomers) which arrive at a surface remain there for a certain length of time and hop from one surface site to another until there occurs collisions between surface atoms to form atomic pairs or a surface atom collides with an already existing cluster of atoms. The process is described by the following rate equations:

$$\frac{dy_1}{dt} = 1 - 2\epsilon\, y_1^2 - y_1 y_s, \qquad y_1(0) = 0$$

$$\frac{dy_s}{dt} = \epsilon y_1^2, \qquad y_s(0) = 0$$

Here $\epsilon = 0.15$ is a constant related to the binding energy of the clusters, y_1 is a scaled variable proportional to the monomer density per site, y_s is a scaled variable proportional to the density of stable clusters and t is a scaled time variable. Solve this system of equations and plot y_1, y_1', and y_s vs time t for $0 \leq t \leq 5.0$.

▶ **16.** **(Population Dynamics)** A population model describing the interaction between two interacting species is given by the simultaneous nonlinear equations

$$\frac{dN_1}{dt} = r_1 N_1 \frac{(\gamma_1 - N_1 - \alpha N_2)}{\gamma_1}, \qquad N_1(0) = 10$$

$$\frac{dN_2}{dt} = r_2 N_2 \frac{(\gamma_2 - N_2 - \beta N_1)}{\gamma_2}, \qquad N_2(0) = 10,$$

where N_1, N_2 are the populations of species, r_1, r_2 are rates of growth, γ_1, γ_2 are saturation values, and α, β represent inhibitory effects of one species on the other. Numerically solve the above system of differential equations for the cases given below and determine which of N_1 or N_2 is the surviving species or whether coexistence between the species exists.

Use the constant values $r_1 = r_2 = 1$, $\gamma_1 = \gamma_2 = 20$ and consider the cases

$$case\ 1: \alpha = 0.75,\ \beta = 1.5 \qquad case\ 3: \alpha = \beta = 0.75$$

$$case\ 2: \alpha = \beta = 1.5 \qquad case\ 4: \alpha = 1.5,\ \beta = 0.75$$

▶ **17.** **(Runge-Kutta Method)**

(a) Construct a second-order Runge-Kutta method to numerically solve the third-order system

$$\frac{du}{dx} = F_1(x, u, v, w), \qquad u(x_0) = A$$

$$\frac{dv}{dx} = F_2(x, u, v, w), \qquad v(x_0) = B$$

$$\frac{dw}{dx} = F_3(x, u, v, w), \qquad w(x_0) = C$$

(b) Represent the differential equation

$$y''' + p(x)y'' + q(x)y' + r(x)y = f(x)$$

in the form of the third-order system of first-order equations.

360

▶ **18.** (**Undetermined Coefficients**) Use the method of undetermined coefficients to derive the following algorithms for solving the initial value problem

$$\frac{dy}{dx} = f(x,y), \quad y(x_0) = y_0$$

(a) Adams-Bashforth (3 step method)

$$x_{n+1} = x_n + h$$
$$y_{n+1} = y_n + \frac{h}{12}\left[23y_n' - 16y_{n-1}' + 5y_{n-2}'\right] + \frac{3}{8}y^{(iv)}(\xi)h^3$$

Three starting values are required.

(b) Adams-Bashforth (5 step method)

$$x_{n+1} = x_n + h$$
$$y_{n+1} = y_n + \frac{h}{720}\left[1901y_n' - 2774y_{n-1}' + 2616y_{n-2}' - 1274y_{n-3}' + 251y_{n-4}'\right] + \frac{95}{288}y^{(6)}(\xi)h^5$$

Five starting values are required.

(c) Adams-Moulton (4 step method)

$$x_{n+1} = x_n + h$$
$$y_{n+1} = y_n + \frac{h}{720}\left[251y_{n+1}' + 646y_n' - 264y_{n-1}' + 106y_{n-2}' - 19y_{n-3}'\right] - \frac{3}{160}y^{(6)}(\xi)h^5$$

This is an implicit method and requires four starting values.

▶ **19.** Many rate equations from chemistry exhibit unusual behavior. Solve the system of rate equations

$$\frac{dy_1}{dt} = -k_1 y_1 \qquad\qquad y_1(0) = 300.0$$
$$\frac{dy_2}{dt} = k_1 y_1 - k_2 y_2 - k_3 y_2 y_3^2 \qquad y_2(0) = 1.0$$
$$\frac{dy_3}{dt} = k_2 y_2 + k_3 y_2 y_3^2 - k_4 y_3 \qquad y_3(0) = 1.0$$
$$\frac{dy_4}{dt} = k_4 y_3 \qquad\qquad y_4(0) = 0.0$$

for $0 \le t \le 600$, where $k_1 = 0.002$, $k_2 = 0.08$, $k_3 = 0.9$ and $k_4 = 0.8$. Plot y_2, y_3 vs t and y_1, y_4 vs t on separate graphs.

▶ **20.** Solve the Lorenz system of differential equations

$$\frac{dx}{dt} = -\alpha(x - y), \quad \frac{dy}{dt} = -xz = \beta x - y, \quad \frac{dz}{dt} = xy - \gamma z$$

subject to the initial conditions $x(0) = y(0) = 6$ and $z(0) = 20$. Use the values $\alpha = 10$, $\beta = 28$ and $\gamma = 8/3$. Examine the projection curve $\{z(t) \text{ vs } x(t)\}$ for $0 \le t \le 40$.

▶ **21.** Construct a numerical solution to the differential equation

$$\frac{dy}{dt} = -\frac{y}{t} + \frac{1}{t^2} \quad y(1) = 1 \qquad 1 \le t \le 10.$$

Compare your solution with the exact solution $y = \dfrac{1 + \ln t}{t}$.

▶ **22.** Construct several (at least three) numerical methods to solve numerically the differential equation

$$\frac{dy}{dx} + 3(\cos 3x)\, y = e^{-3\sin 3x}, \qquad y(0) = 1, \qquad 0 \le x \le 20.$$

Use step sizes of $h = 0.2, 0.1, 0.05, 0.025, 0.01, 0.005$ and compare the different methods you have selected. Compare all answers with the exact solution which you can verify is given by $y = y(x) = (x + 1)e^{-\sin 3x}$.

▶ **23.**

Solve the system of differential equations

$$\frac{dx}{dt} = -y - 0.01x^3, \qquad x(0) = 5.0$$

$$\frac{dy}{dt} = 3x - y, \qquad y(0) = 0.0$$

over the interval $0 \le t \le 4\pi$.

(a) Construct a plot of $x(t)$ vs $y(t)$ for $0 \le t \le 4\pi$.

(b) Construct a plot of $x(t)$ vs t for $0 \le t \le 4\pi$.

(c) Construct a plot of $y(t)$ vs t for $0 \le t \le 4\pi$.

▶ **24.**

Find out everything you can about the functions y_1, y_2 defined by the system of differential equations and initial conditions

$$\frac{dy_1}{dt} = y_2, \qquad y_1(0) = 0$$

$$\frac{dy_2}{dt} = -y_1, \qquad y_2(0) = 1$$

after first solving the system numerically for $0 \le t \le 4\pi$.

362

▶ **25.** Solve the given system over the domain $0 \le t \le 14$. Then find out everything you can about the functions y_1, y_2, y_3 which satisfy the differential equations and initial conditions

$$\frac{dy_1}{dt} = y_2 y_3, \qquad y_1(0) = 0$$

$$\frac{dy_2}{dt} = -y_1 y_3, \qquad y_2(0) = 1$$

$$\frac{dy_3}{dt} = -k^2 y_1 y_2, \qquad y_3(0) = 1$$

where k is a constant $0 \le k \le 1$. These are the Jacobi elliptic functions given by $y_1 = sn(t, k)$, $y_2 = cn(t, k)$, $y_3 = dn(t, k)$.

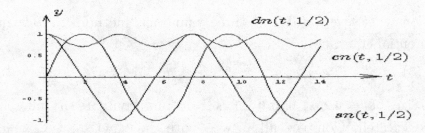

▶ **26.**

(a) Show that when the modified Euler predictor-corrector method

$$y_{n+1,p} = y_n + hf(x_n, y_n), \qquad y_{n+1,c} = y_n + \frac{h}{2}\left(f(x_{n+1}, y_{n+1,p}) + f(x_n, y_n)\right)$$

is applied to the differential equation $\frac{dy}{dx} = \lambda y$ one obtains the recursive relation

$$y_{n+1} = \left(1 + h\lambda + \frac{1}{2}(h\lambda)^2\right) y_n = G y_n.$$

(b) Show the stability region is given by the following figure.

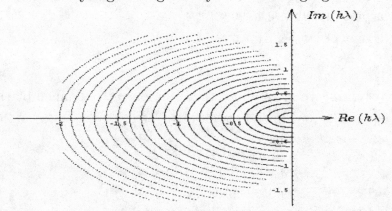

"One must learn by doing the thing; though you think you know it, you have no certainty until you try."

Sophocles (495-406)BCE

Chapter 9
Partial Differential Equations

A linear second order partial differential equation having a dependent variable u and two independent variables x, y, can be defined in terms of the linear partial differential operator $L(u)$ given by

$$L(u) = A(x,y)\frac{\partial^2 u}{\partial x^2} + 2B(x,y)\frac{\partial^2 u}{\partial x \partial y} + C(x,y)\frac{\partial^2 u}{\partial y^2} + D(x,y)\frac{\partial u}{\partial x} + E(x,y)\frac{\partial u}{\partial y} + F(x,y)u$$

where A, B, C, D, E, F are coefficients which are real valued functions of the variables x and y. These coefficients are assumed to possess second derivatives which are continuous over a region R where the solution is desired. Linear second order partial differential equations of the form $L(u) = 0$, $x, y \in R$ are called homogeneous equations and linear second order partial differential equations of the form $L(u) = G(x,y)$, $x, y \in R$ are called nonhomogeneous equations. Partial differential equation of the form

$$A(x,y)\frac{\partial^2 u}{\partial x^2} + 2B\frac{\partial^2 u}{\partial x \partial y} + C\frac{\partial^2 u}{\partial y^2} = F(x,y,u,\frac{\partial u}{\partial x},\frac{\partial u}{\partial y}) \quad x, y \in R$$

are called second order quasilinear partial differential equations. Note in the above definitions it is sometimes desirable, because of the physical problem being considered, to replace one of the variables x or y by the time variable t.

Some examples of linear second order partial differential equations are the Laplace equation

$$L(u) = \frac{\partial^2 u}{\partial x^2} + \frac{\partial^2 u}{\partial y^2} = 0 \quad u = u(x,y) \qquad x, y \in R, \tag{9.1}$$

the one-dimensional heat equation

$$L(u) = \frac{\partial u}{\partial t} - \kappa\frac{\partial^2 u}{\partial x^2} = G(x,t) \quad u = u(x,t), \quad t > 0, \quad 0 \leq x \leq L \tag{9.2}$$

where κ is a constant, and the one-dimensional wave equation

$$L(u) = \frac{\partial^2 u}{\partial t^2} - c^2\frac{\partial^2 u}{\partial x^2} = G(x,t) \quad u = u(x,t), \quad t > 0, \quad 0 \leq x \leq L \tag{9.3}$$

where c is a constant.

Canonical Forms

Associated with both quasilinear and linear second order partial differential equations are canonical forms. These are special forms that the general form assumes under certain variable changes. The canonical forms are classified by the discriminant $\Delta = B^2 - AC$ formed from the A, B, C coefficients which multiply the highest ordered derivatives in the second order linear or quasilinear partial differential equation. In the partial differential equations that we shall consider, the discriminant is assumed to have a constant sign for all x, y in a region R of interest. The partial differential equation is called parabolic if $B^2 - AC = 0$ for all $x, y \in R$, it is called hyperbolic if $B^2 - AC > 0$ for all $x, y \in R$ and it is called elliptic if $B^2 - AC < 0$ for all $x, y \in R$. For example, the Laplace equation (9.1) has discriminant $\Delta = -1 < 0$ and is one of the canonical forms associated with elliptic partial differential equations. The heat equation (9.2) with discriminant $\Delta = 0$ is one of the canonical forms associated with parabolic equations and the wave equation (9.3) with discriminant $\Delta = c^2 > 0$ is one of the canonical forms associated with hyperbolic equations.

Boundary and Initial Conditions

Boundary conditions associated with a linear second order partial differential equation

$$L(u) = G(x, y) \quad \text{for} \quad x, y \in R$$

can be written in the operator form

$$B(u) = f(x, y) \quad \text{for} \quad x, y \in \partial R,$$

where ∂R denotes the boundary of the region R and $f(x, y)$ is a given function of x and y. If the boundary operator $B(u) = u$ the boundary condition represents the dependent variable being specified on the boundary. These type of boundary conditions are called Dirichlet conditions. If the boundary operator $B(u) = \frac{\partial u}{\partial n} = \operatorname{grad} u \cdot \hat{n}$ denotes a normal derivative, then the boundary condition is that the normal derivative at each point of the boundary is being specified. These type of boundary conditions are called Neumann conditions. Neumann conditions require the boundary to be such that one can calculate the normal derivative $\frac{\partial u}{\partial n}$ at each point of the boundary of the given region R. This requires that the unit exterior normal vector $\hat{n}$ be known at each point of the boundary. If the boundary operator is a linear combination of the Dirichlet and Neumann boundary conditions, then the boundary operator has the form $B(u) = \alpha \frac{\partial u}{\partial n} + \beta u$, where α and β are constants. These type of boundary conditions are said to be of the Robin type. The partial differential equation together with a Dirichlet boundary condition is sometimes referred to as a boundary value problem of the first kind. A partial differential equation with a Neumann boundary condition is sometimes referred to as a boundary value problem of the second kind. A boundary value problem of the third kind is a partial differential equation with a Robin type boundary condition. A partial differential equation with a boundary condition of the form

$$B(u) = \begin{cases} u, & \text{for } x, y \in \partial R_1 \\ \frac{\partial u}{\partial n}, & \text{for } x, y \in \partial R_2 \end{cases} \qquad \partial R_1 \cap \partial R_2 = \phi \quad \partial R_1 \cup \partial R_2 = \partial R$$

is called a mixed boundary value problem. If time t is one of the independent variables in a partial differential equation, then a given condition to be satisfied at the time $t = 0$ is referred to as an initial condition. A partial differential equation subject to both boundary and initial conditions is called a boundary-initial value problem.

The Heat Equation

The modeling of the one-dimensional heat flow in a thin rod of length L is accomplished as follows. Denote by $u = u(x,t)$ the temperature in the rod at position x and time t having units of [°C]. Assume the cross sectional area A of the rod is constant and consider an element of volume $d\tau = A\,\Delta x$ located between the positions x and $x + \Delta x$ in the rod as illustrated.

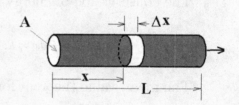

We assume the rod is of a homogeneous material and the surface of the rod is insulated so that heat flows only in the x-direction. Conservation of energy requires that the rate of change of heat energy associated with the volume element must equal the rate of heat energy flowing across the ends of the volume element plus any heat energy produced inside the element of volume. The physical properties of the rod are represented by the quantities:

c, the specific heat of the material with units [cal/g°C]

ϱ, the density of the material with units [g/cm³]

k, the thermal conductivity of the material with units [cal/cm² sec °C/cm]

A, cross sectional area with units [cm²]

The rate of change of heat stored in the volume element is given by

$$H_s = \frac{\partial}{\partial t} \int_x^{x+\Delta x} c\varrho A\, u(x,t)\, dx = \int_x^{x+\Delta x} c\varrho A\, \frac{\partial u(x,t)}{\partial t}\, dx.$$

Here $e = c\varrho u$ represents the thermal energy density of the volume element with units of [cal/cm^3]. The heat loss from the left and right ends of the volume element is found using the Fourier's law of heat flow which states that the heat flow normal to a surface is proportional to the gradient of the temperature. This heat loss can be represented

$$H_\ell = kA\left[\frac{\partial u(x+\Delta x,t)}{\partial x} - \frac{\partial u(x,t)}{\partial x}\right],$$

where k is the thermal conductivity of the material. Let $H(x,t)$ denote the heat generated within the volume element with units of [cal/cm^3], then the heat generated by a source within the volume element can be represented

$$H_g = A\int_x^{x+\Delta x} H(x,t)\, dx.$$

The conservation of energy requires that $H_s = H_\ell + H_g$ or

$$\int_x^{x+\Delta x} c\varrho A\, \frac{\partial u(x,t)}{\partial t}\, dx = kA\left[\frac{\partial u(x+\Delta x,t)}{\partial x} - \frac{\partial u(x,t)}{\partial x}\right] + A\int_x^{x+\Delta x} H(x,t)\, dx. \qquad (9.4)$$

The mean value theorem for integrals

$$\int_x^{x+\Delta x} f(x)\, dx = f(x+\theta\Delta x)\Delta x, \qquad 0 < \theta < 1$$

enables one to express the equation (9.4) in the form

$$kA\left[\frac{\partial u(x+\Delta x,t)}{\partial x} - \frac{\partial u(x,t)}{\partial x}\right] + AH(x+\theta_1\Delta x,t)\Delta x = c\varrho A\frac{\partial u(x+\theta_2\Delta x,t)}{\partial t}\Delta x. \qquad (9.5)$$

Now divide by Δx and take the limit as $\Delta x \to 0$ to obtain the heat equation

$$k\frac{\partial^2 u}{\partial x^2} + H(x,t) = c\varrho\frac{\partial u}{\partial t}, \qquad u = u(x,t), \quad 0 < x < L \qquad (9.6)$$

or

$$\frac{\partial u}{\partial t} = \kappa\frac{\partial^2 u}{\partial x^2} + Q(x,t), \quad \text{where} \quad Q = \frac{H}{c\varrho}, \qquad (9.7)$$

and $\kappa = \dfrac{k}{c\varrho}$ is known as the coefficient of thermal diffusivity with units of cm^2/sec. The following table gives approximate values for the specific heats, density, thermal conductivity and thermal diffusivity of selected materials at 100 °C.

Coefficient	Fe	Al	Cu	Ni	Zn	Ag	Au
c=heat capacity [cal/$g\,°$C]	0.117	0.230	0.101	0.120	0.097	0.058	0.032
ρ= density [g/cm^3]	7.83	2.70	8.89	8.60	7.10	10.6	19.3
k=thermal conductivity [cal/sec cm$^2\,°$C/cm]	0.107	0.490	0.908	0.138	0.262	0.089	0.703
κ=Thermal diffusivity [cm^2/sec]	1.168	0.789	1.011	0.134	0.381	0.145	1.138

A special case of equation (9.6) is when $H = 0$. One then obtains

$$k\frac{\partial^2 u}{\partial x^2} = c\varrho\,\frac{\partial u}{\partial t} \tag{9.8}$$

where $u = u(x,t)$ and k, c, ϱ are constants. This is known as the heat or diffusion equation. This type of partial differential equation arises in the study of diffusion type processes. It is classified as a parabolic equation.

An initial condition associated with the modeling of heat flow in a rod is written as $u(x,0) = f(x)$ where $f(x)$ is a prescribed initial temperature distribution over the rod. Dirichlet boundary conditions for the rod would be to specify the temperature at the ends of the rod and are written $u(0,t) = T_0$ and $u(L,t) = T_1$, where T_0 and T_1 are specified temperatures. (Recall the lateral surface of the rod is assumed to be insulated.) Boundary conditions of the Neumann type are written $-\frac{\partial u(0,t)}{\partial x} = g_0(t)$ and $\frac{\partial u(L,t)}{\partial x} = g_1(t)$ where $g_0(t)$ and $g_1(t)$ are specified functions of time t representing the heat flow across the boundary. If $g_0(t) = 0$, the boundary condition is said to be insulated so that no heat flows across the boundary. Robin type boundary conditions for the rod are expressed

$$-\alpha\frac{\partial u(0,t)}{\partial x} + \beta u(0,t) = g_0(t) \qquad \text{and} \qquad \alpha\frac{\partial u(L,t)}{\partial x} + \beta u(L,t) = g_1(t)$$

where again $g_0(t)$ and $g_1(t)$ are given functions of time t. These type of boundary conditions represent cooling or evaporation at the boundary.

Example 9-1. (Boundary-initial value problem)

As an example of a boundary-initial value problem consider the heat equation without heat sources which models the temperature distribution in a long thin rod of length L which is insulated along its length so that there is no heat loss. We must solve the partial differential equation (PDE)

$$\frac{\partial u}{\partial t} - \kappa \frac{\partial^2 u}{\partial x^2} = 0, \quad u = u(x,t), \quad t > 0, \quad 0 < x < L, \quad K \text{ constant}$$

subject to both boundary and initial conditions.

A Dirichlet boundary-initial value problem for the heat equation has the form

$$\text{PDE:} \quad \frac{\partial u}{\partial t} - \kappa \frac{\partial^2 u}{\partial x^2} = 0, \quad t > 0, \quad 0 < x < L$$
$$\text{BC:} \quad u(0,t) = T_0, \quad u(L,t) = T_L$$
$$\text{IC:} \quad u(x,0) = f(x)$$

Here the boundary conditions (BC) are the temperatures T_0 and T_L being specified at the ends of the rod for all values of the time t. The initial condition (IC) is that the initial temperature distribution $f(x)$ through the rod is being specified at time $t = 0$.

A Neumann boundary-initial value problem for the heat equation has the form

$$\text{PDE:} \quad \frac{\partial u}{\partial t} - \kappa \frac{\partial^2 u}{\partial x^2} = 0, \quad t > 0, \quad 0 < x < L$$
$$\text{BC:} \quad -\frac{\partial u(0,t)}{\partial x} = \phi_0, \quad \frac{\partial u(L,t)}{\partial x} = \phi_L$$
$$\text{IC:} \quad u(x,0) = f(x)$$

Here the unit normal vectors to the ends of the rod are $\hat{n} = \hat{i}$ at $x = L$ and $\hat{n} = -\hat{i}$ at $x = 0$ and consequently the normal derivative at the end point boundaries are

$$\frac{\partial u}{\partial n}\bigg|_{x=0} = \text{grad}\, u \cdot \hat{n}\bigg|_{x=0} = -\frac{\partial u}{\partial x}\bigg|_{x=0} \quad \text{and} \quad \frac{\partial u}{\partial n}\bigg|_{x=L} = \text{grad}\, u \cdot \hat{n}\bigg|_{x=L} = \frac{\partial u}{\partial x}\bigg|_{x=L}$$

These terms represent the temperature gradient across the boundaries. Sometimes these are referred to as heat flow across the boundary since the heat flow is proportional to the gradient of the temperature. If u represents temperature [°C], x represents distance [cm], then $\frac{\partial u}{\partial n}$ has units of [°C/cm]. Note that the condition $\frac{\partial u}{\partial n} = 0$ denotes an insulated boundary. The Neumann boundary-initial value problem specifies the temperature gradients ϕ_0, ϕ_L across the boundaries as well as specifying the initial temperature distribution within the rod.

A Robin boundary-initial value problem for the heat equation has the form

$$\text{PDE:} \qquad \frac{\partial u}{\partial t} - \kappa \frac{\partial^2 u}{\partial x^2} = 0, \quad t > 0, \quad 0 < x < L$$

$$\text{BC:} \qquad -\frac{\partial u(0,t)}{\partial x} + hu(0,t) = \psi_0, \quad \frac{\partial u(L,t)}{\partial x} + hu(L,t) = \psi_L$$

$$\text{IC:} \qquad u(x,0) = f(x)$$

The boundary conditions represent the heat loss from the boundaries with h a heat loss coefficient which is a constant and dependent upon the rod material. In terms of diffusion processes the boundary conditions represent evaporation processes ψ_0, ψ_L specified at the ends $x = 0$ and $x = L$.

■

The wave equation

An example of a hyperbolic equation is the homogeneous partial differential equation

$$\frac{\partial^2 u}{\partial t^2} - c^2 \frac{\partial^2 u}{\partial x^2} = 0 \tag{9.9}$$

where $u = u(x,t)$ and c is a constant. This is the wave equation of mathematical physics. It arises in the modeling of longitudinal and transverse wave motion. Some application areas where it arises are in the study of vibrating strings, electric and magnetic waves, and sound waves.

A mathematical model of a vibrating string is constructed by using Newton's laws and summing the forces acting on an element of string. Consider a section of string between x and $x + \Delta x$ as illustrated. Denote by $u = u(x,t)$ the string displacement [cm], at time t [sec], and introduce the additional symbols: ϱ [g/cm] to denote the lineal

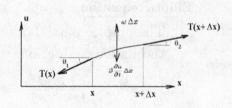

string density, $T(x)$ [dynes] the tension in the string at position x, ω [dynes/cm] an external force per unit length acting on the string. Further assume there exists a damping force proportional to the velocity $\frac{\partial u}{\partial t}$ of the string. This force is represented $\beta \frac{\partial u}{\partial t} \Delta x$ where β [dynes-sec/cm^2] denotes a linear velocity damping force per unit length. Assume there is equilibrium of forces in the horizontal direction. This requires

$$T(x + \Delta x) \cos \theta_2 = T(x) \cos \theta_1 = T_0 = \text{a constant.}$$

In the vertical direction we sum forces and apply Newton's second law to obtain

$$\varrho\Delta x\frac{\partial^2 u(x+\frac{\Delta x}{2},t)}{\partial t^2} = T(x+\Delta x)\sin\theta_2 - T(x)\sin\theta_1 + \omega\Delta x - \beta\Delta x\frac{\partial u(x+\frac{\Delta x}{2},t)}{\partial t}$$

$$= T_0(\tan\theta_2 - \tan\theta_1) + \omega\Delta x - \beta\Delta x\frac{\partial u(x+\frac{\Delta x}{2},t)}{\partial t} \qquad (9.10)$$

Note that $\tan\theta_1 = \dfrac{\partial u(x,t)}{\partial x}$ and $\tan\theta_2 = \dfrac{\partial u(x+\Delta x,t)}{\partial x}$ so that when equation (9.10) is divided by Δx and one takes the limit as Δx approaches zero, there results the equation of motion of the vibrating string

$$\varrho\frac{\partial^2 u}{\partial t^2} = \frac{\partial}{\partial x}\left(T_0\frac{\partial u}{\partial x}\right) + \omega - \beta\frac{\partial u}{\partial t}$$

In the special case $\beta = 0$, $\omega = 0$ and T_0 is constant, this reduces to the one-dimensional wave equation

$$\frac{\partial^2 u}{\partial t^2} - c^2\frac{\partial^2 u}{\partial x^2} = 0 \quad 0 < x < L, \quad t > 0 \qquad (9.11)$$

where $c^2 = T_0/\varrho$. is a constant. Here c has the units of velocity and denotes the wave speed. The wave equation (9.11) can be subjected to Dirichlet, Neumann or Robin type boundary conditions. Dirichlet conditions occur whenever one specifies the string displacements at the ends at $x = 0$ and $x = L$. Neumann conditions occur whenever the derivatives are specified at the ends of the string and Robin conditions occur when some linear combination of displacement and slope is specified at the string ends.

Elliptic equation

The elliptic partial differential equations that occur most frequently are the Laplace equation

$$\frac{\partial^2 u}{\partial x^2} + \frac{\partial^2 u}{\partial y^2} = 0 \quad\text{or}\quad \nabla^2 u = 0, \qquad u = u(x,y), \quad x,y \in R,$$

the Poisson equation

$$\frac{\partial^2 u}{\partial x^2} + \frac{\partial^2 u}{\partial y^2} = g(x,y) \quad\text{or}\quad \nabla^2 u = g(x,y) \qquad u = u(x,y), \quad x,y \in R,$$

(which is the nonhomogeneous form of Laplace's equation) and the Helmholtz equation

$$\frac{\partial^2 u}{\partial x^2} + \frac{\partial^2 u}{\partial y^2} + f(x,y)u = g(x,y) \quad\text{or}\quad \nabla^2 u + f(x,y)u = g(x,y) \qquad u = u(x,y), \quad x,y \in R.$$

These equations occur in a variety of applied disciplines.

Examine the two-dimensional heat equation

$$\frac{\partial u}{\partial t} = \kappa \left(\frac{\partial^2 u}{\partial x^2} + \frac{\partial^2 u}{\partial y^2} \right) \quad u = u(x,y,t) \quad x,y \in R$$

under steady state conditions where $\frac{\partial u}{\partial t} = 0$. If u does not change with time, then $u = u(x,y)$ and so the steady state heat equation is described by the Laplace equation subject to boundary conditions of the Dirichlet, Neumann or Robin type. Additional areas where the Laplace, Poisson and Helmholtz equation occur are potential theory, the study of torsion in cylindrical bars, and in the study of harmonic functions.

Numerical solution of the Laplace equation

Consider the boundary value problem to solve either the Laplace equation

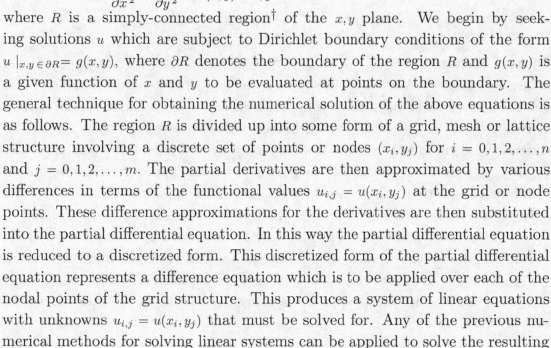

$$\nabla^2 u = \frac{\partial^2 u}{\partial x^2} + \frac{\partial^2 u}{\partial y^2} = 0, \quad x,y \in R$$

or the Poisson equation

$$\nabla^2 u = \frac{\partial^2 u}{\partial x^2} + \frac{\partial^2 u}{\partial y^2} = h(x,y) \quad x,y \in R$$

where R is a simply-connected region[†] of the x,y plane. We begin by seeking solutions u which are subject to Dirichlet boundary conditions of the form $u\mid_{x,y \in \partial R} = g(x,y)$, where ∂R denotes the boundary of the region R and $g(x,y)$ is a given function of x and y to be evaluated at points on the boundary. The general technique for obtaining the numerical solution of the above equations is as follows. The region R is divided up into some form of a grid, mesh or lattice structure involving a discrete set of points or nodes (x_i, y_j) for $i = 0, 1, 2, \ldots, n$ and $j = 0, 1, 2, \ldots, m$. The partial derivatives are then approximated by various differences in terms of the functional values $u_{i,j} = u(x_i, y_j)$ at the grid or node points. These difference approximations for the derivatives are then substituted into the partial differential equation. In this way the partial differential equation is reduced to a discretized form. This discretized form of the partial differential equation represents a difference equation which is to be applied over each of the nodal points of the grid structure. This produces a system of linear equations with unknowns $u_{i,j} = u(x_i, y_j)$ that must be solved for. Any of the previous numerical methods for solving linear systems can be applied to solve the resulting

---

[†] A simply-connected region is such that any simple closed curve within the region can be continuously shrunk to a point without leaving the region.

system of equations. Having obtained the solution values $u_{i,j}$, these discrete values can be used to approximate the true solution $u = u(x, y)$ of the given partial differential equation. If values of $u = u(x, y)$ are desired at nonlattice points, then one must use interpolation to obtain the approximated values.

Sometimes it is assumed that the addition of more lattice or mesh points, to produce a finer grid structure, will give an improved approximation to the true solution. However, this is not necessarily a true statement. As the grid structure gets finer and finer there are produce many more nodal points to solve for and the resulting system of equations can become quite large. This requires more computation to calculate the solution and consequently there is the increased risk of round off error build up. Somewhere between a coarse and fine grid structure lies the optimal grid structure which minimizes both approximation error and round off error in computing the solution.

In general, given a partial differential equation, where the solution is desired over a region R which is rectangular in shape, one can divide the region into a rectangular grid by defining a Δx and Δy spacing given by

$$ h = \Delta x = \frac{b-a}{n}, \qquad k = \Delta y = \frac{d-c}{m}. $$

One can then write $x_i = x_0 + ih$ and $y_j = y_0 + jk$ for $i = 1, 2, \ldots, n$ and $j = 1, 2, \ldots, m$ where $x_0 = a$, $y_0 = c$. We use the notation $u_{i,j} = u(x_i, y_j)$ to denote the value of u at the point (x_i, y_j) and then develop approximations to the various partial derivatives $\frac{\partial u}{\partial x}$, $\frac{\partial u}{\partial y}$, $\frac{\partial^2 u}{\partial x^2}$, $\frac{\partial^2 u}{\partial x \partial y}$ and $\frac{\partial^2 u}{\partial y^2}$ and higher derivatives, occurring in the partial differential equation. We evaluate the given partial differential equation at the point (x_i, y_j) and then substitute the partial derivative approximations to obtain a difference equation. The various partial derivative approximations involve the step sizes h and k and combinations of the grid points in the neighborhood of (x_i, y_j). These derivative approximations can be developed by manipulation of the various Taylor series expansion of a function of two variables about the point (x_i, y_j). It is left as an exercise to derive these various derivative approximations. A variety of these derivative approximations can be found in the exercises at the end of this chapter.

For the Laplace equation we begin with a 5-point formula to approximate the derivatives $\frac{\partial^2 u}{\partial x^2}$ and $\frac{\partial^2 u}{\partial y^2}$ and leave more complicated approximations for the exercises.

Example 9-2. (5-Point formula)

Approximate the derivatives $\dfrac{\partial^2 u}{\partial x^2}$ and $\dfrac{\partial^2 u}{\partial y^2}$ at the point (x_i, y_j) using the value $u_{i,j} = u(x_i, y_j)$ and one or more of the four nearest neighbor points $u_{i+1,j},\ u_{i-1,j},\ u_{i,j+1},\ u_{i,j-1}$

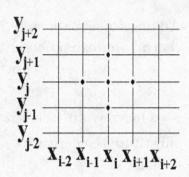

Solution: We use the operator notation

$$D_x u = \frac{\partial u}{\partial x}, \quad D_x^2 = \frac{\partial^2 u}{\partial x^2}, \quad D_x^3 u = \frac{\partial^3 u}{\partial x^3}, \quad \cdots$$

$$D_y u = \frac{\partial u}{\partial y}, \quad D_y^2 u = \frac{\partial^2 u}{\partial y^2}, \quad D_y^3 u = \frac{\partial^3 u}{\partial y^3}, \quad \cdots$$

and write the two-dimensional Taylor series expansion of a function $u(x, y)$ about the point (x_i, y_j) in terms of operators having the form $(\alpha D_x + \beta D_y)^n u$ where all derivatives are to be evaluated at the point (x_i, y_j). The two-dimensional Taylor series expansion can be represented in the form

$$u(x_i + \alpha, y_j + \beta) = u(x_i, y_j) + (\alpha D_x + \beta D_y)\, u + (\alpha D_x + \beta D_y)^2 \frac{u}{2!} + \cdots \tag{9.12}$$

where all derivatives are understood to be evaluated at the point (x_i, y_j). Two special cases of equation (9.12) are when $\beta = 0,\ \alpha = h$ and $\beta = 0,\ \alpha = -h$. These values of α and β are selected to obtain the Taylor series expansions

$$u_{i+1,j} = u_{ij} + h D_x u + \frac{h^2}{2!} D_x^2 u + \frac{h^3}{3!} D_x^3 u + \frac{h^4}{4!} D_x^4 u + \cdots$$

$$u_{i-1,j} = u_{ij} - h D_x u + \frac{h^2}{2!} D_x^2 u - \frac{h^3}{3!} D_x^3 u + \frac{h^4}{4!} D_x^4 u - \cdots \tag{9.13}$$

Now add the equations (9.13) and rearrange terms to obtain the second derivative central approximation in the x-direction

$$\frac{\partial^2 u}{\partial x^2} = \frac{1}{h^2}\left(u_{i+1,j} - 2u_{i,j} + u_{i-1,j}\right) + \mathcal{O}(h^2) \tag{9.14}$$

In a similar manner we construct the two special cases of equation (9.12) where $\alpha = 0,\ \beta = k$ and $\alpha = 0,\ \beta = -k$ to obtain the equations

$$u_{i,j+1} = u_{i,j} + k D_y u + \frac{k^2}{2!} D_y^2 u + \frac{k^3}{3!} D_y^3 u + \frac{k^4}{4!} D_y^4 u + \cdots$$

$$u_{i,j-1} = u_{i,j} - k D_y u + \frac{k^2}{2!} D_y^2 u - \frac{k^3}{3!} D_y^3 u + \frac{k^4}{4!} D_y^4 u - \cdots \tag{9.15}$$

Adding the equations (9.15) and rearranging terms gives the second derivative central approximation in the y-direction

$$\frac{\partial^2 u}{\partial y^2} = \frac{1}{k^2}\left(u_{i,j+1} - 2u_{i,j} + u_{i,j-1}\right) + \mathcal{O}(k^2) \tag{9.16}$$

One way to discretized the Laplace equation $\nabla^2 u = \dfrac{\partial^2 u}{\partial x^2} + \dfrac{\partial^2 u}{\partial y^2} = 0$ is to substitute the derivative approximations from the equations (9.16) and (9.14) to obtain

$$\nabla^2 u(x_i, y_j) = \frac{1}{h^2}\left(u_{i+1,j} - 2u_{i,j} + u_{i-1,j}\right) + \frac{1}{k^2}\left(u_{i,j+1} - 2u_{i,j} + u_{i,j-1}\right) = 0. \tag{9.17}$$

In the special case of equal spacing in the x and y-directions one can set $k = h$ to obtain the simpler approximation

$$\nabla^2 u\Big|_{(x_i, y_j)} = \frac{\partial^2 u}{\partial x^2} + \frac{\partial^2 u}{\partial y^2}\Big|_{(x_i, y_j)} = \frac{1}{h^2}\left(u_{i,j+1} + u_{i,j-1} + u_{i-1,j} + u_{i,j+1} - 4u_{i,j}\right) = 0 \tag{9.18}$$

which can be represented by the pictorial operator

$$\nabla^2 u\Big|_{(x_i, y_j)} = \frac{1}{h^2}\left\{\begin{matrix} & 1 & \\ 1 & -4 & 1 \\ & 1 & \end{matrix}\right\} u_{i,j} = 0 \tag{9.19}$$

indicating the weight assigned to the central value $u_{i,j}$ and also the weights assigned to the dependent variable evaluated at the nearest neighbor points surrounding (x_i, y_j). This pictorial operator is sometimes referred to as a computational molecule. Computational molecules can become complicated in the vicinity of an irregular boundary.

Example 9-3. (Irregular spacing)
Consider the case of irregular spacing in the neighborhood of point (x_i, y_j) near an irregular boundary where the spacing between grid points changes as illustrated. Here we use α_1, α_2, α_3, α_4 as scale factors of the step size h to represent the nonuniform spacing.

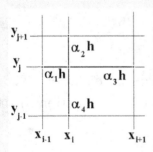

One can use forward difference approximations to approximate the first partial derivatives in the x and y-directions at the midpoints of the x and y spacing

$$\left.\frac{\partial u}{\partial x}\right|_{(x_{i-1/2}, y_j)} = \frac{u_{i,j} - u_{i-1,j}}{\alpha_1 h} \qquad \left.\frac{\partial u}{\partial x}\right|_{(x_{i+1/2}, y_j)} = \frac{u_{i+1,j} - u_{i,j}}{\alpha_3 h}$$

to obtain

$$\left.\frac{\partial u}{\partial y}\right|_{(x_i, y_{j+1/2})} = \frac{u_{i,j+1} - u_{i,j}}{\alpha_2 h} \qquad \left.\frac{\partial u}{\partial y}\right|_{(x_i, y_{j-1/2})} = \frac{u_{i,j} - u_{i,j-1}}{\alpha_4 h} \tag{9.20}$$

The first derivative approximations can now be used to construct the second derivative approximations

$$
\frac{\partial^2 u}{\partial x^2}\bigg|_{(x_i,y_j)} = \frac{\frac{\partial u}{\partial x}\big|_{(x_{i+1/2},y_j)} - \frac{\partial u}{\partial x}\big|_{(x_{i-1/2},y_j)}}{\frac{1}{2}(\alpha_1+\alpha_3)h} = \frac{2}{h^2}\left[\frac{u_{i+1,j}-u_{i,j}}{\alpha_3(\alpha_1+\alpha_3)} - \frac{u_{i,j}-u_{i-1,j}}{\alpha_1(\alpha_1+\alpha_3)}\right]
$$

$$
\frac{\partial^2 u}{\partial y^2}\bigg|_{(x_i,y_j)} = \frac{\frac{\partial u}{\partial y}\big|_{(x_i,y_{j+1/2})} - \frac{\partial u}{\partial y}\big|_{(x_i,y_{j-1/2})}}{\frac{1}{2}(\alpha_2+\alpha_4)h} = \frac{2}{h^2}\left[\frac{u_{i,j+1}-u_{i,j}}{\alpha_2(\alpha_2+\alpha_4)} - \frac{u_{i,j}-u_{i,j-1}}{\alpha_4(\alpha_2+\alpha_4)}\right]
$$

(9.21)

The Laplacian can then be discretized by substituting these derivative approximations to obtain

$$
\nabla^2 u = \frac{2}{h^2}\left[\frac{u_{i-1,j}}{\alpha_1(\alpha_1+\alpha_3)} + \frac{u_{i,j+1}}{\alpha_2(\alpha_2+\alpha_4)} + \frac{u_{i+1,j}}{\alpha_3(\alpha_1+\alpha_3)} + \frac{u_{i,j-1}}{\alpha_4(\alpha_2+\alpha_4)} - \left(\frac{1}{\alpha_1\alpha_3} + \frac{1}{\alpha_2\alpha_4}\right)u_{i,j}\right]
$$

which produces the Laplace equation computational molecule

$$
\nabla^2 u = \frac{\partial^2 u}{\partial x^2} + \frac{\partial^2 u}{\partial y^2} = \frac{2}{h^2}\left\{\begin{array}{ccc} & \frac{1}{\alpha_2(\alpha_2+\alpha_4)} & \\[4pt] \frac{1}{\alpha_1(\alpha_1+\alpha_3)} & -\left(\frac{1}{\alpha_1\alpha_3}+\frac{1}{\alpha_2\alpha_4}\right) & \frac{1}{\alpha_3(\alpha_1+\alpha_3)} \\[4pt] & \frac{1}{\alpha_4(\alpha_2+\alpha_4)} & \end{array}\right\} u_{i,j} = 0 \quad (9.22)
$$

Observe that when all the scale factors are unity, then the computational molecule given by equation (9.22) reduces to the previous computational molecule of equation (9.19) associated with equal spacing. Irregular shaped regions with nonuniform grid spacing along a boundary have scale factors α which are constantly changing. Consequently irregular shaped regions usually require much more additional work in setting up the equations to be solved.

∎

Example 9-4. (Boundary conditions.)

For illustrative purposes we consider a point on the right boundary of a rectangular region where $x = x_n$. You will have to modify the following discussions to apply to the other sides of the rectangle.

Dirichlet boundary conditions have the form $u_{n,j} = u(x_n, y_j) = f(x_n, y_j)$ where $f(x,y)$ is a given function. For Dirichlet boundary conditions one would apply the computational molecule given by equation (9.19) to all the interior points of the region R. If the values at the boundary points are specified, then one would not use the boundary points as the center point of the computational molecule as this would bring into play unknowns $u_{n+1,j}$ exterior to the boundary.

Note that if the functional values are specified at a boundary, then the values $u(x_n, y_j) = u_{n,j} = f(x_n, y_j)$ are known. Observe that when you apply the computational molecule of equation (9.19) to the center point (x_{n-1}, y_j) there results the equation

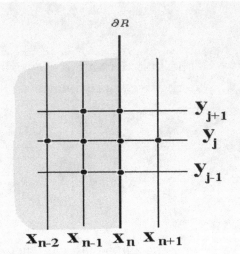

$$u_{n-1,j+1} + u_{n,j} + u_{n-1,j-1} + u_{n-2,j} - 4u_{n-1,j} = 0.$$

Into this equation one would replace the boundary value $u_{n,j}$ by its specified value. This is done to all equations where a boundary point occurs. The known values are then moved to the right-hand side of the resulting system of equations.

Note that in order for the Laplace partial differential equation to have a solution with a Neumann boundary condition, it is required that the line integral of the normal derivative around the boundary be zero. This condition requires that $\int_{\partial R} \frac{\partial u}{\partial n}\, ds = 0$. If this condition is not satisfied, the system of equations that result from applying the computational molecule to all applicable lattice points, will be singular.

If derivative boundary conditions $\frac{\partial u}{\partial n} = F(x, y)$, for $x, y \in \partial R$ are required to be satisfied along a boundary line, say $x = x_n$ for example, one can extend the grid lattice in the x-direction beyond the boundary. One can then use a central difference approximation to write

$$\frac{\partial u}{\partial n}\bigg|_{(x_n, y_j)} = \frac{\partial u}{\partial x}\bigg|_{(x_n, y_j)} = \frac{u_{n+1,j} - u_{n-1,j}}{2h} = F(x_n, y_j), \tag{9.23}$$

which is accurate to $\mathcal{O}(h^2)$. Neumann boundary conditions require that the computational molecule of equation (9.19) be applied to boundary points. As an example, applying this computational molecule to the boundary point (x_n, y_j) there results the equation

$$u_{n,j+1} + u_{n-1,j-1} + u_{n,j-1} + u_{n+1,j} - 4u_{n,j} = 0. \tag{9.24}$$

Solve the boundary condition equation (9.23) for $u_{n+1,j}$ to obtain

$$u_{n+1,j} = u_{n-1,j} + 2hF(x_n, y_j)$$

and observe that this value will simplify the equation (9.24) to the form

$$u_{n,j+1} + 2u_{n-1,j} + u_{n,j-1} + 2hF(x_n, y_j) - 4u_{n,j} = 0.$$

In this way the unknown value $u_{n+1,j}$ is eliminated by making use of the known boundary condition expressed in terms of one of the interior points.

The substitution techniques associated with boundary conditions of the Dirichlet and Neumann types can also be applied to the Robin type boundary condition

$$\alpha \frac{\partial u}{\partial n} + \beta u = F(x, y).$$

Treat the normal derivative at the boundary in the same way it was treated in the Neumann case and treat the value of the dependent variable at the boundary in the same way it was treated in the Dirichlet case.

Similar modifications to the boundary terms must be applied to the top, left-hand side and bottom of a rectangular region. These modifications produce specific values for certain unknowns occurring in the equations resulting from an application of the computational molecule near boundaries. Additional work is required for irregular boundaries. The ideas are the same but the computational molecules change in the neighborhood of a boundary.

∎

Example 9-5. (Dirichlet problem.)

Find the steady state temperature distribution $u = u(x, y)$ satisfying

$$\nabla^2 u = \frac{\partial^2 u}{\partial x^2} + \frac{\partial^2 u}{\partial y^2} = 0, \qquad x, y \in R$$

where R is the square region $0 \le x \le 1, 0 \le y \le 1$ which is subject to the boundary conditions

$$u(x, 1) = 100x, \quad 0 \le x \le 1$$
$$u(x, 0) = 0, \quad 0 \le x \le 1$$
$$u(0, y) = 0, \quad 0 \le y \le 1$$
$$u(1, y) = 100y, \quad 0 \le y \le 1$$

Solution: We divide the area inside the square into a grid with equal spaced Δx and Δy distances between the node points of the grid. Let $\Delta x = h = 1/4$ and $\Delta y = k = 1/4$ so that the interior nodes have the coordinates (x_i, y_j) given by

$x_i = ih$ for $i = 1, 2, 3$ and $y_j = jh$ for $j = 1, 2, 3$. The boundary points of the grid are given and so one can verify the specified temperatures are

$$u(0, y_1) = 0 = T_1 \quad u(x_1, 1) = 25 = T_4 \quad u(1, y_1) = 25 = T_7 \quad u(x_1, 0) = 0 = T_{12}$$

$$u(0, y_2) = 0 = T_2 \quad u(x_2, 1) = 50 = T_5 \quad u(1, y_2) = 50 = T_8 \quad u(x_2, 0) = 0 = T_{11}$$

$$u(0, y_3) = 0 = T_3 \quad u(x_3, 1) = 75 = T_6 \quad u(1, y_3) = 75 = T_9 \quad u(x_3, 0) = 0 = T_{10}$$

The situation is illustrated in the figure 9-1.

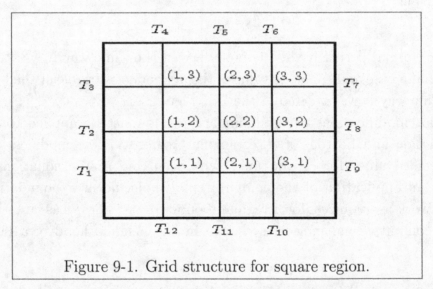

Figure 9-1. Grid structure for square region.

We now apply the computational molecule associated with the partial differential equation to each interior point of the grid. This transforms the given partial differential equation into a system of equations to be solved. Start at the bottom row of the grid and move left to right applying the computational molecule of equation (9.19). By creating a pattern to our movement over the grid we create a pattern to the resulting equations.

nodal point	Resulting Equation
(x_1, y_1)	$T_1 + u_{1,2} + u_{2,1} + T_{12} - 4u_{1,1} = 0$
(x_2, y_1)	$u_{1,1} + u_{2,2} + u_{3,1} + T_{11} - 4u_{2,1} = 0$
(x_3, y_1)	$u_{2,1} + u_{3,2} + T_9 + T_{10} - 4u_{3,1} = 0$

(9.25)

Now move to the next row and apply the computational molecule to the interior nodes to obtain

nodal point	Resulting Equation	
(x_1, y_2)	$T_2 + u_{1,3} + u_{2,2} + u_{1,1} - 4u_{1,2} = 0$	
(x_2, y_2)	$u_{1,2} + u_{2,3} + u_{3,2} + u_{2,1} - 4u_{2,2} = 0$	(9.26)
(x_3, y_2)	$u_{2,2} + u_{3,3} + T_8 + u_{3,1} - 4u_{3,2} = 0$	

Now move up to the next higher row and apply the computational molecule to the interior nodes to obtain

nodal point	Resulting Equation	
(x_1, y_3)	$T_3 + T_4 + u_{2,3} + u_{1,2} - 4u_{1,3} = 0$	
(x_2, y_3)	$u_{1,3} + T_5 + u_{3,3} + u_{2,2} - 4u_{2,3} = 0$	(9.27)
(x_3, y_3)	$u_{2,3} + T_6 + T_7 + u_{3,2} - 4u_{3,3} = 0$	

The equations (9.25), (9.26) and (9.27) can be written in the matrix form

$$
\begin{bmatrix}
-4 & 1 & 0 & 1 & 0 & 0 & 0 & 0 & 0 \\
1 & -4 & 1 & 0 & 1 & 0 & 0 & 0 & 0 \\
0 & 1 & -4 & 0 & 0 & 1 & 0 & 0 & 0 \\
1 & 0 & 0 & -4 & 1 & 0 & 1 & 0 & 0 \\
0 & 1 & 0 & 1 & -4 & 1 & 0 & 1 & 0 \\
0 & 0 & 1 & 0 & 1 & -4 & 0 & 0 & 1 \\
0 & 0 & 0 & 1 & 0 & 0 & -4 & 1 & 0 \\
0 & 0 & 0 & 0 & 1 & 0 & 1 & -4 & 1 \\
0 & 0 & 0 & 0 & 0 & 1 & 0 & 1 & -4
\end{bmatrix}
\begin{bmatrix}
u_{1,1} \\ u_{2,1} \\ u_{3,1} \\ u_{1,2} \\ u_{2,2} \\ u_{3,2} \\ u_{1,3} \\ u_{2,3} \\ u_{3,3}
\end{bmatrix}
=
\begin{bmatrix}
-T_1 - T_{12} \\ -T_{11} \\ -T_9 - T_{10} \\ -T_2 \\ 0 \\ -T_8 \\ -T_3 - T_4 \\ -T_5 \\ -T_6 - T_7
\end{bmatrix}
$$

Using any of the solution techniques previously considered we find the solution of this system produces the temperatures

$$u_{1,3} = 75/4 \qquad u_{2,3} = 75/2 \qquad u_{3,3} = 225/4$$

$$u_{1,2} = 50/4 \qquad u_{2,2} = 50/2 \qquad u_{3,2} = 150/4$$

$$u_{1,1} = 25/4 \qquad u_{2,1} = 25/2 \qquad u_{3,1} = 75/4$$

In this simple example we have used only a 3×3 interior grid to obtain 9-equations in 9-unknowns. Theoretically one could construct a $N \times N$ grid and obtain N^2-equations in N^2-unknowns. The larger the value for N the more calculations are needed and there will be more round off error the larger the value for N.

Note also that in solving for the steady state temperature the thermal properties of the material are not needed. Also note the Laplace equation is independent of scale and the step size h did not enter into the calculations

380

because we selected equal spacing in both the x and y directions. However, with unequal spacings in the x and y directions, the step size h in the x-direction and step size k in the y-direction will enter into the computational molecule as one can see by examining the equation (9.17). As a final note, observe that the corner temperatures do not enter into the resulting equations.

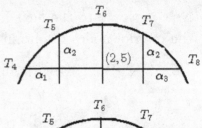

Example 9-6.   (Dirichlet problem.)
Find the steady state temperature distribution $u = u(x,y)$ satisfying

$$\nabla^2 u = \frac{\partial^2 u}{\partial x^2} + \frac{\partial^2 u}{\partial y^2} = 0, \qquad x, y \in R$$

where R is the square region defined by $0 \le x \le 1, 0 \le y \le 1$ with semicircle of radius $1/2$ attached to the top of the square. The region R is subject to the boundary conditions

$$u(x,0) = 0, \qquad 0 \le x \le 1$$
$$u(0,y) = 100y, \qquad 0 \le y \le 1$$
$$u(1,y) = 100y, \quad 0 \le y \le 1$$

with the boundary of the semicircle is maintained at a temperature of 100 degrees.

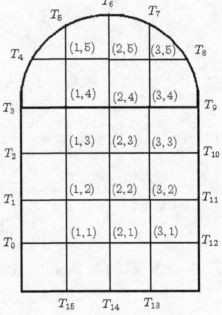

Solution:
We proceed exactly as we did in the previous problem and construct a grid for the region using an equal step size in the x and y-directions of $h = k = 1/4$. It is readily verified that the boundary temperatures are given by

$$T_0 = 25, \; T_1 = 50, \; T_2 = 75, \; T_3 = 100$$

$$T_4 = T_5 = T_6 = T_7 = T_8 = 100$$

$$T_9 = 100, \; T_{10} = 75, \; T_{11} = 50, \; T_{12} = 25$$

$$T_{13} = T_{14} = T_{15} = 0$$

The geometry of the problem produces interior nodal points at $(1,5)$ and $(3,5)$ where the grid spacing must be modified by scale factors α_1, α_2 and α_3 to obtain the position of the boundary points. We use the appropriate computational molecule applied to each of the interior points to replace the given partial differential equation by a system of equations.

Nodal Point	Resulting equation
	$T_0 + u_{1,2} + u_{2,1} + T_{15} - 4u_{1,1} = 0$
(x_1, y_1)	$u_{11} + u_{22} + u_{31} + T_{14} - 4u_{21} = 0$
(x_2, y_1)	$u_{21} + u_{3,2} + T_{12} + T_{13} - 4u_{3,1} = 0$
(x_3, y_1)	$T_1 + u_{1,3} + u_{2,2} + u_{1,1} - 4u_{1,2} = 0$
(x_1, y_2)	$u_{1,2} + u_{2,3} + u_{3,2} + u_{2,1} - 4u_{2,2} = 0$
(x_2, y_2)	$u_{2,2} + u_{3,3} + T_{11} + u_{3,1} - 4u_{3,2} = 0$
(x_3, y_2)	$T_2 + u_{1,4} + u_{2,3} + u_{1,2} - 4u_{1,3} = 0$
(x_1, y_3)	$u_{1,3} + u_{2,4} + u_{3,3} + u_{2,2} - 4u_{2,3} = 0$
(x_2, y_3)	$u_{2,3} + u_{3,4} + T_{10} + u_{3,2} - 4u_{3,3} = 0$
(x_3, y_3)	$T_3 + u_{1,5} + u_{2,4} + u_{1,3} - 4u_{1,4} = 0$
(x_1, y_4)	$u_{1,4} + u_{2,5} + u_{3,4} + u_{2,3} - 4u_{2,4} = 0$
(x_2, y_4)	$u_{2,4} + u_{3,5} + T_9 + u_{33} - 4u_{3,4} = 0$
(x_3, y_4)	
(x_1, y_5)	$\frac{1}{\alpha_1(\alpha_1+1)}T_4 + \frac{1}{\alpha_2(\alpha_2+1)}T_5 + \frac{1}{\alpha_1+1}u_{2,5} + \frac{1}{\alpha_2+1}u_{1,4} - \left(\frac{1}{\alpha_1}+\frac{1}{\alpha_2}\right)u_{1,5} = 0$
(x_2, y_5)	$u_{1,5} + T_6 + u_{3,5} + u_{2,4} - 4u_{2,5} = 0$
(x_3, y_5)	$\frac{1}{1+\alpha_3}u_{2,5} + \frac{1}{\alpha_2(1+\alpha_2)}T_7 + \frac{1}{\alpha_3(1+\alpha_3)}T_8 + \frac{1}{\alpha_2+1}u_{3,4} - \left(\frac{1}{\alpha_3}+\frac{1}{\alpha_2}\right)u_{3,5} = 0$

This system of equations can also be written in the block matrix form

$$\begin{bmatrix} A & I & 0 & 0 & 0 \\ I & A & I & 0 & 0 \\ 0 & I & A & I & 0 \\ 0 & 0 & I & A & I \\ 0 & 0 & 0 & B & C \end{bmatrix} \begin{bmatrix} U_1 \\ U_2 \\ U_3 \\ U_4 \\ U_5 \end{bmatrix} = \begin{bmatrix} V_1 \\ V_2 \\ V_3 \\ V_4 \\ V_5 \end{bmatrix}$$

where

$$A = \begin{bmatrix} -4 & 1 & 0 \\ 1 & -4 & 1 \\ 0 & 1 & -4 \end{bmatrix}, \quad I = \begin{bmatrix} 1 & 0 & 0 \\ 0 & 1 & 0 \\ 0 & 0 & 1 \end{bmatrix}, \quad 0 = \begin{bmatrix} 0 & 0 & 0 \\ 0 & 0 & 0 \\ 0 & 0 & 0 \end{bmatrix}$$

$$B = \begin{bmatrix} \frac{1}{\alpha_2+1} & 0 & 0 \\ 0 & 1 & 0 \\ 0 & 0 & \frac{1}{\alpha_2+1} \end{bmatrix}, \quad C = \begin{bmatrix} -\left(\frac{1}{\alpha_1}+\frac{1}{\alpha_2}\right) & \frac{1}{\alpha_1+1} & 0 \\ 1 & -4 & 1 \\ 0 & \frac{1}{\alpha_3+1} & -\left(\frac{1}{\alpha_3}+\frac{1}{\alpha_2}\right) \end{bmatrix}$$

with

$$U_1 = \begin{bmatrix} u_{1,1} \\ u_{2,1} \\ u_{3,1} \end{bmatrix}, \quad U_2 = \begin{bmatrix} u_{1,2} \\ u_{2,2} \\ u_{3,2} \end{bmatrix}, \quad U_3 = \begin{bmatrix} u_{1,3} \\ u_{2,3} \\ u_{3,3} \end{bmatrix}, \quad U_4 = \begin{bmatrix} u_{1,4} \\ u_{2,4} \\ u_{3,4} \end{bmatrix}, \quad U_5 = \begin{bmatrix} u_{1,5} \\ u_{2,5} \\ u_{3,5} \end{bmatrix}$$

and the right-hand side of the system is given by

$$V_1 = \begin{bmatrix} -T_0 - T_5 \\ -T_{14} \\ -T_{12} - T_{13} \end{bmatrix}, \quad V_2 = \begin{bmatrix} -T_1 \\ 0 \\ -T_{11} \end{bmatrix}, \quad V_3 = \begin{bmatrix} -T_2 \\ 0 \\ -T_{10} \end{bmatrix},$$

$$V_4 = \begin{bmatrix} -T_3 \\ 0 \\ -T_9 \end{bmatrix}, \quad V_5 = \begin{bmatrix} -\frac{1}{\alpha_1(\alpha_1+1)}T_4 - \frac{1}{\alpha_2(\alpha_2+1)}T_5 \\ -T_0 \\ -\frac{1}{\alpha_2(\alpha_2+1)}T_7 - \frac{1}{\alpha_3(\alpha_3+1)}T_8 \end{bmatrix}$$

From the geometry of the problem one can verify the scaling parameters are given by $\alpha_1 = \alpha_2 = \alpha_3 = 0.732051 = \sqrt{3} - 1$ and that the resulting system of equations has the solution set illustrated in the figure 9-2.

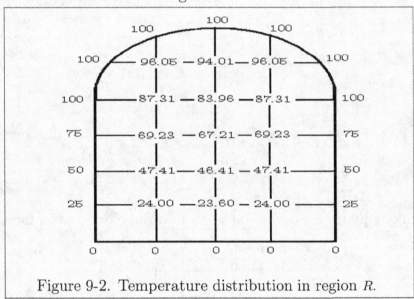

Figure 9-2. Temperature distribution in region R.

Numerical solution of the heat equation

Consider the initial-boundary value problem to solve for the temperature distribution $u = u(x,t)$ in a long thin laterally insulated rod where the temperature is governed by the partial differential equation

$$\frac{\partial^2 u}{\partial x^2} = \alpha^2 \frac{\partial u}{\partial t} \tag{9.28}$$

over the solution domain $0 < x < L, \quad t > 0$. Here $\alpha^2 = 1/\kappa$ is a constant. The solution is subject to the boundary conditions

$$u(0,t) = T_0 \text{ and } u(L,t) = T_L$$

and initial condition $u(0,t) = f(x)$.

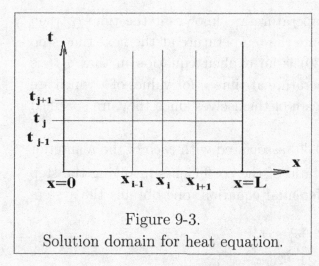

Figure 9-3.

Solution domain for heat equation.

The solution domain is divided up into some type of lattice or grid structure by selecting a step size in the x-direction, $h = \Delta x = L/(N+1)$ and a step size $k = \Delta t$ in the time direction as illustrated. There are $N + 1$ divisions in order to produce N discrete interior points for x. The coordinates (x_i, t_j) within the grid structure are determined from the relations

$$x_i = ih \qquad \text{and} \qquad t_j = jk$$

for $i = 0, 1, 2, \ldots, N+1$ and $j = 0, 1, 2, \ldots$. Denote the value of the temperature at the coordinates (x_i, t_j) by the notation $u_i^j = u(x_i, t_j)$ where the subscript denotes the x-position and the superscript denotes the time.

The partial differential equation can be discretized by substituting difference approximations for the derivatives. Note that different types of difference approximations can produce various types of computational procedures. An explicit computational method is produced using the difference approximations

$$\frac{\partial^2 u}{\partial x^2} = \frac{u_{i+1}^j - 2u_i^j + u_{i-1}^j}{h^2} + \mathcal{O}(h^2)$$
$$\frac{\partial u}{\partial t} = \frac{u_i^{j+1} - u_i^j}{k} + \mathcal{O}(k)$$

$$(9.29)$$

Substituting the derivative approximations from the equations (9.29) into the partial differential equation (9.28) gives

$$\frac{u_{i+1}^j - 2u_i^j + u_{i-1}^j}{h^2} = \alpha^2 \left(\frac{u_i^{j+1} - u_i^j}{k} \right).$$

Now rearrange terms and solve for u_i^{j+1} and show there results the computational procedure

$$u_i^{j+1} = \frac{k}{\alpha^2 h^2}(u_{i+1}^j + u_{i-1}^j) + \left(1 - \frac{2k}{\alpha^2 h^2} \right) u_i^j$$

$$(9.30)$$

Observe that if the values of the temperature are known at the time t_j, then the equation (9.30) tells how to compute the temperature at the next time step t_{j+1} at position x_i. The equation (9.30) is an explicit equation in that u_i^{j+1} is calculated from known values of temperature at time t_j for values of x which are interior points. The end points take care of themselves since they are specified by the boundary conditions.

One can write out the equation (9.30) associated with each of the N interior points. Letting $r = \dfrac{k}{\alpha^2 h^2}$ denote the ratio of the coefficients involving the step sizes and constants of the partial differential equation, one obtains the system of equations

$$u_1^{j+1} = (1 - 2r)u_1^j + ru_2^j$$
$$u_2^{j+1} = ru_1^j + (1 - 2r)u_2^j + ru_3^j$$
$$\vdots$$
$$u_N^{j+1} = ru_{N-1}^j + (1 - 2r)u_N^j$$

which can be written in the tridiagonal matrix form

$$
\begin{bmatrix} u_1^{j+1} \\ u_2^{j+1} \\ \vdots \\ u_{N-1}^{j+1} \\ u_N^{j+1} \end{bmatrix}
=
\begin{bmatrix}
(1-2r) & r & & & \\
r & (1-2r) & r & & \\
& & \ddots & & \\
& & r & (1-2r) & r \\
& & & r & (1-2r)
\end{bmatrix}
\begin{bmatrix} u_1^{j} \\ u_2^{j} \\ \vdots \\ u_{N-1}^{j} \\ u_N^{j} \end{bmatrix}
\tag{9.31}
$$

which can also be written in the simpler matrix form

$$u^{j+1} = Au^j \tag{9.32}$$

The vector u^0 is the vector of initial values at time $t = 0$. These values are known and so one can start calculating in the time direction to obtain

$$u^1 = Au^0$$
$$u^2 = Au^1 = A^2 u^0$$
$$u^3 = Au^2 = A^3 u^0$$
$$\vdots$$
$$u^n = Au^{n-1} = A^n u^0$$

It has been demonstrated in chapter 3 that for stability the eigenvalues of the matrix A must be less than one in absolute value. In chapter 6 we found the eigenvalues of the matrix A are given by

$$\lambda_n = 1 - 4r\sin^2\frac{n\pi}{2(N+1)} \quad \text{for} \quad n = 1, 2, \ldots, N. \tag{9.33}$$

In order to have stability for the explicit method it is required that the ratio r be selected to satisfy the inequalities

$$-1 \le 1 - 4r\sin^2\beta \le 1 \quad \text{where} \quad \beta = \frac{n\pi}{2(N+1)}. \tag{9.34}$$

If $1 - 4r\sin^2\beta \le 1$, then $4r\sin^2\beta > 0$ which is true for all $r > 0$. If $-1 \le 1 - 4r\sin^2\beta$, then $r\sin^2\beta \le 1/2$ which is satisfied if $r \le 1/2$. This condition places restrictions upon the h and k mesh sizes in order to have stability for the numerical method given by equation (9.30).

Stability

There are three methods for investigating stability of a finite difference scheme associated with partial differential equations. The matrix method, illustrated above, the energy method and the Fourier method (sometimes referred to as the Von Neumann method). The energy method can become messy and so it is left for study in a more advanced numerical methods text. To illustrate the Fourier method we write the difference equation (9.30) associated with a mesh point (ph, qk) and obtain

$$u_p^{q+1} = (1 - 2r)u_p^q + r(u_{p+1}^q + u_{p-1}^q) \quad \text{where} \quad r = \frac{k}{\alpha^2 h^2}. \tag{9.35}$$

Assume that

$$u_p^q = U_p^q + \epsilon_p^q \tag{9.36}$$

where U_p^q is an exact solution of the difference equation (9.35) and ϵ_p^q denotes an error. Substituting the assumed solution (9.36) into the difference equation (9.35) we find

$$U_p^{q+1} + \epsilon_p^{q+1} = (1 - 2r)\left(U_p^q + \epsilon_p^q\right) + r\left(U_{p+1}^q + \epsilon_{p+1}^q + U_{p-1}^q + \epsilon_{p-1}^q\right)$$

which simplifies to

$$\epsilon_p^{q+1} = (1 - 2r)\epsilon_p^q + r\left(\epsilon_{p+1}^q + \epsilon_{p-1}^q\right) \tag{9.37}$$

since by hypothesis U_p^q identically satisfies the difference equation (9.35). The equation (9.37) shows that the local errors satisfy the same difference equation as the true solution. The Fourier method assumes a solution for the error ϵ_p^q of the form

$$\epsilon_p^q = e^{i\beta_n ph}e^{\alpha qk} = e^{i\beta_n ph}\xi^q \quad \text{where} \quad \xi = e^{\alpha k} \tag{9.38}$$

with $i^2 = -1$, $\beta_n = n\pi/L$, $(N+1)h = L$ and α is some complex constant. The term ξ^q determines whether the error term grows or decays with increasing time $t = qk$ since the term $e^{i\beta_n ph}$ remains bounded for all values of p. Note that this form of the solution satisfies the relation $\epsilon_p^{q+1} = \xi\epsilon_p^q$, where the ξ term is known as an amplification factor. Substituting the assumed solution (9.38) into the difference equation (9.37) one obtains

$$e^{i\beta_n ph}\xi^{q+1} = (1 - 2r)e^{i\beta_n ph}\xi^q + r\left(e^{i\beta_n(p+1)h} + e^{i\beta_n(p-1)h}\right)\xi^q$$

Now multiply this equation through by $e^{-i\beta_n ph}\xi^{-q}$ to obtain

$$\begin{aligned}
\xi &= (1 - 2r) + r(e^{i\beta_n h} + e^{-i\beta_n h}) \\
\xi &= (1 - 2r) + 2r\cos\beta_n h
\end{aligned} \tag{9.39}$$

The trigonometric identity $1 - \cos\theta = 2\sin^2\frac{\theta}{2}$ further simplifies the equation (9.39) to the form

$$\xi = 1 - 4r\left(\frac{1 - \cos\beta_n h}{2}\right) = 1 - 4r\sin^2\frac{\beta_n h}{2}. \tag{9.40}$$

This gives the result that for stability we require $|\xi| \leq 1$ which requires that $r \leq 1/2$. This is the same result obtained by the matrix method. It can be shown that the solutions given by equation (9.38) represent terms associated with a Fourier series representation of the error $\epsilon(x, t)$.

Other boundary conditions

In the case of Neumann or Robin boundary conditions one must apply the computational molecule to the boundary points at $x = 0$ and $x = L$. This brings in additional node points from outside the solution domain. These exterior points can be removed from the equations by applying a boundary condition which expresses the exterior point in terms of an interior point. For example, assume the Robin boundary condition

$$\left.\frac{\partial u}{\partial x}\right|_{(x_n, y_j)} = \beta(u - u_0)\left.\right|_{(x_n, y_j)}$$

where β and u_0 are constants. Using a forward difference approximation for the derivative gives

$$\left.\frac{\partial u}{\partial x}\right|_{(x_n, y_j)} = \frac{u_{n+1,j} - u_{n,j}}{h} = \beta(u_{n,j} - u_0) \tag{9.41}$$

Observe that the value of the dependent variable evaluated at the exterior point (x_{n+1}, y_j) can now be evaluated from the boundary condition equation (9.41). The resulting value can then be substituted into the computational molecule difference equation formed at the boundary point (x_n, y_j).

Crank-Nicolson method

The derivative approximations

$$\frac{\partial^2 u}{\partial x^2} = \frac{1}{2}\left[\frac{u_{i+1}^j - 2u_i^j + u_{i-1}^j}{h^2} + \frac{u_{i+1}^{j+1} - 2u_i^{j+1} + u_{i-1}^{j+1}}{h^2}\right]$$
$$\frac{\partial u}{\partial t} = \frac{u_i^{j+1} - u_i^j}{k}, \tag{9.42}$$

where the second derivative is represented as an average of the central second derivative approximations at the times t_j and t_{j+1}, produces a discretization known as the Crank-Nicolson method. Substituting the derivative approximations from the equations (9.42) into the heat equation (9.28) produces the discretization

$$\frac{1}{2}\left[\frac{u_{i+1}^j - 2u_i^j + u_{i-1}^j}{h^2} + \frac{u_{i+1}^{j+1} - 2u_i^{j+1} + u_{i-1}^{j+1}}{h^2}\right] = \alpha^2\left(\frac{u_i^{j+1} - u_i^j}{k}\right) \tag{9.43}$$

Rearrange the terms in equation (9.43) and define $r = \dfrac{k}{\alpha^2 h^2}$ to obtain

$$-ru_{i-1}^{j+1} + (2+2r)u_i^{j+1} - ru_{i+1}^{j+1} = ru_{i-1}^j + (2-2r)u_i^j + ru_{i+1}^j \tag{9.44}$$

Here u_i^{j+1} is not expressed explicitly in terms of known values and so the resulting computational molecule is termed implicit. Applying this computational molecule to each of the interior points gives the equations

$$-rT_0 + (2+2r)u_1^{j+1} - ru_2^{j+1} = rT_0 + (2-2r)u_1^j + ru_2^j$$
$$-ru_1^{j+1} + (2++2r)u_2^{j+1} - ru_3^{j+1} = ru_1^j + (2-2r)u_2^j + ru_3^j$$
$$\vdots \qquad\qquad \vdots$$
$$-ru_{N-1}^{j+1} + (2+2r)u_N^{j+1} - rT_L = u_{N-1}^j + (2-2r)u_N^j + rT_L$$

These equations can be expressed in the matrix form

$$Au^{j+1} = Bu^j + C$$

where

$$Au^{j+1} = \begin{bmatrix} (2+2r) & -r & & & \\ -r & (2+2r) & -r & & \\ & & \ddots & & \\ & & -r & (2+2r) & -r \\ & & & -r & (2+2r) \end{bmatrix} \begin{bmatrix} u_1^{j+1} \\ u_2^{j+1} \\ \vdots \\ u_{N-1}^{j+1} \\ u_N^{j+1} \end{bmatrix}$$

$$Bu^{j} = \begin{bmatrix} (2-2r) & r & & & \\ r & (2-2r) & r & & \\ & & \ddots & & \\ & & r & (2-2r) & r \\ & & & r & (2-2r) \end{bmatrix} \begin{bmatrix} u_1^{j} \\ u_2^{j} \\ \vdots \\ u_{N-1}^{j} \\ u_N^{j} \end{bmatrix}$$

and $C = [2rT_0, 0, 0, \ldots, 0, 2rT_L]^T$.

One can write $u^{j+1} = A^{-1}Bu^j + A^{-1}C$ and for stability the eigenvalues

$$\lambda_n = \frac{2 - 4r\sin^2\frac{n\pi}{2(N+1)}}{2 + 4r\sin^2\frac{n\pi}{2(N+1)}} \quad \text{for} \quad n = 1, 2, \ldots, N$$

of the coefficient matrix $A^{-1}B$ must be less than unity. This is true for all values of the parameter r and so the Crank-Nicolson implicit method is always stable.

The heat equation is usually scaled before any numerical computations are performed. Define a time scale $t_0 = \frac{L^2}{\kappa}$ in terms of the length scale L, then the change of variables

$$\xi = \frac{x}{L}, \qquad \tau = \frac{t}{t_0}$$

will reduce the heat equation to the dimensionless form

$$\frac{\partial^2 u}{\partial \xi^2} = \frac{\partial u}{\partial \tau}$$

over the solution domain $0 \le \xi \le 1$, $\tau > 0$. This is equivalent to letting $\alpha = 1$ in the above discussions.

Example 9-7. (Nonlinear heat equation)

Consider the nonlinear heat or diffusion equation

$$\frac{\partial^2 u}{\partial x^2} = D\left(t, x, u, \frac{\partial u}{\partial x}\right)\frac{\partial u}{\partial t}$$

where the diffusion coefficient D changes with position and time as well as being dependent upon the concentration u and changes in concentration with position. It is assumed that the form of the diffusion coefficient D is a known function. One can treat this nonlinear partial differential equation in exactly the same way as the heat equation if one is careful to include changes in the diffusion coefficient. For example, letting $u_i^j = u(x_i, t_j)$ with $h = \Delta x$ and $k = \Delta t$, one can construct a central difference approximation for $\frac{\partial^2 u}{\partial x^2}$ and a forward difference approximation for $\frac{\partial u}{\partial t}$ to obtain the implicit scheme

$$\frac{u_{i+1}^{j+1} - 2u_i^{j+1} + u_{i-1}^{j+1}}{h^2} = D_i^j \left(\frac{u_i^{j+1} - u_i^j}{k} \right).$$

Here the superscript $j+1$ values are unknown and so one obtains the implicit scheme

$$u_{i+1}^{j+1} - (2 + rD_i^j)u_i^{j+1} + u_{i-1}^{j+1} = -rD_i^j u_i^j \qquad \text{with} \qquad r = h^2/k$$

where the D_i^j values are assumed known. This produces a tridiagonal matrix as before with appropriate modification of the diagonal elements.

Alternatively, one can construct a much more difficult system of nonlinear equations by using the difference scheme

$$\frac{u_{i+1}^{j+1} - 2u_i^{j+1} + u_{i-1}^{j+1}}{h^2} = D_i^{j+1} \left(\frac{u_i^{j+1} - u_i^j}{k} \right)$$

where now D_i^{j+1} also contains unknown values. This gives rise to a system of nonlinear equations which must be solved in order to advance the solution in time. These additional computations can be very time consuming. However, if all other methods fail, then sometimes one must try more difficult techniques in order to obtain a solution.

■

Numerical solution of the wave equation

Consider the initial-boundary value problem describing the equation of motion of a vibrating string. We wish to solve the wave equation

$$\frac{\partial^2 u}{\partial t^2} = c^2 \frac{\partial^2 u}{\partial x^2}, \qquad u = u(x,t), \qquad 0 < x < L, \quad t > 0 \qquad (9.45)$$

subject to the boundary conditions $u(0,t) = 0$, $u(L,t) = 0$, denoting that the ends of the string are fixed. The wave equation (9.45) is also subject to the initial conditions $u(x,0) = f(x)$, $\frac{\partial u(x,0)}{\partial t} = g(x)$ where $f(x)$ is an initial shape for the string and $g(x)$ represents an initial velocity.

One can verify that an analytical solution to the above problem is given by the D'Alembert solution

$$u = u(x,t) = \frac{1}{2}[f(x+ct) + f(x-ct)] + \frac{1}{2c}\int_{x-ct}^{x+ct} g(\xi)\,d\xi \qquad (9.46)$$

The characteristic curves associated with the wave equation are given by the family of lines $x - ct = $ constant and $x + ct = $ constant. The solution of the wave equation at time t_0 and position $x = x_0$ is determined by the region between the lines $x + ct = x_0 + ct_0$, $x - ct = x_0 - ct_0$, which pass through the point (x_0, t_0), where $t > 0$. This region is called the domain of dependence. The first term in the D'Alembert solution represents an average of two moving waves. The initial wave shape is $f(x)$ and $f(x - ct)$ represents this wave shape moving to the right and $f(x + ct)$ represents the initial wave shape moving to the left. The second term in the D'Alembert solution represents an integration of the initial velocity over the domain of dependence.

To numerically solve the wave equation boundary-initial value problem we divided the domain $0 < x < L$, $t > 0$ into a grid, mesh or lattice network by selecting a step size $h = L/(N+1)$ in the x-direction and a step size k in the time direction so that the mesh point (x_i, t_j) can be represented using the equations

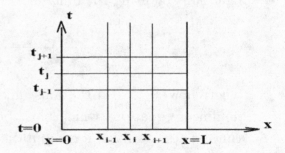

$$x_i = ih, \quad i = 0,1,2,\ldots,N+1 \quad \text{and} \quad t_j = jk, \quad j = 0,1,2,\ldots$$

Let $u_i^j = u(x_i, t_j)$ and substitute the central difference approximations

$$\left.\frac{\partial^2 u}{\partial t^2}\right|_{(x_i,t_j)} = \frac{u_i^{j+1} - 2u_i^j + u_i^{j-1}}{k^2} + \mathcal{O}(k^2)$$

$$\left.\frac{\partial^2 u}{\partial x^2}\right|_{(x_i,t_j)} = \frac{u_{i+1}^j - 2u_i^j + u_{i-1}^j}{h^2} + \mathcal{O}(h^2) \qquad (9.47)$$

for the second derivatives into the equation (9.45). This produces the discrete approximation

$$\frac{u_i^{j+1} - 2u_i^j + u_i^{j-1}}{k^2} = c^2 \left(\frac{u_{i+1}^j - 2u_i^j + u_{i-1}^j}{h^2} \right) \qquad (9.48)$$

Let $r = ck/h$ and write equation (9.48) in the form

$$u_i^{j+1} = 2(1 - r^2)u_i^j + r^2(u_{i+1}^j + u_{i-1}^j) - u_i^{j-1} \qquad (9.49)$$

so that the value of the dependent variable at position x_i and time t_{j+1} is given in terms of the values $u_{i-1}^j, u_i^j, u_{i+1}^j$ at time t_j and u_i^{j-1} at time t_{j-1}. The starting values $u_i^0 = f(x_i)$ and the boundary conditions $u_0^j = 0$, $u_{N+1}^j = 0$ are given. The only difficulty lies in determining a starting value for u_i^{-1}, which is needed when $j = 0$ in equation (9.49), to start marching in the time direction by use of the computational molecule. This starting value can be determined from the initial velocity. In the special case $t_0 = 0$ or $j = 0$, one can employ the central difference approximation

$$\frac{\partial u(x_i, 0)}{\partial t} = \frac{u_i^1 - u_i^{-1}}{2\Delta t} = g(x_i)$$

which implies

$$u_i^{-1} = u_i^1 - 2g(x_i)\Delta t. \qquad (9.50)$$

Thus, in the special case where $t = 0$, one can substitute equation (9.50) into equation (9.49), when $j = 0$, to obtain after simplification

$$u_i^1 = \frac{1}{2}\left[2(1 - r^2)u_i^0 + r^2(u_{i+1}^0 + u_{i-1}^0) - 2g(x_i)\Delta t\right]$$

for the starting values in the row where $t_1 = \Delta t$ or $j = 1$. In this equation the u_i^0 initial values are known for $i = 1, \ldots, N$.

The computational model given by equation (9.49) can now be applied to the interior points x_i at time t_j to obtain the equations

$$u_1^{j+1} = 2(1 - r^2)u_1^j + r^2(u_2^j + u_0^j) - u_1^{j-1}$$
$$u_2^{j+1} = 2(1 - r^2)u_2^j + r^2(u_3^j + u_1^j) - u_2^{j-1}$$
$$\vdots$$
$$u_{N-1}^{j+1} = 2(1 - r^2)u_{N-1}^j + r^2(u_N^j + u_{N-2}^j) - u_{N-1}^{j-1}$$
$$u_N^{j+1} = = 2(1 - r^2)u_N^j + r^2(u_{N+1}^j + u_{N-1}^j) - u_N^{j-1}$$

The boundary conditions require that $u_0^j = 0$ and $u_{N+1}^j = 0$ for all values of j and the initial condition requires that $u_i^0 = f(x_i)$. The above equations can now be written in the matrix form

$$
\begin{bmatrix} u_1^{j+1} \\ u_2^{j+1} \\ \vdots \\ u_{N-1}^{j-1} \\ u_N^{j+1} \end{bmatrix} = \begin{bmatrix} 2(1-r^2) & r^2 & & & \\ r^2 & 2(1-r^2) & r^2 & & \\ & & \ddots & & \\ & & r^2 & 2(1-r^2) & r^2 \\ & & & r^2 & 2(1-r^2) \end{bmatrix} \begin{bmatrix} u_1^j \\ u_2^j \\ \vdots \\ u_{N-1}^j \\ u_N^j \end{bmatrix} - \begin{bmatrix} u_1^{j-1} \\ u_2^{j-1} \\ \vdots \\ u_{N-1}^{j-1} \\ u_N^{j-1} \end{bmatrix}
$$

This equation is not of the form $U^{j+1} = AU^j$ and so a matrix stability analysis is not immediately suggested. Instead, we apply the Fourier method to analyze the difference equation given by equation (9.49). Recall that the error ϵ_p^q satisfies the same difference equation as the true solution and so we can substitute $\epsilon_p^q = e^{i\beta_n ph}\xi^q$ into the difference equation

$$
\epsilon_p^{q+1} = 2(1-r^2)\epsilon_p^q + r^2(\epsilon_{p+1}^q + \epsilon_{p-1}^q) - \epsilon_p^{q-1} \tag{9.51}
$$

and obtain after simplification that ξ must satisfy the equation

$$
\xi^2 - 2\alpha\xi + 1 = 0 \qquad \text{where} \qquad \alpha = 1 - 2r^2\sin^2\frac{\beta_n h}{2} \tag{9.52}
$$

Solving for ξ we find there are two values

$$
\xi_1 = \alpha - \sqrt{\alpha^2 - 1} \qquad \text{and} \qquad \xi_2 = \alpha + \sqrt{\alpha^2 - 1} \tag{9.53}
$$

The condition for stability is that $|\xi| \le 1$. If $\alpha < -1$, then $|\xi_1| > 1$ and so the difference equation would become unstable. If $-1 \le \alpha \le 1$, then stability results since $\xi_1 = \alpha - i\sqrt{1 - \alpha^2}$ and $\xi_2 = \alpha + i\sqrt{1 - \alpha^2}$ so that

$$
|\xi_1| = |\xi_2| = \sqrt{\alpha^2 + 1 - \alpha^2} = 1.
$$

These are the only cases to consider since α cannot be bigger than 1. Therefore, for stability we require that

$$
-1 \le 1 - 2r^2\sin^2\frac{\beta_n h}{2} \le 1
$$

which implies $r^2 \le 1$.

The wave equation is usually scaled before any numerical calculations are performed. One can define the change of variables

$$
\xi = \frac{x}{L}, \qquad \tau = \frac{t}{L/c}
$$

and verify that the wave equation (9.45) reduces to the dimensionless form

$$
\frac{\partial^2 u}{\partial \tau^2} = \frac{\partial^2 u}{\partial \xi^2}.
$$

This is equivalent to letting the wave speed have the value $c = 1$ in the above discussions.

Alternating-Direction Implicit Scheme

Consider the two dimensional heat equation

$$\frac{\partial u}{\partial t} = \alpha^2\left(\frac{\partial^2 u}{\partial x^2} + \frac{\partial^2 u}{\partial y^2}\right), \qquad u = u(x,y,t), \qquad x,y \in R, \quad t > 0 \qquad (9.54)$$

where α^2 is constant and x,y is restricted to the rectangular region R defined by $0 \le x \le a$ and $0 \le y \le b$ for $t > 0$. The partial differential equation is subject to boundary conditions along the edges of the region R and required to satisfy the initial condition $u(x,y,0) = f(x,y)$.

The alternating-direction implicit (ADI) method begins by constructing a discrete version of the partial differential equation by replacing the derivatives by finite difference approximations. The region R is divided into a Δx and Δy grid structure. Let $\Delta x = a/(N+1)$ with $x_i = i\Delta x$ for $i = 0,1,\ldots,N+1$ and $\Delta y = b/(M+1)$ with $y_j = j\Delta y$ for $j = 0,1,\ldots,M+1$. This grid structure is now used to construct discrete approximations of the spatial derivatives. The alternating-direction implicit scheme (ADI) uses two time steps of length $\Delta t/2$ to march a distance Δt. Let $u_{i,j}^k = u(x_i, y_j, t_k)$ and consider the first time step from t_k to $t_{k+1/2}$ where the time derivative is a forward difference approximation and the spatial derivatives are central difference approximations. One obtains the discrete form of equation (9.54) written as

$$\frac{u_{i,j}^{k+1/2} - u_{i,j}^k}{\Delta t/2} = \alpha^2\left[\frac{u_{i+1,j}^k - 2u_{i,j}^k + u_{i-1,j}^k}{\overline{\Delta x}^2} + \frac{u_{i,j+1}^{k+1/2} - 2u_{i,j}^{k+1/2} + u_{i,j-1}^{k+1/2}}{\overline{\Delta y}^2}\right] \qquad (9.55)$$

where the second derivative $\frac{\partial^2 u}{\partial x^2}$ is represented explicitly in terms of known values along the t_k time grid. Note that the equation (9.55) defines the values $u_{i,j}^{k+1/2}$ implicitly. Rearranging terms in equation (9.55) one can write

$$-r_2 u_{i,j-1}^{k+1/2} + 2(1 + r_2)u_{i,j}^{k+1/2} - r_2 u_{i,j+1}^{k+1/2} = r_1 u_{i-1,j}^k + 2(1 - r_1)u_{i,j}^k + r_1 u_{i+1,j}^k \qquad (9.56)$$

where $r_1 = \alpha^2 \Delta t/\overline{\Delta x}^2$ and $r_2 = \alpha^2 \Delta t/\overline{\Delta y}^2$. Note that by applying the computational molecule given by equation (9.56) to the nodes in any column of unknowns in the y-direction, as illustrated in the figure 9-4, there results a tridiagonal system of equations to be solved.

394

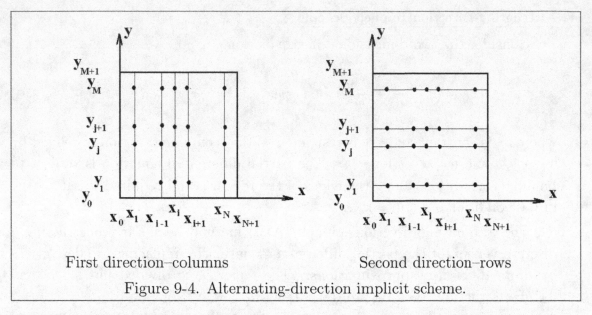

First direction–columns   Second direction–rows

Figure 9-4. Alternating-direction implicit scheme.

For example, apply the computational molecule from equation (9.56) to all the nodes points in the x_i column of values. One obtains, for each value of the index i, a set of equations having the tridiagonal form

$$
\begin{bmatrix}
2(1+r_2) & -r_2 & & \\
-r_2 & 2(1+r_2) & -r_2 & \\
& \ddots & & \\
& & -r_2 & 2(1+r_2)
\end{bmatrix}
\begin{bmatrix}
u_{i,1}^{k+1/2} \\
u_{i,2}^{k+1/2} \\
\vdots \\
u_{i,M}^{k+1/2}
\end{bmatrix}
=
\begin{bmatrix}
b_1 \\
b_2 \\
\vdots \\
b_M
\end{bmatrix}
=
\begin{bmatrix}
r_1 u_{i-1,1}^{k} + 2(1-r_1)u_{i,1}^{k} + r_1 u_{i+1,1}^{k} \\
r_1 u_{i-1,2}^{k} + 2(1-r_1)u_{i,2}^{k} + r_1 u_{i+1,2}^{k} \\
\vdots \\
r_1 u_{i-1,M}^{k} + 2(1-r_1)u_{i,M}^{k} + r_1 u_{i+1,M}^{k}
\end{bmatrix}
$$

where $b_1, \ldots, b_M$ represent known values or values determined from boundary conditions. In this tridiagonal system of equations every term on the right-hand side of the equation is known and so one need only solve these sets of equations for each value of $i = 1, \ldots, N$ and determine all the values $u_{i,j}^{k+1/2}$. This completes the first step of the alternating-direction implicit scheme.

What has been done for columns of unknown values can now be done for rows of unknown values. The second step of the ADI procedure marches from time $t_{k+1/2}$ to t_{k+1} and replaces the partial differential equation (9.54) by the discrete approximation

$$
\frac{u_{i,j}^{k+1} - u_{i,j}^{k+1/2}}{\Delta t/2} = \alpha^2 \left[\frac{u_{i+1,j}^{k+1} - 2u_{i,j}^{k+1} + u_{i-1,j}^{k+1}}{\overline{\Delta x}^2} + \frac{u_{i,j+1}^{k+1/2} - 2u_{i,j}^{k+1/2} + u_{i,j-1}^{k+1/2}}{\overline{\Delta y}^2} \right] \tag{9.57}
$$

where now the terms $u_{i,j}^{k+1}$ represent the implicit values to be determined. The equation (9.57) can be expressed as the computational molecule

$$-r_1 u_{i-1,j}^{k+1} + 2(1+r_1)u_{i,j}^{k+1} - r_1 u_{i+1,j}^{k+1} = r_2 u_{i,j-1}^{k+1/2} + 2(1-r_2)u_{i,j}^{k+1/2} + r_2 u_{i,j+1}^{k+1/2} \qquad (9.58)$$

Applying this computational molecule to the node points in the row where $y = y_j$ produces the system of equations

$$\begin{bmatrix} 2(1+r_1) & -r_1 & & \\ -r_1 & 2(1+r_1) & -r_1 & \\ & & \ddots & \\ & & -r_1 & 2(1+r_1) \end{bmatrix} \begin{bmatrix} u_{1,j}^{k+1} \\ u_{2,j}^{k+1} \\ \vdots \\ u_{N,j}^{k+1} \end{bmatrix} = \begin{bmatrix} b_1^* \\ b_2^* \\ \vdots \\ b_N^* \end{bmatrix} = \begin{bmatrix} r_2 u_{1,j-1}^{k+1/2} + 2(1-r_2)u_{1,j}^{k+1/2} + r_2 u_{1,j+1}^{k+1/2} \\ r_2 u_{2,j-1}^{k+1/2} + 2(1-r_2)u_{2,j}^{k+1/2} + r_2 u_{2,j+1}^{k+1/2} \\ \vdots \\ r_2 u_{N,j-1}^{k+1/2} + 2(1-r_2)u_{N,j}^{k+1/2} + r_2 u_{N,j+1}^{k+1/2} \end{bmatrix}$$

where $b_1^*, \ldots, b_N^*$ represent known values or values determined from boundary conditions. Again we have a tridiagonal system of equations to solve where every term on the right-hand side is known. One must solve these sets of equations for each value of $j = 1, \ldots, M$. The resulting value $u_{i,j}^{k+1}$ for $i = 1, \ldots, N$ and $j = 1, \ldots, M$ are the desired nodal point temperatures at the $(k+1)$st time step. This procedure of alternating directions at each half of a time step gives the method its name.

Systems of Partial Differential Equations

Systems of partial differential equations are usually written in matrix notation. For example, if there are n-unknowns u_i, $i = 1, \ldots, n$, which are dependent variables and assume that there are two independent variables, say x and y, then one can write $u_i = u_i(x,y)$ for $i = 1, \ldots, n$. If these dependent variables are related by a coupled system of n first order partial differential equations of the form

$$\sum_{j=1}^{n} a_{ij}\frac{\partial u_j}{\partial x} + \sum_{j=1}^{n} b_{ij}\frac{\partial u_j}{\partial y} + f_i(x, y, u_1, u_2, \ldots, u_n) = 0 \qquad (9.59)$$

for $i = 1, \ldots, n$ and where $a_{ij} = a_{ij}(x, y, u_1, \ldots, u_n)$, $b_{ij} = b_{ij}(x, y, u_1, \ldots, u_n)$ are coefficients which may be either constants, functions of x and y, or functions of x, y and the dependent variables u_i, $i = 1, \ldots, n$. One can define the column vectors

$$\vec{u} = \text{col}\,[u_1, u_2, \ldots, u_n], \qquad \vec{f} = \text{col}\,[f_1, f_2, \ldots, f_n] \qquad (9.60)$$

and write the system of partial differential equations (9.59) in the matrix form

$$A\frac{\partial \vec{u}}{\partial x} + B\frac{\partial \vec{u}}{\partial y} + \vec{f} = \vec{0} \qquad (9.61)$$

where A,B are coefficient matrices given by

$$A = (a_{ij}) = \begin{bmatrix} a_{11} & a_{12} & \cdots & a_{1n} \\ a_{21} & a_{22} & \cdots & a_{2n} \\ \vdots & \vdots & \ddots & \vdots \\ a_{n1} & a_{n2} & \cdots & a_{nn} \end{bmatrix} \qquad B = (b_{ij}) = \begin{bmatrix} b_{11} & b_{12} & \cdots & b_{1n} \\ b_{21} & b_{22} & \cdots & b_{2n} \\ \vdots & \vdots & \ddots & \vdots \\ b_{n1} & b_{n2} & \cdots & b_{nn} \end{bmatrix} \qquad (9.62)$$

and $\vec{0} = \text{col}[0, 0, \ldots, 0]$ is a column vector of zeros.

Note that equation (9.61) defines a basic form associated with systems of first order partial differential equations in two independent variables. Sometimes it is convenient to replace x by t and y by x in the equations (9.59) or (9.61). This is usually done whenever time t is one of the independent variables. A special case of the system of partial differential equations (9.61) occurs when $A = I$ is the identity matrix. In this special case the system of partial differential equations (9.61) can be written in the form

$$\frac{\partial \vec{u}}{\partial x} + B\frac{\partial \vec{u}}{\partial y} + \vec{f} = \vec{0}. \qquad (9.63)$$

Systems of partial differential equations having the form

$$\frac{\partial \vec{u}}{\partial t} + \frac{\partial}{\partial x}\vec{F}(\vec{u}) = \vec{0} \qquad (9.64)$$

are said to be in a conservative law form. Note that the equation (9.64) can be written in the alternative form

$$\frac{\partial \vec{u}}{\partial t} + B(\vec{u})\frac{\partial \vec{u}}{\partial x} = \vec{0} \qquad (9.65)$$

where $B(\vec{u}) = \left[\frac{\partial F_i}{\partial u_j}\right]$ is the Jacobian matrix of $\vec{F}$. This is a special case of the equation (9.61) with the variables changed.

Example 9-8.   (Cauchy-Riemann equations)

The Cauchy-Riemann equations for solenoidal and irrotational fluid flow is represented by the system of first order partial differential equations

$$\frac{\partial u_1}{\partial x} - \frac{\partial u_2}{\partial y} = 0, \qquad \frac{\partial u_1}{\partial y} + \frac{\partial u_2}{\partial x} = 0, \qquad x, y \in R$$

subject to boundary conditions. This first order system of equations can written in the matrix form of equation (9.61)

$$\begin{bmatrix} \frac{\partial u_1}{\partial x} \\ \frac{\partial u_2}{\partial x} \end{bmatrix} + \begin{bmatrix} 0 & -1 \\ 1 & 0 \end{bmatrix} \begin{bmatrix} \frac{\partial u_1}{\partial y} \\ \frac{\partial u_2}{\partial y} \end{bmatrix} = \begin{bmatrix} 0 \\ 0 \end{bmatrix}$$

where $\vec{u} = \text{col}[u_1, u_2]$, $\vec{f} = \vec{0}$ and $B = \begin{bmatrix} 0 & -1 \\ 1 & 0 \end{bmatrix}$ is a constant matrix.

Example 9-9. (Euler equations)

The Euler equations of motion describing the isentropic flow of a perfect fluid moving in the x-direction, is governed by the system of first order partial differential equations

$$\frac{\partial u}{\partial t} + u\frac{\partial u}{\partial x} + \frac{c^2}{\varrho}\frac{\partial \varrho}{\partial x} = 0, \qquad u = u(x,t)$$

$$\frac{\partial \varrho}{\partial t} + u\frac{\partial \varrho}{\partial x} + \varrho\frac{\partial u}{\partial x} = 0, \qquad \varrho = \varrho(x,t)$$

for $x,t \in R$, and subject to boundary conditions. In these equations u represents velocity in the x-direction, ϱ represents fluid density, and $c = c(\varrho)$ expresses the speed of sound as a known function of the density ϱ. This system can be represented in the matrix form of equation (9.61)

$$\begin{bmatrix} \frac{\partial u}{\partial t} \\ \frac{\partial \varrho}{\partial t} \end{bmatrix} + \begin{bmatrix} u & c^2/\varrho \\ \varrho & u \end{bmatrix} \begin{bmatrix} \frac{\partial u}{\partial x} \\ \frac{\partial \varrho}{\partial x} \end{bmatrix} = \begin{bmatrix} 0 \\ 0 \end{bmatrix}.$$

Here $\vec{u} = \text{col}[u, \varrho]$, $\vec{f} = 0$, and $B = \begin{bmatrix} u & c^2/\varrho \\ \varrho & u \end{bmatrix}$ is a function of the dependent variables.

Most partial differential equations of second order and higher can be written in the matrix form of equation (9.61).

Example 9-10. (Poisson's equation)

Consider the Poisson equation

$$\frac{\partial^2 u}{\partial x^2} + \frac{\partial^2 u}{\partial y^2} = f(x,y) \tag{9.66}$$

Let $u_1 = u$, $u_2 = \frac{\partial u}{\partial x}$, and $u_3 = \frac{\partial u}{\partial y}$ and represent the Poisson equation (9.66) by the system of partial differential equations

$$\frac{\partial u_1}{\partial x} - u_2 = 0, \qquad \frac{\partial u_1}{\partial y} - u_3 = 0, \qquad \frac{\partial u_2}{\partial x} + \frac{\partial u_3}{\partial y} = f(x,y) \tag{9.67}$$

which can be written in the matrix form

$$\begin{bmatrix} 1 & 0 & 0 \\ 0 & 0 & 0 \\ 0 & 1 & 0 \end{bmatrix} \frac{\partial}{\partial x} \begin{bmatrix} u_1 \\ u_2 \\ u_3 \end{bmatrix} + \begin{bmatrix} 0 & 0 & 0 \\ 1 & 0 & 0 \\ 0 & 0 & 1 \end{bmatrix} \frac{\partial}{\partial y} \begin{bmatrix} u_1 \\ u_2 \\ u_3 \end{bmatrix} + \begin{bmatrix} 0 & -1 & 0 \\ 0 & 0 & -1 \\ 0 & 0 & 0 \end{bmatrix} \begin{bmatrix} u_1 \\ u_2 \\ u_3 \end{bmatrix} = \begin{bmatrix} 0 \\ 0 \\ f(x,y) \end{bmatrix}$$

Example 9-11.   (The heat equation)

The heat equation with source term

$$\frac{\partial u}{\partial t} - \alpha^2 \frac{\partial^2 u}{\partial x^2} = f(x,t)$$

can be written in matrix form. Let $u_1 = u$, $u_2 = \frac{\partial u}{\partial x}$, and replace the heat equation by the system of partial differential equations

$$\frac{\partial u_1}{\partial x} - u_2 = 0, \quad \text{and} \quad \frac{\partial u_1}{\partial t} - \alpha^2 \frac{\partial u_2}{\partial x} = f(x,t)$$

which can be written in the matrix form

$$\begin{bmatrix} 0 & 0 \\ 1 & 0 \end{bmatrix} \frac{\partial}{\partial t} \begin{bmatrix} u_1 \\ u_2 \end{bmatrix} + \begin{bmatrix} 1 & 0 \\ 0 & -\alpha^2 \end{bmatrix} \frac{\partial}{\partial x} \begin{bmatrix} u_1 \\ u_2 \end{bmatrix} + \begin{bmatrix} 0 & -1 \\ 0 & 0 \end{bmatrix} \begin{bmatrix} u_1 \\ u_2 \end{bmatrix} = \begin{bmatrix} 0 \\ f(x,t) \end{bmatrix}.$$

■

Example 9-12.   (The wave equation)

The nonhomogeneous wave equation

$$\frac{\partial^2 u}{\partial t^2} - c^2 \frac{\partial^2 u}{\partial x^2} = f(x,t) \tag{9.68}$$

can be written in a matrix form. Let $u_1 = \frac{\partial u}{\partial t}$, and $u_2 = \frac{\partial u}{\partial x}$ and replace the equation (9.68) be the first order system of partial differential equations

$$\frac{\partial u_2}{\partial t} - \frac{\partial u_1}{\partial x} = 0, \quad \text{and} \quad \frac{\partial u_1}{\partial t} - c^2 \frac{\partial u_2}{\partial x} = f(x,t)$$

which can be written in the matrix form

$$\begin{bmatrix} 0 & 1 \\ 1 & 0 \end{bmatrix} \frac{\partial}{\partial t} \begin{bmatrix} u_1 \\ u_2 \end{bmatrix} + \begin{bmatrix} -1 & 0 \\ 0 & -c^2 \end{bmatrix} \frac{\partial}{\partial x} \begin{bmatrix} u_1 \\ u_2 \end{bmatrix} = \begin{bmatrix} 0 \\ f(x,t) \end{bmatrix}.$$

■

First order systems of partial differential equations in more than two independent variables can be represented by matrix equations of the form

$$\frac{\partial \vec{u}}{\partial t} + A_1 \frac{\partial \vec{u}}{\partial x_1} + A_2 \frac{\partial \vec{u}}{\partial x_2} + \cdots + A_n \frac{\partial \vec{u}}{\partial x_n} = \vec{f} \tag{9.69}$$

where $\vec{u} = \text{col}[u_1, u_2, \ldots, u_m]$ are the dependent variables and $x_1, x_2, \ldots, x_n, t$ are the independent variables.

The numerical techniques that you have applied to scalar partial differential equations can now be extended to systems of partial differential equations. The basic idea remains the same. One replaces continuous derivatives by discrete approximations to obtain difference equations. This is the easy part. The hard part is to determine whether your discrete approximation produces a computational molecule which is valid. Is the discrete form you created even stable? These are important concerns that must always be addressed.

Example 9-13. (Wave equation)

The above concerns can be illustrated using the first order wave equation

$$\frac{\partial u}{\partial t} = -\alpha \frac{\partial u}{\partial x}, \quad \alpha > 0, \text{ and constant} \quad u = u(x,t). \tag{9.70}$$

Introduce a rectangular grid and let $u_i^j = u(x_i, t_j)$, then various derivative approximations substituted into this equation produce different discrete methods and one must be careful about using a method as it might be an unstable method. Some example substitutions for the equation (9.70) are as follows.

(i) First order forward derivative approximations in x and t to obtain Euler's method

$$\frac{u_i^{j+1} - u_i^j}{\Delta t} = -\alpha \frac{u_{i+1}^j - u_i^j}{\Delta x}.$$

This seems reasonable to do, but it turns out the method is unstable.

(ii) First order derivative approximation in t and second order derivative approximation in x, this gives

$$\frac{u_i^{j+1} - u_i^j}{\Delta t} = -\alpha \frac{u_{i+1}^j - u_{i-1}^j}{2\Delta x}.$$

This also looks reasonable, but it too is an unstable method.

(iii) It might seem reasonable to use more accurate approximations for the derivatives, like a central difference approximation in time t and x to obtain

$$\frac{u_i^{j+1} - u_i^{j-1}}{2\Delta t} = -\alpha \left(\frac{u_{i+1}^j - 2u_i^j + u_{i-1}^j}{(\Delta x)^2} \right)$$

However, this method also turns out to be unstable.

(iv) Forward differencing in time t and backward differencing in distance x to obtain

$$\frac{u_i^{j+1} - u_i^j}{\Delta t} = -\alpha \frac{u_{i+1}^j - u_i^j}{\Delta x}$$

This method turns out to be stable if $\alpha > 0$ and $\alpha \Delta t / \Delta x \leq 1$. The method is known as an upwind differencing scheme.

(v) The Lax method produces an averaged approximation to the time derivative by modifying the forward difference approximation in time $\frac{\partial u}{\partial t} = \frac{u_i^{j+1} - u_i^j}{\Delta t}$ by replacing the u_i^j term by an average value $\frac{u_{i+1}^j + u_{i-1}^j}{2}$ to obtain the derivative approximation

$$\frac{\partial u}{\partial t} = \frac{u_i^{j+1} - \frac{u_{i+1}^j + u_{i-1}^j}{2}}{\Delta t}$$

When this derivative approximation is substituted into the equation (9.70) along with a central difference approximation in the x direction there results

$$\frac{u_i^{j+1} - \frac{u_{i+1}^j + u_{i-1}^j}{2}}{\Delta t} = -\alpha \frac{u_{i+1}^j - u_{i-1}^j}{2\Delta x}.$$

This method is stable for $\alpha \Delta t / \Delta x \leq 1$.

(vi) The Lax-Wendroff method uses the Taylor series expansion

$$u(x, t + \Delta t) = u(x, t) + \frac{\partial u}{\partial t}\Delta t + \frac{\partial^2 u}{\partial t^2}\frac{(\Delta t)^2}{2!} + h.o.t$$

where higher order terms (h.o.t.) have been neglected. Taking the derivative of equation (9.70) with respect to time gives

$$\frac{\partial}{\partial t}\frac{\partial u}{\partial t} = \frac{\partial^2 u}{\partial t^2} = -\alpha\frac{\partial}{\partial t}\frac{\partial u}{\partial x} = -\alpha\frac{\partial}{\partial x}\frac{\partial u}{\partial t} = -\alpha\frac{\partial}{\partial x}\left(-\alpha\frac{\partial u}{\partial x}\right) = \alpha^2\frac{\partial^2 u}{\partial x^2}$$

where the order of differentiation has been interchanged. The Taylor series expansion can then be written

$$u(x, t + \Delta t) = u(x, t) + \frac{\partial u}{\partial t}\Delta t + \alpha^2\frac{\partial^2 u}{\partial x^2}\frac{(\Delta t)^2}{2}.$$

Evaluating this Taylor series at x_i, t_j and using difference approximations for the derivatives gives

$$u_i^{j+1} = u_i^j - \alpha\Delta t\left(\frac{u_{i+1}^j - u_{i-1}^j}{2\Delta x}\right) + \frac{\alpha^2}{2}(\Delta t)^2\left(\frac{u_{i+1}^j - 2u_i^j + u_{i-1}^j}{(\Delta x)^2}\right).$$

This iterative scheme turns out to be stable for $\alpha \Delta t / \Delta x \leq 1$.

∎

The above examples illustrate that there are many kinds of difference schemes and averaging techniques that can be applied to the derivatives, of a scalar u or vector $\vec{u}$ in a partial differential equation or system of partial differential equations, from which one can derive algorithms for solving partial differential equations. One must become familiar with various types of norms in order to understand the errors associated with these methods. These techniques are left for more advanced courses in numerical methods for partial differential equations.

Exercises Chapter 9

▶ **1.**

For $u_{i,j} = u(x_i, y_j)$ with $x_i = x_0 + ih$, for $i = 0, 1, 2, \ldots, n$, verify the first partial derivative approximations

(a) $\left. \dfrac{\partial u}{\partial x} \right|_{i,j} = \dfrac{u_{i+1,j} - u_{i,j}}{h} + \mathcal{O}(h)$ forward difference approximation

(b) $\left. \dfrac{\partial u}{\partial x} \right|_{i,j} = \dfrac{u_{i,j} - u_{i-1,j}}{h} + \mathcal{O}(h)$ backward difference approximation

(c) $\left. \dfrac{\partial u}{\partial x} \right|_{i,j} = \dfrac{u_{i+1,j} - u_{i-1,j}}{2h} + \mathcal{O}(h^2)$ central difference approximation

with similar expressions for the partial derivatives in the y-direction.

▶ **2.**

Verify the first partial derivative difference approximations

(a) $\left. \dfrac{\partial u}{\partial x} \right|_{i,j} = \dfrac{-3u_{i,j} + 4u_{i+1,j} - u_{i+2,j}}{2h} + \mathcal{O}(h^2)$

(b) $\left. \dfrac{\partial u}{\partial x} \right|_{i,j} = \dfrac{3u_{i,j} - 4u_{i-1,j} + u_{i-2,j}}{2h} + \mathcal{O}(h^2)$

(c) $\left. \dfrac{\partial u}{\partial x} \right|_{i,j} = \dfrac{u_{i-2,j} - 8u_{i-1,j} + 8u_{i+1,j} - u_{i+2,j}}{12h} + \mathcal{O}(h^4)$

with similar expressions for the partial derivatives in the y-direction.

▶ **3.**

Verify the second partial derivative approximations

(a) $\left. \dfrac{\partial^2 u}{\partial x^2} \right|_{i,j} = \dfrac{u_{i,j} - 2u_{i+1,j} + u_{i+2,j}}{h^2} + \mathcal{O}(h)$

(b) $\left. \dfrac{\partial^2 u}{\partial x^2} \right|_{i,j} = \dfrac{u_{i,j} - 2u_{i-1,j} + u_{i-2,j}}{h^2} + \mathcal{O}(h)$

(c) $\left. \dfrac{\partial^2 u}{\partial x^2} \right|_{i,j} = \dfrac{u_{i+1,j} - 2u_{i,j} + u_{i-1,j}}{h^2} + \mathcal{O}(h^2)$

(d) $\left. \dfrac{\partial^2 u}{\partial x^2} \right|_{i,j} = \dfrac{-u_{i-2,j} + 16u_{i-1,j} - 30u_{i,j} + 16u_{i+1,j} - u_{i+2,j}}{12h^2} + \mathcal{O}(h^4)$

with similar expressions for the partial derivatives in the y-direction.

▶ **4.**

Verify the derivative approximations

(a) $\quad \dfrac{\partial u}{\partial x}\bigg|_{i,j} = \dfrac{u_{i+1,j+1} - u_{i-1,j+1} + u_{i+1,j-1} - u_{i-1,j-1}}{4h} + \mathcal{O}(h^2)$

(b) $\quad \dfrac{\partial^2 u}{\partial x^2}\bigg|_{i,j} = \dfrac{1}{3h^2}\big[\, u_{i+1,j+1} - 2u_{i,j+1} + u_{i-1,j+1}$

$\qquad\qquad\qquad\qquad u_{i+1,j} \quad - 2u_{i,j} \quad + u_{i-1,j}$

$\qquad\qquad\qquad\qquad u_{i+1,j-1} - 2u_{i,j-1} + u_{i-1,j-1} \,\big] + \mathcal{O}(h^2)$

▶ **5.**

Verify the third partial derivative approximations

(a) $\quad \dfrac{\partial^3 u}{\partial x^3}\bigg|_{i,j} = \dfrac{u_{i+3,j} - 3u_{i+2,j} + 3u_{i+1,j} - u_{i,j}}{h^3} + \mathcal{O}(h)$

(b) $\quad \dfrac{\partial^3 u}{\partial x^3}\bigg|_{i,j} = \dfrac{u_{i,j} - 3u_{i-1,j} + 3u_{i-2,j} - u_{i-3,j}}{h^3} + \mathcal{O}(h)$

(c) $\quad \dfrac{\partial^3 u}{\partial x^3}\bigg|_{i,j} = \dfrac{u_{i+2,j} - 2u_{i+1,j} + 2u_{i-1,j} - u_{i-2,j}}{2h^3} + \mathcal{O}(h^2)$

(d) $\quad \dfrac{\partial^3 u}{\partial x^3}\bigg|_{i,j} = \dfrac{-5u_{i,j} + 18u_{i+1,j} - 24u_{i+2,j} + 14u_{i+3,j} - 3u_{i+4,j}}{2h^3} + \mathcal{O}(h^2)$

(e) $\quad \dfrac{\partial^3 u}{\partial x^3}\bigg|_{i,j} = \dfrac{5u_{i,j} - 18u_{i-1,j} + 24u_{i-2,j} - 14u_{i-3,j} + 3u_{i-4,j}}{2h^3} + \mathcal{O}(h^2)$

(f) $\quad \dfrac{\partial^3 u}{\partial x^3}\bigg|_{i,j} = \dfrac{u_{i-3,j} - 8u_{i-2,j} + 13u_{i-1,j} - 13u_{i+1,j} + 8u_{i+2,j} - u_{i+3,j}}{8h^3} + \mathcal{O}(h^4)$

▶ **6.**

Verify the fourth partial derivative approximations

(a) $\quad \dfrac{\partial^4 u}{\partial x^4}\bigg|_{i,j} = \dfrac{u_{i+4,j} - 4u_{i+3,j} + 6u_{i+2j} - 4u_{i+1,j} + u_{i,j}}{h^4} + \mathcal{O}(h)$

(b) $\quad \dfrac{\partial^4 u}{\partial x^4}\bigg|_{i,j} = \dfrac{u_{i,j} - 4u_{i-1,j} + 6u_{i-2,j} - 4u_{i-3,j} + u_{i-4,j}}{h^4} + \mathcal{O}(h)$

(c) $\quad \dfrac{\partial^4 u}{\partial x^4}\bigg|_{i,j} = \dfrac{u_{i+2,j} - 4u_{i+1,j} + 6u_{i,j} - 4u_{i-1,j} + u_{i-2,j}}{h^4} + \mathcal{O}(h^2)$

(d) $\quad \dfrac{\partial^4 u}{\partial x^4}\bigg|_{i,j} = \dfrac{3u_{i,j} - 14u_{i+1,j} + 26u_{i+2,j} - 24u_{i+3,j} + 11u_{i+4,j} - 2u_{i+5,j}}{h^4} + \mathcal{O}(h^2)$

(e) $\quad \dfrac{\partial^4 u}{\partial x^4}\bigg|_{i,j} = \dfrac{3u_{i,j} - 14u_{i-1,j} + 26u_{i-2,j} - 24u_{i-3,j} + 11u_{i-4,j} - 2u_{i-5,j}}{h^4} + \mathcal{O}(h^2)$

(f) $\quad \dfrac{\partial^4 u}{\partial x^4}\bigg|_{i,j} = \dfrac{-u_{i-3,j} + 12u_{i-2,j} - 39u_{i-1,j} + 56u_{i,j} - 39u_{i+1,j} + 12u_{i+2,j} - u_{i+3,j}}{6h^4} + \mathcal{O}(h^4)$

▶ **7.**

Verify the following mixed partial derivative approximations associated with a mesh having a $\Delta x = h$ and $\Delta y = k$ spacing.

(a) $\quad \dfrac{\partial^2 u}{\partial x \partial y}\bigg|_{i,j} = \dfrac{1}{h}\left[\dfrac{u_{i+1,j} - u_{i+1,j-1}}{k} - \dfrac{u_{i,j} - u_{i,j-1}}{k}\right] + \mathcal{O}(h,k)$

(b) $\quad \dfrac{\partial^2 u}{\partial x \partial y}\bigg|_{i,j} = \dfrac{1}{h}\left[\dfrac{u_{i+1,j+1} - u_{i+1,j}}{k} - \dfrac{u_{i,j+1} - u_{i,j}}{k}\right] + \mathcal{O}(h,k)$

(c) $\quad \dfrac{\partial^2 u}{\partial x \partial y}\bigg|_{i,j} = \dfrac{1}{h}\left[\dfrac{u_{i,j} - u_{i,j-1}}{k} - \dfrac{u_{i-1,j} - u_{i-1,j-1}}{k}\right] + \mathcal{O}(h,k)$

(d) $\quad \dfrac{\partial^2 u}{\partial x \partial y}\bigg|_{i,j} = \dfrac{1}{h}\left[\dfrac{u_{i,j+1} - u_{i,j}}{k} - \dfrac{u_{i-1,j+1} - u_{i-1,j}}{k}\right] + \mathcal{O}(h,k)$

(e) $\quad \dfrac{\partial^2 u}{\partial x \partial y}\bigg|_{i,j} = \dfrac{1}{h}\left[\dfrac{u_{i,j+1} - u_{i,j-1}}{2k} - \dfrac{u_{i-1,j+1} - u_{i-1,j-1}}{2k}\right] + \mathcal{O}(h,k^2)$

(f) $\quad \dfrac{\partial^2 u}{\partial x \partial y}\bigg|_{i,j} = \dfrac{1}{2h}\left[\dfrac{u_{i+1,j+1} - u_{i+1,j}}{k} - \dfrac{u_{i-1,j+1} - u_{i-1,j}}{k}\right] + \mathcal{O}(h^2,k)$

(g) $\quad \dfrac{\partial^2 u}{\partial x \partial y}\bigg|_{i,j} = \dfrac{1}{h}\left[\dfrac{u_{i+1,j+1} - u_{i+1,j-1}}{2k} - \dfrac{u_{i,j+1} - u_{i,j-1}}{2k}\right] + \mathcal{O}(h,k^2)$

(h) $\quad \dfrac{\partial^2 u}{\partial x \partial y}\bigg|_{i,j} = \dfrac{1}{2h}\left[\dfrac{u_{i+1,j} - u_{i+1,j-1}}{k} - \dfrac{u_{i-1,j} - u_{i-1,j-1}}{k}\right] + \mathcal{O}(h^2,k)$

(i) $\quad \dfrac{\partial^2 u}{\partial x \partial y}\bigg|_{i,j} = \dfrac{1}{2h}\left[\dfrac{u_{i+1,j+1} - u_{i+1,j-1}}{2k} - \dfrac{u_{i-1,j+1} - u_{i-1,j-1}}{2k}\right] + \mathcal{O}(h^2,k^2)$

▶ **8.**

(a) Derive an implicit backward difference scheme for the heat equation $\dfrac{\partial u}{\partial t} = \alpha^2 \dfrac{\partial^2 u}{\partial x^2}$ by using the backward and central difference approximations

$$\dfrac{\partial u}{\partial t} = \dfrac{u_i^{j+1} - u_i^j}{k} \quad \text{and} \quad \dfrac{\partial^2 u}{\partial x^2} = \dfrac{u_{i+1}^{j+1} - 2u_i^{j+1} + u_{i-1}^{j+1}}{h^2}, \quad 0 < x < 1, \quad t > 0$$

(b) What conditions on h, k must be satisfied for stability if one assumes boundary and initial conditions are given?

▶ **9. Heat equation**

Consider the boundary value problem

$$\frac{\partial u}{\partial t} = \frac{\partial^2 u}{\partial x^2}, \quad u = u(x,t), \quad 0 < x < 1, \ t > 0$$

with boundary conditions $u(0,t) = 0$ and $u(1,t) = 0$. Assume an initial condition $u(x,0) = \sin \pi x$.

(a) Show that with $u_i^j = u(x_i, t_j)$ and $\Delta x = h$, $\Delta t = k$ a forward difference approximation in time and a central difference approximation in distance produces the difference equation

$$\frac{u_i^{j+1} - u_i^j}{k} = \frac{u_{i+1}^j - 2u_i^j + u_{i-1}^j}{h^2}$$

(b) Let $r = k/h^2$ and solve for u_i^{j+1} to produce a computational scheme.

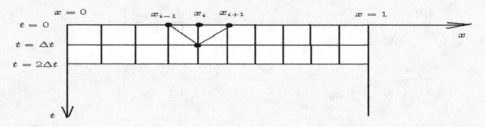

(c) Let $h = \Delta x = 0.1$ with $r = 0.4$ and create a table like the following

$t \backslash x$	0.0	0.1	0.2	0.3	0.4	0.5	0.6	0.7	0.8	0.9	1.0
$t = 0$											
$t = \Delta t$											
$t = 2\Delta t$											
$\vdots$											

   (i) What values go in the row where $t = 0$? What values go in the columns where $x = 0.0$ and $x = 1.0$?

   (ii) Show that each row can be completed, starting with the $x = 0.1$ column and ending with the $x = 0.9$ column, using the computational scheme from part (b).

   (iii) At what row do you terminate your calculations?

(d) Compare your answers to the exact solution $u = u(x,t) = e^{-\pi^2 t} \sin \pi x$.

(e) Solve the above problem using $r = 0.9$ and explain what happens.

405

▶ **10.** **Wave equation**

Consider the boundary value problem

$$\frac{\partial^2 u}{\partial t^2} = \frac{\partial^2 u}{\partial x^2}, \quad u = u(x,t), \quad 0 < x < 1, \ t > 0$$

subject to the boundary conditions $u(0,t) = 0$ and $u(1,t) = 0$. Assume the initial conditions are given by $u(x,0) = \sin \pi x$ and $\frac{\partial u(x,0)}{\partial t} = 0$.

(a) Show that with $u_i^j = u(x_i, t_j)$ and $\Delta x = h$, $\Delta t = k$ a central difference approximation in time and a central difference approximation in distance reduces the wave equation to the difference scheme

$$\frac{u_i^{j+1} - 2u_i^j + u_i^{j-1}}{k^2} = \frac{u_{i+1}^j - 2u_i^j + u_{i-1}^j}{h^2}.$$

(b) Let $r^2 = k^2/h^2$ and solve for u_i^{j+1} to produce a computational scheme.

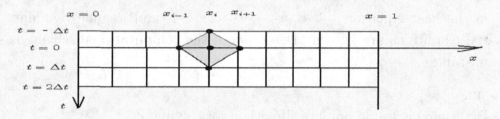

(c) Let $h = \Delta x = 0.1$ with $r = 0.4$ and create a table like the following

$t \backslash x$	0.0	0.1	0.2	0.3	0.4	0.5	0.6	0.7	0.8	0.9	1.0
$t = -\Delta t$	Initial condition for velocity used to determine u_i^{-1} for row $t = \Delta t$										
$t = 0$											
$t = \Delta t$											
$t = 2\Delta t$											
⋮											

What values go in the row where $t = 0$? What values go in the columns where $x = 0.0$ and $x = 1.0$? Note the row $t = -\Delta t$ has entries which are not known. However, one can use the other initial condition to modify the starting values.

(d) Note that a central difference approximation on initial conditions of the form $\frac{\partial u(x_i,0)}{\partial t} = g(x_i)$ gives a difference equation $\frac{u_i^1 - u_i^{-1}}{2k} = g(x_i)$ which gives the result $u_i^{-1} = u_i^1 - 2g(x_i)k$ at $t = 0$. Use this equation to modify your computational scheme when $j = 0$. The initial values in the row $t = 0$ are inserted, then the modified equation, in the special case where $j = 0$, calculates the next row and thereafter each additional row can be completed by employing the computational scheme developed in part (a), starting with the $x = 0.1$ column and ending with the $x = 0.9$ column.

(iii) At what row do you terminate your calculations?

(d) Compare your answers to the exact solution $u = u(x, t) = \sin \pi x \cos \pi t$.

(e) Solve the above problem using $r = 1.5$ and explain what happens.

(f) If an error such as round off error occurs in the second line of the computations, then how does this error propagate from row to row of the calculations?

▶ **11.**   Use the derivative approximations

$$\frac{\partial u}{\partial t} = \frac{u_i^{j+1} - u_i^{j-1}}{2k}, \qquad \frac{\partial^2 u}{\partial x^2} = \frac{u_{i+1}^j - 2u_i^j + u_{i-1}^j}{h^2}$$

in the heat equation $\dfrac{\partial u}{\partial t} = \alpha^2 \dfrac{\partial^2 u}{\partial x^2}$, $0 < x < 1$, and assume boundary and initial conditions are known. Determine if the computational scheme is stable or unstable.

▶ **12.**

Determine a stable numerical method for solving

$$PDE: \qquad \frac{\partial u}{\partial t} = \frac{\partial^2 u}{\partial x^2} - u, \qquad 0 < x < 1, \quad t > 0$$
$$BC: \qquad u(0, t) = 0, \quad u(1, t) = 0$$
$$IC: \qquad u(x, 0) = f(x)$$

Test your numerical method using $f(x) = x(1 - x)$.

▶ **13.**   **Triangular grid.**

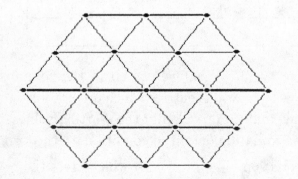
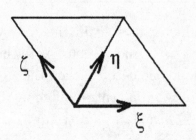

Consider a triangular grid with equal grid spacing between nodal points.

(a) Use rotation of axes $x = \bar{x}\cos\theta - \bar{y}\sin\theta$, $y = \bar{x}\sin\theta + \bar{y}\cos\theta$ and show that under a change of variable $u(x,y) \rightarrow u(\bar{x},\bar{y})$ the partial derivatives are calculated

$$\frac{\partial u}{\partial \bar{x}} = \frac{\partial u}{\partial x}\cos\theta + \frac{\partial u}{\partial y}\sin\theta$$

$$\frac{\partial^2 u}{\partial \bar{x}^2} = \frac{\partial^2 u}{\partial x^2}\cos^2\theta + \frac{\partial^2 u}{\partial x \partial y}2\sin\theta\cos\theta + \frac{\partial^2 u}{\partial y^2}\sin^2\theta$$

(b)

(i) For $\theta = 0$ show $\quad \dfrac{\partial^2 u}{\partial \xi^2} = \dfrac{\partial^2 u}{\partial x^2}$

(ii) For $\theta = \pi/3$ show $\quad \dfrac{\partial^2 u}{\partial \eta^2} = \dfrac{1}{4}\dfrac{\partial^2 u}{\partial x^2} + \dfrac{\sqrt{3}}{2}\dfrac{\partial^2 u}{\partial x \partial y} + \dfrac{3}{4}\dfrac{\partial^2 u}{\partial y^2}$

(iii) For $\theta = 2\pi/3$ show $\quad \dfrac{\partial^2 u}{\partial \zeta^2} = \dfrac{1}{4}\dfrac{\partial^2 u}{\partial x^2} - \dfrac{\sqrt{3}}{2}\dfrac{\partial^2 u}{\partial x \partial y} + \dfrac{3}{4}\dfrac{\partial^2 u}{\partial y^2}$

(c) Show the Laplace equation $\nabla^2 u = \dfrac{\partial^2 u}{\partial x^2} + \dfrac{\partial^2 u}{\partial y^2} = 0$ in a triangular network can be written

$$\nabla^2 u = \frac{2}{3}\left(\frac{\partial^2 u}{\partial \xi^2} + \frac{\partial^2 u}{\partial \eta^2} + \frac{\partial^2 u}{\partial \zeta^2}\right) = 0$$

(d) Use central difference approximations for the derivatives in part (c) and show there results the computational molecule

$$\nabla^2 u = \frac{3}{h^2}\left\{\begin{matrix} & 1 & & 1 & \\ 1 & & -6 & & 1 \\ & 1 & & 1 & \end{matrix}\right\} u = 0$$

for use with a triangular grid.

▶ 14.   **Polar coordinates.**

The Laplace operator $\nabla^2 u$ has the following form in polar coordinates (r,θ)

$$\nabla^2 u = \frac{\partial^2 u}{\partial r^2} + \frac{1}{r}\frac{\partial u}{\partial r} + \frac{1}{r^2}\frac{\partial^2 u}{\partial \theta^2} \qquad u = u(r,\theta).$$

In dealing with polar coordinates one usually constructs a grid structure involving circles and rays. In the Taylor series expansion given by equation (9.12) replace x by r and y by θ with ray spacings of $k = \Delta\theta$ and circular spacing of $h = \Delta r$ as illustrated in the figure. Denote by $u_{i,j} = u(r_i, \theta_j)$ the function $u = u(r,\theta)$ evaluated at the point r_i and θ_j of the grid structure.

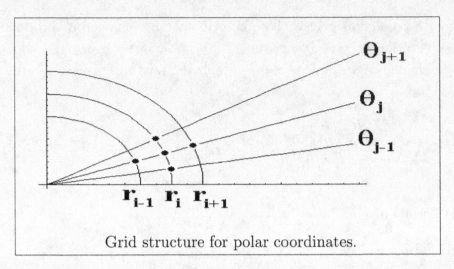

Grid structure for polar coordinates.

(a) Verify the derivative approximations

$$\frac{\partial u}{\partial r}\bigg|_{(r_i,\theta_j)} = \frac{u_{i+1,j} - u_{i-1,j}}{2h}$$

$$\frac{\partial^2 u}{\partial r^2}\bigg|_{(r_i,\theta_j)} = \frac{u_{i+1,j} - 2u_{i,j} + u_{i-1,j}}{h^2}$$

$$\frac{\partial^2 u}{\partial \theta^2}\bigg|_{(r_i,\theta_j)} = \frac{u_{i,j+1} - 2u_{i,j} + u_{i,j-1}}{k^2}$$

(b) Substitute the above derivative approximations into the Laplacian and verify the discrete representation

$$\nabla^2 u = \frac{u_{i+1,j} - 2u_{i,j} + u_{i-1,j}}{h^2} + \frac{1}{r_i}\left(\frac{u_{i+1,j} - u_{i-1,j}}{2h}\right) + \frac{1}{r_i^2}\left(\frac{u_{i,j+1} - 2u_{i,j} + u_{i,j-1}}{k^2}\right)$$

▶ **15.** **Three-dimensional computational molecule.**

Construct an equally spaced lattice structure in three-dimensions and use the notation $u_{i,j,k} = u(x_i, y_j, z_k)$ where $x_i = ih$, $y_j = jh$, and $z_k = kh$, to show that one discrete form of the Laplace equation

$$\nabla^2 u = \frac{\partial^2 u}{\partial x^2} + \frac{\partial^2 u}{\partial y^2} + \frac{\partial^2 u}{\partial z^2} = 0$$

is the approximation

$$\nabla^2 u \Big|_{x_i,y_j,z_k} = \frac{u_{i+1,j,k} - 2u_{i,j,k} + u_{i-1,j,k}}{h^2}$$

$$+ \frac{u_{i,j+1,k} - 2u_{i,j,k} + u_{i,j-1,k}}{h^2}$$

$$+ \frac{u_{i,j,k+1} - 2u_{i,j,k} + u_{i,j,k-1}}{h^2} = 0$$

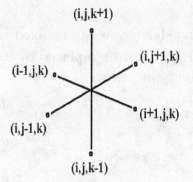

▶ **16.** Consider the one-dimensional heat or diffusion equation

$$\frac{\partial u}{\partial t} = D \frac{\partial^2 u}{\partial x^2}, \quad 0 < x < 1, \quad t \geq 0$$

which is subject to the boundary conditions $u(0,t) = T_0$, $u(1,t) = T_1$, and initial conditions $u(x,0) = 0$, where T_0, T_1 are constant temperatures.

Divide the interval $0 \leq x \leq 1$ into $N+1$ segments by defining

$$\Delta x_i = \frac{1}{N+1}, \quad \text{with} \quad x_i = i\Delta x_i = \frac{i}{N+1}, \quad \text{for} \quad i = 0,1,\ldots,N+1$$

(a) Show that from the Taylor series expansions

$$u(x + \Delta x, t) = u(x,t) + \frac{\partial u}{\partial x}\Delta x + \frac{\partial^2 u}{\partial x^2}\frac{\overline{\Delta x}^2}{2!} + \cdots$$

$$u(x - \Delta x, t) = u(x,t) - \frac{\partial u}{\partial x}\Delta x + \frac{\partial^2 u}{\partial x^2}\frac{\overline{\Delta x}^2}{2!} - \cdots$$

one can obtain the approximation

$$\frac{\partial^2 u}{\partial x^2} = \frac{u(x + \Delta x, t) - 2u(x,t) + u(x - \Delta x, t)}{(\Delta x)^2}$$

(b) Show that for a constant value of x, the heat equation can be approximated by

$$\frac{du}{dt} = \frac{D}{(\Delta x)^2}[u(x + \Delta x, t) - 2u(x,t) + u(x - \Delta x, t)] \qquad (16a)$$

(c) Let

$$u_1 = u(x_1, t), \; u_2 = u(x_2, t), \ldots, u_N = u(x_N, t)$$

and show by assigning fixed values of $x_1, \ldots, x_n$ to the variable x that equation (16(a)) can be replaced by a system of differential equations. In particular, show that when

$$x = x_1, \quad \frac{du_1}{dt} = \frac{D}{(\Delta x)^2}[u_2 - 2u_1 + T_0]$$

$$x = x_2, \quad \frac{du_2}{dt} = \frac{D}{(\Delta x)^2}[u_3 - 2u_2 + u_1]$$

$$x = x_3, \quad \frac{du_3}{dt} = \frac{D}{(\Delta x)^2}[u_4 - 2u_3 + u_2]$$

$$\vdots$$

$$x = x_N, \quad \frac{du_N}{dt} = \frac{D}{(\Delta x)^2}[T_1 - 2u_N + u_{N-1}]$$

(d) By defining $\bar{u} = \text{col}(u_1, u_2, \ldots, u_N)$ show that there results the system of differential equations

$$\frac{d\bar{u}}{dt} = A\bar{u} + \bar{f}, \qquad (16b)$$

where $\bar{f} = \frac{D}{(\Delta x)^2} \text{col}(T_0, 0, \ldots, 0, T_1)$ and A is the $N \times N$ tridiagonal matrix

$$A = \frac{D}{(\Delta x)^2}\begin{bmatrix} -2 & 1 & & & & \\ 1 & -2 & 1 & & & \\ & 1 & -2 & 1 & & \\ & & & \ddots & & \\ & & & 1 & -2 & 1 \\ & & & & 1 & 2 \end{bmatrix}$$

(e) The solution of the differential system (16b) is dependent upon the eigenvalues of A. Find the eigenvalues of the matrix A.

▶ **17.** Show the eigenvalues of the $N \times N$ tridiagonal matrix

$$A = \begin{bmatrix} 1-2r & r & & & & \\ r & 1-2r & r & & & \\ & r & 1-2r & r & & \\ & & & \ddots & & \\ & & & r & 1-2r & r \\ & & & & r & 1-2r \end{bmatrix}$$

are given by $1 - 4r \sin^2(n\pi/2(N+1))$ for $n = 1, 2, \ldots, N$.

"There is no result in nature without a cause; understand the cause and you will have no need for the experiment."

Leonardo da Vinci (1452-1519)

Chapter 10

Monte Carlo Methods

In very broad terms one can say that a computer simulation is the process of designing a model of a real or abstract system and then conducting numerical experiments using the computer to obtain a statistical understanding of the system behavior. That is, sampling experiments are performed upon the model. This requires that certain variables in the model be assign random values associated with certain probability distributions. This sampling from various probability distributions requires the use of random numbers to create a stochastic simulation of the system behavior. This stochastic simulation of system behavior is called a Monte Carlo simulation. Monte Carlo simulations are used to construct theories for observed behavior of complicated systems, predict future behavior of a system, and study effects on final results based upon input and parameter changes within a system. The stochastic simulation is a way of experimenting with a system to find ways to improve or better understand the system behavior.

Monte Carlo methods use the computer together with the generation of random numbers and mathematical models to generate statistical results that can be used to simulate and experiment with the behavior of various business, engineering and scientific systems. Some examples of application areas where Monte Carlo modeling and testing have been used are: the simulation and study of specific business management practices, modeling economic conditions, war games, wind tunnel testing of aircraft, operations research, information processing, advertising, complex queuing situations, analysis of mass production techniques, analysis of complex system behavior, analysis of traffic flow, the study of shielding effects due to radiation, the modeling of atomic and subatomic processes, and the study of nuclear reactor behavior. These are just a few of the numerous applications of Monte Carlo techniques.

Monte Carlo simulations usually employ the application of random numbers which are uniformly distributed over the interval $[0, 1]$. These uniformly distributed random numbers are used for the generation of stochastic variables from various probability distributions. These stochastic variables can then be used to approximate the behavior of important system variables. In this way one can generate sampling experiments associated with the modeling of system behavior. The statistician can then apply statistical techniques to analyze the data collected on system performance and behavior over a period of time. The generation of a system variable behavior from a specified probability distribution involves the use of uniformly distributed random numbers over the interval $[0,1]$. The quantity of these random numbers generated during a Monte Carlo simulation can range anywhere from hundreds to hundreds of thousands. Consequently, the computer time necessary to run a Monte Carlo simulation can take anywhere from minutes to months depending upon both the computer system and the application being simulated. The Monte Carlo simulation produces various numerical data associated with both the system performance and the variables affecting the system behavior. These system variables which model the system behavior are referred to as model parameters. The study of the sensitivity of model parameters and their affect on system performance is a large application area of Monte Carlo simulations. These type of studies involve a great deal of computer time and can be costly. We begin this introduction to Monte Carlo techniques with a discussion of random number generators.

Uniformly distributed random numbers

Examine the built in functions associated with the computer language you use most often. Most computer languages have some form of random number generator which can be used to generate uniformly distributed random numbers between 0 and 1. The majority of these random number generators use modulo arithmetic to generate numbers which appear to be uniformly distributed. Consequently, the random number generators are called pseudorandom number generators because they are not truly random. They only simulate the behavior of a uniformly distributed random number on the interval $[0,1]$. The basic equations used in generating these pseudorandom numbers usually have one of the forms

$$X_{i+1} \equiv AX_i \pmod{M} \qquad \text{A multiplicative congruential generator}$$

or $\qquad X_{i+1} \equiv AX_i + C \pmod{M} \qquad$ A mixed congruential generator

where A,C and M are nonnegative integers. The sequence of numbers $\{X_i\}$, for $i = 0, 1, 2, \ldots$, generated by a congruential generator, needs a starting value X_0. The initial value X_0 is called the seed of the random number generator. The quantity X_{i+1} represents one of the integers from the set $\{0, 1, 2, \ldots, M-1\}$. In general $A \equiv B \pmod{M}$ is read A is congruent to B modulo M, where the quantity A is calculated from the relation $A = B - K * M$, where $K = [B/M]$ denotes the largest positive integer resulting from the truncation of B/M to form an integer. The quantity A represents the remainder when B is divided by M. By using modulo arithmetic, the numbers of the sequence $\{X_i\}$ eventually start to repeat themselves and so only a finite number of distinct integers are generated by the above methods. The number of integers generated by the congruential generator before repetition starts to occur is called the period of the pseudorandom number generator.

Using the mixed congruential generator, with a given seed X_0, one can write

$X_1 = AX_0 + C - MK_1 \qquad K_1$ is some appropriate constant.

$X_2 = AX_1 + C - MK_2 \qquad K_2$ is some appropriate constant.

Substituting for X_1 and simplifying gives

$X_2 = A^2 X_0 + C(1 + A) - M(K_2 + AK_1)$

$X_3 = AX_2 + C - MK_3 \qquad K_3$ is some appropriate constant.

Substituting for X_2 and simplifying gives

$X_3 = A^3 X_0 + C(1 + A + A^2) - M(K_3 + K_2 A + K_1 A^2)$

$\vdots \qquad$ Continuing in this manner one can show

$X_n = A^n X_0 + C(1 + A + A^2 + \cdots + A^{n-1}) - MK$

where $K = K_n + K_{n-1}A + \cdots + K_1 A^{n-1} = \sum_{i=1}^{n} K_i A^{n-i}$ is some new constant. Knowledge of the geometric series enables us to simplify the above result to the form $X_n = A^n X_0 + \dfrac{C(A^n - 1)}{A - 1} \mod M$. Now if for some value of n we have $X_0 = A^n X_0 + \dfrac{C(A^n - 1)}{A - 1} \mod M$, then one can write

$$(A^n - 1)\left[X_0 + \frac{C}{A-1}\right] \equiv 0 \mod M \qquad (10.1)$$

The minimum value of n which satisfies the equation (10.1) is the period of the pseudorandom number generator. The special case where $C = 0$ is easier to understand. In this special case one can employ the Fermat theorem that if M is prime and A is not a multiple of M, then $A^M \equiv 1 \mod M$. Then as a special case of the equation (10.1) $n = M = P$ is the period of the pseudorandom number generator.

The equation (10.1) has been extensively analyzed using number theory and one conclusion is that the quantity M can be selected as a power of 2. The power to which 2 is raised is based upon the number of bits b associated with a computer word size. The largest integer that can be stored using b-bits is $2^b - 1$. Selecting $M = 2^b$ enables the largest possible period to be obtained for the computer system being used. Alternatively, M can be selected as a large prime number compatible with the computer word size. Once M is selected the value of A must satisfy $0 < A < M$. The sequence of values $\{X_i\}$ are then determined by the values M, A, X_0 and C and so the sequence generated will have a period $P \leq M$. That is, $P \leq M$ distinct values will be generated before the sequence starts to repeat itself. For each index i, the uniformly distributed pseudorandom numbers R_i are calculated from the relation $R_i = X_i/M$ and satisfy $0 < R_i < 1$. Optimally, one should select the quantities M, A, X_0, and C to maximize the period of the sequence and reduce the degree of correlation between the numbers generated. However, this is not always achieved.

There are many Fortran language packages created and sold under different brand names. Some Fortran languages treat the random number generator as a subroutine and in others the random number generator is treated as an intrinsic function. In the Fortran examples that follow it is assumed that there exists a subroutine RANDOM(RN) which generates a uniform random number RN between 0 and 1. It is also assumed there is a Fortran subroutine SEED(INTEGER) for setting the seed value for the pseudo random number generator. These subroutines can then be used to create other subroutines which return random variates satisfying various conditions. Some examples are illustrated.

```
      SUBROUTINE RAN1(IX,N)
C     Generate random integer IX
C     Satisfying   0 .LE. IX .LE. N
      CALL RANDOM(RN)
      IX=INT(RN*(N+1))
      RETURN
      END
```

```
      SUBROUTINE RAN2(NMIN,NMAX, N)
C     Generate random integer N
C     satisfying   NMIN .LE. N .LE. NMAX
      CALL RANDOM(RN)
      N=NMIN +INT((NMAX+1-NMIN)*RN)
      RETURN
      END
```

Chi-square χ^2 goodness of fit

The chi-square goodness of fit test is used to compare actual frequencies from sampled data with frequencies from theoretical distributions. The chi-square statistic is calculated from the relation

$$\chi^2 = \sum_{k=1}^{n} \frac{(f_{ok} - f_{ek})^2}{f_{ek}} \tag{10.2}$$

where f_{ok} is the observed frequency of the kth class or interval, f_{ek} is the expected frequency of the kth class or interval due to theoretical considerations, and n is the number of classes or intervals. Let the theoretical distribution function be denoted by $F(x)$. From this theoretical probability distribution one can calculate the probability p_k that a random variable X takes on a value in the kth interval. One finds $f_{ek} = np_k$ as the number of theoretical expected values in the kth class or interval. In using the chi-square goodness of fit test, one is testing the null hypothesis H_0 that there is no significant difference between the frequencies of the sampled data and that expected from theoretical considerations. Ideally, if $\chi^2 = 0$ then the observed frequencies and theoretical frequencies agree exactly. For $\chi^2 > 0$ there is a discrepancy between the observed and theoretical frequencies and one must resort to tables of χ^2 critical values, associated with a significance level and degrees of freedom, to determine if one should accept or reject the null hypothesis. If the computed value of χ^2 is greater than the tabular critical value at some significance level found in tables, then one must reject the null hypothesis.

As an example, apply the chi-square goodness of fit test to the random number generator associated with your computer programming language. Write a computer program to generate 1000 random numbers between 0 and 1. You can then divided the interval 0 to 1 into 10 classes using the intervals $(0, .1), (.1, .2), \cdots, (.9, 1.0)$ and then sort the 1000 random numbers to determine the number in each class. These values are the observed frequencies determined by the experiment. If the pseudorandom number generator is truly uniform, then the theoretical frequency associated with each class would have a value of 100. Note that when using the chi-square goodness of fit test one should always use actual counts for the frequencies. Do not use relative frequencies or percentages. Also the theoretical frequencies associated with each class or interval should number greater than 5. If this is not the case then adjacent intervals or nearest neighbor intervals must be combined into a new class or interval with frequency greater than or equal to 5. The degrees of freedom ν associated a chi-square test is given by $\nu = n - 1 - m$ where n is the number of classes or intervals and m is the number of parameters in the theoretical distribution being tested. For the chi-square goodness of fit test associated with our random number generator test, the value of the degrees of freedom is $\nu = 10 - 1 - 0 = 9$. One would then find a chi-square table of critical values, such as the one on page 497, and look up the critical value associated with a $(1 - \alpha)$ significance level by selecting an appropriate column and then select a row of the table which represents the degrees of freedom. The tabulated value found is then compared with the calculated χ^2-value to determine if the random number generator differs significantly from the theoretical values expected.

Example 10-1. **(Chi-square test.)**

A pseudorandom number generator associated with a certain Fortran computer language was used to generate 1000 random numbers X. The frequencies associated with 10 equal spaced intervals over the range $(0,1)$ where calculated and are given in the accompanying table.

Use a chi-square test to compare the resulting frequencies with theoretical values associated with a uniform distribution of random numbers. The chi-square statistic is found to be

$$\chi^2 = \sum_{k=1}^{10} \frac{(f_{ok} - 100)^2}{100} = 14.90$$

X range	Frequency	Symbol
$X \leq 0.1$	106	f_{o1}
$0.1 < X \leq 0.2$	110	f_{o2}
$0.2 < X \leq 0.3$	79	f_{o3}
$0.3 < X \leq 0.4$	99	f_{o4}
$0.4 < X \leq 0.5$	88	f_{o5}
$0.5 < X \leq 0.6$	98	f_{o6}
$0.6 < X \leq 0.7$	125	f_{o7}
$0.7 < X \leq 0.8$	107	f_{o8}
$0.8 < X \leq 0.9$	97	f_{o9}
$0.9 < X \leq 1.0$	91	f_{o10}

This number is compared with the tabular value of 23.5893 from the $\chi^2_{0.995}$ column and $\nu = 9$ row of the chi-square table of critical values found on page 497. Since the χ^2-value 14.90 is less than the critical value of 23.5893 we can accept the numbers generated by the Fortran computer code as representing a uniform random number generator. ∎

Discrete and continuous distributions

In constructing Monte Carlo simulations it is important that one know how to generate random variables X which come from a specified probability distribution. Let X denote a random variable with cumulative probability distribution function $F(x)$ and let RN, denote a uniformly distributed random number $0 \leq RN \leq 1$. If $RN = F(X)$, then the inverse function gives $X = F^{-1}(RN)$. The situation is illustrated in the figure 10-1.

The computer generation of a random variable X associated with a discrete or continuous distribution can be illustrated graphically. Calculate the distribution function $F(x)$ associated with a discrete or continuous probability function or relative frequency function $f(x)$. Then generate a uniform random number RN, $0 \leq RN \leq 1$ and plot this number on the $F(x)$ axis and then move horizontally until you hit the distribution function curve. Then drop down to obtain the random variable X. This is the inverse function method of generating a random variable X associated with a given distribution.

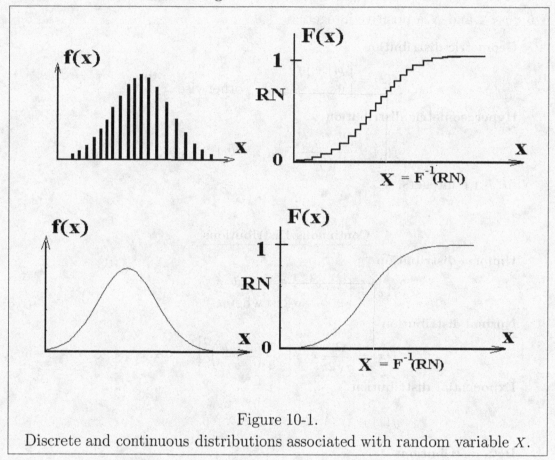

Figure 10-1.
Discrete and continuous distributions associated with random variable X.

The following is a list of some of the more popular discrete and continuous probability distributions that can be used to help model various Monte Carlo simulations.

Discrete Distributions

Discrete uniform distribution

$$f(x) = \begin{cases} \frac{1}{N_2 + 1 - N_1}, & x = N_1, N_1 + 1, \ldots, N_2 \\ 0 & \text{otherwise} \end{cases}$$

N_1, N_2 integers with $N_2 > N_1$.

Poisson distribution

$$f(x) = \begin{cases} e^{-\lambda}\lambda^x/x! & x = 0, 1, 2, \ldots \\ 0 & \text{otherwise} \end{cases} \quad \lambda > 0.$$

Binomial distribution

$$f(x) = \begin{cases} \binom{N}{x}p^x(1-p)^{N-x}, & x = 0, 1, 2, \ldots, N \\ 0 & \text{otherwise} \end{cases}$$

$0 < p < 1$ and N a positive integer.

Geometric distribution

$$f(x) = \begin{cases} p(1-p)^{x-1}, & x = 1, 2, \ldots \\ 0 & \text{otherwise} \end{cases} \quad 0 < p < 1.$$

Hypergeometric distribution

$$f(x) = \frac{\binom{M}{x}\binom{N-M}{J-x}}{\binom{N}{J}} \text{ for } x = 0, 1, 2, \ldots, J,$$

N, M, J are integers.

Continuous Distributions

Uniform distribution

$$f(x) = \begin{cases} \frac{1}{b-a}, & a < x < b \\ 0 & \text{elsewhere} \end{cases}$$

Normal distribution

$$f(x) = \frac{1}{\sigma\sqrt{2\pi}} \exp\left[-\frac{1}{2}\left(\frac{x-\mu}{\sigma}\right)^2\right] \quad -\infty < x < \infty$$

Exponential distribution

$$f(x) = \lambda e^{-\lambda x} \quad x > 0 \text{ and } \lambda > 0.$$

Beta distribution

$$f(x) = \begin{cases} \frac{\Gamma(\alpha+\beta)}{\Gamma(\alpha)\Gamma(\beta)}x^{\alpha-1}(1-x)^{\beta-1} & 0 < x < 1 \\ 0 & \text{elsewhere} \end{cases} \quad \alpha > 0, \beta > 0.$$

Gamma distribution

$$f(x) = \begin{cases} \frac{1}{\Gamma(\alpha)\beta^\alpha}x^{\alpha-1}e^{-x/\beta} & x > 0 \\ 0 & \text{elsewhere} \end{cases}, \quad \alpha > 0, \beta > 0.$$

Weibull distribution

$$f(x) = \begin{cases} \alpha\beta x^{\beta-1}e^{-\alpha x^\beta} & x > 0 \\ 0 & \text{elsewhere} \end{cases} \quad \alpha > 0, \beta > 0.$$

Log-Normal distribution

$$f(x) = \begin{cases} \frac{1}{\beta\sqrt{2\pi}} \frac{1}{x} e^{-(\ln x - \alpha)^2/2\beta^2} & x > 0 \\ 0 & \text{elsewhere} \end{cases} \qquad \beta > 0.$$

Chi-square distribution

$$f(x) = \begin{cases} \frac{1}{2^{\nu/2}\sigma^\nu\Gamma(\nu/2)} x^{(\nu/2)-1} e^{-(x/2\sigma^2)} & x > 0 \\ 0 & x < 0 \end{cases}$$

where ν represents the degrees of freedom.

Student's t-distribution

$$f(x) = \frac{1}{\sqrt{n\pi}} \frac{\Gamma[(n+1)/2]}{\Gamma[n/2]} \left(1 + \frac{x^2}{n}\right)^{-(n+1)/2}$$

with $n = 1, 2, 3, \ldots$ degrees of freedom.

F-distribution

$$f(x) = \begin{cases} \frac{\Gamma[(m+n)/2]}{\Gamma(m/2)\Gamma(n/2)} (m/n)^{m/2} \frac{x^{(m/2)-1}}{[1+(m/n)x]^{(m+n)/2}} & x > 0 \\ 0 & x < 0 \end{cases}$$

with parameters $m = 1, 2, \ldots$ and $n = 1, 2, \ldots$

In the discussions that follow we develop programs to generate random variates from only a few select probability distributions. The more complicated random variate generators are left for more advanced simulation courses. In this introduction to Monte Carlo simulations we consider only discrete empirical distributions, binomial distributions, Poisson distributions, normal distributions and exponential distributions as these distributions are easy to work with and are representative of how one employs various discrete and continuous probability distributions for modeling purposes.

Selected discrete distributions

Recall that associated with a discrete sample is the function

$$f(x) = \begin{cases} f_j & x = x_j \quad j = 1, 2, 3, \ldots \\ 0 & \text{otherwise} \end{cases} \qquad \sum_{k=1}^{\infty} f_j = 1 \qquad (10.3)$$

where f_j are relative frequencies associated with the sample. The function $f(x)$ given by equation (10.3) is also referred to as the probability distribution function of the sample. To calculate the probability $P(a < X \leq b)$ one would calculate

$$P(a < X \leq b) = \sum_{a < x_j \leq b} f(x_j) \qquad (10.4)$$

The function $F(x)$ representing the cumulative relative frequency function is given by

$$F(x) = P(X \leq x) = \sum_{x_j \leq x} f(x_j) \quad \text{with} \quad P(X > x) = 1 - F(x) \tag{10.5}$$

and is called the distribution function associated with the sample. Note that the above definition implies

$$P(a < X \leq b) = P(X \leq b) - P(X \leq a) = F(b) - F(a) \tag{10.6}$$

The mean μ associated with a discrete distribution is defined

$$\mu = \sum_j x_j f(x_j)$$

The variance σ^2 associated with a discrete distribution is defined

$$\sigma^2 = \sum_j (x_j - \mu)^2 f(x_j).$$

The positive square root of the variance gives the standard deviation σ. For X a random variable and $g(X)$ any continuous function, then the mathematical expectation of $g(X)$ is defined for discrete distributions as

$$E(g(X)) = \sum_j g(x_j) f(x_j) \quad (X \text{ discrete}).$$

Empirical distributions associated with a sample are generated by collecting data and calculating the frequency, relative frequency and cumulative relative frequency associated with the data.

The **Binomial probability distribution**, which is sometimes referred to as a Bernoulli distribution, is given by

$$f(x) = \begin{cases} \binom{n}{x} p^x (1-p)^{n-x} & \text{for } x = 0, 1, 2, \ldots, n \\ 0 & \text{otherwise} \end{cases}$$

and contains the two parameters p and n. The parameter p is a probability satisfying $0 \leq p \leq 1$ and the parameter $n = 1, 2, 3, \ldots$. This probability distribution is used in application areas where one of two possible outcomes can result. For example, the number x of successes in n independent repeated events in which the probability of success p for each event is governed by the binomial distribution.

421

This distribution has mean $\mu = np$ and variance $\sigma^2 = np(1-p)$ and is used in studying sampling with replacements.

The **Poisson probability distribution** is given by

$$f(x) = \begin{cases} e^{-\lambda}\lambda^x/x! & \text{for } x = 0, 1, 2, \dots \\ 0 & \text{otherwise} \end{cases}$$

and contains the single parameter $\lambda > 0$ which is a constant. The Poisson probability distribution is used to represent the occurrence of isolated events during some period of time. For example, the number of persons coming to a waiting line during a time interval and the arrival of incoming telephone calls during a time interval are described by a Poisson distribution. The Poisson distribution has mean $\mu = \lambda$ and variance $\sigma^2 = \lambda$.

Selected continuous distributions

A random variable X associated with a probability density function $f(x) > 0$, $-\infty < x < \infty$ which is continuous except for possibly a finite number of jump discontinuities produces a distribution function

$$F(x) = \int_{-\infty}^{x} f(x)\,dx$$

with the property that

$$\int_{-\infty}^{\infty} f(x)\,dx = 1$$

To calculate the probability $P(a < X \le b)$ one needs to calculate

$$P(a < X \le b) = F(b) - F(a) = \int_{a}^{b} f(x)\,dx$$

which represents the area under the probability density function $f(x)$ between the values $x = a$ and $x = b$

The mean value μ associated with a continuous distribution is defined as

$$\mu = \int_{-\infty}^{\infty} x f(x)\,dx$$

The variance σ^2 associated with a continuous distribution is defined

$$\sigma^2 = \int_{-\infty}^{\infty} (x - \mu)^2 f(x)\,dx.$$

The positive square root of the variance is called the standard deviation σ. For X a random variable and $g(X)$ any continuous function, then the mathematical expectation of $g(X)$ for continuous distributions is defined as

$$E(g(X)) = \int_{-\infty}^{\infty} g(x)f(x)\,dx \quad (X \text{ continuous}).$$

A variable X associated with a **Normal distribution** or Gaussian distribution has the probability density function

$$f(x) = \frac{1}{\sigma\sqrt{2\pi}} \exp\left[-\frac{1}{2}\left(\frac{x-\mu}{\sigma}\right)^2\right], \qquad -\infty < x < \infty$$

with the parameters μ, $-\infty < \mu < \infty$ and $\sigma > 0$. This type of distribution is associated with errors in measurements, bell shaped curves, the central limit theorem. It is also the basis for the chi-square, student-t, and F-distributions used in testing of hypothesis, and determining confidence intervals associated with various statistical samples. The normal distribution is symmetric with respect to the line $x = \mu$ and has mean μ and variance σ^2. The distribution function $F(x)$ associated with the normal probability distribution function $f(x)$ can be represented

$$F(x) = \frac{1}{\sigma\sqrt{2\pi}} \int_{-\infty}^{x} \exp\left[-\frac{1}{2}\left(\frac{x-\mu}{\sigma}\right)^2\right]\,dx. \tag{10.7}$$

If a random variable X comes from a normal distribution with mean μ and variance σ^2, then the variable $Z = \frac{X-\mu}{\sigma}$ has a mean of 0 and a variance of 1. The variable Z is called a standardized or normalized variable corresponding to the variable X. The distribution function $F(x)$ can be expressed in terms of the normalized distribution function

$$\Phi(z) = \frac{1}{\sqrt{2\pi}} \int_{-\infty}^{z} e^{-u^2/2}\,du \tag{10.8}$$

where one can verify that with a change of variable

$$F(x) = \Phi\left(\frac{x-\mu}{\sigma}\right). \tag{10.9}$$

Tables for the normalized integral given by equation (10.8) can be found in most elementary statistics books.

The **exponential probability distribution** has a parameter $\lambda > 0$ and can be written

$$f(x) = \begin{cases} \lambda e^{-\lambda x} & \text{for } x > 0 \\ 0 & \text{otherwise} \end{cases}$$

This distribution has a mean $\mu = \frac{1}{\lambda}$ and variance of $\sigma^2 = \frac{1}{\lambda^2}$. Exponential distributions are associated with time intervals between the occurrence of events or waiting time for the next event to occur. For example, waiting for an elevator, or waiting to be served while in a line at a bank or store. This distribution also occurs in studying the life expectancy of electrical equipment, and the length of time for a telephone conversation. The distribution function $F(x)$ associated with this probability distribution is given by

$$F(x) = \int_0^x \lambda e^{-\lambda x}\, dx = 1 - e^{-\lambda x}.$$

There are many other types of discrete and continuous probability density functions that can be associated with a random variable X. The above examples are but a small sampling of the many probability distributions that exist.

Example 10-2.   (Continuous distribution.)

To generate a random variable X from the uniform distribution

$$f(x) = \frac{1}{b-a}, \quad a \le x \le b$$

with mean $\mu = \frac{1}{a+b}$

and variance $\sigma^2 = \frac{1}{12}(b-a)^2$

one can calculate the distribution function

$$F(x) = \begin{cases} 0 & x < a \\ \frac{x-a}{b-a} & a \le x \le b \\ 1 & x > b \end{cases}$$

The generation of a random variable X from this distribution one can use the inverse function of the

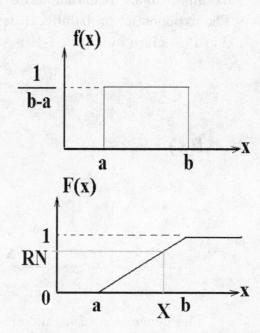

distribution function. In other words, solve the equation $RN = \dfrac{X - a}{b - a}$ for the variable X to obtain

$$X = a + (b - a) * RN$$

where RN denotes a uniform random number between 0 and 1. One can construct a Fortran subroutine for the generation of uniform random numbers between $x = a$ and $x = b$ having the form

```
SUBROUTINE UNIFORM(A,B,X)
Call RANDOM(RN)
X=A+(B-A)*RN
RETURN
END
```

where RANDOM is the internal Fortran call statement to generate uniform random numbers RN satisfying $0 \leq RN \leq 1$.

■

Example 10-3.   (Continuous distribution.)
The exponential probability distribution $f(x) = \lambda e^{-\lambda x}$ has the distribution function $F(x)$ given by

$$F(x) = \int_0^x f(x)\,dx = 1 - e^{-\lambda x}$$

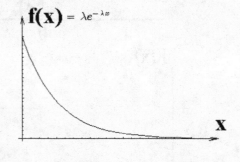

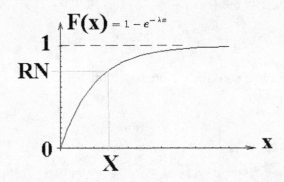

The generation of a random variable X from this distribution can be constructed from the inverse of the distribution function $F(x)$. If RN is a uniform random number which satisfies $RN = 1 - e^{-\lambda X}$, then solving for X we obtain

$X = -\dfrac{1}{\lambda}\ln(1-RN)$. Note that RN is between 0 and 1 and so $1-RN$ is also between 0 and 1 and if RN is random then so is $1-RN$. Consequently, we can replace the above equation with the equivalent result $X = -\dfrac{1}{\lambda}\ln RN$.

A Fortran subroutine for the generation of random variates from an exponential distribution can be constructed having the form

```
        SUBROUTINE EXPONEN(LAMBDA,X)
        REAL LAMBDA
100     CONTINUE
        CALL RANDOM(RN)
        IF( RN .LE. 0.0) GO TO 100
        X=-ALOG(RN)/LAMBDA
        RETURN
        END
```

Many distributions do not have nice inverses and so alternate means of generating random numbers from these distributions are developed. Such is the case with the normal distribution having mean μ and variance σ^2. The Box and Muller method[†] is used to construct a Fortran subroutine for generating random normally distributed numbers $R1$ and $R2$ which represent random variates from a normal distribution with mean μ and standard deviation σ.

```
        SUBROUTINE NORMAL(MU,STD,R1,R2)
        REAL MU
100     CONTINUE
        CALL RANDOM(RN1)
        CALL RANDOM(RN2)
        X1=2.0*RN1-1.0
        X2=2.0*RN2-1.0
        TEST=X1*X1+X2*X2
        IF(TEST .GE. 1.0) GO TO 100
        IF(TEST .LE. 0.0) GO TO 100
        SCALE=SQRT((-2.0*ALOG(TEST))/TEST)
        RAN1=X1*SCALE
        RAN2=X2*SCALE
        R1=MU+RAN1*STD
        R2=MU+RAN2*STD
        RETURN
        END
```

[†] G.E. Box, M.E. Muller, *A note on the Generation of Random Normal Deviates*, Annals of Mathematical Statistics, Vol. 29, 1858.

The Poisson distribution is a discrete distribution and again special steps are needed to obtain the equivalent of an inverse function in order to construct a random variable X from this distribution. Tocher[‡] has developed a method for generating random numbers from a Poisson distribution based upon multiplying uniform random numbers until the following condition is satisfied

$$\prod_{j=0}^{x} RN_i \geq e^{-\lambda} > \prod_{j=0}^{x+1} RN_i.$$

One can use this test to construct the Fortran subroutine

```
        SUBROUTINE POISSON(LAMBDA,X)
        REAL LAMBDA
        X=0.0
        TEST=EXP(-LAMBDA)
        PROD=1.0
100     CONTINUE
        CALL RANDOM(RN)
        PROD=PROD*RN
        IF((PROD-TEST) .LE. 0.0) RETURN
        X=X+1.0
        GO TO 100
        END
```

Monte Carlo examples

The following are some examples of Monte Carlo simulations using some of the above distributions.

Example 10-4. (Discrete Empirical distribution.)

Construct a Fortran subroutine to simulate the rolling of a pair of dice.

Solution: For illustration we investigate a method that employs a discrete probability function and associated cumulative distribution function. We begin by examining all possible outcomes from rolling a pair of honest dice. These events are described by the sample space $S = \{2, 3, 4, 5, 6, 7, 8, 9, 10, 11, 12\}$ from which one can construct the table 10.1 of all possible outcomes.

The table 10.1 gives data for the construction of graphs for the discrete probability function $f(x)$ and cumulative distribution function or cumulative relative frequency function $F(x)$. Graphs of these functions are illustrated in the figures 10-2 and 10-3. Observe that the cumulative distribution function can be used to generate random throws of the dice. One can generate a uniform random

---

[‡] See the reference K.D. Tocher.

number RN, $0 \leq RN \leq 1$ and then plot this point on the $F(x)$ axis. By moving horizontally over to the curve and then dropping down one finds the random variable X representing the sum of the dice.

	Table 10.1			
Sum of dice	Combinations (die 1, die 2)	Frequency	Relative frequency $f(x)$	Cumulative relative frequency $F(x)$
2	(1,1)	1	1/36	1/36
3	(1,2),(2,1)	2	2/36	3/36
4	(1,3),(2,2),(3,1)	3	3/36	6/36
5	(1,4),(2,3),(3,2),(4,1)	4	4/36	10/36
6	(1,5),(2,4),(3,3),(4,2),(5,1)	5	5/36	15/36
7	(1,6),(2,5),(3,4),(4,3),(5,2),(6,1)	6	6/36	21/36
8	(2,6),(3,5),(4,4),(5,3),(6,2)	5	5/36	26/36
9	(3,6),(4,5),(5,4),(6,3)	4	4/36	30/36
10	(4,6),(5,5),(6,4)	3	3/36	33/36
11	(6,5),(5,6)	2	2/36	35/36
12	(6,6)	1	1/36	36/36

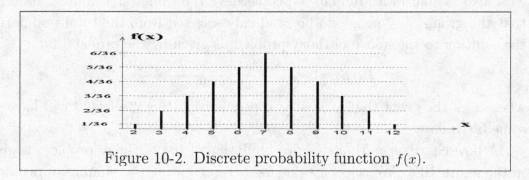
Figure 10-2. Discrete probability function $f(x)$.

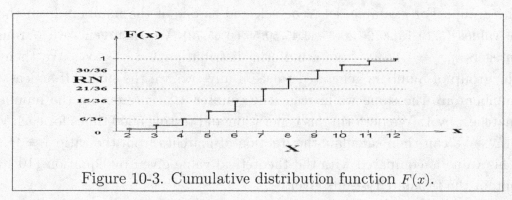
Figure 10-3. Cumulative distribution function $F(x)$.

428

One can use the data generated to construct a Fortran subroutine, similar to the one below, for generating the random throw of dice.

```
      SUBROUTINE DICESUM(X)
      DIMENSION CUMSUM(11)
      INTEGER X
      DATA CUMSUM/1.,3.,6.,10.,15.,21.,26.,30.,33.,35.,36./
      CALL RANDOM(RN)
      X=2
100   CONTINUE
      IF(RN .LE. CUMSUM(X-1)/36.) RETURN
      X=X+1
      GO TO 100
      END
```

■

Example 10-5.   (The birthday problem.)

The following is a simple example of a Monte Carlo simulation. In the study of probability there arises the birthday problem. What is the probability that at least two people in a group of N-people have the same birth date? Assuming that the group of N-people to be randomly selected from the total population, the solution to the above birthday problem is given by the probability

$$P_N(E) = 1 - \frac{(365)(364)\cdots(365-N+1)}{(365)^N} \qquad (10.10)$$

where E is the event that at least two people from the N selected will have the same birth date.

A flow chart for a Monte Carlo simulation of the birthday problem is given in the figure 10-4. We select a value for N representing the number of people in the group. For example, let N be selected as one of the numbers from the set of values $\{5, 10, 15, 20, 25, 30, 35, 40, 45, 50, 55, 60, 65, 70\}$. We then generate N random integers $n_i, i = 1, \ldots, N$ from a uniform distribution $1 \le n_i \le 365$. We then test the group of numbers generated to see if any two are the same. If at least two numbers are the same we increase the counter M, representing the number of matches, by 1. We perform this generation and testing 5000 times for each value of N. We can then calculate the fraction of matches from the ratio $f = M/5000$. This value is compared with the theoretical value given by equation (10.10) to obtain the tabulated results given.

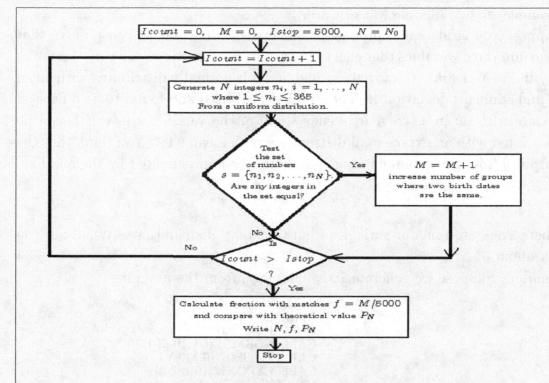

$Icount = 0, \quad M = 0, \quad Istop = 5000, \quad N = N_0$

$Icount = Icount + 1$

Generate N integers n_i, $i = 1, \ldots, N$
where $1 \leq n_i \leq 365$
From a uniform distribution.

Test
the set
of numbers
$s = \{n_1, n_2, \ldots, n_N\}$.
Are any integers in
the set equal?

Yes

$M = M + 1$
increase number of groups
where two birth dates
are the same.

No

Is
$Icount > Istop$
?

No

Yes

Calculate fraction with matches $f = M/5000$
and compare with theoretical value P_N
Write N, f, P_N

Stop

Figure 10-4. Flow chart for Monte Carlo simulation of birthday problem.

N	f	P_N
5	.2520000E-01	.2713555E-01
10	.1220000E+00	.1169482E+00
15	.2474000E+00	.2529013E+00
20	.4202000E+00	.4114384E+00
25	.5718000E+00	.5686997E+00
30	.7106000E+00	.7063162E+00
35	.8106000E+00	.8143833E+00
40	.8944000E+00	.8912318E+00
45	.9448000E+00	.9409759E+00
50	.9686000E+00	.9703736E+00
55	.9886000E+00	.9862623E+00
60	.9948000E+00	.9941227E+00
65	.9984000E+00	.9976831E+00
70	.9996000E+00	.9991596E+00

Results from Monte Carlo simulation of the birthday problem.

You may wish to experiment and do the simulation for 10,000 trials and other values for N. Today's computers do the simulation rather quickly. ∎

Example 10-6. (Simple Monte Carlo)

Suppose you study some process in a manufacturing plant. You observe that there are three variables that effect the output. You decide to call these variables x, y and z. You study the variable x and find it is normally distributed with mean 50 and standard deviation 10. The variable y is tested and found to be a Poisson variate with mean 0.342. A frequency study of the variable z reveals it can be associated with an exponential distribution with mean 0.187. You think that the output g from the manufacturing process can be approximated by the relation

$$g = g(x, y, z) = x * z + y$$

where x, y, z are random variables from the above distributions. What can you say about g?

Solution: Suppose we generate 100 values for g from the program

```
                    DIMENSION g(100)
                    DO 100 I=1,100
                    CALL NORMAL(50.,10.,X,R2)
                    CALL POISSON(0.342,Y)
                    CALL EXPONEN(0.187,Z)
                    g(I)=X*Z+Y
        100         CONTINUE
```

The above example program can be modified and used to test your hypothesis concerning the behavior of the manufacturing process. The results from running this program might be plotted to obtain something like the figure given.

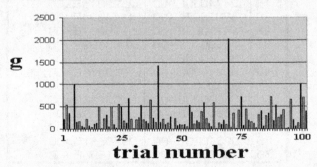

The above program can be modified to perform specific studies. Some example studies are listed.

(i) Studying the effect of the variability of the inputs x, y, z on the output g.

(ii) Studying possible ways to alter variable behavior and outcomes.

(iii) Studying alternative ways to obtain g by computer testing.

(iv) A way to study g and possibly construct a theory for the manufacturing process results.

(v) You could change the parameters used in the above distributions, or even change the distribution associated with a variable, to model future behavior under different circumstances.

(v) Modify the above model for g so that it better describes the manufacturing process.

The central limit theorem states that if we repeat the generating of 100 trials a large number of times, say N, then the average value for g, calculated from each computer run, must be a normal variate having some mean μ and standard deviation σ. The values for μ and σ can be calculated from the data collected from the N computer runs. For illustration, the value of $N = 500$ is selected and the average values are plotted in a frequency distribution to obtain the result illustrated

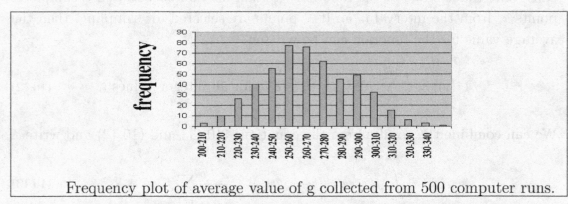

Frequency plot of average value of g collected from 500 computer runs.

For the data collected, the above distribution has a mean $\mu = 266.93$ and a standard deviation of $\sigma = 25.97$. This example illustrates that no matter how complicated a system is, one can employ the central limit theorem and run a Monte Carlo simulation a large number of times to collect many averages. These averages will give normal distributions which can be analyzed to obtain means, variances, standard deviations, and confidence intervals associated with system behavior.

The above program represents an over simplified representation of the fol-

432

lowing idea. If the result g from a manufacturing process or processes can be represented by some functional relationship between the variables involved $x, y, z, \ldots$, then one can write $g = g(x, y, z, \ldots)$. One then has a functional relationship which describes the operating characteristics between the variables within a system. This would be an example of a deterministic relationship between the variables of a system as opposed to a stochastic relationship between the variables. One can then devise computer studies with the objective of improving or better understanding the system behavior.

■

Example 10-7.   (Integration)

The integral $I = \int_a^b f(x)\, dx$ can be evaluated by generating uniform random numbers from the interval (a, b). The average value of the function $f(x)$ defined over the interval (a, b) is given by

$$\langle f(x) \rangle = \frac{1}{b-a} \int_a^b f(x)\, dx. \tag{10.11}$$

Note that the average of the function $f(x)$ can also be based upon a sample of points x_i from the interval (a, b). If N points are selected for sampling, then the average value for the function can be written

$$\langle f(x) \rangle = \frac{1}{N} \sum_{i=1}^{N} f(x_i) \qquad x_i \in (a, b) \text{ for all integer values } i. \tag{10.12}$$

We can combine the results from the equations (10.11) and (10.12) and write

$$I = \int_a^b f(x)\, dx = (b-a)\langle f(x) \rangle = \frac{b-a}{N} \sum_{i=1}^{N} f(x_i). \tag{10.13}$$

Note that polynomial interpolation can be interpreted as a sample the points x_i in a small equally spaced manner which are used to obtain an approximation for the integral. Whenever this is not possible, then the Monte Carlo sampling is an alternative.

As an example we use a uniform random number generator to evaluate the integral $I = \int_0^{\pi} \sin x\, dx$ using the equation (10.13) with $f(x) = \sin x$. We used the time variable from our digital machine multiplied by a large prime number as input to the SEED subroutine to change the starting values of the uniform random numbers generated. We then generated 100 uniformly distributed random numbers in the interval $(0, \pi)$ and calculated the results from equation (10.13). The program was executed 15 times with the following results for the integral I.

{2.021276, 1.829944, 1.705557, 2.172054, 2.046782, 1.829049, 2.131321, 2.047764,

1.95345, 2.116943, 1.95207, 2.020918, 2.029641, 2.013587, 2.164814}

Descriptive statistics on the above results produced the following table of values.

Mean	2.002345
Standard error	0.03404
Median	2.021276
Standard deviation	0.131837
Variance	0.017381

You can perform other Monte Carlo experiments with the above integral and can compare your results with the exact value of $I = 2$.

■

Queuing Theory

As one would expect, queuing theory deals with all kinds of problems associated with queuing and waiting for a service. It can include waiting for failure of a piece of electronic equipment or waiting for a part or piece of machinery to fail. Most individuals are familiar with queues associated with banks and supermarkets where they have to wait in line.

Queues form because there are not enough resources to go around. Some terminology associated with queues are "customers", (customers is used as a generic term and can denote people, machine parts, cars, etc.) , "waiting time", "service", (service is used as a generic term indicating how customers are handled), "service time", "arrival process", "arrival time", and "service time distribution". Queues are also classified by how customers are treated. Queues can be FIFO (First-In-First-Out), or FCFS (First-Come-First-Served). Queues can be finite or infinite in length. Various types of queuing systems are illustrated in the figure 10-5.

The Kendall queue notation has the form

$$A/S/s/c/p/Q$$

and is used to classify queuing systems. In the above notation, A denotes an arrival process description with options of M,D or G, where M is for Markovian (Poisson arrival distribution or exponential interarrival distribution), D is for deterministic or constant value and G is for general distribution with a known mean and variance. The second letter S is for a service process description representing a probability distribution. The lower case s,c and p denote respectively the number of servers, the capacity of the queue and the population being served. The default values for c and p are infinite. The last term Q is for the queuing discipline. The default value for Q is FIFO. For example, the simplest single queue is denoted M/M/1 which represents a single line FIFO queue with a single server where the arrival and service times are Markovian.

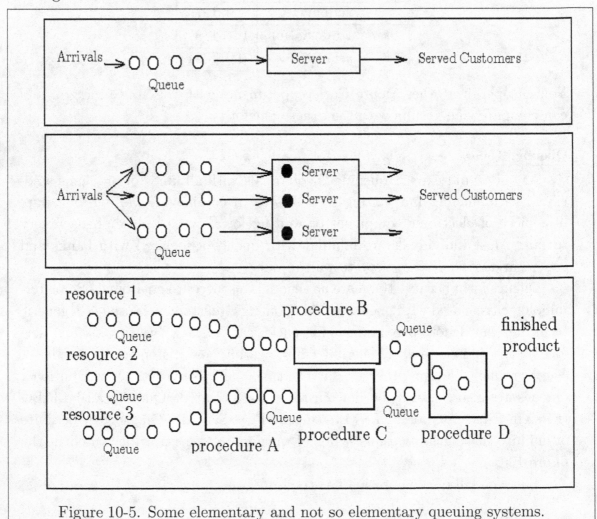

Figure 10-5. Some elementary and not so elementary queuing systems.

Example 10-8.   (Waiting time)

The problem to be simulated using random numbers is the waiting time associated with visits to a doctors office. We present an elementary model for simplicity. You can add complications later on. The following assumptions are made.

(i) Appointment times for morning and afternoon sessions are 15 minutes apart. There are 16 morning appointments beginning at 8:00AM and ending at 11:45AM. There are fourteen afternoon appointments beginning at 1:15PM and ending at 4:30PM. We assume that all patients keep their appointments. That is, we do not consider such things as last minute cancellations, or patients who can't wait any longer and have to leave for another appointment.

(ii) Patients arrive X minutes early for their appointments where X is a random number from a Poisson distribution, with parameter $\lambda = 3.8$.

(iii) There is a 2 minute interval of time before the doctor sees the next patient. (i.e. The 2 minute interval, for this example, is fixed. However, later on one might want to associate this time with a normal distribution about 2 minutes. This time is used to simulate other work that has to be done which is associated with the office routine.)

(iv) The service time (ST) is the time the doctor spends with the patient. It is assumed to be normally distributed with mean 15 minutes and standard deviation of 1.0 minute, with the constraint that if $ST < 12$, then set $ST = 12$ minutes. (i.e. It is assumed that the minimum service time is 12 minutes.)

Using the above assumptions one can construct a subroutine to simulate such things as the waiting time associated with each patient, the average number of customers in the queue at any time, the average time that a patient spends in the waiting room. One can set a dollar figure on each apart of the queuing process and then try to figure out how to minimize the cost associated with the overall handling of patients. One can try to simulate a "better" way to accomplish the objectives of serving the patients. The figure 10-6 depicts a flowchart for a simple subroutine for simulating the waiting time. A sample result from simulating just one day of activity is given in the accompanying tables.

Morning appointments	
Scheduled Appointment Time	Waiting Time (minutes)
8:00	2.30
8:15	4.71
8:30	7.47
8:45	6.90
9:00	16.21
9:15	13.36
9:30	13.21
9:45	17.61
10:00	15.88
10:15	25.78
10:30	22.56
10:45	29.45
11:00	28.08
11:15	29.24
11:30	29.91
11:45	31.79

Afternoon appointments	
Scheduled Appointment Time	Waiting Time (minutes)
1:15	4.00
1:30	6.12
1:45	10.30
2:00	13.61
2:15	11.91
2:30	13.44
2:45	16.23
3:00	15.20
3:15	17.21
3:30	18.45
3:45	22.91
4:00	19.95
4:15	21.27
4:30	25.58

The simulation of one week of activity gives a frequency diagram similar to the figure illustrated.

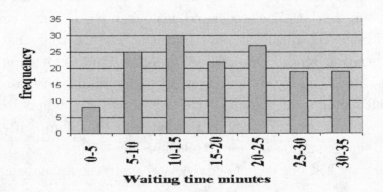

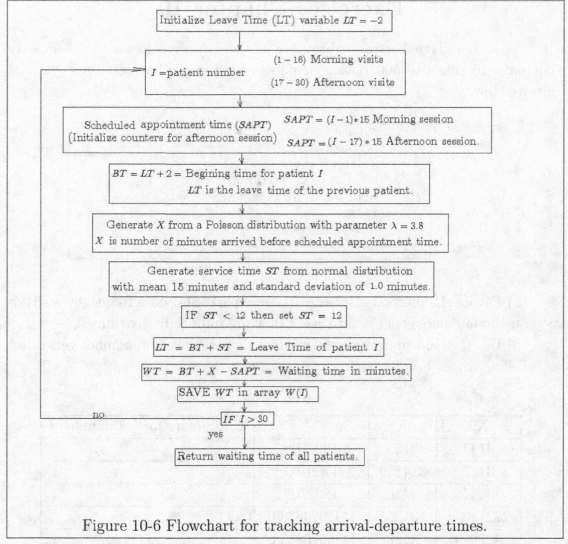

Figure 10-6 Flowchart for tracking arrival-departure times.

This example illustrates some of the necessary bookkeeping required to study queues. One must keep track of arrival times, service times, and waiting times. The model can be modified in many ways. For example, the model can be made more complicated by considering cancellations and handling of emergency cases. The parameters associated with the distributions can be adjusted. The scheduling times of appointments can be adjusted. One can do simulations over many weeks of activity. The data collected can then be analyzed statistically. The more complicated the model the more bookkeeping is required in order to do the necessary statistics on many sample runs of the model.

Exercises Chapter 10

▶ **1.** Use the given Fortran subroutine or its equivalent in another computer language to generate 500 random numbers for the parameter values $X0, A, M$ given below.

```
        SUBROUTINE URANNUM(X0,A,M,RN)
        INTEGER X0,X1,A,M
        X1=MOD(X0*A,M)
        RN=FLOAT(X1)/FLOAT(M)
C       REPLACE X0 BY X1 FOR NEXT CALL
        X0=X1
        RETURN
        END
```

(a) Divide the interval $(0,1)$ into 10 equal parts and do a frequency analysis on the numbers generated to see if they are uniformly distributed.

(b) Use the chi-square test to determine a "best" random number generator. Define what "best" means.

X0	A	M	Chi-Square for 500 numbers
3147	3125	1299709	
3147	314159	1299709	
3147	899	1299709	
27211871	3125	343597383337	
427989	3125	343597383337	

▶ **2.**

(a) Use the sample space $S = \{1,2,3,4,5,6\}$ and plot a graph representing the probability function $f(x)$ associated with the throw of a single honest die.

(b) Plot a graph of the corresponding distribution function $F(x)$.

(c) If you generate a random number RN, $0 \le RN \le 1$ and plot this point on the $F(x)$ axis and then move horizontally until you hit the graph and then drop down you obtain a random variable X. What does X represent?

(d) How would you use the results from part (c) to construct the throw of two honest dice?

▶ **3.**   (Computer Problem) Use the following subroutine (or something equivalent to it in another computer language) for generating random variates from a binomial distribution.

```
                 SUBROUTINE BINOMIAL(N,P,X)
        C        P=probability of success
        C        N is the number of trials
        C        X is the number of successes in N trials
                 X=0.0
                 DO 100 I=1,N
                 CALL RANDOM(RN)
                 IF((RN-P) .LE. 0.0) X=X+1.0
        100      CONTINUE
                 RETURN
                 END
```

Use the Chi-square goodness of fit test to evaluate this subroutine. Write out in detail how you will set up and test the above subroutine. Discuss the results from the execution of your testing.

The binomial probability distribution can be written

$$b = b(x; n, p) = \frac{n!}{x!(n-x)!}p^x(1-p)^x \qquad \text{for } x = 1, 2, 3, \ldots$$

Note that the binomial distribution has two parameters n and p. The parameter n is an integer $1, 2, 3, \ldots$ representing the number of independent trials (sometimes referred to as Bernoulli trials) and the parameter p represents a probability $0 < p < 1$. Binomial distributions occurs in areas where one of two possible outcomes are possible. For example, (yes,no),(success,failure),(left,right),(use of item, no use of item). The quantity $b = b(x; n, p)$ represents the probability of getting x successes in n trials, where p is the probability of success in any single trial.

▶ **4.**   (Computer Problem)
Use the chi-square goodness of fit test to evaluate the
        SUBROUTINE NORMAL(MU,STD,R1,R2).
given in the text. Write out in detail how you will set up and test the above subroutine. Discuss the results from the execution of your testing.

The normal probability distribution or normal probability density is given by

$$f(x; \mu, \sigma) = \frac{1}{\sigma\sqrt{2\pi}}e^{-\frac{1}{2}\left(\frac{x-\mu}{\sigma}\right)^2}, \qquad -\infty < x < \infty.$$

The normal probability function has two parameters, μ the mean and σ the standard deviation. This type of distribution occurs in errors of measurement. It is sometimes referred to as a bell shaped curve. It is the theoretical basis for the chi-square, student-t and F-distributions used for testing of hypothesis and establishing confidence intervals associated with independent sampling.

▶ **5.** **(Computer Problem)**

Use the chi-square goodness of fit test to evaluate the

SUBROUTINE POISSON(LAMBDA,X).

given in the text. Write out in detail how you will set up and test the above subroutine. Discuss the results from the execution of your testing.

The Poisson probability distribution

$$p(x; \lambda) = \begin{cases} \frac{\lambda^x}{x!} e^{-\lambda}, & \text{for } x = 0, 1, 2, 3, \ldots \\ 0, & \text{otherwise} \end{cases}$$

has a single parameter λ. This type of distribution represents the probability of occurrence of isolated events over a period of time. For example, the number of telephone lines in use, the number of customers waiting in a line, the number of incoming phone calls during a time interval, the number of cars passing an intersection during a time interval. In general, the Poisson distribution is used to represent the probability of x successes during a time interval.

▶ **6.** **(Computer Problem)**

Use the chi-square goodness of fit test to evaluate the

SUBROUTINE EXPONEN(LAMBDA,X).

given in the text. Write out in detail how you will set up and test the above subroutine. Discuss the results from the execution of your testing.

The exponential probability distribution or exponential density is given by

$$f(x; \lambda) = \begin{cases} \lambda e^{-\lambda x}, & \text{for } x > 0, \\ 0, & \text{otherwise} \end{cases}$$

It has one parameter $\lambda > 0$. It occurs in life testing of electronic equipment (time for failure of a light bulb, computer, transistor, etc.), waiting time between arrivals. In general, it occurs in representing the waiting time for a next event to occur.

► 7.

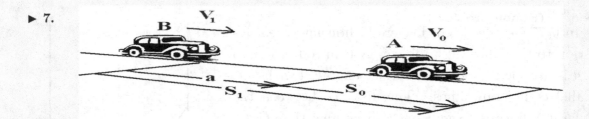

We consider a previous problem from chapter 2. The stopping distance of a car is a function of reaction time and reaction time distance plus braking distance. Stopping under rainy conditions or during night time driving conditions adds to this stopping distance. Data from the department of motor vehicles gives the following approximate formula for stopping distance S in feet of a vehicle moving with velocity of V miles per hour under rain and night time conditions $S = -2.0 + 2.4V + 0.059V^2 + 0.000028V^3$, for $20 \leq V \leq 70$.

$V_0 = 40\,\text{mph}$	
a (feet)	V_1 (mph)
10	
20	
30	
40	
50	

You are in car B traveling at V_1 miles per hour and approach the slower moving car A which is traveling at V_0 miles per hour. Assume that when the distance between cars A and B is a-feet, car A suddenly hits its brakes.

We modify this problem by adding a random variable X to the stopping distance S and assume the stopping distance is given by

$$S = X - 2.0 + 2.4V + 0.59V^2 + 0.000028V^3, \quad 20 \leq V \leq 70$$

where X is generated from a normal distribution with mean 5 and standard deviation of 1.

(a) Simulate the stopping distances of 10 cars using the distances given in the table and determine the average minimum velocity V_1 for an accident to occur if V_0 is 40 miles per hour. Perform this simulation many times to see if the averages follow a normal distribution.

(b) Make up tables similar to the one illustrated for cases where V_0 is greater than 40 mph, say 50 mph and 60 mph.

(c) Assume the stopping distance of the first car has added to it a random variable from a normal distribution. Assume some values for the means and standard deviation and create a Monte Carlo simulation where the stopping distances of both cars have a certain variability. Do averages on many runs and report your results.

442

▶ **8.** (**Buffon problem**)
In 1777 Georges Louis Leclerc-Buffon gave the solution to the following problem. A thin rod of length 2ℓ is dropped upon a flat surface engraved with parallel equidistant lines. The distance between the parallel lines is $2a$ and one can assume that $\ell < a$. Take the parallel line nearest to the center c of the rod and call it the y-axis and then construct an x-axis through the center of the rod. The situation is illustrated in the figure. Find the probability that the rod will intersect one of the parallel lines.

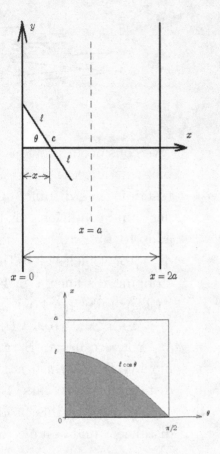

Let x denote the distance of the point c from the origin and make use of symmetries that exist in the problem. An examination of the accompanying figure shows that the rod will only intersect the parallel line if $x < \ell \cos \theta$. Think of x as a random variable between 0 and a and let θ denote a random variable between 0 and $\pi/2$. Then the number pair (θ, x) can be thought of as a point somewhere inside the rectangle illustrated. Graph the function $x = \ell \cos \theta$ on this figure also. An examination of the rectangular figure shows that the probability that x is less than $\ell \cos \theta$ is given by

$$P\{x < \ell \cos \theta\} = \frac{\text{Area under } x = \ell \cos \theta \text{ curve between 0 and } \pi/2.}{\text{Total area}}$$

$$P\{x < \ell \cos \theta\} = \frac{\ell}{\frac{\pi}{2}a} = \frac{2\ell}{\pi a}$$

(a) Select values for ℓ and a and generate uniform random numbers for x and θ in the appropriate ranges. Then perform a simulation for the dropping of a thin rod 1000 times. Keep count of the number of intersections that occur and compare the ratio of (number of intersections)/(total number of drops) with the theoretical probability given above. Examine the cases $\ell = a$ and $\ell < a$.

(b) Perform the simulation 10,000 times and calculate the probability.

(c) Perform the simulation in part (a) 10 times and calculate the average probability.

(d) Perform the simulation in part (c) 100 times and do a frequency analysis on the averages. Find the mean and standard deviation for your results.

► **9.** **(Random walk)**

A random walker steps a unit length either East, South, West or North at each step. Assume that at each step she marches North with a probability of 0.4, East with probability 0.3, South with probability 0.1 and West with probability 0.2. Assume the walker starts at the origin. Set up a simulation to determine the average distance of the walker from the origin

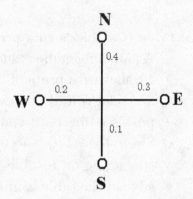

after 100 steps. Can you determine an average vector pointing to where the walker will be after 100 steps? Can you put confidence intervals on the position of the walker after 100 steps?

► **10.** **(Random walk)**

(a) A random walk in two dimensions can be simulated by generating a uniformly distributed random number between 0 and 2π, which represents a random direction θ. It is assumed that each step taken can then be represented by a vector of length $\ell = $ a constant, in the direction θ. One can then write

$$\text{x new position} = \text{x old position} + \ell \cos \theta$$

$$\text{y new position} = \text{y old position} + \ell \sin \theta$$

One can then draw vectors representing each step taken from the previous position. Assume the walker takes random steps of length $\ell = 1/2$ in a random direction starting at the origin. Set up a simulation to determine the walkers average distance from the origin after 100 random steps?

(b) A random walk in three dimensions can be simulated by generating a uniformly distributed random number θ between 0 and π and then generating a uniformly distributed random number ϕ between 0 and 2π. One can then model the vector step in three dimensions using the equations

$$\text{x new position} = \text{x old position} + \ell \sin \theta \cos \phi$$

$$\text{y new position} = \text{y old position} + \ell \sin \theta \sin \phi$$

$$\text{z new position} = \text{z old position} + \ell \cos \theta$$

Assume the walker takes random steps of length $\ell = 1/2$ in a random direction

$$\vec{\Omega} = \sin \theta \cos \phi \, \hat{i} + \sin \theta \sin \phi \, \hat{j} + \cos \theta \, \hat{k}.$$

Assuming the walker starts at the origin, then set up a simulation to determine the walkers average distance from the origin after 100 steps.

444

▶ **11.** (**Gambler's ruin problem**)

A gambling game is such that you can win 1 dollar with probability ν and lose 1 dollar with probability $(1-\nu)$ at each turn. Let P_n denote the probability of financial ruin if you currently have n dollars. Let N denote the total money possessed by both you and your opponent.

(a) Show $P_n = \nu P_{n+1} + (1-\nu)P_{n-1}$. Notice this game is equivalent to a drunk staggering on a table top which is divided into N equal segments. The left edge of the table is marked 0 and the right-edge of the table is marked N and the segments in between are labeled consecutively $1, 2, 3, \ldots, n, \ldots, N-1$. At time $t=0$, the drunk is at position n, and at each successive time interval, he staggers (random walks) either toward N with probability ν or toward 0 with probability $(1-\nu)$. Here P_n denotes the probability a drunk, starting at n, falls off the table at 0 .

(b) Solve the difference equation and find P_n.

Hint: For $n=0$, $P_0 = 1$, and for $n = N$, $P_N = 0$.

(c) What is the solution in the case $\nu = 1/2$?

(d) Write a Monte Carlo simulation to test your answers.

▶ **12.**

Write a Monte Carlo simulation to calculate the volume of the spherical cap inside the sphere $x^2+y^2+z^2 = 1$ and bounded by the plane $x+y+z = 1$. The situation is illustrated in the accompanying figure. This example illustrates that Monte Carlo simulations can be used to calculate areas and volumes associated with more complicated configurations. (Exact answer $= 0.482129$)

Hint: Construct cube and find distance of random point from plane and sphere.

▶ **13.**

Write a Monte Carlo simulation to calculate the area bounded by the curves $y = e^{-x^2/2}$ and $y = 0.4$ as illustrated in the accompanying figure.

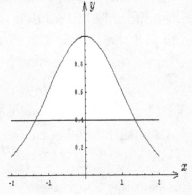

"What we know is not much. What we do not know is immense. "

Pierre Simon De Laplace (1749-1827)

Chapter 11
Miscellaneous Numerical Methods

The following sections contain a very brief introduction into selected miscellaneous topics and examples from some specialty areas of numerical methods and analysis. The topics selected reflect upon current research of the time. We give a very brief look at parallel computing and the message passing interface language MPI. We take a passing glance at such areas as integral equations, Bézier curves, and B-splines. These are some areas of numerical methods and analysis that you might want to study in more depth. The material presented is far from complete and should be viewed only as introductory examples which are intended to serve as motivation for some readers to get more involved in the subject areas.

Parallel Computer Systems

Imagine that you have a computer code to run and instead of a single computer executing your computer program you had N ($N > 1$) computers hooked up in parallel so that they all could communicate with each other and exchange information. How would you make use of all this extra computing power? This concept of a cluster of computers all doing calculations and communicating with each other is called parallel computing. The concept of serial and parallel computing is illustrated in the figure 11-1. We will be interested in situations where computer codes can be broken up so that the computations can be divided among the many computers available. The information calculated by the cluster of parallel computers can then be shared to compute the desired output.

Assuming that all the computers in serial and parallel modes are the same, then what kind of reduction in computational speed can one expect when using parallel computing? It turns out that most computer codes run faster in a parallel mode as compared to a serial mode. The reduction in computing time depends upon the type and number of parallel computers, the kind of code being run and how much message passing is done between computers within the cluster. For example, if a computer code, which runs in a serial mode, is modified to run in a parallel mode, then you can define the ratio

$$\frac{\text{Parallel Computing Time}}{\text{Serial Computing Time}} = f,$$

where all computers are the same. In comparing the run times of the computer codes, one will probably obtain an approximation like $f = .3 + \beta/N < 1$ for some positive constant β. Don't think that because you have N computers in parallel, that f will be around $1/N$ as this is unrealistic. It is the message passing that slows things down.

Not all computer codes are amenable for conversion to run under a parallel structure. Sometimes it is wiser to spend your money on a good serial computer. The kind of computer structure "best" for you will depend heavily upon the type of applications you are using your computer system to solve.

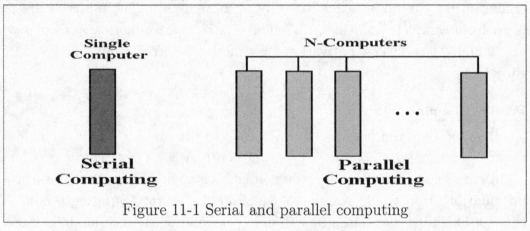

Figure 11-1 Serial and parallel computing

Building a parallel computer system is not hard. There are many individuals who collect old throw away computers and hook them up in a parallel structure. Loosely speaking any collection of computers that is capable of running a parallel code is called a Beowulf system or Beowulf cluster. (There are many definitions of a Beowulf system.) Hardware for connecting the computers comes in a variety of forms from inexpensive, with slow communication between computers, to expensive, with high speed communication between computers. The good news is that most software for multi-computer structures is free. For example, the MPI (Message-Passing Interface) libraries consist of Fortran, C and C++ language macros, functions and subroutines that can be inserted into Fortran, C and C++ computer codes to convert them to run within a parallel structure. There are many predefined constants built into the MPI libraries and one should note that variable names, function names and subroutine names beginning with MPI_ are protected and you should not change them. The following is a brief introduction into MPI programming for the Fortran computer language as this seems to be the language of choice of most individuals involved in scientific computing. As you begin your study of parallel computing you might think about how parallel computer systems can be employed to speed up numerical procedures. You will find by examining the scientific literature that parallel computing is being applied to almost all areas of numerical computing and new ideas are welcomed.

I cannot tell you how to compile and execute your MPI parallel code. At this time the commands for compiling and executing a parallel computer code varies widely from one computer system to another. Compiling and running codes for a parallel cluster also depends upon how environment variables are set up for your computer system. I suggest you contact the administrator of the parallel system you will be using to find out the compile and execute commands associated with your parallel computer system. On some unix systems the compile and run commands have the form

f77 -o compiledprogram fortranprogram.f -lmpi

mpirun -np 6 compiledprogram

where $-np$ is a mpirun processor flag which is followed by an integer representing the number of processor required to run the compiled code. The number 6 was selected as an example.

The following Fortran programs contain a small selection of MPI commands with a brief explanation as to their function. I will not go into detail about the function of each MPI command, but will leave it to the reader to purchase a text on MPI programming to find out the details associated with all the message passing between computers. The following computer codes are given strictly as an introduction for the purpose of illustrating selected applications and usages of parallel computing. The illustrative examples are just to get you started into some of the introductory aspects of parallel computing by way of examples. I will leave the complexities associated with parallel computing for additional study.

Example 11-1. (Getting started.)
The first example program is a variation of the 'HELLO WORLD' program found in

many introductory MPI manuals. In this first example, and all of the subsequent examples, the MPI library commands are written in bold face type so that you take notice of them. In the first example program it is assumed that it is compiled to run on 6-processors. That is, a copy of the program is placed upon each of the processors. What each processor does is controlled by the program.

The first MPI command to recognize is the

INCLUDE 'mpif.h'

command which is the first directive to be placed at the beginning of the Fortran code. The file mpif.h contains all the necessary information for compiling an MPI Fortran code. At some place after this first directive and before any MPI commands are issued, there must occur the initialization statement

CALL MPI_INIT(IERR)

where IERR is an error code with 0 denoting that there is no error. At the end of the program, after all MPI library usage has ended, there must occur the finalization statement

CALL MPI_FINIALIZE(IERR).

This statement will clear up any MPI commands that have not been completed during the execution of the program.

The commands

CALL MPI_COMM_RANK(MPI_COMM_WORLD, ID-PROC, IERR)

CALL MPI_COMM_SIZE(MPI_COMM_WORLD, NP, IERR)

determines the rank of each processor (ID-PROC) and number of processors (NP). The parameter MPI_COMM_WORLD is a predefined integer variable which is associated with information about all the processors being used in the program execution. It is referred to as a communicator argument. The parameter ID-PROC is the identification number assigned to each processor at start up. For the first example program, we use 6 processors and so each processors is assigned its own identification number of 0,1,2,3,4,5. Sometimes the processor 0 is referred to as the master processor. The command MPI_COMM_SIZE returns the integer variable NP denoting the number of processors being used.

Always remember that a copy of the code goes to each processor and that each processor has its own identification number. The first example code tests to see if the correct number of processors are being used. It then has each processor write out its identification number. Note that the order of the output depends upon which processor finishes first. The bottom half of the program has the message 'GOOD NIGHT WORLD' sent from the processor with identification rank 0 to the other processors and illustrates the use of the commands

CALL MPI_SEND(SENDDATA, ICOUNT, DATATYPE, IDESTIN, ITAG, ICOMM, IERR)

CALL MPI_RECV(RECVDATA, ICOUNT, DATATYPE, ISOURCE, ITAG, ICOMM, STATUS, IERR)

used to send and receive information. The parameters SENDDATA, RECVDATA represent the data being sent or received. The parameters ICOUNT and DATATYPE are used for system identification of the data sent and received. The MPI data types are

MPI_INTEGER	MPI_COMPLEX
MPI_REAL	MPI_LOGICAL
MPI_DOUBLE_PRECISION	MPI_CHARACTER

The parameters IDESTIN and ISOURCE are integers specifying the ranks of the receiving and sending processors. The ITAG parameter and MPI_COMM_WORLD parameters represent an integer tag and communicator. The parameter STATUS has dimension 2 and contains information regarding the data actually received. The first element of STATUS represents the source and the second element represents the tag.

After each processor writes out its processor number and value of NP, the code has an IF statement which tests the processor identification number. If this number is zero, then processor 0 creates the message 'GOOD NIGHT WORLD. A do-loop is then created to send this message to the processors 1,2,3,4,5. Note that each processor is executing the program. When a processor encounters the IF statement testing the processor number, then if the processor identification number is not zero, the processor is told that it should set up to receive the message being sent from processor zero. Always remember, a copy of the Fortran code is sent to each processor and each processor is executing the same code. The code then tells each processor what it should do. In the first half of the example program each processor was told to write its identification number and the value of NP (total number of processors). The second half of the program tells processor zero to create a message and then send the message to the other processors. All the processors with rank greater than zero are told to get ready to receive a message. The parameters within the MPI_SEND and MPI_RECV commands are used to distinguish between data being sent and received. This is the message-passing part of MPI that slows down the over-all run time of the code. After receiving the message that was sent, each processor is told to write out its processor identification number, the value of NP and the message that was received. The MPI_FINALIZE(IERR) is a must statement at the end of the MPI usage.

The MPI library has around 125 commands built in. The type of commands and number of commands are changing year to year. The six basic commands

MPI_INIT	MPI_SEND
MPI_COMM_RANK	MPI_RECV
MPI_COMM_SIZE	MPI_FINALIZE

are all that is needed for many of the elementary uses of parallel processors running MPI Fortran codes.

```
        PROGRAM EXAMPLE1
C           Assume 6 processors are being used
        CHARACTER *16 MESSAGE
        INCLUDE 'mpif.h'
        INTEGER STATUS
C
        CALL MPI_INIT(IERR)
        CALL MPI_COMM_RANK(MPI_COMM_WORLD, ID-PROC, IERR)
        CALL MPI_COMM_SIZE(MPI_COMM_WORLD, NP, IERR)
        IF(NP .NE. 6) STOP 'WRONG NUMBER OF PROCESSORS'
        WRITE(*,100) ID-PROC,NP
100     FORMAT(1X,' hello world! I am processor number ',I3,
     1  ' OUT OF A TOTAL OF ',I3, ' PROCESSORS')
        TAG=5
        IF(ID-PROC .EQ. 0) THEN
        MESSAGE='GOOD NIGHT WORLD'
        IDNO=NP-1
        DO 10 I=1,IDNO
        CALL MPI_SEND(MESSAGE, 16, MPI_CHARACTER, I, TAG,
     1      MPI_COMM_WORLD, IERR)
10      CONTINUE
        ELSE
        CALL MPI_RECV(MESSAGE, 16, MPI_CHARACTER, 0, TAG),
     1      MPI_COMM_WORLD, STATUS, IERR)
        ENDIF
        WRITE(*,101) ID-PROC,NP,MESSAGE
101     FORMAT(1X,'I am finished with processor number ',I3,
     1    ' OUT OF A TOTAL OF ',I3, ' PROCESSORS',1x,A16)
C
        CALL MPI_FINALIZE(IERR)
        STOP
        END
```

Example 11-2. (Another example.)

Another example of parallel programing, using the above commands, is given in the computer program example2. The example2 code assumes the use of 4 processors where each processor is told to run a different computer code.

```
      PROGRAM EXAMPLE2
C         Assume 4 processors are being used
      INCLUDE 'mpif.h'
C

      CALL MPI_INIT(IERR)
      CALL MPI_COMM_RANK(MPI_COMM_WORLD, ID, IERR)
      CALL MPI_COMM_SIZE(MPI_COMM_WORLD, NP, IERR)
C         ID is the processor number
C         NP is the number of processors
      IF(NP .NE. 4) STOP 'WRONG NUMBER OF PROCESSORS'
C         ID=0,1,2,3 are the processor numbers
C         job 1 done on processor 0, job 2 done on processor 1, etc.
      IF(ID .EQ. 0) THEN
      {put job number 1 here}
      ELSE IF (ID .EQ. 1) THEN
      {put job number 2 here}
      ELSE IF (ID .EQ. 2) THEN
      {put job number 3 here}
      ELSE
      {put job number 4 here}
      END IF
      CALL MPI_FINALIZE(IERR)
      STOP 'FINISHED'
      END
```

Example 11-3.

The next example evaluates the integral $I = \int_a^b f(x)\,dx$ using 32 processors. Therefore, we divide the integral I into 32 parts and write

$$I = \int_{a_0}^{b_0} f\,dx + \int_{a_1}^{b_1} f\,dx + \cdots + \int_{a_i}^{b_i} f\,dx + \cdots + \int_{a_{31}}^{b_{31}} f\,dx$$

where $a_0 = a$ and $b_{31} = b$. The ith integration $I_i = \int_{a_i}^{b_i} f(x)\,dx$ is to be performed on processor number i, $i = 0, 1, 2, \ldots, 31$. Here the distance $b_i - a_i$ is the same for each integral I_i and is given by

$$b_i - a_i = \frac{b-a}{32}.$$

Therefore, each processor can use its own identification number to calculate the local limits of integration a_i and b_i. The local starting value a_i is found from the relation

$$a_i = a + ID(b-a)/32$$

where ID is the processor identification number. The local upper limit b_i is found from the relation

$$b_i = a_i + (b-a)/32$$

for $i = 1, 2, \ldots, 31$. We can define a local step size $h = (b_i - a_i)/N$ for each processor and calculate the local integral I_i using the trapezoidal rule. For purposes of illustration we select to use $N = 30$ panels for each local integral. After each processor calculates the local integral the value obtained is sent back to processor zero where all the information is summed.

In the program example3, we have selected $a = 10$ and $b = 100$ for the lower and upper limit, these numbers can be changed. The subroutine for calculating the local area using the trapezoidal rule and the function being integrated needs to be defined and append to the program example3. The following subroutine and function statement are representative of these required items.

```
      SUBROUTINE TRAP(A, B, N, H, AREA)
C        Trapezoidal rule for area under curve FUN(x)
      EXTERNAL FUN
      Area=(FUN(A)+FUN(B))/2.
      x=A
      DO 100 I=1,N-1
      x=x+H
      AREA=AREA+FUN(X)
100      CONTINUE
      AREA=AREA*H
      RETURN
      END


      FUNCTION FUN(X)
C        Function to be integrated
      FUN={put your function here}
      RETURN
      END
```

```
      PROGRAM EXAMPLE3
C         Assume 32 processors are being used
      INTEGER STATUS
      PARAMETER( a=10 , b= 100 )
      PARAMETER(N=30, ITAG=10, IDESTIN=0)
      INCLUDE 'mpif.h'
      CALL MPI_INIT(IERR)
      CALL MPI_COMM_RANK(MPI_COMM_WORLD, ID, IERR)
      CALL MPI_COMM_SIZE(MPI_COMM_WORLD, NP, IERR)
C         ID=0,1,2,3,...,31 are the processor numbers
C         Local limits of integration are given by
      ai=a+ID*(b-a)/NP
      bi=ai+(b-a)/NP
C         Local step size H is
      H=(bi-ai)/N
C         Have each processor calculate local area
      CALL TRAP(ai, bi, N, H, AREAi)
C         Have processor 0 get ready to sum results and receive information
C         TAREA is total area which is sum of areas AREAi
      IF(ID .EQ. 0) THEN
      TAREA=AREAi
C         Get ready to receive information from other processors
      DO 100 ISOURCE=1,NP-1
      CALL MPI_RECV(AREAi, 1, MPI_REAL, ISOURCE, ITAG,
     1 MPI_COMM_WORLD, STATUS, IERR)
      TAREA=TAREA+Areai
100       CONTINUE
      ELSE
C         If processor ID greater than 0, then send results to processor 0
      CALL MPI_SEND(AREAi, 1, MPI_REAL, IDESTIN, ITAG,
     1 MPI_COMM_WORLD, IERR)
      END IF
C         Write out results
      IF(ID .EQ. 0) THEN
      WRITE(*,200) TAREA
200       FORMAT(1x,'AREA BY TRAPEZOIDAL RULE = ',E14.7)
      END IF
      CALL MPI_FINIALIZE(IERR)
      STOP
      END
```

Example 11-4.

The next example program gets a little more complicated. The program
example4 is designed to test the transfer of an array using the MPI_SEND and
MPI_RECV commands. The example assumes the use of 4 processors.

```fortran
      PROGRAM EXAMPLE4
C        Assume 4 processors are being used-fill array with integer values 1-48
      DIMENSION A(3,4,4)
      INTEGER STATUS
      INCLUDE 'mpif.h'
      CALL MPI_INIT(IERR)
      CALL MPI_COMM_RANK(MPI_COMM_WORLD, ID, IERR)
      CALL MPI_COMM_SIZE(MPI_COMM_WORLD, NP, IERR)
C        Each processor fills in part of the array
      J=ID+1
      DO 20 I=1,3
      DO 21 K=1,4
      A(I,J,K)=I+3.*(J-1)+12*(K-1)
21        CONTINUE
20        CONTINUE
C        Define new array ANEWA describing blocks, block length, stride
      CALL MPI_TYPE_VECTOR(4, 3, 12, MPI_REAL, ANEWA, IERR)
C        You must commit this new array to all processors
      CALL MPI_TYPE_COMMIT(ANEWA, IERR)
C        Note in sending new array – only position of first element need be specified.
      IF(ID .NE. 0) THEN
      k=ID+1
      CALL MPI_SEND(A(1,k,1), 1, ANEWA, 0, 30, MPI_COMM_WORLD, IERR)
      ELSE
      DO 40 I=1,3
      K=I+1
      ISOURCE=I
      CALL MPI_RECV(A(1,K,1), 1, ANEWA, ISOURCE, 30,
     1 MPI_COMM_WORLD, STATUS, IERR)
40        CONTINUE
      END IF
      IF(ID .EQ. 0) THEN
C        Write out array and indexing
      OPEN(4,FILE='EX4.DAT',STATUS='UNKNOWN')
      DO 50 I=1,3
      DO 60 J=1,4
      DO 70 K=1,4
      INT=I+3*(J-1)+12*(K-1)
      WRITE(4,90)INT,I,J,K,A(I,J,K)
90        FORMAT(1x,4(1x,I3),1x,F5.1)
70        CONTINUE
60        CONTINUE
50        CONTINUE
      close (4)
      END IF
      CALL MPI_FINALIZE(IERR)
      STOP
      END
```

Note that arrays are stored as vectors. For example, if you have a variable called A with dimension A(NX,NY,NZ), then this array is stored as a column vector in the form

$$A(1) = A(1, 1, 1)$$
$$A(2) = A(2, 1, 1,)$$
$$\vdots$$
$$A(I + NX * (J - 1) + NX * NY * (K - 1)) = A(I, J, K)$$
$$\vdots$$
$$A(NX * NY * NZ) = A(NX, NY, NZ)$$

For example, consider the array A with dimension $A(3, 4, 4)$. This is stored as a column vector with length 48. If we write an MPI code to fill up this array using, say 4 processors, the code might have a statement something like the following.

```
C          Assume 4 processors
C          ID is processor identification number
C          ID=0,1,2,3
           J=ID+1
C          Note J=1,2,3, or 4 depending on processor ID number
           DO 20 I=1,3
           DO 21 k=1,4
           A(I,J,K)=I+3.*(J-1)+12*(K-1)
   21      CONTINUE
   20      CONTINUE
```

The above commands tell each processor to fill up part of the array. Eventually we want the numbers $1, 2, 3, \ldots, 48$ to be placed in the $A(1), A(2), A(3), \ldots, A(48)$ storage locations respectively. Processor 0 fills in the storage locations

$$A(1), A(2), A(3), \quad A(13), A(14), A(15), \quad A(25), A(26), A(27), \quad A(37), A(38), A(29)$$

because these are the storage locations where $J = 1$ in the array $A(I, J, K)$. The processor 1 fills in the storage locations

$$A(4), A(5), A(6), \quad A(16), A(17), A(18), \quad A(28), A(29), A(30), \quad A(40), A(41), A(42)$$

because these are the storage locations where $J = 2$ in the array $A(I, J, K)$. Similarly, the processor 2 fills in the storage locations

$$A(7), A(8), A(9), \quad A(19), A(20), A(21), \quad A(31), A(32), A(33), \quad A(43), A(44), A(45)$$

corresponding the value $J = 3$ in the array $A(I, J, K)$ and the processor 3 fills in the storage locations

456

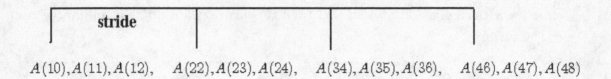

$A(10), A(11), A(12), \quad A(22), A(23), A(24), \quad A(34), A(35), A(36), \quad A(46), A(47), A(48)$

corresponding to $J = 4$ in the array $A(I, J, K)$. That is each processor calculates 4 blocks, with 3 elements in each block, with 12 steps (the stride) between the starting element number of each block. In order to transfer this information from processors 1,2,3 to processor 0, it is required to define a new datatype for use in communicating between processors. The MPI command

MPI_TYPE_VECTOR(ICOUNT, IBLOCK, ISTRIDE, IOLD, INEW, IERR)

is used to define the new datatype for communication and the MPI command

MPI_TYPE_COMMIT(IDATATYPE, IERR)

is needed to commit the new data type to all processors. Once the new data type is defined you need only specify the position of the first element in the array when sending or receiving this datatype.

■

Example 11-5.

This example illustrates complications that can arise if all processors are not used in the same way. For this example assume there are five processors which are used to fill up the two-dimensional and four-dimensional arrays defined by the dimension statement

    DIMENSION A(3,4), B(3,4,5,6)

Note that Fortran allows up to seven dimensions

    ARRAY(N1,N2,N3,N4,N5,N6,N7)

Arrays are stored in column major format so the element ARRAY(i1,i2,i3,i4,i5,i6,i7) is stored in ARRAY(I) where the index I is given by

I=i1+N1*(i2-1)+N1*N2*(i3-1)+N1*N2*N3*(i4-1)+N1*N2*N3*N4*(i5-1)

$$+N1*N2*N3*N4*N5*(i6-1)+N1*N2*N3*N4*N5*N6*(i7-1)$$

Note that when the indices (i1,i2,i3,i4,i5,i6,i7)=(N1,N2,N3,N4,N5,N6,N7), then the index I has the value

$$I=N1*N2*N3*N4*N5*N6*N7.$$

In the following example we store the associated column major index I in the corresponding array locations $A(i1, i2)$, and $B(i1, i2, i3, i4)$. We use the first four processors to fill the array A as follows. If ID denotes the processor identification number 0,1,2, or 3, then processor ID fills in the elements $A(j, \text{ID}+1)$ for $j = 1, 2, 3$. We use all five processors to fill in the array elements $B(j, m, \text{ID}+1, n)$ for $j = 1, 2, 3$, $m = 1, 2, 3, 4$ and $n = 1, 2, 3, 4, 5, 6$ where ID is the processor identification number of 0,1,2,3 or 4. The problem is to transfer back to processor 0 all the information calculated on the other processors. The given Fortran code illustrates the following basic idea. There are certain MPI commands, where MPI expects all processors to participate in the command. MPI can hang up and wait forever for a certain processors information, even though that processor was not involved in the calculation of information. This is called a deadlock situation where everything comes to a stand still. A situation to be avoided.

A communicator like MPI_COMM_WORLD is associated with a group of processors. In this example an additional communicator and group are introduced to handle the array A which was constructed using only four of the five processors. In this way deadlock is avoided. In this example we send the calculated arrays A and B to the processor 0, using the MPI_GATHER command which has the form

MPI_GATHER(SENDA, ISCOUNT, ISTYPE, RECA, IRCOUNT, IRANK, ICOMM, IERR)

where SENDA is the starting address of the sending material, ISCOUNT is the integer send count (number of elements to send), ISTYPE is the data type of the send elements, RECA is the address of the receive material, IRCOUNT is the number of elements to receive, IRANK is the rank of the receiving processor, ICOMM is the communicator, IERR is the returned error code.

Note in the first call to MPI_GATHER we use $12 = N1 * N2$ as the stride between the elements and MPI_COMM_WORLD is the communicator used so that all processors participate in the gather command. In contrast, the second call to MPI_GATHER uses the stride 3 between elements and the communicator ICOMMNEW is used, because only 4 out of the 5 processors participate in the gathering of the data. In using the communicator ICOMMNEW only 4 processors are expected to be heard from. If you had used the communicator MPI_COMM_WORLD in the second call to MPI_GATHER, then deadlock would have occurred as MPI would be expecting to hear from 5 processors.

458

```
      PROGRAM EXAMPLE5
C     Assume 5 processors are being used
      DIMENSION A(3,4), B(3,4,5,6)
      PARAMETER (N1=3,N2=4,N3=5,N4=6)
      INCLUDE 'mpif.h'
C     Define communicator and groups
      INTEGER ICOMMNEW, IGROUPNEW, ICOMMWORLD,IGROUPWORLD, IPROC(1)
C     set up global communicator which uses 5 processors
      CALL MPI_INIT(IERR)
      CALL MPI_COMM_RANK(MPI_COMM_WORLD, ID, IERR)
      CALL MPI_COMM_SIZE(MPI_COMM_WORLD, NP, IERR)
C     Set up an additional communicator
      ICOMMWORLD=MPI_COMM_WORLD
      CALL MPI_COMM_GROUP(ICOMMWORLD, IGROUPWORLD, IERR)
      IPROC(1)=NP-1
C     exclude IPROC(1) components from IGROUPWORLD
      CALL MPI_GROUP_EXCL(IGROUPWORLD, 1, IPROC, IGROUPNEW, IERR)
      CALL MPI_COMM_CREATE(ICOMMWORLD, IGROUPNEW, ICOMMNEW, IERR)
C     fill in the A-array elements using 4 processors
      K=ID+1
      DO 10 I1=1,N1
      IF(K .NE. 5) THEN
      A(I1,K)=I1+N1*(K-1)
      END IF
10       CONTINUE
C     fill in B-array using all 5 processors
      DO 15 I1=1,N1
      DO 14 I2=1,N2
      DO 13 I4=1,N4
      B(I1,I2,K,I4)=I1+N1*(I2-1)+N1*N2*(K-1)+N1*N2*N3*(I4-1)
13       CONTINUE
14       CONTINUE
15       CONTINUE
C     transfer portions of arrays back to processor 0 and then write out arrays
      DO 16 I4=1,N4
      CALL MPI_GATHER(B(1,1,K,I4), 12, MPI_REAL,B(1,1,1,I4),12, MPI_REAL,
     1 0, MPI_COMM_WORLD, IERR)
16       CONTINUE
      IF(K .NE. 5) THEN
      CALL MPI_GATHER(A(1,K), 3, MPI_REAL, A(1,1), 3, MPI_REAL,
     1 0, ICOMMNEW, IERR)
      END IF
```

```
C         program example5 continued
C         write out results
          IF(ID .EQ. 0) THEN
          OPEN(4,FILE='OUTPUT.DAT',STATUS='UNKNOWN')
          DO 20 I1=1,N1
          DO 21 I2=1,N2
          INDEX=I1+N1*(I2-1)
          WRITE(4,5) INDEX,I1,I2,A(I1,I2)
5            FORMAT(1x,I4,1x,'A(',I1,',',I1,')=',F7.1)
21           CONTINUE
20           CONTINUE
          DO 50 I1=1,N1
          DO 60 I2=1,N2
          DO 70 I3=1,N3
          DO 80 I4=1,N4
          INDEX=I1+N1*(I2-1)+N1*N2*(I3-1)+N1*N2*N3*(I4-1)
          WRITE(4,6)INDEX,I1,I2,I3,I4,B(I1,I2,I3,I4)
6            FORMAT(1x,I4,1X,'B(',I1,',',I1,',',I1,',',I1,')=',F7.1)
80           CONTINUE
70           CONTINUE
60           CONTINUE
50           CONTINUE
          CLOSE(4)
          END IF
          CALL MPI_FINALIZE(IERR)
          STOP 'FINISHED'
          END
```

∎

A complete list of MPI commands and their function can be obtained from the reference

Message Passing Interface Forum, MPI: *A Message-Passing Interface Standard*, International Journal of Supercomputer Applications, vol 8, nos 3/4, 1994. This reference is available for download from the

Computer Science Department, University of Tennessee, Knoxville, TN. Internet address: **www.cs.utk.edu/~ library/1994.html**, Report CS-94-230.

Bézier Curves

Consider a polygonal line connecting the $n+1$ data points

$$P_0 = (x_0, y_0), \quad P_1 = (x_1, y_1), \quad \ldots, P_{n-1} = (x_{n-1}, y_{n-1}), \quad P_n = (x_n, y_n)$$

called vertices. The polygonal line connecting the vertices results in a jagged representation of a curve associated with the data points. A Bézier curve associated with the above data points is a parametric curve which allows the interior points of the curve to act as parameters which affects the shape of the curve. Only the first and last vertex points will lie on the Bézier curve associated with the data points.

Define the polynomial basis functions

$$\mathcal{B}_{n,m}(t) = \binom{n}{m} t^m (1-t)^{n-m}, \qquad \binom{n}{m} = \frac{n!}{m!(n-m)!} \tag{11.1}$$

where m represents an integer associated with the data point $P_m = (x_m, y_m)$ and n is the degree of the polynomial represented by the equation (11.1). One can define a nth ordered polynomial associated with the $n+1$ data points $P_0, P_1, \ldots, P_n$ by the equation

$$\mathcal{P}(t) = \sum_{m=0}^{n} P_m \mathcal{B}_{n,m}(t) = (x(t), y(t)), \qquad 0 < t < 1 \tag{11.2}$$

where $P_m = (x_m, y_m)$ is to be treated as a vector with components x_m and y_m representing the vertices of the mth data point. The parametric curves $x = x(t)$ and $y = y(t)$ represents the Bézier curve through the data points.

Example 11-6.

Construct a second degree Bézier curve associated with the data points

$$P_0 = (x_0, y_0), \quad P_1 = (x_1, y_1), \quad P_2 = (x_2, y_2).$$

Solution: By definition

$$\mathcal{P}(t) = \sum_{m=0}^{2} P_m \mathcal{B}_{2,m}(t), \qquad 0 < t < 1$$

so that

$$\mathcal{P}(t) = P_0 \mathcal{B}_{2,0}(t) + P_1 \mathcal{B}_{2,1}(t) + P_2 \mathcal{B}_{2,2}(t)$$

where

$$\mathcal{B}_{2,0}(t) = (1-t)^2, \qquad \mathcal{B}_{2,1}(t) = 2t(1-t), \qquad \mathcal{B}_{2,2}(t) = t^2.$$

This gives the parametric representation

$$\mathcal{P}(t) = P_0(1-t)^2 + P_1 2t(1-t) + P_2 t^2$$
$$\mathcal{P}(t) = (x_0, y_0)(1-t)^2 + (x_1, y_1)2t(1-t) + (x_2, y_2)t^2$$
$$\mathcal{P}(t) = \left(x_0(1-t)^2 + 2x_1 t(1-t) + x_2 t^2, y_0(1-t)^2 + 2y_1 t(1-t) + y_2 t^2 \right) = (x(t), y(t))$$

which produces the parametric curves

$$x(t) = x_0(1-t)^2 + 2x_1 t(1-t) + x_2 t^2$$
$$y(t) = y_0(1-t)^2 + 2y_1 t(1-t) + y_2 t^2$$

(11.3)

To show the affect of the middle data point on the shape of the curve we select the data points

$$(x_0, y_0) = (1,1), \qquad (x_1, y_1) = (2, y), \qquad (x_2, y_2) = (3, 2)$$

and let y have the values $y = 0, 1, 2, 3, 4, 5$. The resulting Bézier quadratic curves are illustrated in the figure 11-2.

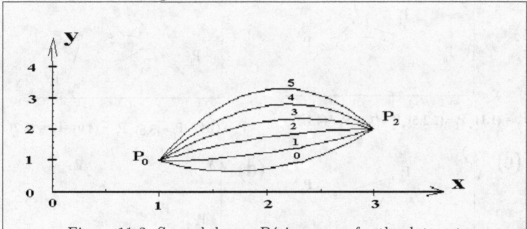

Figure 11-2. Second degree Bézier curves for the data set
$$P_0 = (1,1), \ P_1 = (2, y), \ P_2 = (3, 2)$$
for $y = 0, 1, 2, 3, 4, 5$.

Example 11-7.

Construct a third degree Bézier curve associated with the data points

$$P_0 = (x_0, y_0), \quad P_1 = (x_1, y_1), \quad P_2 = (x_2, y_2), \quad P_3 = (x_3, y_3).$$

Solution: The Bézier curve is defined $\mathcal{P}(t) = \displaystyle\sum_{m=0}^{3} P_m \mathcal{B}_{3,m}(t)$ where

$$\mathcal{B}_{3,0}(t) = (1-t)^3, \quad \mathcal{B}_{3,1}(t) = 3t(1-t)^2, \quad \mathcal{B}_{3,2}(t) = 3t^2(1-t), \quad \mathcal{B}_{3,3}(t) = t^3$$

so that

$$\mathcal{P}(t) = P_0(1-t)^3 + P_1 3t(1-t)^2 + P_2 3t^2(1-t) + P_3 t^3 = (x(t), y(t)).$$

462

This gives the parametric curve $\mathcal{P}(t) = (x(t), y(t))$ where

$$x(t) = x_0(1-t)^3 + 3x_1 t(1-t)^2 + 3x_2 t^2(1-t) + x_3 t^3$$
$$y(t) = y_0(1-t)^3 + 3y_1 t(1-t)^2 + 3y_2 t^2(1-t) + y_3 t^3$$

The figures 11-3(a),(b),(c),(d) illustrate various Bézier curves together with the polygonal line associated with the given data points.

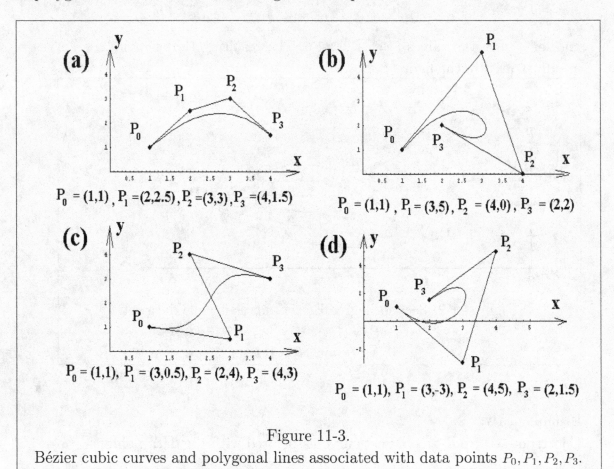

Figure 11-3.

Bézier cubic curves and polygonal lines associated with data points P_0, P_1, P_2, P_3.

B-Splines

Recall that the Heaviside unit step function $H(\xi)$ is defined

$$H(\xi) = \begin{cases} 0, & \text{if } \xi < 0 \\ 1, & \text{if } \xi > 0 \end{cases}$$

That is, the Heaviside unit step function has a value of zero if the argument of the function is negative and it has a value of unity if the argument of the function is positive. We shall employ the Heaviside unit step function to construct functions $B_{i,n}(t)$ of a single variable t, where the t axis is partitioned as

The points t_i are referred to as knots. Note that the partitioning is such that the knots are not necessarily evenly spaced. Define on each interval (t_i, t_{i+1}) the impulse function

$$B_{i,0}(t) = H(t - t_i) - H(t - t_{i+1})$$

which is illustrated in the figure 11-4.

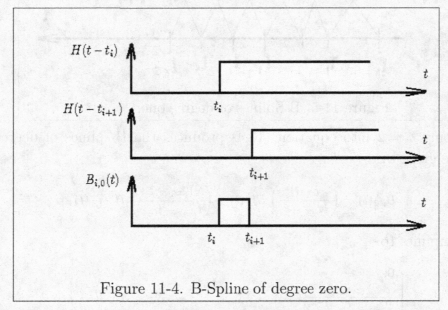

Figure 11-4. B-Spline of degree zero.

The function $B_{i,0}(t)$ is called a B-spline of degree or order zero. Higher ordered B-splines are defined by the recursive relation

$$B_{i,n}(t) = \left(\frac{t - t_i}{t_{i+n} - t_i} \right) B_{i,n-1}(t) + \left(\frac{t_{i+n+1} - t}{t_{i+n+1} - t_{i+1}} \right) B_{i+1,n-1}(t) \qquad (11.4)$$

for $n = 1, 2, 3, \ldots$ and $i = 0, \pm 1, \pm 2, \pm 3, \ldots$. Letting $n = 1$ in equation (11.4) gives the

464

B-splines of degree 1 associated with each knot

$$B_{i,1}(t) = \left(\frac{t - t_i}{t_{i+1} - t_i}\right)[H(t - t_i) - H(t - t_{i+1})] + \left(\frac{t_{i+2} - t}{t_{i+2} - t_{i+1}}\right)[H(t - t_{i+1}) - H(t - t_{i+2})]$$

which can also be expressed in the form

$$B_{i,1}(t) = \begin{cases} 0, & t < t_i \\ \frac{t - t_i}{t_{i+1} - t_i}, & t \in [t_i, t_{i+1}] \\ \frac{t_{i+2} - t}{t_{i+2} - t_{i+1}}, & t \in [t_{i+1}, t_{i+2}] \\ 0, & t > t_{i+2} \end{cases} \qquad (11.5)$$

Several of these B-splines are illustrated in the figure 11-5.

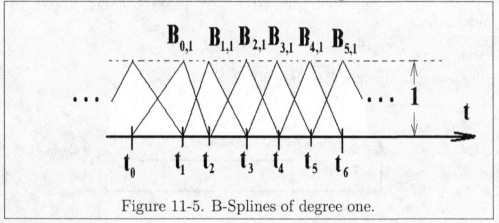

Figure 11-5. B-Splines of degree one.

Substituting $n = 2$ into equation (11.4) produces the B-splines of degree two which are given by

$$B_{i,2}(t) = \left(\frac{t - t_i}{t_{i+2} - t_i}\right)B_{i,1}(t) + \left(\frac{t_{i+3} - t}{t_{i+3} - t_{i+1}}\right)B_{i+1,1}(t)$$

which simplifies to

$$B_{i,2}(t) = \begin{cases} 0, & t < t_i \\ \frac{(t - t_i)^2}{(t_{i+2} - t_i)(t_{i+1} - t_i)}, & t \in [t_i, t_{i+1}] \\ \frac{(t - t_i)(t_{i+2} - t)}{(t_{i+2} - t_i)(t_{i+2} - t_{i+1})} + \frac{(t_{i+3} - t)(t - t_{i+1})}{(t_{i+3} - t_{i+1})(t_{i+2} - t_{i+1})}, & t \in [t_{i+1}, t_{i+2}] \\ \frac{(t_{i+3} - t)^2}{(t_{i+3} - t_{i+1})(t_{i+3} - t_{i+2})}, & t \in [t_{i+2}, t_{i+3}] \\ 0, & t > t_{i+3} \end{cases} \qquad (11.6)$$

Several of these B-splines are illustrated in the figure 11-6.

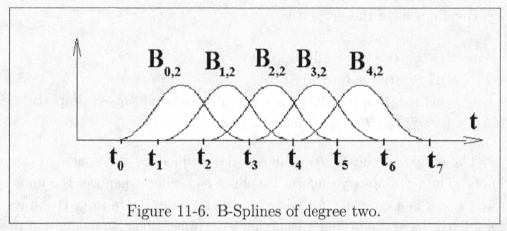

Figure 11-6. B-Splines of degree two.

Continuing in this manner one can construct B-Splines of any degree by using the recursion relation given by equation (11.4). The first couple of B-splines are illustrated in the figure 11-7. The support of the B-spline functions is defined as the set of points t where $B_{i,n}(t) \neq 0$. Observe in figure 11-7 that the support region increases as the degree of the spline increases.

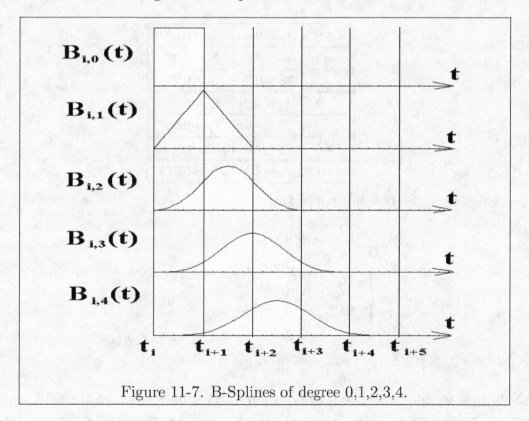

Figure 11-7. B-Splines of degree 0,1,2,3,4.

B-splines have the properties

(i) $B_{i,n}(t) \geq 0$, for all values of t.

(ii) $B_{i,n}(t) = 0$, for $t \notin [t_i, t_{i+n}]$.

(iii) $B_{i,n}(t)$ is $(n-2)$ times differentiable (smoothness property)

(iv) $\sum_{i=-\infty}^{\infty} B_{i,n}(t) = 1$ for all x and $n \geq 0$.

The shape of the $B_{i,n}(t)$ splines is determined by the spacing between the knots. The terminology uniform B-splines is applied whenever the knot spacing $t_{j+1} - t_j = h$ is a constant for all values of the index j. Uniform B-splines $B_{i,n}(t)$, for n fixed, have the property that they are translations of one another or shifted copies of one another.

Example 11-8.

Any of the B-splines can be used for interpolation. Consider the representation of a curve through the data points

$i = 0$	$t_0 = 0.0$	$f_0 = 0.000$
$i = 1$	$t_1 = 0.4$	$f_1 = 0.389$
$i = 2$	$t_2 = 0.8$	$f_2 = 0.717$
$i = 3$	$t_3 = 1.2$	$f_3 = 0.932$
$i = 4$	$t_4 = 1.6$	$f_4 = 0.999$
$i = 5$	$t_5 = 2.0$	$f_5 = 0.909$
$i = 6$	$t_6 = 2.4$	$f_6 = 0.675$

Construct the B-splines of order one

$$B_0(t) = \frac{t - t_1}{t_1 - t_0}, \quad t_0 \leq t \leq t_1$$

$$B_i(t) = \begin{cases} 0, & t < t_{i-1} \\ \frac{t - t_{i-1}}{t_i - t_{i-1}}, & t_{i-1} \leq t \leq t_i \\ \frac{t_{i+1} - t}{t_{i+1} - t_i}, & t_i \leq t \leq t_{i+1} \\ 0, & t > t_{i+1} \end{cases} \qquad \text{for } i = 1, 2, 3, 4, 5$$

$$B_6(t) = \frac{t - t_5}{t_6 - t_5}, \quad t_5 \leq t \leq t_6$$

Note that we have modified the notation for B-splines since we are only working with first degree splines. These B-splines are illustrated in the figure 11-8(a).

Note that these B-splines have the property that $B_j(t_j) = 1$ for $j = 0, 1, \ldots, 6$. We can make use of this fact and write an interpolating function having the form

$$f(t) = \sum_{j=0}^{6} C_j B_j(t).$$

For collocation of the given data points we want

$$f(t_m) = f_m = \sum_{j=0}^{6} C_j B_j(t_m) = C_m, \quad \text{for } m = 0, 1, 2, \ldots, 6,$$

since the only nonzero term in the summation, when $t = t_m$, occurs when $j = m$ and one obtains $C_m B_m(t_m) = C_m$. The resulting interpolating function can be expressed

$$f(t) = \sum_{j=0}^{6} f_j B_j(t).$$

The above set of data points was created from the curve $f = f(t) = \sin t$ and so we can compare the actual curve with this B-spline interpolation function. The comparison is illustrated in the figure 11-8(b).

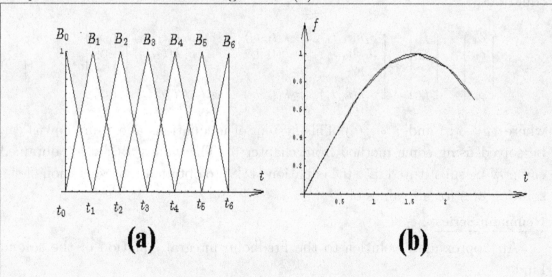

Figure 11-8. (a) B-Splines used for interpolation function $f(t)$.
(b) Interpolating function and sine function.

Fredholm integral equation

The Fredholm integral equation of the second kind

$$\phi(x) = f(x) + \lambda \int_a^b k(x,t)\phi(t)\, dt \qquad a < x < b \tag{11.7}$$

presents the problem of determining $\phi(x)$ when $f(x)$, $k(x,t)$ are known functions and a, b, λ are known constants. The Fredholm integral equation (11.7) has a definite integral to be evaluated and so one can select an integration formula and evaluate the given integral. In this way the integral equation (11.7) can be reduced to a system of equations. For example, if a Gaussian integration formula is employed to evaluate the integral in equation (11.7), there results an equation of the form

$$\phi(x) = f(x) + \beta_1 k(x,t_1)\phi(t_1) + \beta_2 k(x,t_2)\phi(t_2) + \cdots + \beta_n k(x,t_n)\phi(t_n) \tag{11.8}$$

where $\beta_1, \beta_2, \ldots, \beta_n$ and $t_1, t_2, \ldots, t_n$ are known constants. The equation (11.8) can now be evaluated at the points $x = t_1$, $x = t_2$, $\cdots$, $x = t_n$ to obtain the system of equations.

$$
\begin{bmatrix} \phi_1 \\ \phi_2 \\ \vdots \\ \phi_n \end{bmatrix} = \begin{bmatrix} f_1 \\ f_2 \\ \vdots \\ f_n \end{bmatrix} + \begin{bmatrix} \beta_1 k(t_1,t_1) & \beta_2 k(t_1,t_2) & \cdots & \beta_n k(t_1,t_n) \\ \beta_1 k(t_2,t_1) & \beta_2 k(t_2,t_2) & \cdots & \beta_n k(t_2,t_n) \\ \vdots & \vdots & \ddots & \vdots \\ \beta_1 k(t_n,t_1) & \beta_2 k(t_n,t_2) & \cdots & \beta_n k(t_n t_n) \end{bmatrix} \begin{bmatrix} \phi_1 \\ \phi_2 \\ \vdots \\ \phi_n \end{bmatrix} \tag{11.9}
$$

where $\phi_i = \phi(t_i)$ and $f_i = f(t_i)$. This system of n-equations and n-unknowns can be solved using some method from chapter 3. The values $(\phi_1, \ldots, \phi_n)$ obtained can now be substituted into the equation (11.8) to obtain a representation of the solution $\phi(x)$ for all values of x.

Neumann Series

An approximate solution to the Fredholm integral equation of the second kind

$$\phi(x) = f(x) + \lambda \int_a^b k(x,t)\phi(t)\, dt \qquad a < x < b$$

can be obtained by using an iterative method $\phi_{n+1}(x) = f(x) + \lambda \int_a^b k(x,t)\phi_n(t)\, dt$ for $n = 0, 1, 2, 3, \ldots$ using a starting value of $\phi_0(x) = f(x)$. One finds that

$\phi_1(x) = f(x) + \lambda \int_a^b k(x,t) f(t)\, dt$, $\phi_2(x) = f(x) + \lambda \int_a^b k(x,t)\phi_1(t)\, dt$ or in expanded form $\phi_2(x) = f(x) + \lambda \int_a^b k(x,t) f(t)\, dt + \lambda^2 \int_a^b \int_a^b k(x,t)k(t,s)f(s)\, ds\, dt$. Continuing in this manner one can show the solution can be represented

$$\phi_n(x) = \sum_{i=0}^{n} \lambda^i u_i(x)$$

where $u_0(x) = f(x)$, $u_1(x) = \int_a^b k(x,t)f(t)\, dt$, $u_2(x) = \int_a^b \int_a^b k(x,t)k(t,s)f(s)\, ds\, dt, \ldots$.

Volterra integral equation

The Volterra integral equation of the second kind

$$\phi(x) = f(x) + \lambda \int_a^x k(x,t)\phi(t)\, dt \qquad a \le x < b \tag{11.10}$$

has the known quantities $f(x)$, $k(x,t)$, λ, a and b and one must determine the unknown $\phi = \phi(x)$. An infinite series is obtained if one iteratively substitutes for $\phi(t)$ the value defined by equation (11.10). One then obtains a Neumann expansion similar to the previous results cited. Alternatively, an approximate solution can be obtained over a region $a \le x \le b$ by defining points x_i such that

$$a = x_0 < x_1 < x_2 < \ldots < x_N = b$$

One can then define the first order B-splines

$$B_j(x) = \begin{cases} \frac{x - x_{j-1}}{x_j - x_{j-1}} & \text{for } x_{j-1} \le x \le x_j \\ \frac{x_{j+1} - x}{x_{j+1} - x_j} & \text{for } x_j \le x \le x_{j+1} \qquad j = 0, 1, 2, \ldots, N \\ 0 & \text{elsewhere} \end{cases} \tag{11.11}$$

which have the property $B_j(x_j) = 1$. The first order B-spline is illustrated in the figures 11-8 and 11-9.

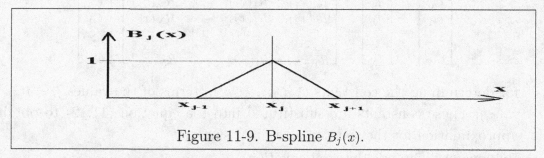

Figure 11-9. B-spline $B_j(x)$.

We assume a solution to equation (11.10) over the region $a \le x \le b$ having the form of a finite sum given by

$$\phi(x) = \sum_{j=0}^{N} C_j B_j(x), \qquad a \le x \le b \tag{11.12}$$

where C_j are constants to be determined. Substituting the assumed solution from equation (11.12) into the Volterra integral equation (11.10) and rearranging terms one obtains

$$\sum_{j=0}^{N} C_j B_j(x) = f(x) + \lambda \int_a^x k(x,t) \sum_{j=0}^{N} C_j B_j(t)\, dt$$
$$\sum_{j=0}^{N} C_j B_j(x) = f(x) + \sum_{j=0}^{N} C_j \int_a^x \lambda k(x,t) B_j(t)\, dt \tag{11.13}$$

Define the quantities

$$R_j(x) = \int_a^x \lambda k(x,t)\, B_j(t)\, dt, \qquad j = 0,1,\ldots,N \tag{11.14}$$

and write equation (11.13) in the form

$$\sum_{j=0}^{N} C_j B_j(x) = f(x) + \sum_{j=0}^{N} C_j R_j(x). \tag{11.15}$$

Now evaluate the equation (11.15) at each of the discrete points $x = x_i$ for the values $i = 0,1,\ldots,N$. Note that at $x = x_m$, there results the equations

$$C_m = f(x_m) + \sum_{j=0}^{N} C_j R_j(x_m), \qquad m = 0,1,\ldots,N$$

which can be expressed as the system of equations

$$\begin{bmatrix} C_0 \\ C_1 \\ \vdots \\ C_N \end{bmatrix} = \begin{bmatrix} f_0 \\ f_1 \\ \vdots \\ f_N \end{bmatrix} + \begin{bmatrix} R_0(x_1) & R_1(x_1) & \cdots & R_N(x_1) \\ R_0(x_2) & R_1(x_2) & \cdots & R_N(x_2) \\ \vdots & \vdots & \ddots & \vdots \\ R_0(x_N) & R_1(x_N) & \cdots & R_N(x_N) \end{bmatrix} \begin{bmatrix} C_0 \\ C_1 \\ \vdots \\ C_N \end{bmatrix} \tag{11.16}$$

for determining the constants $C_0, C_1, \ldots, C_N$ in terms of the values $f_i = f(x_i)$ and $R_i(x_j)$. These constants are substituted into the equation (11.12) to obtain an approximation for the solution as a finite sum.

Boltzmann differential-integral equation

The Boltzmann differential-integral equation

$$\frac{\partial \phi(x,E)}{\partial x} + \sigma(E)\phi(x,E) = \sum_m \int_E^{E/\alpha} \sigma_m(E,E')\phi(x,E')\, dE' + g(x,E), \tag{11.17}$$

for $0 \leq x \leq x_N$, $E_0 \leq E \leq E_N$, which describes the neutron fluence at position x with energy E within a shielding material, can also be solved using B-splines. Assume the quantities $\sigma(E), \sigma_m(E, E'), g(x, E)$ are all known, then one can partition the energy domain $E_0 < E_1 < \ldots < E_N$ and define the B-splines

$$B_j(E) = \begin{cases} \frac{E-E_{j-1}}{E_j-E_{j-1}} & \text{for } E_{j-1} \leq E \leq E_j \\ \frac{E_{j+1}-E}{E_{j+1}-E_j} & \text{for } E_j \leq E \leq E_{j+1} \quad j = 0,1,2,\ldots,N \\ 0 & \text{elsewhere} \end{cases} \tag{11.18}$$

Assume a solution to the Boltzmann equation (11.17) of the form

$$\phi = \phi(x, E) = \sum_{j=1}^{N} U_j(x) B_j(E) \tag{11.19}$$

and substitute this solution into the Boltzmann equation to obtain

$$\sum_{j=1}^{N} \frac{dU_j}{dx} B_j(E) + \sigma(E) \sum_{j=0}^{N} U_j(x) B_j(E) = \sum_m \int_E^{E/\alpha} \sigma_m(E, E') \sum_{j=0}^{N} U_j(x) B_j(E') \, dE' + g(x, E) \tag{11.20}$$

Now evaluate the equation (11.20) at the discrete values $E = E_i$, for $i = 0, 1, \ldots, N$. In this way the Boltzmann equation can be reduced to studying a system of ordinary differential equations subject to specified boundary conditions. The resulting system of differential equations has the form

$$\frac{dU_0}{dx} + \sigma(E_0) U_0(x) = \sum_{j=0}^{N} C_j(E_0) U_j(x) + g(x, E_0)$$

$$\frac{dU_1}{dx} + \sigma(E_1) U_1(x) = \sum_{j=0}^{N} C_j(E_1) U_j(x) + g(x, E_1) \tag{11.21}$$

$$\vdots \qquad \vdots$$

$$\frac{dU_N}{dx} + \sigma(E_N) U_N(x) = \sum_{j=0}^{N} C_j(E_N) U_j(x) + g(x, E_N)$$

where

$$C_j(E) = \sum_m \int_E^{E/\alpha} \sigma_m(E, E') B_j(E') \, dE'. \tag{11.22}$$

The system of differential equations (11.21) are numerically solved for the coefficients $\{U_0(x_i), U_1(x_i), \ldots, U_N(x_i)\}$ for $i = 0, 1, \ldots, N$. The numerical form of equation (11.19) then gives a representation of the solution $\phi(x_i, E_j)$ at the selected nodal points (x_i, E_j).

Bibliography

- Abramowitz, M. , Stegun, A. **Handbook of Mathematical Functions**, 10th edition, New York: Dover Publications, 1972.

- Ames, W.F., **Numerical Methods for Partial Differential Equations**, Barnes & Noble, Inc., New York, 1969.

- Anderson, D.A., Tannehill, J.C., Fletcher,, R.H., **Computational Fluid Mechanics and Heat Transfer**, Hemisphere Publishing Corporation, New York, 1984.

- Atkinson, K.E., **An Introduction to Numerical Analysis**, John Wiley and Sons, New York, 1978.

- Bronson, R., **Matrix Operations**, Schaum's Outline Series, McGraw-Hill Book Company, New York, 1989.

- Burden, R.L., Faires, J. D., **Numerical Analysis**, Fourth Edition, PWS-Kent Publishing Company, Boston, 1989.

- Burden, R.L., Faires, J. D., **Numerical Methods**, PWS-Kent Publishing Company, Boston, 1993.

- Chapra, S.C., Canale, R.P., **Numerical Methods for Engineers**, McGraw-Hill Book Company, New York, 1988.

- Cheney, W., Kincaid, D., **Numerical Mathematics and Computing**, Second Edition, Brooks/Cole Publishing Company, Monterey, California, 1985.

- Conte, S.D., de Boor, C., **Elementary Numerical Analysis**, McGraw-Hill Book Company, New York, 1980.

- Dahlquist, G., Björck, Åke, **Numerical Methods**, Prentice-Hall, Inc, Inglewood Cliffs, New Jersey, 1974.

- Davis, P.J., **Interpolation and Approximation**, Dover Publications, Inc, New York, 1975.

- Fishman, G.S., **Principles of Discrete Event Simulation**, John Wiley & Sons, New York, 1978.

- Gerald, C.F., **Applied Numerical Analysis**, Second Edition, Addison-Wesley Publishing Company, Reading, Massachusetts, 1980.

- Golub, G.H., Ortega, J.M., **Scientific Computing and Differential Equations**, Academic Press, Inc., Boston, 1992.

- Gordon, G., **System Simulation**, Second Edition, Prentice-Hall, Inc., Englewood Cliffs, New Jersey, 1978.

- Greenspan, D., Casulli, V., **Numerical Analysis for Applied Mathematics, Science and Engineering**, Addison-Wesley Publishing Company, Reading, Massachusetts, 1988.

- Hall, C.A., Porsching, T.A., **Numerical Analysis of Partial Differential Equations**, Prentice Hall, Englewood Cliffs, New Jersey, 1990.

- Hamming, R.W., **Numerical Methods for Scientists and Engineers**, Second Edition, Dover Publications, New York, 1973.

- Hildebrand, F.B., **Introduction to Numerical Analysis**, Second Edition, Dover Publications, New York, 1974.

- Hoffman, J.D., **Numerical Methods for Engineers and Scientists**, McGraw Hill Book Compay, New York, 1992.

- Hoffman, K.A., Chiang, S.T., **Computational Fluid Dynamics for Engineers**, Vols. 1 and 2, A publication of Engineering Education System , Wichita, Kansas, 1993.

- Hornbeck, R.W., **Numerical Methods**, Quantum Publishers, Inc, New York, 1975.

- Hovanessian, S.A., Pipes, L.A., **Digital Computer Methods in Engineering**, McGraw-Hill Book Company, New York, 1969.

- Hultquist, P.F., **Numerical Methods for Engineers and Computer Scientists**, The Benjamin/Cummings Publishing Company, Menlo Park, California, 1988.

- Jensen, J.A., Rowland, J.H., **Methods of Computation**, Scott, Foresman and Company, Glenview, Illinois, 1975.

- Johnson, L.H., **Nomography and Empirical Equations**, John Wiley & Sons, New York, 1952.

- Johnson, L.W., Riess, R.D., **Numerical Analysis**, Second Edition, Addision-Wesley Publishing Company, Reading, Massachusetts, 1982.

- Kunz, Kaiser S., **Numerical Analysis**, McGraw-Hill Book Company, New York, 1957.

- Kuo, S.S., **Computer Applications of Numerical Methods**, Addison-Wesley Publishing Company, Inc, Reading Massachusetts, 1972.

- Lambert, J.D., **Computational Methods in Ordinary Differential Equations**, John Wiley & Sons, New York, 1973.

- Lapidus, L., Schiesser, W.E., **Numerical Methods for Differential Systems**, Academic Press Inc, New York, 1976.

- Lopez, R.J., **Advanced Engineering Mathematics**, Addison Wesley Publishing Company, Inc., Boston, 2001.

- Maron, M.J., **Numerical Analysis: A Practical Approach**, Macmillan Publishing Company, New York, 1982.

- Mendenhall, W., **Introduction to Linear Models and The Design and Analysis of Experiments**, Wadsworth Publishing Company, Belmont, California, 1968.

- Message Passing Interface Forum, MPI:*A Message-Passing Interface Standard*, International Journal of Supercomputer Applications, vol 8, nos 3/4, 1994. (Available for download from the Computer Science Department, University of Tennessee, Knoxville, TN. Internet address **www.cs.utk.edu/$\sim$ library/1994.html**, Report CS-94-230.)

Bibliography

- Mitchell, A.R., **Computational Methods in Partial Differential Equations**, John Wiley & Sons, New York, 1969.
- Moore, R.E., **Mathematical Elements of Scientific Computing**, Holt, Rinehart and Winston, New York, 1975.
- Pipes, L.A., Hovanessian, S.A., **Matrix-Computer Methods in Engineering**, John Wiley and Sons, New York, 1969.
- Plybon, B.F., **An Introduction to Applied Numerical Analysis**, PWS-Kent Publishing Company, Boston, Massachusetts, 1992.
- Press, W.H., Flannery, B.P., Teukolsky, S.A., Vetterling, W.T., **Numerical Recipes The Art of Scientific Computing**, Cambridge University Press, London, 1986.
- Ralston, A., Rabinowitz, P., **A First Course in Numerical Analysis**, McGraw-Hill Book Company, New York, 1978.
- Richtmyer, R.D., Morton, K.W., **Difference Methods for Initial-Value Problems**, Second Edition, Interscience Publishers, New York, 1957.
- Rogers, D.F., Adams, J.A., **Mathematical Elements for Computer Graphics**, McGraw Hill Book Co., New York, 1976.
- Rubinstein, R.Y., **Simulation and the Monte Carlo Method**, John Wiley & Sons, New York, 1981.
- Scarborough, J.B., **Numerical Mathematical Analysis**, The Johns Hopkins Press, Baltimore, Maryland, 1955.
- Scheid, F., **Theory and Problems of Numerical Analysis**, Schaum's Outline Series, McGraw-Hill Book Company, New York, 1968.
- Shannon, R.E., **Systems Simulation the art and science**, Prentice Hall, Inc., Englewood Cliffs, New Jersey, 1975.
- Tocher, K.D., **The Art of Simulation**, D. Van Nostrand Co., Inc, Princeton, N.J. 1963.
- Smith, G.D., **Numerical Solution of Partial Differential Equations**, Oxford University Press, New York, 1965.
- Smith, J.M., **Mathematical Modeling & Digital Simulation for Engineers and Scientists**, John Wiley & Sons, New York, 1977.
- Twizell, E.H., **Computational Methods for Partial Differential Equations**, Ellis Horwood Limited, Chichester, West Sussex, England, 1984.
- Vemure, V., Karplus, W.J., **Digital Computer Treatment of Partial Differential Equations,** Prentice-Hall, Inc., Englewood Cliffs, New Jersey, 1981.
- Young, D.M., Gregory, R.T., **A Survey of Numerical Mathematics**, Vols. 1 and 2, Addison-Wesley Publishing Company, Inc., Reading, Massachusetts, 1973.
- Zeigler, B.P., **Theory of Modelling and Simulation**, John Wiley & Sons, New York, 1976.

Bibliography

APPENDIX A
Units of Measurement

The following units, abbreviations and prefixes are from the
Système International d'Unitès   (designated SI in all Languages.)

Prefixes.

Prefix	Abbreviations Multiplication factor	Symbol
tera	10^{12}	T
giga	10^{9}	G
mega	10^{6}	M
kilo	10^{3}	K
hecto	10^{2}	h
deka	10	da
deci	10^{-1}	d
ccnti	10^{-2}	c
milli	10^{-3}	m
micro	10^{-6}	μ
nano	10^{-9}	n
pico	10^{-12}	p

Basic Units.

Basic units of measurement Unit	Name	Symbol
Length	meter	m
Mass	kilogram	kg
Time	second	s
Electric current	ampere	A
Temperature	degree Kelvin	$^{\circ}$K
Luminous intensity	candela	cd

Supplementary units Unit	Name	Symbol
Plane angle	radian	rad
Solid angle	steradian	sr

DERIVED UNITS		
Name	Units	Symbol
Area	square meter	m^2
Volume	cubic meter	m^3
Frequency	hertz	Hz (s^{-1})
Density	kilogram per cubic meter	kg/m^3
Velocity	meter per second	m/s
Angular velocity	radian per second	rad/s
Acceleration	meter per second squared	m/s^2
Angular acceleration	radian per second squared	rad/s^2
Force	newton	N $(kg \cdot m/s^2)$
Pressure	newton per square meter	N/m^2
Kinematic viscosity	square meter per second	m^2/s
Dynamic viscosity	newton second per square meter	$N \cdot s/m^2$
Work, energy, quantity of heat	joule	J $(N \cdot m)$
Power	watt	W (J/s)
Electric charge	coulomb	C $(A \cdot s)$
Voltage, Potential difference	volt	V (W/A)
Electromotive force	volt	V (W/A)
Electric force field	volt per meter	V/m
Electric resistance	ohm	Ω (V/A)
Electric capacitance	farad	F $(A \cdot s/V)$
Magnetic flux	weber	Wb $(V \cdot s)$
Inductance	henry	H $(V \cdot s/A)$
Magnetic flux density	tesla	T (Wb/m^2)
Magnetic field strength	ampere per meter	A/m
Magnetomotive force	ampere	A

Physical Constants:

- $4 \arctan 1 = \pi = 3.14159\,26535\,89793\,23846\,2643\ldots$
- $\lim_{n \to \infty} \left(1 + \frac{1}{n}\right)^n = e = 2.71828\,18284\,59045\,23536\,0287\ldots$
- Euler's constant $\quad \gamma = 0.57721\,56649\,01532\,86060\,6512\ldots$
- $\gamma = \lim_{n \to \infty} \left(1 + \frac{1}{2} + \frac{1}{3} + \cdots + \frac{1}{n} - \log n\right)$
- Speed of light in vacuum $= 2.997925(10)^8\,m\ s^{-1}$
- Electron charge $= 1.60210(10)^{-19}\,C$
- Avogadro's constant $= 6.02252(10)^{23}\,mol^{-1}$
- Plank's constant $= 6.6256(10)^{-34}\,J\,s$
- Universal gas constant $= 8.3143\,J\,K^{-1}\,mol^{-1} = 8314.3\,J\,Kg^{-1}\,K^{-1}$
- Boltzmann constant $= 1.38054(10)^{-23}\,J\,K^{-1}$
- Stefan–Boltzmann constant $= 5.6697(10)^{-8}\,W\,m^{-2}\,K^{-4}$
- Gravitational constant $= 6.67(10)^{-11}\,N\,m^2kg^{-2}$

Appendix A

APPENDIX B
Solutions to Selected Exercises

Solutions Chapter 1

▶ 1. (a) $f'(x) = 2x \cos x^2$ (b) $f'(x) = -x^2 \sin x + 2x \cos x$

▶ 2. (a) $f'(x) = \dfrac{2x}{\sqrt{1 - x^4}}$ (c) $f'(x) = (1 - 2x^2)e^{-x^2}$

▶ 3. $I_a = \displaystyle\int_0^\pi \sin x\, dx = 2$ $I_b = \displaystyle\int_0^\infty e^{-\alpha x}\, dx = \dfrac{1}{\alpha}$

▶ 4. $I_a = \displaystyle\int_0^\pi x \cos x^2\, dx = \dfrac{1}{2}\sin \pi^2$ $I_b = \displaystyle\int_0^\infty xe^{-\alpha x}\, dx = \dfrac{1}{\alpha^2}$

▶ 5. (a) $\xi = 7/2$ (b) $\xi = 4$

▶ 6.   All series converge.

▶ 7. (a) $|x| < 1$ (b) $|x| < 2$

▶ 8. $\ln(1 - x) = -x - \dfrac{x^2}{2} - \dfrac{x^3}{3} - \dfrac{x^4}{4} - \cdots$

▶ 9. $\ln(1 + x) = x - \dfrac{x^2}{2} + \dfrac{x^3}{3} - \dfrac{x^4}{4} + \cdots$

▶ 10. $\dfrac{1}{1 - x} = 1 + x + x^2 + x^3 + \cdots$ $|x| < 1$

▶ 11.

π		e		$\sqrt{2}$	
3.14159	0.000084	2.71828	0.000067	1.41421	0.000251

▶ 12. (a) $f = xy^2 = 1 + (x-1) + 2(y-1) + \dfrac{1}{2!}\left[4(x-1)(y-1) + 2(y-1)^2\right] + \dfrac{1}{3!}\left[6(x-1)(y-1)^2\right]$

▶ 14. $e^{-2} = 0.135335,$ $e^{-4} = 0.0183156,$ $e^{-6} = 0.00247875$

▶ 15. (a) $N_t = 10,$ $9.995 < N_a < 10.005,$ $N_t = 1000,$ $999.5 < N_a < 1000.5$

▶ 16. $\dfrac{\Delta y}{y} = \epsilon\left[\dfrac{1}{x_1} + \dfrac{1}{x_2} + \cdots + \dfrac{1}{x_n}\right]$

▶ 17.   Maximum error occurs when Δx_2 is negative.

▶ 18. $0.1 = 0.000\overline{1100}_2$ $0.01 = (0.010.000000101000111101\ldots)_2$

▶ 19.   Approximate values are $0.1 \approx 0.06_8$ $0.01 \approx 0.005_8$

▶ 20.   Approximate values are $0.1 \approx 0.2_{16}$ $0.01 \approx 0.03_{16}$

▶ 21.   Approximate values are $0.1 \approx 0.0022_3$ $0.01 \approx 0.000021_3$

▶ 24. (a) (i) $\sqrt{5} \approx 0.707107$ (ii) $\sqrt{1.5} \approx 1.22474$

▶ 25. (a) $\xi = \arcsin(2/\pi) \approx 0.690107$

▶ 26. $\epsilon/3$

▶ 30.
$$N_a = 0.00011001100110011001100_2 = 0.09999990463256836$$
$$N_a + \epsilon_1 = 0.00011001100110011001101_2 = 1.000000238418579$$
$$N_a - \epsilon_2 = 0.00011001100110011001011_2 = 0.9999978542327881$$

Solutions Chapter 2

► 1.

(a)	$x = 0.01682279781$	(c)	$x = 1.029866535$
(b)	$x = 1.074569932$	(d)	$x = 1.406929669$

► 5.   18/105,   11/42,   17/30

► 6.

$V_0 = 50$ mph	
10	51.1650
20	52.3105
30	53.4374
40	54.5465
50	55.6387

► 7.

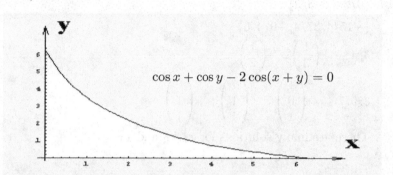

$$\cos x + \cos y - 2\cos(x+y) = 0$$

► 8.

Curves intersect at
$x = -0.404495$ and $x = 2.07854$

► 9.

Curves intersect at
$y = 2.36063$ and $y = -0.707452$

► 12.   2.55,   4.22,   7.15

► 14. $\tan x = x$ for
$x_1 = 0.0104761,$ $x_2 = 4.49341,$ $x_3 = 7.72525,$ $x_4 = 10.9041,$ $x_5 = 14.0662$

► 15. $x = 0.619061,$ $x = 1.51213$

► 18. $h = 3.6479$

► 20.

(a) $x = -0.703467$

(b) $x = 0.259171,$ $x = 2.54264$

(c) $x = 0.666239$

► 22.

(a) $x = -0.732051,$ $x = 1.0,$ $x = 2.73205$

(b) $x = -2.53209,$ $x = -1.3473,$ $x = 0.879385$

► 23.   (a) $x = 0.785398,$ (b) $x = 0.539785$

► 24. $V = 1.345$ [m^3/mol]

480

Solutions Chapter 3

▶ 1. $\{-40, -60, -60\}$

▶ 2. $\{5, 6\}$

▶ 3. $\{4/3, 16/3\}$

▶ 4. $\{3/2, 4, -3/2\}$

▶ 7. $\{-15/4, 37/4, -9/4, 5\}$

▶ 8. $3, 4$ $\begin{pmatrix} 1 \\ 3 \end{pmatrix}$, $\begin{pmatrix} 1 \\ 2 \end{pmatrix}$

▶ 10. $3, 5, 7$ $\begin{pmatrix} 1 \\ 0 \\ 1 \end{pmatrix}$, $\begin{pmatrix} 1 \\ 1 \\ 1 \end{pmatrix}$, $\begin{pmatrix} 0 \\ 1 \\ 1 \end{pmatrix}$

▶ 12. There are many solutions for x_1, x_2 and x_3

x1	x2	x3
0.320761	1.13003	-0.74717
-0.477404	1.40141	-0.555672
-0.231724	0.0509845	0.119164
0.144342	-0.189602	0.207334
1.72138	-0.412756	0.68867
1.33333	0.416667	1
0.917351	0.768654	1.11449
0.898993	-0.514243	2.38098
1.09635	-0.560201	2.4314

▶ 13. $(1, 0)$ and $(-0.215044, 0.976604)$

▶ 14. $\rho(A) = 16$, $\sqrt{\rho(A^T A)} = 89.5779$

▶ 15. $x_1 = -45$, $x_2 = 31$

▶ 22. $\vec{x} = \begin{pmatrix} x_1 \\ x_2 \end{pmatrix} = c_1 \begin{pmatrix} -1 \\ 1 \end{pmatrix} e^{-3t} + c_2 \begin{pmatrix} 1 \\ 1 \end{pmatrix} e^{-t}$

▶ 23. $x_1 = 13/4$, $x_2 = 9/4$

▶ 24. $\vec{x} = \text{col}\{8, -3, 8\}$

▶ 25. $x = 0.771845$, $x = 0.419643$

▶ 26. $\vec{x} = \text{col}\{3, -4, 2, -1\}$

▶ 28. $\vec{x} = \text{col}\{1, 3, 5, 7\}$

▶ 29. $\vec{x} = \text{col}\{5/3, 4/3, -1\}$

▶ 30. $\vec{x} = \text{col}\{1, 1, 1\}$

▶ 31. $\{-5.5, 6.75, 7.75\}$, $\{6, 1, 2\}$, $\{9, -44, 44\}$

Solutions Chapter 4

▶ 1. $\quad P_3(x) = \dfrac{1}{24}(-105 - 61x + 81x^2 - 11x^3)$

▶ 2. $\quad$ (c) $\quad P_2(x) = 20 - 26x + 9x^2 \qquad$ (d) $\quad P_2(2.5) = 11.25$

▶ 3. $\quad P_{0,1}(x) = \dfrac{1}{2}x^3 - \dfrac{3}{2}x + 3 \qquad P_{1,2}(x) = -\dfrac{1}{2}(x-1)^3 + \dfrac{3}{2}(x-1)^2 + 2$

▶ 4. $\quad P_{0,1}(x) = -\dfrac{13}{4}x^3 + \dfrac{21}{4}x^2 + 2 \qquad\qquad P_{1,2}(x) = \dfrac{11}{4}(x-1)^3 - \dfrac{9}{2}(x-1)^2 + \dfrac{3}{4}(x-1) + 4$

▶ 7. $\quad$ (b) $\quad P(x) = 25x^2 - 73x + 73, \qquad$ (d) $\quad P(2.5) = 46.75$

▶ 8. $\quad$ (b) $\quad P(x) = 11 + 30(x-2) + 8(x-2)(x-3)$

▶ 10. $\quad F(2.3, 4.7) = 4.38764$

▶ 12. $\quad R_{3,2}(x) = \dfrac{x - \frac{7}{60}x^3}{1 + \frac{1}{20}x^2}$

▶ 14. $\quad R_{4,3}(x) = \dfrac{1 - \frac{7}{15}x^2 + \frac{1}{40}x^4}{1 + \frac{1}{30}x^2}$

▶ 16.

$$(a) \quad f(x) = -\frac{4}{3}P_0(x) + \frac{8}{5}P_1(x) - \frac{8}{3}P_2(x) + \frac{2}{5}P_3(x)$$

$$(c) \quad f(x) = -U_0(x) + \frac{3}{4}U_1(x) - U_2(x) + \frac{1}{8}U_3(x)$$

$$(e) \quad f(x) = -2H_0(x) + \frac{5}{4}H_1(x) - H_2(x) + \frac{1}{8}H_3(x)$$

▶ 17. $\quad b_n = \dfrac{2L}{\pi}\dfrac{(-1)^{n+1}}{n}, \qquad n = 1, 2, 3, \ldots$

▶ 18. $\quad a_0 = \dfrac{L}{2}, \qquad a_n = \dfrac{2L}{n^2\pi^2}[(-1)^n - 1], \qquad n = 1, 2, 3, \ldots$

▶ 20. $\quad y_0 = 9/2, \quad y_1 = 14, \quad y_0' = -1, \qquad y_1' = 4$

$$L_{1,0}(x) = \frac{x - x_1}{x_0 - x_1} \quad L_{1,1}(x) = \frac{x - x_0}{x_1 - x_0}$$

$$L_{1,0}'(x) = \frac{1}{x_0 - x_1} \quad L_{1,1}'(x) = \frac{1}{x_1 - x_0}$$

$$U_0(x) = [1 - 2L_{1,0}'(x_0)(x-1)]L_{1,0}^2(x) \quad V_0(x) = (x-1)L_{1,0}^2(x)$$

$$U_1(x) = [1 - 2L_{1,1}'(x_1)(x-2)]L_{1,1}^2(x) \quad V_1(x) = (x-2)L_{1,1}^2(x)$$

$$h(x) = y_0 U_0(x) + y_0' V_0(x) + y_1 U_1(x) + y_1' V_1(x)$$

482

Solutions Chapter 5

▶ 1.

(a) $\quad a + bx = \dfrac{x}{y} = Y$

(b) $\quad \ln(y - c) = \ln a + b \ln x, \qquad Y = \ln(y - c), \quad X = \ln x$

(c) $\quad \ln(y - d) = \ln(ae^c) + bx, \qquad Y = \ln(y - d), \quad \alpha = \ln(ae^c)$

(d) $\quad \text{plot} \quad \dfrac{y - y_0}{x - x_0} \text{ vs } x - x_0$

(e) $\quad \dfrac{x - x_0}{y - y_0} = \alpha + \beta x, \qquad \alpha = (a + bx_0), \quad \beta = (a + bx_0)(b/a)$

(f) $\quad Y = mx + \beta, \qquad m = 1/a, \qquad \beta = b/a$

▶ 8. $\quad \beta_1 = 1/2, \quad \beta_2 = 3$

▶ 9. $\quad y = 3x^{1.5}$

▶ 10. $\quad y = 2.5 + 3.5 \ln x$

▶ 11. $\quad y = 3e^{-0.3x}$

▶ 12. $\quad y = 1.5 + \dfrac{2}{x}$

▶ 13. $\quad R = 0.072 + 0.0003\,T$

▶ 14. $\quad y(x) = 3e^{-x} + 2\sin x - 5\sin 2x$

▶ 16.

(a) $\quad y = \dfrac{177}{79} + \dfrac{129}{158}x$

(b) $\quad y = \dfrac{158}{53} + \dfrac{30}{53}x$

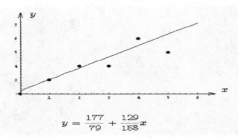

$y = \dfrac{177}{79} + \dfrac{129}{158}x$

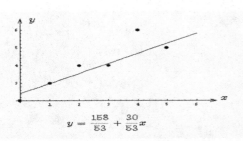

$y = \dfrac{158}{53} + \dfrac{30}{53}x$

Solutions Chapter 6

► 2.

$$(a) \quad y_n = c_1 4^n$$

$$(e) \quad y_n = c_1 - \frac{n}{2} + \frac{n^2}{2}$$

$$(c) \quad y_n = c_1 a^n + \frac{b}{1-a} \qquad a \neq 1$$

$$(g) \quad y_n = c_1 + 3n^2$$

► 3.

$$(a) \quad \Delta y_n = 2n+1, \qquad \Delta^2 y_n = 2$$

$$(c) \quad \Delta y_n = n \, n!, \qquad \Delta^2 y_n = (n^2 + n + 1) \, n!$$

$$(e) \quad \Delta y_n = (n+2) \, 2^n, \qquad \Delta^2 y_n = (n+4) \, 2^n$$

► 4.

$$(a) \quad y_n = c_1 3^n + c_2 2^n$$

$$(b) \quad y_n = 2^n \left(A \cos \frac{2n\pi}{3} + B \sin \frac{2n\pi}{3} \right)$$

► 5. $y_{n+2} - 6y_{n+1} + 8y_n = 0$

► 8.

$$(a) \quad f(n) = n^{[4]} + 9n^{[3]} + 18n^{[2]} + 6n^{[1]} - 5$$

$$(c) \quad f(n) = n^{[3]} + 3n^{[2]} + n^{[1]} - 2$$

$$(e) \quad f(n) = n^{[4]} + 6n^{[3]}$$

► 9.

$$(a) \quad P_n = \frac{1}{n!} (\lambda \tau)^n P_0$$

$$(b) \quad P_0 = e^{-\lambda \tau}$$

► 11.

$$(a) \quad y_k = c_1 4^k + c_2 2^k - 3^k$$

$$(c) \quad y_k = c_1 4^k + c_2 2^k + \frac{1}{3} k^{[2]} + \frac{11}{9} k^{[1]} + \frac{38}{27}$$

► 12. $(a) \quad U_n = \frac{1}{10}(5 + \sqrt{5}) \left(\frac{1 + \sqrt{5}}{2} \right)^n + \frac{\sqrt{5}}{10}(-1 + \sqrt{5}) \left(\frac{1 - \sqrt{5}}{2} \right)^n$

► 13. $y_n = \frac{1}{3} 2^n + \frac{2}{3}(-1)^n$

► 14. $y_{46} = 2^{46} - 1$

► 16. $(b) \quad P_n = \frac{r^N - r^n}{r^N - 1}, \qquad (c) \quad P_n = 1 - \frac{n}{N}, \quad \nu = 1/2$

► 18. $(a) \quad y_n = y_0 A^n + B \frac{(1-A)}{1-A}, \quad A \neq 1, \qquad (b) \quad y_n = y_0 + Bn, \quad A = 1$

► 19. $(a) \quad \frac{A_n}{A_0} = 1 + rn, \qquad (b) \quad \frac{A_n}{A_0} = (1 + i)^n$

484

▶ 24. (a) $y_k = \dfrac{1}{k+1}$, (b) $y_k = k!\left(1 + \dfrac{1}{1!} + \dfrac{2^1}{2!} + \dfrac{2^2}{3!} + \cdots + \dfrac{2^k}{k!}\right)$

▶ 25.

$$(a) \quad y_n = c_1 2^n + c_2 n 2^n + c_3 3^n$$

$$(c) \quad y_n = A\cos\dfrac{n\pi}{2} + B\sin\dfrac{n\pi}{2}$$

▶ 26. (a) $y_n = c_1 3^n + c_2 4^n + 5^n$

▶ 27.

$$(a) \quad y_n = (c_1 + c_2 n)\cos\dfrac{2\pi n}{3} + (c_3 + c_4 n)\sin\dfrac{2\pi n}{3}$$

$$(c) \quad y_n = c_1 + c_2 n + c_3 2^n + c_4 n 2^n + c_5 3^n + c_6 n 3^n$$

$$(e) \quad y_n = c_1 3^n + c_2 n 3^n + \dfrac{1}{18}n^2 3^n$$

▶ 33. $\mathcal{Z}\{H(n-k)\} = \displaystyle\sum_{m=0}^{\infty} H(m) z^{-m-k} = \dfrac{1}{z^k}\left(\dfrac{1}{1-1/z}\right)$

▶ 35. $\displaystyle\int_z^\infty \dfrac{F(z)}{z}\,dz = \sum_{n=0}^{\infty} f(n)\int_z^\infty z^{-n-1}\,dz = \sum_{n=0}^{\infty} \dfrac{f(n)/n}{z^n} = \mathcal{Z}\{\dfrac{f(n)}{n}\}$

▶ 37.

$$\dfrac{Y(z)}{z} = \dfrac{5/4}{z-3} + \dfrac{3/4}{z-1} + \dfrac{1/2}{(z-1)^2}$$

$$y_n = \dfrac{5}{4}3^n + \dfrac{3}{4}(1)^n + \dfrac{1}{2}n$$

▶ 38.

$$(a) \qquad \mathcal{Z}\{\sin kx\} = \dfrac{z\sin x}{z^2 - 2z\cos x + 1}$$

$$\mathcal{Z}\{a^k \sin kx\} = (z/a)\dfrac{\sin x}{(z/a)^2 - 2(z/a)\cos x + 1} = G(z)$$

$$\mathcal{Z}\{\sum_{k=0}^{n} a^k \sin kx\} = \dfrac{z}{z-1}G(z) = H(z)$$

$$\lim_{z\to 1}(z-1)H(z) = \dfrac{a\sin x}{a^2 - 2a\cos x + 1} = \sum_{k=0}^{\infty} a^k \sin kx$$

$$(b) \qquad \sum_{k=0}^{\infty} a^k \cos kx = \dfrac{1 - a\cos x}{a^2 - 2a\cos x + 1}$$

$$(c) \qquad \sum_{k=0}^{\infty} \dfrac{(-1)^k x^{k+1}}{k+1} = \ln(1+x)$$

▶ 39. (b) $\mathcal{Z}\{n^3\} = \dfrac{6z}{(z-1)^4} + \dfrac{6z}{(z-1)^3} + \dfrac{z}{(z-1)^2}$

▶ 40.

$$(c) \qquad \mathcal{Z}\{n^2 + n\} = \frac{2z}{(z-1)^3} + \frac{2z}{(z-1)^2}$$

$$\mathcal{Z}\{\sum_{m=0}^{n} (m^2 + m)\} = F(z) = \frac{2z^2}{(z-1)^4} + \frac{2z^2}{(z-1)^3}$$

$$\mathcal{Z}^{-1}\{\frac{F(z)}{z}\} = f(n) = \frac{1}{3}n^{[3]} + n^{[2]} = \frac{1}{3}(n-1)n(n+1)$$

$$\mathcal{Z}^{-1}\{F(z)\} = f(n+1) = \frac{1}{3}n(n+1)(n+2)$$

▶ 43. $\quad y_n = -3H(n) - 2n + 32^n$

▶ 44. $\quad 6$

▶ 46. $\quad \dfrac{1}{z}$

▶ 53. $\quad y_n = (-4 + 2\sqrt{3})^n + (-4 - 2\sqrt{3})^n$

▶ 55.

$$(a) \qquad S_m = (\cos\theta)^{m+1}\frac{\sin m\theta}{\sin\theta}$$

$$(b) \qquad S_m = a^{m+1}\left[\frac{a\sin m\theta - \sin(m+1)\theta}{a^2 - 2a\cos\theta + 1}\right] + \frac{a\sin\theta}{a^2 - 2a\cos\theta + 1}$$

$$(c) \qquad S_m = \frac{a^{m+1} - a}{a - 1}, \qquad a \neq 1$$

▶ 57.

Let $\omega = \frac{2\pi(\ell-n)}{N}$, then $S_N = \sum_{k=0}^{N-1} e^{ik\omega}$ is a geometric series

$$S_N = \sum_{k=0}^{N-1} e^{ik\omega} = \frac{1 - \cos 2\pi(\ell - n)}{1 - e^{i\omega}}$$

$S_N = 0$ for ℓ and n integers and for $\ell = n$ we have $\omega = 0$ and $S_N = N$.

Therefore, one can write

$$\sum_{k=0}^{N-1} F(k)e^{ik\ell 2\pi/N} = \sum_{n=0}^{N-1} f(n) \sum_{k=0}^{N-1} e^{i2\pi k(\ell-n)/N}$$

$$\sum_{k=0}^{N-1} F(k)e^{ik\ell 2\pi/N} = \sum_{n=0}^{N-1} f(n)N\delta_{n\ell} = f(\ell)N$$

Hence, $\qquad f(\ell) = \frac{1}{N} \sum_{k=0}^{N-1} e^{i2\pi k\ell/N}$

Solutions Chapter 6

Solutions Chapter 7

▶ 7. $I = 4.42527$,   Exact value is $I = 4\arctan(2)$

▶ 8. $I = 4.42856$,   Exact value is $I = 4\arctan(2)$

▶ 9.   Exact value is π

▶ 10.   Exact value is -2

▶ 11.   Exact value is π

▶ 12.   Exact value is -2

▶ 13.   Exact value is $\dfrac{1}{2}(1 - 1/e)$

▶ 14.   Exact value is $\arctan(2)$

▶ 15.   Exact value is $2 - 5/e$ $x = \frac{1}{2}(1 + t)$, $dx = \frac{1}{2}dt$

$$I_1 = \beta_1 f(x_1) = 0.151633$$
$$I_2 = \beta_1 f(x_1) + \beta_2 f(x_2) = 0.15941$$
$$I_3 = \beta_1 f(x_1) + \beta_2 f(x_2) + \beta_3 f(x_3) = 0.160595$$
$$I_4 = \beta_1 f(x_1) + \beta_2 f(x_2) + \beta_3 f(x_3) + \beta_4 f(x_4) = 0.160603$$

▶ 16.   Exact value is $11/6$ $x = \frac{1}{2}(1 + t)$, $dx = \frac{1}{2}dt$
$I_1 = 7/4$, $I_2 = I_3 = I_4 = 1.83333$

▶ 17.   Exact value is $1/2$ $f(x) = \sin x$ with

$$I_1 = 0.842471, \quad I_2 = -.471916, \quad I_3 = 0.49603, \quad I_4 = 0.504889$$

▶ 18.   Exact value is $1/2$ $f(x) = \cos x$ with

$$I_1 = 0.540302, \quad I_2 = 0.5648, \quad I_3 = 0.476521, \quad I_4 = 0.502547$$

▶ 19. $I_1 = 0.78538$, $I_2 = 1.98013$, $I_3 = 2.98299$, $I_4 = 3.97746$

▶ 20. $I_1 = 0.0$, $I_2 = 0.66292$, $I_3 = 0.911799$, $I_4 = 1.21918$

▶ 28.   (b) $I = 3.10438$

▶ 29.   (d) (i) $I_1 = 252$, (ii) $I_2 = 2\pi$

▶ 30.   (e) (i) $I_1 = 1.9541$, (ii) $I_2 = 4.66667$

▶ 33.   Exact values $I_1 = 2\pi/\sqrt{3}$ and $I_2 = \pi\ln(9/2)$

▶ 34. $I_1 = 3.6276$, $I_2 = 4.7252$

▶ 37. $I_1 = 1.45256$, $I_2 = 2.02285$

▶ 39. $I_1 = 0.886227$, $I_2 = 1.43485$

Solutions Chapter 8

▶ 1. **Euler Method**

(a) $y_{exact} = e^{-x^2/2}$ (b) $y_{exact} = (x+1)^2$

x	y	y_{exact}	y'
0	1	1	0
0.05	1	0.9987508	-0.05
0.1	0.9975	0.9950125	-0.09975
0.15	0.9925125	0.988813	-0.148877
0.2	0.9850687	0.9801987	-0.197014
0.25	0.975218	0.9692332	-0.243804
0.3	0.9630277	0.9559975	-0.288908
0.35	0.9485823	0.9405881	-0.332004
0.4	0.9319821	0.9231163	-0.372793
0.45	0.9133425	0.9037071	-0.411004
0.5	0.8927923	0.8824969	-0.446396
0.55	0.8704725	0.8596328	-0.47876
0.6	0.8465345	0.8352702	-0.507921
0.65	0.8211385	0.8095716	-0.53374
0.7	0.7944515	0.7827045	-0.556116
0.75	0.7666457	0.7548396	-0.574984
0.8	0.7378964	0.726149	-0.590317
0.85	0.7083806	0.6968048	-0.602123
0.9	0.6782744	0.6669768	-0.610447
0.95	0.6477521	0.6368316	-0.615364
1	0.6169838	0.6065307	-0.616984

x	y	y_{exact}	y'
0.00000	1.00000	1.00000	2.00000
0.05000	1.10000	1.10250	2.09524
0.10000	1.20476	1.21000	2.19048
0.15000	1.31429	1.32250	2.28571
0.20000	1.42857	1.44000	2.38095
0.25000	1.54762	1.56250	2.47619
0.30000	1.67143	1.69000	2.57143
0.35000	1.80000	1.82250	2.66667
0.40000	1.93333	1.96000	2.76190
0.45000	2.07143	2.10250	2.85714
0.50000	2.21429	2.25000	2.95238
0.55000	2.36190	2.40250	3.04762
0.60000	2.51429	2.56000	3.14286
0.65000	2.67143	2.72250	3.23810
0.70000	2.83333	2.89000	3.33333
0.75000	3.00000	3.06250	3.42857
0.80000	3.17143	3.24000	3.52381
0.85000	3.34762	3.42250	3.61905
0.90000	3.52857	3.61000	3.71429
0.95000	3.71429	3.80250	3.80952
1.00000	3.90476	4.00000	3.90476

488

▶ 2. **Taylor Series Method**

(a) $y_{exact} = 2e^{x^2/2} - 1$

Sample results

x	y	y_{exact}	y'	y''
0.00000	1.00000	1.00000	0.00000	2.00000
0.02500	1.00063	1.00063	0.05002	2.00188
0.05000	1.00250	1.00250	0.10013	2.00751
0.07500	1.00563	1.00563	0.15042	2.01691
0.10000	1.01002	1.01003	0.20100	2.03012
0.12500	1.01568	1.01569	0.25196	2.04718
0.15000	1.02262	1.02263	0.30339	2.06813
0.17500	1.03085	1.03086	0.35540	2.09305
0.20000	1.04039	1.04040	0.40808	2.12201
0.22500	1.05126	1.05127	0.46153	2.15510
0.25000	1.06347	1.06349	0.51587	2.19243
0.27500	1.07705	1.07707	0.57119	2.23413
0.30000	1.09203	1.09206	0.62761	2.28031
0.32500	1.10843	1.10846	0.68524	2.33113
0.35000	1.12629	1.12633	0.74420	2.38676
0.37500	1.14564	1.14569	0.80462	2.44737
0.40000	1.16652	1.16657	0.86661	2.51316
0.42500	1.18897	1.18903	0.93031	2.58435
0.45000	1.21304	1.21311	0.99587	2.66118
0.47500	1.23876	1.23884	1.06341	2.74389
0.50000	1.26621	1.26630	1.13310	2.83276
0.52500	1.29542	1.29552	1.20510	2.92810
0.55000	1.32646	1.32657	1.27955	3.03022
0.57500	1.35940	1.35952	1.35665	3.13947
0.60000	1.39430	1.39443	1.43658	3.25624
0.62500	1.43123	1.43138	1.51952	3.38093
0.65000	1.47027	1.47044	1.60568	3.51396
0.67500	1.51151	1.51170	1.69527	3.65582
0.70000	1.55504	1.55524	1.78853	3.80701
0.72500	1.60094	1.60117	1.88568	3.96806
0.75000	1.64932	1.64957	1.98699	4.13957
0.77500	1.70029	1.70056	2.09272	4.32215
0.80000	1.75396	1.75426	2.20317	4.51649
0.82500	1.81045	1.81077	2.31862	4.72331
0.85000	1.86989	1.87024	2.43941	4.94339
0.87500	1.93242	1.93281	2.56587	5.17756
0.90000	1.99819	1.99861	2.69837	5.42672

► 3.   **Taylor Series Method**

(b) $y_{exact} = e^{\sin x}$

x	y	y_{exact}	y'	y''	y'''
0.00000	1.00000	1.00000	1.00000	1.00000	0.00000
0.05000	1.05125	1.05125	1.04994	0.99608	-0.16005
0.10000	1.10499	1.10499	1.09947	0.98366	-0.34025
0.15000	1.16118	1.16118	1.14815	0.96173	-0.54037
0.20000	1.21978	1.21978	1.19547	0.92931	-0.75969
0.25000	1.28070	1.28070	1.24089	0.88546	-0.99695
0.30000	1.34383	1.34383	1.28381	0.82934	-1.25030
0.35000	1.40903	1.40902	1.32361	0.76021	-1.51721
0.40000	1.47613	1.47612	1.35961	0.67745	-1.79455
0.45000	1.54492	1.54491	1.39112	0.58064	-2.07846
0.50000	1.61516	1.61515	1.41744	0.46957	-2.36446
0.55000	1.68657	1.68655	1.43784	0.34425	-2.64745
0.60000	1.75884	1.75882	1.45163	0.20497	-2.92177
0.65000	1.83161	1.83159	1.45812	0.05232	-3.18134
0.70000	1.90452	1.90450	1.45666	-0.11281	-3.41975
0.75000	1.97714	1.97712	1.44665	-0.28920	-3.63044
0.80000	2.04904	2.04901	1.42758	-0.47529	-3.80687
0.85000	2.11974	2.11971	1.39899	-0.66921	-3.94273
0.90000	2.18877	2.18874	1.36056	-0.86878	-4.03214
0.95000	2.25563	2.25560	1.31206	-1.07156	-4.06987
1.00000	2.31981	2.31978	1.25340	-1.27484	-4.05159

▶ 4. **Runge Kutta Method**

(b) $y_{exact} = 3e^x - x^2 - 2x - 2$

x	y	y_{exact}	k_1	k_2	k_3
0.00000	1.00000	1.00000	0.05000	0.05128	0.05275
0.05000	1.05131	1.05131	0.05269	0.05416	0.05585
0.10000	1.10551	1.10551	0.05578	0.05745	0.05936
0.15000	1.16300	1.16300	0.05928	0.06116	0.06330
0.20000	1.22421	1.22421	0.06321	0.06532	0.06771
0.25000	1.28957	1.28958	0.06760	0.06995	0.07259
0.30000	1.35957	1.35958	0.07248	0.07507	0.07799
0.35000	1.43470	1.43470	0.07786	0.08071	0.08391
0.40000	1.51547	1.51547	0.08377	0.08690	0.09040
0.45000	1.60243	1.60244	0.09025	0.09366	0.09748
0.50000	1.69616	1.69616	0.09731	0.10102	0.10517
0.55000	1.79725	1.79726	0.10499	0.10902	0.11352
0.60000	1.90635	1.90636	0.11332	0.11768	0.12254
0.65000	2.02411	2.02412	0.12233	0.12705	0.13229
0.70000	2.15125	2.15126	0.13206	0.13715	0.14280
0.75000	2.28849	2.28850	0.14255	0.14802	0.15410
0.80000	2.43661	2.43662	0.15383	0.15971	0.16623
0.85000	2.59642	2.59644	0.16595	0.17225	0.17925
0.90000	2.76879	2.76881	0.17894	0.18569	0.19319
0.95000	2.95461	2.95463	0.19286	0.20008	0.20810
1.00000	3.15482	3.15485	0.20774	0.21547	0.22403

▶ 6.

(d) Error is sum of previou errors

(e) $e_{n+1} \leq e_n(1-\alpha) + \dfrac{h^2}{2}y''(\xi)$, $e_0 = 0$

$e_n \leq \dfrac{h^2}{2}\left[y''(\xi)(1-\alpha)^{n-1} + \cdots\right]$, $|1-\alpha| < 1$

▶ 8.   **Milne's Method**

(b) $y_{exact} = 2e^x - x - 1$

x	$y_{predictor}$	$y_{corrector}$	y_{exact}
0.000000	0.000000	1.000000	1.000000
0.050000	0.000000	1.052542	1.052542
0.100000	0.000000	1.110342	1.110342
0.150000	0.000000	1.173668	1.173668
0.200000	1.242805	1.242806	1.242806
0.250000	1.318050	1.318050	1.318051
0.300000	1.399717	1.399718	1.399718
0.350000	1.488134	1.488135	1.488135
0.400000	1.583649	1.583649	1.583649
0.450000	1.686624	1.686624	1.686624
0.500000	1.797442	1.797443	1.797443
0.550000	1.916505	1.916506	1.916506
0.600000	2.044237	2.044238	2.044238
0.650000	2.181081	2.181081	2.181082
0.700000	2.327505	2.327505	2.327505
0.750000	2.483999	2.484000	2.484000
0.800000	2.651081	2.651082	2.651082
0.850000	2.829293	2.829293	2.829294
0.900000	3.019206	3.019206	3.019206
0.950000	3.221418	3.221419	3.221419
1.000000	3.436563	3.436564	3.436564

▶ 10.   (b) The function $y = e^{3x}$ is a parasitic solution.

▶ 15.

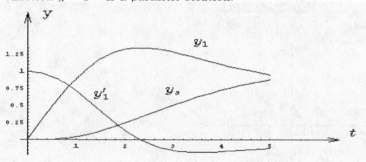

▶ 16.

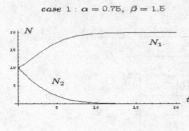

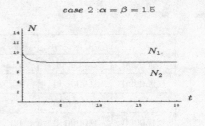

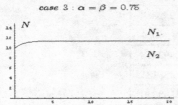

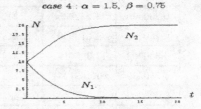

▶ 19.

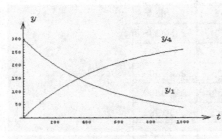

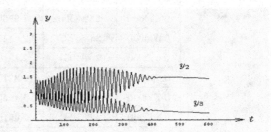

▶ 20.

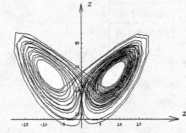

▶ 23.

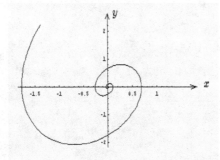

Solutions Chapter 8

Solutions Chapter 9

▶ 8.   Stable for all values of r.

▶ 9.   Sample results rounded after calculations.

t \ x	0.000	0.100	0.200	0.300	0.400	0.500	0.600	0.700	0.800	0.900	1.000
0.000	0.000	0.309	0.588	0.809	0.951	1.000	0.951	0.809	0.588	0.309	0.000
0.004	0.000	0.297	0.565	0.777	0.914	0.961	0.914	0.777	0.565	0.297	0.000
0.008	0.000	0.285	0.543	0.747	0.878	0.923	0.878	0.747	0.543	0.285	0.000
0.012	0.000	0.274	0.521	0.718	0.844	0.887	0.844	0.718	0.521	0.274	0.000
0.016	0.000	0.263	0.501	0.690	0.811	0.852	0.811	0.690	0.501	0.263	0.000
0.020	0.000	0.253	0.481	0.663	0.779	0.819	0.779	0.663	0.481	0.253	0.000
0.024	0.000	0.243	0.463	0.637	0.748	0.787	0.748	0.637	0.463	0.243	0.000
0.028	0.000	0.234	0.444	0.612	0.719	0.756	0.719	0.612	0.444	0.234	0.000
0.032	0.000	0.224	0.427	0.588	0.691	0.726	0.691	0.588	0.427	0.224	0.000
0.036	0.000	0.216	0.410	0.565	0.664	0.698	0.664	0.565	0.410	0.216	0.000
0.040	0.000	0.207	0.394	0.543	0.638	0.671	0.638	0.543	0.394	0.207	0.000
0.044	0.000	0.199	0.379	0.521	0.613	0.644	0.613	0.521	0.379	0.199	0.000
0.048	0.000	0.191	0.364	0.501	0.589	0.619	0.589	0.501	0.364	0.191	0.000
0.052	0.000	0.184	0.350	0.481	0.566	0.595	0.566	0.481	0.350	0.184	0.000
0.056	0.000	0.177	0.336	0.462	0.544	0.572	0.544	0.462	0.336	0.177	0.000
0.060	0.000	0.170	0.323	0.444	0.522	0.549	0.522	0.444	0.323	0.170	0.000
0.064	0.000	0.163	0.310	0.427	0.502	0.528	0.502	0.427	0.310	0.163	0.000
0.068	0.000	0.157	0.298	0.410	0.482	0.507	0.482	0.410	0.298	0.157	0.000
0.072	0.000	0.151	0.286	0.394	0.463	0.487	0.463	0.394	0.286	0.151	0.000
0.076	0.000	0.145	0.275	0.379	0.445	0.468	0.445	0.379	0.275	0.145	0.000
0.080	0.000	0.139	0.264	0.364	0.428	0.450	0.428	0.364	0.264	0.139	0.000
0.084	0.000	0.134	0.254	0.350	0.411	0.432	0.411	0.350	0.254	0.134	0.000
0.088	0.000	0.128	0.244	0.336	0.395	0.415	0.395	0.336	0.244	0.128	0.000
0.092	0.000	0.123	0.235	0.323	0.380	0.399	0.380	0.323	0.235	0.123	0.000
0.096	0.000	0.118	0.225	0.310	0.365	0.383	0.365	0.310	0.225	0.118	0.000
0.100	0.000	0.114	0.217	0.298	0.350	0.368	0.350	0.298	0.217	0.114	0.000
0.104	0.000	0.109	0.208	0.286	0.337	0.354	0.337	0.286	0.208	0.109	0.000
0.108	0.000	0.105	0.200	0.275	0.323	0.340	0.323	0.275	0.200	0.105	0.000
0.112	0.000	0.101	0.192	0.264	0.311	0.327	0.311	0.264	0.192	0.101	0.000
0.116	0.000	0.097	0.185	0.254	0.299	0.314	0.299	0.254	0.185	0.097	0.000

494

▶ 10.   Sample results rounded after calculations.

t \ x	0.000	0.100	0.200	0.300	0.400	0.500	0.600	0.700	0.800	0.900	1.000
0.000	0.000	0.309	0.588	0.809	0.951	1.000	0.951	0.809	0.588	0.309	0.000
0.040	0.000	0.307	0.583	0.803	0.944	0.992	0.944	0.803	0.583	0.307	0.000
0.080	0.000	0.299	0.569	0.784	0.921	0.969	0.921	0.784	0.569	0.299	0.000
0.120	0.000	0.287	0.547	0.753	0.885	0.930	0.885	0.753	0.547	0.287	0.000
0.160	0.000	0.271	0.516	0.710	0.834	0.877	0.834	0.710	0.516	0.271	0.000
0.200	0.000	0.250	0.476	0.656	0.771	0.810	0.771	0.656	0.476	0.250	0.000
0.240	0.000	0.226	0.430	0.591	0.695	0.731	0.695	0.591	0.430	0.226	0.000
0.280	0.000	0.198	0.376	0.518	0.608	0.640	0.608	0.518	0.376	0.198	0.000
0.320	0.000	0.166	0.317	0.436	0.512	0.539	0.512	0.436	0.317	0.166	0.000
0.360	0.000	0.133	0.252	0.347	0.408	0.429	0.408	0.347	0.252	0.133	0.000
0.400	0.000	0.097	0.184	0.253	0.298	0.313	0.298	0.253	0.184	0.097	0.000
0.440	0.000	0.059	0.113	0.155	0.183	0.192	0.183	0.155	0.113	0.059	0.000
0.480	0.000	0.021	0.040	0.055	0.065	0.068	0.065	0.055	0.040	0.021	0.000
0.520	0.000	-0.018	-0.034	-0.046	-0.054	-0.057	-0.054	-0.046	-0.034	-0.018	0.000
0.560	0.000	-0.056	-0.107	-0.147	-0.173	-0.181	-0.173	-0.147	-0.107	-0.056	0.000
0.600	0.000	-0.094	-0.178	-0.245	-0.288	-0.303	-0.288	-0.245	-0.178	-0.094	0.000
0.640	0.000	-0.130	-0.247	-0.339	-0.399	-0.419	-0.399	-0.339	-0.247	-0.130	0.000
0.680	0.000	-0.164	-0.311	-0.428	-0.504	-0.530	-0.504	-0.428	-0.311	-0.164	0.000
0.720	0.000	-0.195	-0.371	-0.511	-0.600	-0.631	-0.600	-0.511	-0.371	-0.195	0.000
0.760	0.000	-0.224	-0.425	-0.585	-0.688	-0.723	-0.688	-0.585	-0.425	-0.224	0.000
0.800	0.000	-0.248	-0.473	-0.650	-0.765	-0.804	-0.765	-0.650	-0.473	-0.248	0.000
0.840	0.000	-0.269	-0.512	-0.705	-0.829	-0.872	-0.829	-0.705	-0.512	-0.269	0.000
0.880	0.000	-0.286	-0.544	-0.749	-0.881	-0.926	-0.881	-0.749	-0.544	-0.286	0.000
0.920	0.000	-0.299	-0.568	-0.782	-0.919	-0.966	-0.919	-0.782	-0.568	-0.299	0.000
0.960	0.000	-0.306	-0.582	-0.802	-0.942	-0.991	-0.942	-0.802	-0.582	-0.306	0.000
1.000	0.000	-0.309	-0.588	-0.809	-0.951	-1.000	-0.951	-0.809	-0.588	-0.309	0.000
1.040	0.000	-0.307	-0.584	-0.804	-0.945	-0.993	-0.945	-0.804	-0.584	-0.307	0.000
1.080	0.000	-0.300	-0.571	-0.786	-0.924	-0.971	-0.924	-0.786	-0.571	-0.300	0.000
1.120	0.000	-0.289	-0.549	-0.756	-0.888	-0.934	-0.888	-0.756	-0.549	-0.289	0.000
1.160	0.000	-0.273	-0.519	-0.714	-0.839	-0.882	-0.839	-0.714	-0.519	-0.273	0.000

APPENDIX C
Tables

Area Under Standard Normal Curve

$$Area = \frac{1}{\sqrt{2\pi}} \int_{-\infty}^{z} e^{-\xi^2/2} \, d\xi$$

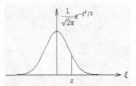

z	Area	z	Area	z	Area	z	Area	z	Area	z	Area	z	Area
0.00	.5000	.50	.6915	1.00	.8413	1.50	.9332	2.00	.9772	2.50	.9938	3.00	.9987
.01	.5040	.51	.6950	1.01	.8438	1.51	.9345	2.01	.9778	2.51	.9940	3.01	.9987
.02	.5080	.52	.6985	1.02	.8461	1.52	.9357	2.02	.9783	2.52	.9941	3.02	.9987
.03	.5120	.53	.7019	1.03	.8485	1.53	.9370	2.03	.9788	2.53	.9943	3.03	.9988
.04	.5160	.54	.7054	1.04	.8508	1.54	.9382	2.04	.9793	2.54	.9945	3.04	.9988
.05	.5199	.55	.7088	1.05	.8531	1.55	.9394	2.05	.9798	2.55	.9946	3.05	.9989
.06	.5239	.56	.7123	1.06	.8554	1.56	.9406	2.06	.9803	2.56	.9948	3.06	.9989
.07	.5279	.57	.7157	1.07	.8577	1.57	.9418	2.07	.9808	2.57	.9949	3.07	.9989
.08	.5319	.58	.7190	1.08	.8599	1.58	.9429	2.08	.9812	2.58	.9951	3.08	.9990
.09	.5359	.59	.7224	1.09	.8621	1.59	.9441	2.09	.9817	2.59	.9952	3.09	.9990
.10	.5398	.60	.7257	1.10	.8643	1.60	.9452	2.10	.9821	2.60	.9953	3.10	.9990
.11	.5438	.61	.7291	1.11	.8665	1.61	.9463	2.11	.9826	2.61	.9955	3.11	.9991
.12	.5478	.62	.7324	1.12	.8686	1.62	.9474	2.12	.9830	2.62	.9956	3.12	.9991
.13	.5517	.63	.7357	1.13	.8708	1.63	.9484	2.13	.9834	2.63	.9957	3.13	.9991
.14	.5557	.64	.7389	1.14	.8729	1.64	.9495	2.14	.9838	2.64	.9959	3.14	.9992
.15	.5596	.65	.7422	1.15	.8749	1.65	.9505	2.15	.9842	2.65	.9960	3.15	.9992
.16	.5636	.66	.7454	1.16	.8770	1.66	.9515	2.16	.9846	2.66	.9961	3.16	.9992
.17	.5675	.67	.7486	1.17	.8790	1.67	.9525	2.17	.9850	2.67	.9962	3.17	.9992
.18	.5714	.68	.7517	1.18	.8810	1.68	.9535	2.18	.9854	2.68	.9963	3.18	.9993
.19	.5753	.69	.7549	1.19	.8830	1.69	.9545	2.19	.9857	2.69	.9964	3.19	.9993
.20	.5793	.70	.7580	1.20	.8849	1.70	.9554	2.20	.9861	2.70	.9965	3.20	.9993
.21	.5832	.71	.7611	1.21	.8869	1.71	.9564	2.21	.9864	2.71	.9966	3.21	.9993
.22	.5871	.72	.7642	1.22	.8888	1.72	.9573	2.22	.9868	2.72	.9967	3.22	.9994
.23	.5910	.73	.7673	1.23	.8907	1.73	.9582	2.23	.9871	2.73	.9968	3.23	.9994
.24	.5948	.74	.7704	1.24	.8925	1.74	.9591	2.24	.9875	2.74	.9969	3.24	.9994
.25	.5987	.75	.7734	1.25	.8944	1.75	.9599	2.25	.9878	2.75	.9970	3.25	.9994
.26	.6026	.76	.7764	1.26	.8962	1.76	.9608	2.26	.9881	2.76	.9971	3.26	.9994
.27	.6064	.77	.7794	1.27	.8980	1.77	.9616	2.27	.9884	2.77	.9972	3.27	.9995
.28	.6103	.78	.7823	1.28	.8997	1.78	.9625	2.28	.9887	2.78	.9973	3.28	.9995
.29	.6141	.79	.7852	1.29	.9015	1.79	.9633	2.29	.9890	2.79	.9974	3.29	.9995
.30	.6179	.80	.7881	1.30	.9032	1.80	.9641	2.30	.9893	2.80	.9974	3.30	.9995
.31	.6217	.81	.7910	1.31	.9049	1.81	.9649	2.31	.9896	2.81	.9975	3.31	.9995
.32	.6255	.82	.7939	1.32	.9066	1.82	.9656	2.32	.9898	2.82	.9976	3.32	.9995
.33	.6293	.83	.7967	1.33	.9082	1.83	.9664	2.33	.9901	2.83	.9977	3.33	.9996
.34	.6331	.84	.7995	1.34	.9099	1.84	.9671	2.34	.9904	2.84	.9977	3.34	.9996
.35	.6368	.85	.8023	1.35	.9115	1.85	.9678	2.35	.9906	2.85	.9978	3.35	.9996
.36	.6406	.86	.8051	1.36	.9131	1.86	.9686	2.36	.9909	2.86	.9979	3.36	.9996
.37	.6443	.87	.8078	1.37	.9147	1.87	.9693	2.37	.9911	2.87	.9979	3.37	.9996
.38	.6480	.88	.8106	1.38	.9162	1.88	.9699	2.38	.9913	2.88	.9980	3.38	.9996
.39	.6517	.89	.8133	1.39	.9177	1.89	.9706	2.39	.9916	2.89	.9981	3.39	.9997
.40	.6554	.90	.8159	1.40	.9192	1.90	.9713	2.40	.9918	2.90	.9981	3.40	.9997
.41	.6591	.91	.8186	1.41	.9207	1.91	.9719	2.41	.9920	2.91	.9982	3.41	.9997
.42	.6628	.92	.8212	1.42	.9222	1.92	.9726	2.42	.9922	2.92	.9982	3.42	.9997
.43	.6664	.93	.8238	1.43	.9236	1.93	.9732	2.43	.9925	2.93	.9983	3.43	.9997
.44	.6700	.94	.8264	1.44	.9251	1.94	.9738	2.44	.9927	2.94	.9984	3.44	.9997
.45	.6736	.95	.8289	1.45	.9265	1.95	.9744	2.45	.9929	2.95	.9984	3.45	.9997
.46	.6772	.96	.8315	1.46	.9279	1.96	.9750	2.46	.9931	2.96	.9985	3.46	.9997
.47	.6808	.97	.8340	1.47	.9292	1.97	.9756	2.47	.9932	2.97	.9985	3.47	.9997
.48	.6844	.98	.8365	1.48	.9306	1.98	.9761	2.48	.9934	2.98	.9986	3.48	.9997
.49	.6879	.99	.8389	1.49	.9319	1.99	.9767	2.49	.9936	2.99	.9986	3.49	.9998

Critical Values for the
chi-square distribution with ν degrees of freedom

$$\int_0^{\chi^2_{(1-\alpha)}} \frac{1}{2^{\nu/2}\Gamma(\nu/2)} x^{(\nu/2)-1} \exp(-x/2)\, dx = 1 - \alpha$$

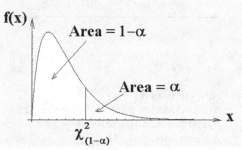

α	0.005	0.010	0.025	0.050	0.100
ν	$\chi^2_{0.995}$	$\chi^2_{0.990}$	$\chi^2_{0.975}$	$\chi^2_{0.950}$	$\chi^2_{0.900}$
1	7.87944	6.63479	5.02389	3.84146	2.70554
2	10.5966	9.21033	7.37776	5.99145	4.60517
3	12.8381	11.3447	9.3484	7.81473	6.25138
4	14.8602	13.2765	11.1433	9.48773	7.77944
5	16.7494	15.0863	12.8325	11.0705	9.23636
6	18.5475	16.8119	14.4494	12.5916	10.6446
7	20.2777	18.4753	16.0128	14.0671	12.017
8	21.9546	20.0902	17.5345	15.5073	13.3616
9	23.5893	21.6659	19.0228	16.919	14.6837
10	25.1882	23.2092	20.4831	18.307	15.9872
11	26.7568	24.725	21.92	19.6751	17.275
12	28.2995	26.2169	23.3367	21.0261	18.5493
13	29.8192	27.6882	24.7356	22.362	19.8119
14	31.3193	29.1412	26.1189	23.6848	21.0641
15	32.8013	30.5777	27.4884	24.9958	22.3071
16	34.2671	31.9999	28.8454	26.2962	23.5418
17	35.7178	33.4087	30.191	27.5871	24.769
18	37.1565	34.8053	31.5264	28.8693	25.9894
19	38.5823	36.1908	32.8522	30.1435	27.2036
20	39.9968	37.5661	34.1696	31.4104	28.412
21	41.4009	38.9321	35.4789	32.6706	29.6151
22	42.7957	40.2893	36.7807	33.9244	30.8133
23	44.1813	41.6384	38.0755	35.1724	32.0069
24	45.5583	42.9796	39.364	36.415	33.1962
25	46.9278	44.3141	40.6463	37.6525	34.3816
26	48.2899	45.6414	41.9232	38.8851	35.5632
27	49.6449	46.9629	43.1945	40.1133	36.7412
28	50.993	48.2782	44.4607	41.3371	37.9159
29	52.3356	49.5879	45.7223	42.557	39.0874
30	53.6719	50.8921	46.9792	43.773	40.256
40	66.7645	63.6899	59.1969	31.0267	32.8753
50	79.4898	76.1539	71.4202	67.5048	63.0789
60	91.9516	88.3794	83.2977	79.0819	74.397
70	104.214	100.425	95.0232	90.5311	85.527
80	116.32	112.329	106.629	101.879	96.5782
90	128.299	124.116	118.136	113.145	107.565
100	140.169	135.806	129.561	124.342	118.498

Appendix C

Index

Printed in the United States
By Bookmasters